AF598014

NEUROMETHODS

For further volumes:
http://www.springer.com/series/7657

Methods for the Discovery and Characterization of G Protein-Coupled Receptors

Edited by

Craig W. Stevens

Department of Pharmacology and Physiology, College of Osteopathic Medicine, Oklahoma State University-Center for Health Sciences, Tulsa, OK, USA

Editor
Craig W. Stevens, Ph.D
Department of Pharmacology and Physiology
College of Osteopathic Medicine
Oklahoma State University-Center for Health Sciences
Tulsa, OK
USA
cw.stevens@okstate.edu

ISSN 0893-2336 e-ISSN 1940-6045
ISBN 978-1-61779-178-9 e-ISBN 978-1-61779-179-6
DOI 10.1007/978-1-61779-179-6
Springer New York Dordrecht Heidelberg London

Library of Congress Control Number: 2011930757

Printed on acid-free paper

Humana Press is part of Springer Science+Business Media (www.springer.com)

Preface to the Series

Under the guidance of its founders Alan Boulton and Glen Baker, the Neuromethods series by Humana Press has been very successful since the first volume appeared in 1985. In about 17 years, 37 volumes have been published. In 2006, Springer Science+Business Media made a renewed commitment to this series. The new program will focus on methods that are either unique to the nervous system and excitable cells or which need special consideration to be applied to the neurosciences. The program will strike a balance between recent and exciting developments like those concerning new animal models of disease, imaging, *in vivo* methods, and more established techniques. These include immunocytochemistry and electrophysiological technologies. New trainees in neurosciences still need a sound footing in these older methods in order to apply a critical approach to their results. The careful application of methods is probably the most important step in the process of scientific inquiry. In the past, new methodologies led the way in developing new disciplines in the biological and medical sciences. For example, Physiology emerged out of Anatomy in the nineteenth century by harnessing new methods based on the newly discovered phenomenon of electricity. Nowadays, the relationships between disciplines and methods are more complex. Methods are now widely shared between disciplines and research areas. New developments in electronic publishing also make it possible for scientists to download chapters or protocols selectively within a very short time of encountering them. This new approach has been taken into account in the design of individual volumes and chapters in this series.

Wolfgang Walz

Preface

The prospect of assembling a volume on "G Protein-Coupled Receptors" for the *Neuromethods* series was a daunting task. The topics of investigation for the characterization and discovery of G protein-coupled receptors, or GPCRs as they are commonly abbreviated, are widespread and numerous methodologies exist for their study. GPCRs are arguably the most important class of signaling molecules in humans and other vertebrates as they are the largest class of membrane receptor proteins coded for in the genome. They are the primary targets of most current medications and will likely remain so for new drugs acting on presently uncharacterized GPCRs, quaintly known as "orphan" GPCRs. This volume presents the very latest on the methods and technology used to characterize and discover novel mechanisms of GPCRs and in many cases can be used directly to design experiments for the reader's particular GPCR of interest and the specific avenue of investigation.

Given the enormity of information on GPCRs, an organizing principle based on the life-cycle of GPCRs was imposed such that this volume is divided into four parts, each with five chapters. The first part is *The G Protein-Coupled Receptor in the Genome* and includes chapters on the *in silico* identification of GPCRs, alternative splicing of GPCRs, GPCR polymorphisms, transcription of GPCRs, and the evolution of vertebrate GPCRs. Part two is entitled *The Trafficking of G Protein-Coupled Receptors* and comprises chapters on visualization of endogenous GPCRs, the post-translational regulation of GPCRs, motifs involved in the export of GPCRs from the endoplasmic reticulum, protein partners of GPCRs acting as escorts, chaperones, and tethers, as well as a novel method for determining the kinetics of GPCR trafficking. The third part of the volume is *The G Protein-Coupled Receptor on the Membrane* with the topics including the characterization of GPCRs in transfected cells, novel assays for the discovery of drugs targeting GPCRs, cell type-specific phosphorylation of GPCRs, quantifying allosteric modulation of GPCRs, and studies designed to determine the receptor–receptor interactions of GPCR activation. The volume concludes with *The Regulation of G Protein-Coupled Receptors* featuring the agonist-selective mechanisms of GPCR desensitization, the role of arrestins in GPCR regulation, mobility of GPCRs within the membrane, using RNA interference to downregulate GPCRs, and upregulating GPCRs with receptor antagonists.

Each chapter was written by scientists with international expertise in their field and with emphasis on the methodology used to characterize GPCRs in the specific aspect of receptor research being investigated. An important section called Future Directions is present in each chapter, which gives the reader an insight into advances in each area soon to be realized. Finally, many of the authors focused on presenting examples of data and methodology used to characterize one of the more famous types of GPCRs, the opioid receptors, so researchers in this area will find this book especially useful.

Tulsa, OK *Craig W. Stevens*

Contents

Contributors

LAMIA ACHOUR • *Institut Cochin, Université Paris Descartes, CNRS-INSERM, Paris, France*
ELISA ALVAREZ-CURTO • *Institute of Neurosciences and Psychology, University of Glasgow, Glasgow, UK*
CHRIS P. BAILEY • *Department of Pharmacy and Pharmacology, University of Bath, Bath, UK*
NICOLAS BEAUDET • *Department of Physiology and Biophysics, Centre des Neurosciences de Sherbrooke, Université de Sherbrooke, Sherbrooke, QC, Canada*
VALÉRIE BÉGIN-LAVALLÉE • *Department of Physiology and Biophysics, Centre des Neurosciences de Sherbrooke, Université de Sherbrooke, Sherbrooke, QC, Canada*
LAURA M. BOHN • *Department of Molecular Therapeutics, The Scripps Research Institute, Jupiter, FL, USA*
ADRIAN J. BUTCHER • *Department of Cell Physiology and Pharmacology, University of Leicester, Leicester, UK*
MARC-ANDRÉ DANSEREAU • *Department of Physiology and Biophysics, Centre des Neurosciences de Sherbrooke, Université de Sherbrooke, Sherbrooke, QC, Canada*
MATTHEW N. DAVIES • *SGDP, Institute of Psychiatry, King's College London, London, UK*
CHUNMIN DONG • *Department of Pharmacology and Experimental Therapeutics, Louisiana State University Health Sciences Center, New Orleans, LA, USA*
LOUIS DORÉ-SAVARD • *Department of Physiology and Biophysics, Centre des Neurosciences de Sherbrooke, Université de Sherbrooke, Sherbrooke, QC, Canada*
FREDERICK J. EHLERT • *Department of Pharmacology, School of Medicine, University of California, Irvine, CA, USA*
DARREN R. FLOWER • *School of Life and Health Sciences, Aston University, Birmingham, UK*
DAVID E. GLORIAM • *Department of Medicinal Chemistry, University of Copenhagen, Copenhagen, Denmark*
STEVEN GRINNELL • *Molecular Pharmacology and Chemistry Program, Memorial Sloan-Kettering Cancer Center, New York City, NY, USA*
CHEOL KYU HWANG • *Department of Pharmacology, University of Minnesota Medical School, Minneapolis, MN, USA*
EAMONN KELLY • *Department of Physiology and Pharmacology, University of Bristol, Bristol, UK*
BRIGITTE L. KIEFFER • *Département Neurobiologie et Génétique, Institut de Génétique et de Biologie Moléculaire et Cellulaire, Illkirch, France*

MARY JEANNE KREEK • *Laboratory of the Biology of Addictive Diseases, The Rockefeller University, New York City, NY, USA*
KOK CHOI KONG • *Department of Cell Physiology and Pharmacology, University of Leicester, Leicester, UK*
PING-YEE LAW • *Department of Pharmacology, University of Minnesota Medical School, Minneapolis, MN, USA*
HORACE H. LOH • *Department of Pharmacology, University of Minnesota Medical School, Minneapolis, MN, USA*
LILIANA PARDO LOPEZ • *Instituto de Biotechnology, Universitad Nacional Autonoma de Mexico, Cuernavaca Morelos, Mexico*
PATRICIA H. MCDONALD • *Department of Molecular Therapeutics, The Scripps Research Institute, Jupiter, FL, USA*
GRAEME MILLIGAN • *Institute of Neurosciences and Psychology, University of Glasgow, Glasgow, UK*
STEFANO MARULLO • *Institut Cochin, Université Paris Descartes, CNRS-INSERM, Paris, France*
YU MING • *Department of Clinical Neuroscience, Karolinska Institute, Stockholm, Sweden*
YING-XIAN PAN • *Molecular Pharmacology and Chemistry Program, Memorial Sloan-Kettering Cancer Center, New York City, NY, USA*
GAVRIL W. PASTERNAK • *Molecular Pharmacology and Chemistry Program, Memorial Sloan-Kettering Cancer Center, New York City, NY, USA*
DENNIS PAUL • *Department of Pharmacology and Experimental Therapeutics, Louisiana State University Health Sciences Center, New Orleans, LA, USA*
PAUL L. PRATHER • *Department of Pharmacology and Toxicology, University of Arkansas for Medical Sciences, Little Rock, AR, USA*
DMITRI PROUDNIKOV • *Laboratory of the Biology of Addictive Diseases, The Rockefeller University, New York City, NY, USA*
YU QIU • *Department of Pharmacology, University of Minnesota Medical School, Minneapolis, MN, USA*
PHILIPPE SARRET • *Department of Physiology and Biophysics, Centre des Neurosciences de Sherbrooke, Université de Sherbrooke, Sherbrooke, QC, Canada*
GREGORY W. SAWYER • *Department of Biochemistry and Microbiology, College of Osteopathic Medicine, Oklahoma State University-Center for Health Sciences, Tulsa, OK, USA*
GRÉGORY SCHERRER • *Department of Physiology and Cell Biophysics, Columbia University, New York City, NY, USA*
KATHRYN A. SEELY • *Department of Pharmacology and Toxicology, University of Arkansas for Medical Sciences, Little Rock, AR, USA*
CRAIG W. STEVENS • *Department of Pharmacology and Physiology, College of Osteopathic Medicine, Oklahoma State University-Center for Health Sciences, Tulsa, OK, USA*
HINAKO SUGA • *Department of Pharmacology, School of Medicine, University of California Irvine, Irvine, CA, USA*
LARS TERENIUS • *Department of Clinical Neuroscience, Karolinska Institute, Stockholm, Sweden*

PASCAL TÉTREAULT • *Department of Physiology and Biophysics, Centre des Neurosciences de Sherbrooke, Université de Sherbrooke, Sherbrooke, QC, Canada*
ANDREW B. TOBIN • *Department of Cell Physiology and Pharmacology, University of Leicester, Leicester, UK*
ELLEN M. UNTERWALD • *Department of Pharmacology and Center for Substance Abuse Research, Temple University School of Medicine, Philadelphia, PA, USA*
VLADANA VUKOJEVIĆ • *Department of Clinical Neuroscience, Karolinska Institute, Stockholm, Sweden*
RICHARD J. WARD • *Institute of Neuroscience and Psychology, University of Glasgow, Glasgow, UK*
LI-NA WEI • *Department of Pharmacology, University of Minnesota Medical School, Minneapolis, MN, USA*
GUANGYU WU • *Department of Pharmacology and Toxicology, Georgia Health Science University, Augusta, GA, USA*
VADIM YUFEROV • *Laboratory of the Biology of Addictive Diseases, The Rockefeller University, New York City, NY, USA*

Part I

The G Protein-Coupled Receptor in the Genome

Chapter 1

In Silico Identification of Novel G Protein-Coupled Receptors

Matthew N. Davies, David E. Gloriam, and Darren R. Flower

Abstract

The G protein-coupled receptors (GPCRs) form the largest and most multi-functional protein superfamilies known. From a drug discovery and pharmaceutical industry perspective, the GPCRs are among the most commercially and economically important groups of proteins yet identified, since they have so many vital metabolic functions and interact with such a diversity of ligands. Many distinct methodologies have been proposed to classify the GPCRs: motif-based techniques, machine learning, and several alignment-free techniques have all been used successful in this regard. This chapter reviews the available methodologies for classifying GPCRs. In particular, we allude to several innate problems in developing such approaches, such as the lack of sequence similarity between the six GPCR classes and the low sequence similarity of many newly identified family members to other GPCRs.

Key words: Receptor classification, Genomics, Machine learning, In silico

1. Introduction

The G protein-coupled receptors (GPCRs) form a large grouping of integral membrane proteins implicated in an extensive series of physiological tasks (1, 2). They turn an assortment of endogenous, extracellular signals into a restricted number of intracellular responses. A bewildering diversity of ligands can bind to GPCRs. Such ligands include ions, hormones, neurotransmitters, peptides, and proteins, as well as light, in the form of photons. Since GPCRs are implicit in physiological processes as diverse as for example, neurotransmission, cellular metabolism, secretion, inflammatory responses, and cellular differentiation (3), they have become a consistent target for the development of medicines. Roughly 50% of all marketed drugs target a GPCR (4).

Craig W. Stevens (ed.), *Methods for the Discovery and Characterization of G Protein-Coupled Receptors*, Neuromethods, vol. 60, DOI 10.1007/978-1-61779-179-6_1, © Springer Science+Business Media, LLC 2011

Most drugs targeting GPCRs have been derived through the inherently haphazard processes which permeate medicinal chemistry. Driven originally by the whims and caprices of synthetic chemistry, rather than the focussed, rationality of structure-based design, GPCR drug discovery and design is now able to make full use of the entire battery of sequence analysis tools and relevant crystal structure information, allowing *in silico* approaches to tell us much about GPCR sequences, including the potential function of newly discovered sequences.

Despite the diversity of the superfamily, there are many commonalities among these proteins. Every member of the GPCR superfamily contains seven highly conserved transmembrane segments (of 25–35 consecutive residues), each displaying a high degree of hydrophobicity. Rather than forming a perfect circle or regular ellipse, the seven membrane-crossing α-helical segments (TM1-7) form a flattened two-layer structure known as the transmembrane bundle, which is thought to be common to all GPCRs (5).

Compared to many apparently similar sets of proteins, the GPCRs exhibit a far greater conservation of structure than of sequence. Here we outline various approaches that have been used to develop GPCR classification algorithms and attempt to highlight the strengths and weaknesses of the various approaches. The approaches have important applications not only in discovering and characterising novel protein sequences but also in better understanding the relations within the GPCR superfamily.

2. Brief Overview of Nomenclature, Classification, and Repertoires

There are several difficulties in producing a comprehensive classification system for protein superfamilies (6). In this regard, the GPCRs have proved especially contentious, as family members are so numerous, and the relations between members are so multifarious and complicated. Definitive evolutionary relationships between GPCR groups remain cryptic and unclear: some receptors may have arisen through convergent evolution to adopt a particular structural scaffold, and may not even be homologous. Other GPCR families appear to have arisen through gene duplication, as demonstrated by Stevens in Chap. 5 of this book.

One of the first GPCR superfamily classification systems was introduced by Kolakowski for the now defunct GCRDb database (7), and further developed by Vriend et al. for the GPCRDB database (8–10). GPCRDB divides the superfamily into six classes. The first of which is Class A, the so-called Rhodopsin-like GPCRs, accounting for over 80% of family members across species. There are around 300 human non-olfactory Class A receptors mostly binding peptides, biogenic amines, or lipids (11). The structure of bovine rhodopsin, published in 2000, has been followed recently by those of ligand-bound avian and human β_1- and β_2-adrenoceptors, and the human A_{2A} adenosine receptor, which were determined in inactive, ligand-bound conformations (12–16).

The second class is Class B or Secretin-like GPCRs; as a group they have only weak similarity at the sequence level to Class A receptors, despite a presumed similarity of more significant proportions at the structural and functional level (17). The group is also rather smaller, with only 15 members; they bind large endogenous peptides such as glucagon or glucagon-like peptide 1 (GLP-1). Class B receptors have a large N-terminal extracellular domain of 100–160 residues, which undertakes a crucial role in ligand binding.

The third class is Class C and comprises the Metabotropic glutamate-like receptors (mGluRs). These excitatory neurotransmitter receptors are activated via an indirect metabotropic process (18). In humans, mGluRs are found principally within pre- and postsynaptic neurons in the hippocampus, cerebellum, and the cerebral cortex, as well as other regions of the brain and in the periphery (19).

The fourth class is Class D, which contains about 20 distinct proteins, and comprises highly divergent receptors for peptide pheromones (20, 21). Class D GPCRs are split between two major subfamilies: Ste2 and Ste3. There is no obvious sequence similarity between these two subfamilies, and as a group these receptors lack many features characteristic of Class A GPCRs. They have no ERY or DRY motif on TM3, no NPxxY motif on TM7, and no disulfide between the extracellular end of TM3 and loop 2.

The fifth class of GPCRs is Class E, comprising cAMP receptors from the protozoan amoeba *Dictyostelium discoideum*, which form part of several chemotactic signalling systems (18). Compared to other lower eukaryotes with sequenced genomes, *Dictyostelium* has over 55 GPCRs: including four receptors for extracellular cAMP (22, 23). In addition to class D and E, other groups of GPCRs are only found exclusively outside the subphylum vertebrata, such as the large family of nematode chemosensory receptors (24).

Finally, the sixth class is Class F, which contains Frizzled/smoothened receptors from *Drosophila*, which are necessary for Wnt binding and the mediation of hedgehog signalling respectively (25). This recently identified group of 7 TM receptors are considered the most highly divergent, especially with respect to rhodopsin (26).

An alternative, and potentially superior, sequence-based classification system has been proposed for the GPCR family (27, 28). The GRAFS classification system was developed using phylogenetic analysis (29). GRAFS divides the GPCR superfamily into the *Glutamates*, *Rhodopsins*, *Adhesions*, *Frizzled/Taste 2*, and *Secretin* families, from which the acronym GRAFS is derived. The authors of GRAFS were able successfully to differentiate pseudogenes from functional genes, and were also able to classify all human GPCR leading to the identification of several new GPCRs (30–37).

2.1. GPCR Repertoire

In spite of the high degree of structural similarity within the GPCR superfamily, a proper, unambiguous phylogenetic analysis of these proteins is next to impossible. The lack of overt sequence similarity between GPCR families makes a putative common origin very much an open question. Of the classes, adhesion and secretin families are most likely to have originated together (38). Previous best guesses put the number of GPCRs within the human genome at approximately 1% of total genes, with other estimations putting the number of GPCRs involved in olfaction at an inaccurate and unlikely additional 1,000–2,000. An early analysis by Fredriksson et al. (28) put the total number of human GPCR genes at 802, while Niimura and Nei have put the current number of olfactory receptor (OR) genes at 388 and pseudogenes at 414 (39). Subsequently, Fredriksson and co-workers, and indeed several other groups as well, have been able to identify many new rhodopsin-like and adhesion-like GPCRs in the burgeoning suite of genomes available for study. In light of these findings, the size of any genome and the number of GPCRs within it must remain educated guesses. While both will alter, particularly as the genomes of individual humans are sequenced, we can be reasonably confident that the majority genes and most GPCRs have been discovered. The Human Genome Project used a combination of protein families and protein domains to estimate that there are 616 GPCR sequences belonging to Classes A, B, and C. A motif-based approach was used whereby InterPro estimated the total number of Rhodopsin-like GPCRs to be 569 (40). Takeda and colleagues extracted approximately 950 open reading frames from the human genome that had 200–1,500 amino acid residues similar to those of GPCRs (41). The GPCR repertoires of several other species have also been published, including mouse (29), rat (34), chicken (33), pufferfish (42), among several others. The recently determined GPCR repertoire within the dog genome was shown to be more similar to that found in humans than that found in rodents (30).

2.2. GPCR Training and Test Sets for Classification Algorithms

In the machine learning scenario, a classification algorithm is trained with examples (i.e. GPCRs) with known classes and the classification model discovered from this set is used to predict the classes of further examples drawn from a separate test set, which were unseen during training (43). Issues of clarity, precision, and bias are faced when we try to define the training and test sets to be used in GCPR classification. It is clearly worth ventilating some of the more apposite issues here. Usually, one would expect verification through the use of independent test data to be ideal; however, things can be deceptive. In general, the choice of both training and testing examples is important. Predicting examples very similar to training data is typically a much easier prospect than predicting instances which are wildly unalike. Consider the classification task of predicting whether or not a protein is a GPCR. In this task the training set would contain, as positive examples, proteins known to be GPCR, while the negative examples would be proteins known not to be GPCRs. It is possible to create data sets containing positive and negative examples which will favour good validation statistics. For example, if we have a valid positive set of GPCRs, we could choose very different sequences – say small globular proteins or sequences with extreme amino acid compositions or whatever – as negative examples. However, if one chooses as negative examples proteins which are similar to GPCRs – membrane proteins of a similar size composition and a similar number of transmembrane helices – then the task would seem to become very much demanding. What can be done to circumvent such problems? We can propose that a cascade of different negative sets of increasing difficulty is likely to be a more reliable and accurate test of a method's effectiveness. Independent tests should, if possible, be conducted in a double blind fashion, since almost invariably when an author is party to an evaluation (and thus influences the choice of the test and the way that the test is conducted) it is never truly independent.

The above discussion considered the classification task of discriminating GPCRs from non-GPCRs. However, other types of classification problems can be defined for GPCRs. In particular, one can have a training set where all examples are known to be GPCRs, and then try to predict to which class a given GPCR protein belongs. For example, when developing GPCR classification algorithms, Davies et al. (44) built as large and comprehensive a dataset of GPCR sequences as possible with which to train and test the classifier. All protein sequences for the dataset were obtained from Entrez (45) using text-based searching and these were used to construct each GPCR sub-family and Class level dataset. Only human proteins sequences were incorporated, with the exception of Class D proteins, which are found only in fungi, and Class E, which is only found in *Dictyostelium*. Atypically short, and probably incomplete, GPCR sequences less than 280 amino acids were removed, as were all duplicate sequences. Thus the construction of this data set, which relied on accumulated annotations extant within database, also relied on the insight, and the bias, of the many investigators who worked on the problem over the decades.

3. Discovery of GPCRs

3.1. Full-Length Sequence Searching Approaches to the Discovery and Annotation of GPCRs

The most obvious and straightforward approach to characterising a protein sequence usually involves searching a sequence database – which contains within it sets of previously annotated sequences – using a pair-wise similarity tool, such as FastA (46) or BLAST (Basic Local Alignment Search Tool) (47). BLAST searches typically reveal obvious similarities between the query and one or more sequences in the database, as determined from pair-wise alignments along with concomitant statistical significance. Proteins are listed, as ranked by expectation or "E" values. Such values are a measure of the reliability of the similarity calculated by the method. Low E values are more significant, implying the greater reliability of the identified relatedness between the two sequences.

For a GPCR query, most proteins sitting at the top of the list, and thus evincing high sequence identity, will be true GPCRs. BLAST searches have often identified new GPCR proteins, mainly where there is detectable sequence similarity to other GPCR sequences, a situation which is becoming increasingly uncommon. Currently, this kind of obvious similarity is harder and harder to find, as bioinformaticians find themselves increasingly working at the margins.

In the present era, BLAST, while well used, is often of limited value for hunting out new members of the GPCR superfamily, since there is often a low degree of sequence similarity between the six families and between outliers within groups. An ideal result will show unambiguous similarity to a well-characterised protein over the full length of the query. However, often outputs contain no significant hits. Obviously, a more typical state of affairs would fall between such extremes, affording a list of incomplete matches to a wide variety of proteins. Many of these hits will be uncharacterised or have dubious or contradictory annotations (48). The difficulty then lies in the reliable inference of homology (the verification of a divergent evolutionary relationship) and, from this, the extrapolation to biological function.

However, as the size of sequence databases rises inexorably, and is increasingly contaminated by populations of poor quality or partial sequences, the probability of making high-scoring yet actually random matches will also rise. Moreover, if not appropriately masked, hits matching atypical regions may swamp and thus obscure search outputs. Many sections of protein sequences are atypical: some have repetitious sequences where the same pattern is repeated many times over. Others have what is called low sequence complexity, where one or two residue types are used to the exclusion of all others (49). This contrasts with normal protein sequences where the usage and repetition of each of the 20 amino acids varies little from the perfect average of 5%.

The modular or multi-domain structure of many proteins is also a problem. It may not be obvious, when matching to many concurrent domains, which corresponds correctly to the query. Even when the correct domain has been identified, direct transfer of extant functional annotation may not be appropriate, since the function of the domain may be quite different. Even if a wholly correct and validated match can be discovered, pair-wise similarity struggles to distinguish orthologues from paralogues. Thus, in sum and to lapse into the vernacular, BLAST is very much a blunt instrument, particularly for the fine-detail analysis of large and/or complex protein families.

3.2. Motif-Based Approaches to the Discovery and Annotation of GPCRs

To a first approximation, BLAST generates generic, full-length similarities between sequences, while so-called motif-based approaches focus on specific, length-restricted traits unique to families or sub-families. Many protein family databases – most famously typified by PROSITE (50) or PFAM (51), and latterly subsumed by InterPro (52), a system combining sequence profiles from several databases – are built on such an approach. They use multiple alignments to identify highly conserved regions that can form the basis of characteristic, and even diagnostic, motifs for family or subfamily membership.

Of available approaches – single motifs through to HMM models of entire sequences – perhaps the more informative are so-called "fingerprints". GPCR fingerprints have been developed using patterns of common conservation within the seven transmembrane regions (53–55). Rather than identifying a single, lone motif, fingerprinting looks at many short yet conserved regions within the sequence group. Sub-family and sub-sub-family level fingerprints are derived from segments within the TM regions, parts of the loops and parts of the N- and C-termini. False positives are readily determined since typically sequences will lack one or more of the motifs.

The PRINTS database system (55) contains within it hundreds of GPCR fingerprints. Individual motifs within such fingerprint can reflect structurally or functionally important sections of sequence, say a TM domain or a ligand-binding site. PRINTS has been demonstrated to identify similarities between receptors with low sequence similarity: it allows a user to find the GPCR superfamily to which a particular query sequence belongs (i.e. at the level of rhodopsin-like vs. secretin-like, etc.); the family to which it belongs (e.g. muscarinic vs. adrenergic, etc.); and also its subtype (i.e. muscarinic M1, M2, etc.). However, as known members become more numerous, it becomes ever harder to define fingerprints with synoptic precision. Nor can very atypical GPCR sequences be easily identified using the fingerprint method or indeed other methods.

Holden and Freitas (56) classified GPCRs using three different kinds of motifs: PROSITE patterns, PRINTS fingerprints, and

InterPro (52) entries. Three different GPCR datasets were created. Each dataset used a different set of attributes: 338 proteins and 281 attributes were derived from PRINTS; 194 proteins and 127 attributes from PROSITE; and 584 proteins and 448 attributes from InterPro. Holden and Freitas used a swarm intelligence algorithm (57) for GPCR classification. Their algorithm induced sets of IF-THEN classification rules. These took the form: *IF <set of motifs is present> THEN <predict a certain class>*. The motifs forming these sets could come from either PROSITE, PRINTS, or InterPro. The goal of this work was to find the most discriminating set of motifs which formed the most accurate rule. PRINTS motifs performed best (89.6% classification accuracy at the family level), InterPro marginally worse (86.3% classification accuracy at the family level), while PROSITE patterns performed poorly. Substantially lower accuracy rates were obtained for sub-families and below.

3.3. Machine Learning and Statistical Pattern Recognition Approaches to the Discovery and Annotation of GPCRs: Artificial Neural Networks, Hidden Markov Models, and Support Vector Machines

In many cases, conventional bioinformatics techniques, such as global sequence searching and/or motif matching, can determine useful information from a sequence through pair-wise alignment or by comparing the sequence to previously determined motifs. Although such an alignment or motif-based approach is without question valid, it may not always be optimal when trying to identify GPCRs. First, the sequence of the GPCR superfamily varies between 290 and 834 amino acids in length, meaning that many of the subfamilies cannot be effectively aligned without significant and subjective manual intervention. One should also remember that conventional biochemically based GPCR Classification schemes were created using the identity of the ligand to which the receptor binds not sequence similarity. A more computationally sophisticated if not necessarily a more effective approach to the GPCR classification problem is through use of techniques based on Machine Learning, a branch of Artificial Intelligence or statistical pattern recognition.

An example of machine learning in the analysis of GPCR data is the use of Self-Organising Maps (SOMs) (58). SOMs perform unsupervised learning (in this case, clustering) to discriminate protein families from each other. Sequences from the same family are expected to form a cluster although it cannot be assumed that the clusters will be visually recognised on the SOM output map. The overall performance of the map can be assessed using the sensitivity and specificity values as well as calculating the total accuracy of the clustering. Otaki et al. (59) reported a 97.4% precision at clustering 12 Class A sub-families using SOMs.

A Hidden Markov model or HMM is a statistical model where the system being modelled is assumed to be a Markov process with unknown parameters. In a Markov process, the probability distribution describing future states depends solely on the present state not on states prior to that: the future depends upon the present not the past. In a regular Markov model, the state is seen by the observer and thus only state transition probabilities are parameters. In an HMM, the state is not visible, although variables influenced by that state can be seen, and so the aim is to determine the hidden parameters from the observable parameters. HMMs have gained significant currency, particularly when used for sequence alignment (60).

Support Vector Machines (SVMs) are machine-learning algorithms based on statistical learning theory (61). In two-class problems, an SVM maps two sets of distinct data representing sequence descriptions onto a multi-dimensional feature space and then sets about constructing a division between the classes. The optimal division is one with a maximum distance to the closest data point from each of the two classes. Finding this optimal division is important since should another data point be added, it is easier to classify it correctly when there is a significant separation between classes. The data points nearest to the optimal division are termed support vectors. Although SVMs are more commonly used to solve 2-class problems, this technique has been applied to the classification of GPCR data with more than two classes by running the algorithm multiple times (once for each class) (62).

3.4. Alignment-Free Methods Approaches to the Discovery and Annotation of GPCRs: Proteochemometrics, Properties, and Statistics

Rather than aligning sequences, and from such alignments deducing pseudo-evolutionary relationships, alignment-independent classification systems use the physiochemical properties of amino acids to give insight into functionally or structurally important differences between sequences. To enable this process, we need to turn the symbolic structure of the protein into a set of numbers. Proteochemometrics is an example of such an approach; it has been applied to the classification of the GPCR superfamily. Proteochemometrics uses Wold's five Z values which encode key properties of the 20 biogenic amino acids (63–68). Z1 values account for amino acid lipophilicity: a large negative value corresponds to a lipophilic amino acid, and vice versa. Size or, more properly, volume properties are accounted for by Z2 values. Large negative values correspond to low volume amino acids while large positive numbers indicate amino acids with large volume and surface area. Z3 values describe amino acid polarity. Polar or hydrophilic amino acids have large positive values, while non-polar amino acids have large negative values. Recondite electronic effects are described by the Z4 and Z5 values.

Replacing each amino acid in the sequence with these five Z values, and then transforming it in some manner, reduces a protein sequence to the required numerical description. The resulting, normalised matrix is analysed using Principal Component

Analysis (PCA) and Partial Least Squares (PLS), generating a classification model. Using the proteochemometrics method, Lapnish et al. developed a model with an accuracy of 0.76 for a diverse set of amine GPCRs (67).

Kim et al. also developed a physico-chemically based classification method (69), which separates sequences into specific categories using a linear discriminant function, called a Quasi-predictor Feature Classifier (QFC) algorithm, within a statistically defined "feature space". The resulting model was used to screen databases for novel GPCRs. The QFC approach was trained on 750 GPCRs from the GPCRDB and 1,000 randomly chosen non-GPCR proteins of 200–1,000 amino acids in length. Several amino acid property scales were examined and the values normalised using a sliding window. Windows comprising 13–16 amino acids were more effective than those of 32 or 64 amino acids. Test sets of 100 GPCRs and 100 non-GPCRs were classified with a 99% accuracy vs. 530 ion channels (non-GPCR transmembrane proteins), QFC was 96.4% accurate. QFC had a higher false positive rate than many motif-based techniques, which is consistent with the approach needing more filtering.

Huang (70) used Quinlan's C4.5 algorithm (71) to induce a decision tree partitioning 4,395 GPCR sequences into 5 Classes, 39 sub-families, 93 sub-sub-families, and types. Each protein was represented as a vector comprising its normalised composition. C4.5 chooses to split data by selecting the composition feature that best discriminates the classes to be predicted. Division continues until a defined stopping criterion is attained. The technique was 86.9% accurate at the sub-family level and 81.5% accurate for sub-sub-families.

4. Prediction Servers

Servers that identify and classify GPCR family membership from their sequence now abound, and we describe many of these in coming paragraphs (see Table 1). GPCRHMM uses an HMM to recognise GPCRs specifically (72). Models are estimated using a maximum likelihood and a discriminative method. TMHMM (Transmembrane Hidden Markov Model) also predicts transmembrane helices by using an HMM. It partitions a protein sequence into the most probable distribution compared to known GPCRs (73), but has a high false positive rate, and many proteins with seven transmembrane helices are incorrectly predicted as possessing six or eight TM regions.

Pred-GPCR (74) combined FFT (Fast Fourier Transforms) and SVMs to leverage sequence hydrophobicity in the identification of GPCRs. Four hundred and three sequences from 17

Table 1
Servers for GPCR function prediction

Server name	URL	Reference
TMHMM	www.cbs.dtu.dk/services/TMHMM/	(73)
GPCRHMM	noble.gs.washington.edu/~lukall/gpcrhmm/	(74)
Pred-GPCR	athina.biol.uoa.gr/bioinformatics/PRED-GPCR/	(74)
GPCRsClass	www.imtech.res.in/raghava/gpcrsclass/	(75)
GPCRPred	www.imtech.res.in/raghava/gpcrpred/	(77)
GPCRTree	igrid-ext.cryst.bbk.ac.uk/gpcrtree/	(78)
7TMHMM	tp12.pzr.uni-rostock.de/~moeller/7tmhmm/	–

sub-families from GPCR Classes B, C, D, and F were used to train the program. Optimal performance reached an accuracy of 93.3%, and the accuracies for different subfamilies varied between 66.7 and 100%. One should bear in mind that 105 of the 403 sequences came from the atypical frizzled/smoothened family.

GPCRsClass (75) is an SVM-based server that focuses on Class A GPCRs. GPCRsClass is 99.7% accurate at dividing amine from non-GPCRs, and 92% accurate when splitting sequences into sub-subfamilies. A similar program, GPCRPred, first determines if a sequence is a GPCR, then which class it belongs to, and finally, assuming it is a Class A GPCR, to which subfamily it belongs (76). GPCR vs. non-GPCR sequences had 99.5% accuracy, the Class prediction was 97.3% accurate, and the sub-family was on average 85% accurate. The hierarchical approach to GPCR classification developed by Secker, Davies, co-workers has also been made available freely over the World Wide Web, implemented within the webserver GPCRTree (77). Certain other servers, such as the GPCR Subfamily Classifier, have now been retired from active service, while others, including a variant of the TMHMM program, called 7TMHMM, are only available for download.

5. Future Directions

Orphan GPCRs may their very nature have relatively low sequence similarity to well-characterised GPCRs with known functions and/or known ligands; thus inferring information about their function can be problematic. It may be that many such orphan receptors have ligand-independent properties, such as the constitutive regulation of cell surface GPCRs (78, 79), as suggested by a study of the Class C metabotropic γ-aminobutyric acid B ($GABA_B$) receptor, which indicated that it was a heterodimer composed of two subunits, B1 and B2 (80). $GABA_{B1}$ was responsible for the binding of the ligand while the $GABA_{B2}$ subunit promotes the efficient transport of $GABA_{B1}$.

It is also possible that many of the orphan receptors are also responsible for the regulation of non-orphan GPCR cell surface expression, in either a positive (81) or a negative way (78). If this is true then the relative expression of orphan and non-orphan GPCR proteins could be an important factor for the regulation of cell signalling. There has also been considerable interest in the tendency of GPCRs to form higher order oligomers in living cells (82). Dimeric ligands linked by spacer arms have been used to identify the importance of co-expression of certain GPCR subtypes, indicating that the formation of these oligomers is a crucial part of GPCR signalling, although the extent to which oligomerisation occurs across the whole GPCR superfamily remains uncertain.

The search for new GPCRs in a newly studied genome is typically confounded by issues that arise from the complex nature of multi-gene families: database search techniques cannot easily differentiate between proteins that have arisen by a process of speciation (so-called orthologues, where the functional counterpart of a sequence is found in another species) and those that have arisen via intra-species duplication and divergence (so-called paralogues, which may undertake related yet distinct functions within the same organism).

Examination of the current literature shows that no real consensus exists for tackling the problem of *in silico* GPCR Classification. GPCR prediction is a complicated problem that may go beyond conventional bioinformatics techniques. Classification models based upon motifs are both simple and comprehensible to the user, allowing the user to understand why a GPCR falls within a particular group. Such methods can have unacceptably high false positive and false negative rates however. Models constructed by SVMs (Support Vector Machines) or ANNs are typically totally opaque to the typical user: that is a non-computer science literate biologist with a burning interest in GPCRs but only a causal interest in search systems and protocols. Nonetheless, such techniques can be very effective. The alignment-independent methods, while showing some of the highest overall accuracy, do not allow the user to infer any information about the protein sequence other than to which family it likely belongs. Therefore, there is arguably a trade-off between the accuracy of the predictive technique and the comprehensibility of its results (83).

6. Conclusions

While many of the algorithms described here show significant accuracy, often techniques have not been assessed independently. Further benchmarking making use of several different GPCR datasets seems an obligatory next step. Moreover, a technique that can discriminate GPCRs from non-GPCRs may be markedly less successful at identifying the class, sub-family, or sub-sub-family level. Different approaches could therefore be employed at each classification level. Furthermore, all the predictive techniques have as yet been assessed using the GPCRDB Classification system. Future work in this field may need to be directed towards training algorithms based upon alternative classification systems, such as GRAFS, in order to determine the most comprehensive approach to classifying the GPCR superfamily.

Caveats aside, analysis of the GPCRs has been a success story for bioinformatics and comparative sequence analysis. It remains a key test bed for new technical approaches to classification, since the quality of the data is so much cleaner and better understood than is the case for many other families. In this regard, the GPCRs represent a strong platform on which to build.

References

1. Bissantz C (2003) Conformational changes of G protein-coupled receptors during their activation by agonist binding. J Recept Signal Transduct Res **23**:123–153
2. Tuteja N (2009) Signaling through G protein coupled receptors. Plant Signal Behav **4**:942–947
3. Hebert TE, Bouvier M (1998) Structural and functional aspects of G protein-coupled receptor oligomerization. Biochem Cell Biol **76**:1–11
4. Flower DR (1999) Modelling G-protein-coupled receptors for drug design. Biochim Biophys Acta **1422**:207–234
5. Yeagle PL, Albert AD (2007) G-protein coupled receptor structure. Biochim Biophys Acta **1768**:808–824
6. Cheng BY, Carbonell JG, Klein-Seetharaman J (2005) Protein classification based on text document classification techniques. Proteins **58**:955–970
7. Kolakowski LF (1994) GCRDb: a G-protein-coupled receptor database. Receptors Channels **2**:1–7
8. Horn F, Bettler E, Oliveira L et al (2003) GPCRDB information system for G protein-coupled receptors. Nucleic Acids Res **31**:294–297
9. Horn F, Vriend G, Cohen FE (2001) Collecting and harvesting biological data: the GPCRDB and NucleaRDB information systems. Nucleic Acids Res **29**: 346–349
10. Horn F, Weare J, Beukers MW et al (1998) GPCRDB: an information system for G protein-coupled receptors. Nucleic Acids Res **26**: 275–279
11. Fridmanis D, Fredriksson R, Kapa I et al (2007) Formation of new genes explains lower intron density in mammalian Rhodopsin G protein-coupled receptors. Mol Phylogenet Evol **43**:864–880
12. Bokoch MP, Zou Y, Rasmussen SG et al (2010) Ligand-specific regulation of the extracellular surface of a G-protein-coupled receptor. Nature **463**:108–112
13. Cherezov V, Rosenbaum DM, Hanson MA et al (2007) High-resolution crystal structure of an engineered human *beta*2-adrenergic G protein-coupled receptor. Science **318**: 1258–1265
14. Rosenbaum DM, Cherezov V, Hanson MA et al (2007) GPCR engineering yields high-resolution structural insights into *beta*2-adrenergic receptor function. Science **318**: 1266–1273

15. Day PW, Rasmussen SG, Parnot C et al (2007) A monoclonal antibody for G protein-coupled receptor crystallography. Nat Methods **4**:927–929
16. Rasmussen SG, Choi HJ, Rosenbaum DM et al (2007) Crystal structure of the human *beta*2 adrenergic G-protein-coupled receptor. Nature **450**:383–387
17. Parthier C, Reedtz-Runge S, Rudolph R et al (2009) Passing the baton in class B GPCRs: peptide hormone activation via helix induction? Trends Biochem Sci **34**:303–310
18. Pin JP, Galvez T, Prezeau L (2003) Evolution, structure, and activation mechanism of family 3/C G-protein-coupled receptors. Pharmacol Ther **98**:325–354
19. Brauner-Osborne H, Wellendorph P, Jensen AA (2007) Structure, pharmacology and therapeutic prospects of family C G-protein coupled receptors. Curr Drug Targets **8**:169–184
20. Burkholder AC, Hartwell LH (1985) The yeast *alpha*-factor receptor: structural properties deduced from the sequence of the STE2 gene. Nucleic Acids Res **13**:8463–8475
21. Eilers M, Hornak V, Smith SO et al (2005) Comparison of class A and D G protein-coupled receptors: common features in structure and activation. Biochemistry **44**:8959–8975
22. Eichinger L, Noegel AA (2005) Comparative genomics of *Dictyostelium discoideum* and *Entamoeba histolytica*. Curr Opin Microbiol **8**:606–611
23. Eichinger L, Pachebat JA, Glockner G et al (2005) The genome of the social amoeba *Dictyostelium discoideum*. Nature **435**:43–57
24. Troemel ER, Chou JH, Dwyer ND et al (1995) Divergent seven transmembrane receptors are candidate chemosensory receptors in *C. elegans*. Cell **83**:207–218
25. Prabhu Y, Eichinger L (2006) The *Dictyostelium* repertoire of seven transmembrane domain receptors. Eur J Cell Biol **85**:937–946
26. Chou KC, Elrod DW (2002) Bioinformatical analysis of G-protein-coupled receptors. J Proteome Res **1**:429–433
27. Schioth HB, Fredriksson R (2005) The GRAFS classification system of G-protein coupled receptors in comparative perspective. Gen Comp Endocrinol **142**:94–101
28. Fredriksson R, Lagerstrom MC, Lundin LG et al (2003) The G-protein-coupled receptors in the human genome form five main families. Phylogenetic analysis, paralogon groups, and fingerprints. Mol Pharmacol **63**:1256–1272
29. Bjarnadottir TK, Gloriam DE, Hellstrand SH et al (2006) Comprehensive repertoire and phylogenetic analysis of the G protein-coupled receptors in human and mouse. Genomics **88**:263–273
30. Haitina T, Fredriksson R, Foord SM et al (2009) The G protein-coupled receptor subset of the dog genome is more similar to that in humans than rodents. BMC Genomics **10**:24
31. Gloriam DE, Fredriksson R, Schioth HB (2007) The G protein-coupled receptor subset of the rat genome. BMC Genomics **8**:338
32. Nordstrom KJ, Mirza MA, Larsson TP et al (2006) Comprehensive comparisons of the current human, mouse, and rat RefSeq, Ensembl, EST, and FANTOM3 datasets: identification of new human genes with specific tissue expression profile. Biochem Biophys Res Commun **348**:1063–1074
33. Lagerstrom MC, Hellstrom AR, Gloriam DE et al (2006) The G protein-coupled receptor subset of the chicken genome. PLoS Comput Biol **2**:e54
34. Gloriam DE, Schioth HB, Fredriksson R (2005) Nine new human Rhodopsin family G-protein coupled receptors: identification, sequence characterisation and evolutionary relationship. Biochim Biophys Acta **1722**: 235–246
35. Gloriam DE, Bjarnadottir TK, Yan YL et al (2005) The repertoire of trace amine G-protein-coupled receptors: large expansion in zebrafish. Mol Phylogenet Evol **35**:470–482
36. Gloriam DE, Bjarnadottir TK, Schioth HB et al (2005) High species variation within the repertoire of trace amine receptors. Ann N Y Acad Sci **1040**:323–327
37. Bjarnadottir TK, Fredriksson R, Hoglund PJ et al (2004) The human and mouse repertoire of the adhesion family of G-protein-coupled receptors. Genomics **84**:23–33
38. Nordstrom KJ, Lagerstrom MC, Waller LM et al (2009) The secretin GPCRs descended from the family of adhesion GPCRs. Mol Biol Evol **26**:71–84
39. Niimura Y, Nei M (2003) Evolution of olfactory receptor genes in the human genome. Proc Natl Acad Sci USA **100**:12235–12240
40. Lander ES, Linton LM, Birren B et al (2001) Initial sequencing and analysis of the human genome. Nature **409**:860–921
41. Takeda S, Kadowaki S, Haga T et al (2002) Identification of G protein-coupled receptor genes from the human genome sequence. FEBS Lett **520**:97–101
42. Metpally RP, Sowdhamini R (2005) Genome wide survey of G protein-coupled receptors in *Tetraodon nigroviridis*. BMC Evol Biol **5**:41
43. Davies MN, Gloriam DE, Secker A et al (2007) Proteomic applications of automated GPCR classification. Proteomics **7**:2800–2814

44. Davies MN, Secker A, Freitas AA et al (2007) On the hierarchical classification of G protein-coupled receptors. Bioinformatics **23**:3113–3118
45. Wheeler DL, Barrett T, Benson DA et al (2008) Database resources of the National Center for Biotechnology Information. Nucleic Acids Res **36**:D13–21
46. Lipman DJ, Pearson WR (1985) Rapid and sensitive protein similarity searches. Science **227**:1435–1441
47. Altschul SF, Madden TL, Schaffer AA et al (1997) Gapped BLAST and PSI-BLAST: a new generation of protein database search programs. Nucleic Acids Res **25**:3389–3402
48. Altschul SF, Boguski MS, Gish W et al (1994) Issues in searching molecular sequence databases. Nat Genet **6**:119–129
49. Wootton JC (1994) Non-globular domains in protein sequences: automated segmentation using complexity measures. Comput Chem **18**:269–285
50. Sigrist CJ, Cerutti L, de Castro E et al (2010) PROSITE, a protein domain database for functional characterization and annotation. Nucleic Acids Res **38**:D161–166
51. Finn RD, Mistry J, Tate J et al (2010) The Pfam protein families database. Nucleic Acids Res **38**:D211–222
52. Hunter S, Apweiler R, Attwood TK et al (2009) InterPro: the integrative protein signature database. Nucleic Acids Res **37**:D211–215
53. Attwood TK (2001) A compendium of specific motifs for diagnosing GPCR subtypes. Trends Pharmacol Sci **22**:162–165
54. Flower DR, Attwood TK (2004) Integrative bioinformatics for functional genome annotation: trawling for G protein-coupled receptors. Sem Cell Dev Biol **15**:693–701
55. Mulder NJ, Apweiler R, Attwood TK et al (2007) New developments in the InterPro database. Nucleic Acids Res **35**:D224–228
56. Holden N, Freitas AA (2005) A hybrid particle swarm/ant colony algorithm for the classification of hierarchical biological data. 2005 IEEE Swarm Intelligence Symp: 100–107
57. Holden N, Freitas AA (2009) Hierarchical classification of protein function with ensembles of rules and particle swarm optimisation. Soft Computing **13**:259–272
58. Yan AX (2006) Application of self-organizing maps in compounds pattern recognition and combinatorial library design. Comb Chem High Throughput Screen **9**:473–480
59. Otaki JM, Mori A, Itoh Y et al (2006) Alignment-free classification of G-protein-coupled receptors using self-organizing maps. J Chem Inform Mod **46**:1479–1490
60. Gollery M (2008) Handbook of hidden Markov models in bioinformatics. CRC Press, Boca Raton
61. Cristianini N, Shawe-Taylor J (2000) An introduction to Support Vector Machines and other kernel-based learning methods. Cambridge University Press, Cambridge
62. Lorena AC, de Carvalho ACPLF (2004) Comparing techniques for multiclass classification using binary SVM predictors. Adv Artifical Intell **2972**:272–281
63. Strombergsson H, Prusis P, Midelfart H et al (2006) Rough set-based proteochemometrics modeling of G-protein-coupled receptor-ligand interactions. Prot Struct Func Bioinf **63**:24–34
64. Lapinsh M, Prusis P, Uhlen S et al (2005) Improved approach for proteochemometrics modeling: application to organic compound-amine G protein-coupled receptor interactions. Bioinformatics **21**:4289–4296
65. Lapinsh M, Prusis P, Lundstedt T et al (2002) Proteochemometrics modeling of the interaction of amine G-protein coupled receptors with a diverse set of ligands. Mol Pharm **61**:1465–1475
66. Lapinsh M, Prusis P, Gutcaits A et al (2001) Development of proteo-chemometrics: a novel technology for the analysis of drug-receptor interactions. Biochimica Et Biophysica Acta-Gen Subj **1525**:180–190
67. Lapinsh M, Gutcaits A, Prusis P et al (2002) Classification of G-protein coupled receptors by alignment-independent extraction of principal chemical properties of primary amino acid sequences. Protein Sci **11**:795–805
68. Freyhult E, Prusis P, Lapinsh M et al (2005) Unbiased descriptor and parameter selection confirms the potential of proteochemometric modelling. BMC Bioinformatics **6**:50.
69. Kim J, Moriyama EN, Warr CG et al (2000) Identification of novel multi-transmembrane proteins from genomic databases using quasi-periodic structural properties. Bioinformatics **16**:767–775
70. Huang Y, Cai J, Ji L, Li Y (2004) Classifying G-protein coupled receptors with bagging classification tree. Comput Biol Chem **28**:275–280
71. Quinlan JR (1993) C4.5: programs for machine learning. Morgan Kaufmann, San Mateo
72. Wistrand M, Kall L, Sonnhammer EL (2006) A general model of G protein-coupled receptor sequences and its application to detect remote homologs. Protein Sci **15**:509–521

73. Inoue Y, Yamazaki Y, Shimizu T (2005) How accurately can we discriminate G-protein-coupled receptors as 7-TM protein sequences from other sequences? Biochem Biophys Res Commun **338**:1542–1546
74. Papasaikas PK, Bagos PG, Litou ZI et al (2004) PRED-GPCR: GPCR recognition and family classification server. Nucleic Acids Res **32**:W380–382
75. Bhasin M, Raghava GP (2005) GPCRsclass: a web tool for the classification of amine type of G-protein-coupled receptors. Nucleic Acids Res **33**:W143–147
76. Bhasin M, Raghava GP (2004) GPCRpred: an SVM-based method for prediction of families and subfamilies of G-protein coupled receptors. Nucleic Acids Res **32**:W383–389
77. Davies MN, Secker A, Halling-Brown M et al (2008) GPCRTree: online hierarchical classification of GPCR function. BMC Res Notes **1**:67
78. Levoye A, Dam J, Ayoub MA et al (2006) Do orphan G-protein-coupled receptors have ligand-independent functions? New insights from receptor heterodimers. EMBO Rep 7:1094–1098
79. Levoye A, Jockers R, Ayoub MA et al (2006) Are G protein-coupled receptor heterodimers of physiological relevance? Focus on melatonin receptors. Chronobiol Int **23**:419–426
80. Pin JP, Kniazeff J, Liu J et al (2005) Allosteric functioning of dimeric class C G-protein-coupled receptors. FEBS J **272**:2947–2955
81. Milasta S, Pediani J, Appelbe S et al (2006) Interactions between the Mas-related receptors MrgD and MrgE alter signalling and trafficking of MrgD. Mol Pharmacol **69**: 479–491
82. Casado V, Cortes A, Mallol J et al (2009) GPCR homomers and heteromers: a better choice as targets for drug development than GPCR monomers? Pharmacol Ther **124**: 248–257
83. Freitas AA, Weiser DC, Appweiler R (2010) On the importance of comprehensible classification model for protein function prediction. IEEE/ACM Trans Comp Biol Bioinform 7:10

Chapter 2

Alternative Pre-mRNA Splicing of G Protein-Coupled Receptors

Ying-Xian Pan, Steven Grinnell, and Gavril W. Pasternak

Abstract

Alternative pre-mRNA splicing involves editing of a gene to generate a number of different mRNAs and proteins. It provides a mechanism for only 20,000 genes to generate hundreds of thousands of proteins. Like other proteins, it is estimated that 50% of G protein-coupled receptors undergo alternative splicing. While most commonly involving either the N-terminus or C-terminus, some variants have modifications in the interior of the receptor. Alternative splicing generates functionally distinct variants, due to an intrinsic difference in transduction or location. These features are well illustrated by the *mu* opioid receptor gene, *OPRM1*, which undergoes extensive alternative splicing.

Key words: Alternative splicing, MOR-1, OPRM1, Splice variants, *Mu* opioid receptor, Truncated variants

1. Introduction

One of the surprises of the molecular biology revolution was the realization that the human genome contains only approximately 20,000 genes, far fewer than early estimates and not that many more than found in lower species. On the surface, this appears to be inadequate to generate the number of proteins found in humans, but through the mechanism of alternative pre-mRNA splicing in higher eukaryotic organisms are able to generate hundreds of thousands or more proteins from a far more limited set of genes. More than 90% of human genes produce more than one splice variant, including about 52% of G protein-coupled receptors (GPCRs) (1).

GPCRs are among the most abundant class of receptors, with hundreds having been cloned or implied from genomic studies. A very large percentage of the current drugs used in medicine

Craig W. Stevens (ed.), *Methods for the Discovery and Characterization of G Protein-Coupled Receptors*, Neuromethods, vol. 60, DOI 10.1007/978-1-61779-179-6_2, © Springer Science+Business Media, LLC 2011

target these receptors, emphasizing their importance both physiologically and pharmacologically. Structurally, these receptor are often termed serpentine, since they traverse the membrane 7 times, with the N-terminus on the outside of the cell and the C-terminus inside. About half of the GPCR genes contain a single coding exon and thus generate a single protein. However, the remainders of the GPRC genes have the potential of producing multiple proteins through the mechanism of alternative splicing.

1.1. Splicing Mechanisms

Genes are composed of exons and introns. Pre-mRNA splicing involves removing introns and joining exons together to produce mature mRNA. Pre-mRNA splicing is carried out by the spliceosome, a large complex composed of five small nuclear RNAs and more than 100 proteins (2, 3). In higher eukaryotes, many genes undergo alternative splicing to create protein diversity. For example, over 90% of genes in human undergo alternative splicing, producing a number of proteins far greater than the number of genes. There are many different patterns of alternative splicing. These include exon inclusion/skipping, alternative 5′ splicing and/or 3′ splicing, intron retention, mutually exclusive exons, alternative promoters, and alternative polyadenylation sites (Fig. 1) (4). All these patterns have been described in alternative

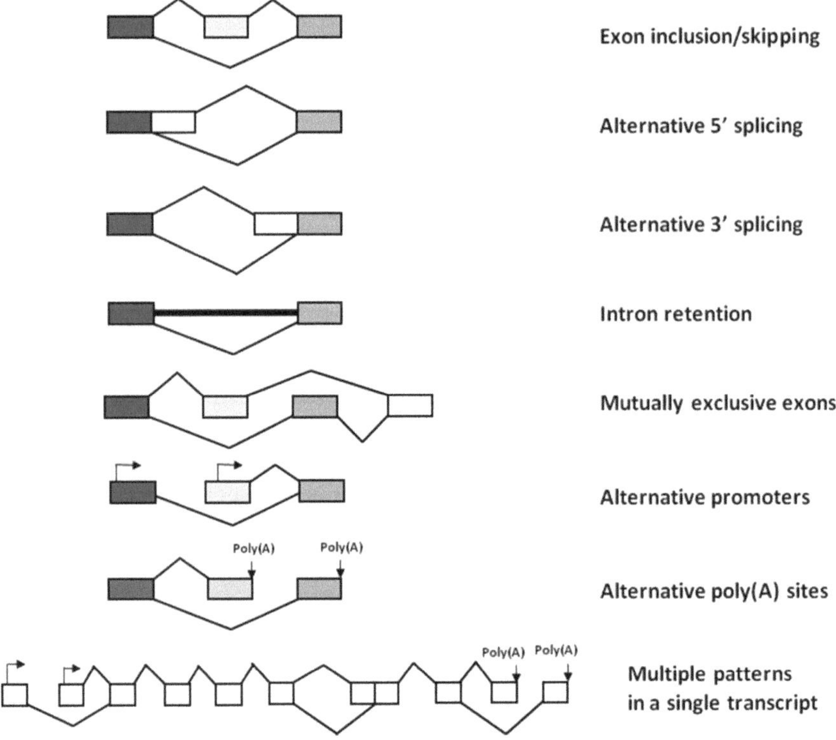

Fig. 1. Schematic on alternative splicing. Exons are indicated by *boxes* and introns are shown by *horizontal line* or *blanked*. Splicing is shown by connecting lines among exons. Promoters and polyA sites are indicated by *arrows*. Adapted from (4).

splicing of GPCRs. Alternative splicing can lead to generating multiple protein products or isoforms from single gene. These protein products can encode truncated or novel proteins by losing segments of the protein or gaining new protein sequences, or changing the amino acid sequence through a reading-frame shifting or an early stop codon. Alternative splicing of GPCRs often leads to truncations at the N-terminus, C-terminus and transmembrane segments, or generation of new N-terminal and C-terminal sequences. Different protein isoforms generated through alternative splicing have diverse functions, which can be related or not to the original protein. In GPCRs, alternative splicing has been demonstrated to modulate receptor structures related to ligand binding, G protein-coupling, receptor membrane targeting, and internalization. Alternative splicing is also highly regulated in different tissues and cell types, as well as in various developmental stages, providing their functions in spatial and temporal fashion. Alternative splicing provides a mechanism for on/off switching of gene regulation through truncation or nonsense-mediated decay (NMD).

Detailed steps and mechanisms involving intron removal and exon ligation have been well established (2–4). However, splicing is not uniform from cell to cell, with some cells making one set of isoforms and other cells a different set. Unfortunately, the mechanisms modulating alternative splicing remain largely unknown. Alternative splicing in a number of genes such as the c-src (5–8), Troponin T (9–11), and FGFR2 (12–14) has been extensively studied (9), which provides great insights to understanding the mechanisms of alternative splicing. Isolation of a number of trans-acting factors that regulate alternative splicing, such as the serine/arginine-rich (SR) family proteins (15–17), hnRNPs, the CELF protein family (18–20), and the neuro-oncological ventral antigen (NOVA) family (21–23), has greatly facilitated our understanding of alternative splicing mechanisms. Generally, alternative splicing involves coordination of multiple components, including cis-acting elements located within the exons and proximal or distal introns, trans-acting factors that interact with the cis-acting elements, and their interactions with the basal spliceosome in a particular cell environment or in response to an extracellular stimulus.

1.2. Alternative Splicing of GPCRs: An Overview

One of the first examples of alternative splicing of a GPCR was the dopamine D_2 receptor (24, 25). Here, a 29 amino acid insertion in the third intracellular loop distinguishes between D_{2L} and D_{2S}. This is interesting in that this region of the receptor has been implicated with coupling to G proteins. Since then, a wide range of GPCRs have been noted to have splice variants (1).

GPCR splicing involves a series of distinct forms (1). Splicing can occur at the N-terminus through the use of different promoters, exon inclusion, alternative donor/acceptor sites, or exon deletion. These changes may, or may not, impact binding. More commonly, receptor isoforms are generated from splicing at the C-terminus (Table 1). Some involve changes in the length of the C-terminus while others change the sequences. These variants may have different signaling properties based upon the changes in the C-terminus. In others, the insertion of a stop codon results in truncation of the C-terminus or even the loss of transmem-

Table 1
Splicing patterns of GPCRs

N-terminus	Intracellular loops
Corticotropin-releasing hormone receptor type 2	Corticotropin-releasing hormone (CRH-R1) receptor type 1
Parathyroid (PTH) receptor	D_2 dopamine receptor
Mu opioid receptor MOR-1	H_3 histamine receptor
C-terminus	Cholecystokinin-B (CCK_B) receptor
Prostaglandin EP_3 receptor	Pituitary adenylate cyclase-activating peptide (PAC_1) receptor
Prostaglandin $F_{2\alpha}$ receptor	**Extracellular loops**
Parathyroid hormone (PTH) receptor	Calcitonin (CT) receptor
α_{1A}-adrenoreceptor	D3 dopamine receptor
$GABA_B$ receptor	Orphanin FQ/nociceptin (OFQ/N) receptor
mGlu receptor	**Shortened TM7**
Mu opioid receptor	Corticotropin-releasing hormone (CRH-R1) receptor type 1
Serotonin or 5-hydroxytryptamine (5HT) receptor	Calcitonin (CT) receptor
Somatostatin receptor	Parathyroid (PTH) receptor
Thyrotropin-releasing hormone (TRH) receptor type 1	VIP/Pituitary $VPAC_2$ receptor
Neurokinin-1 (NK_1) receptor	**Soluble isoforms**
Gonadotropin-releasing hormone (GRH) receptor	Luteinizing hormone receptor
Follicle stimulating hormone (FSH) receptor[a]	Metabotropic glutamate (mGlu) receptor
Thromboxane	Corticotropin-releasing hormone (CRH-R1) receptors
Neuropeptide S receptor	

[a]Truncation converts the receptor from a GPCR to a growth factor receptor (1)

brane domains. Some of these truncated versions act as dominant negatives, blocking the actions of the active receptor.

Other types of splicing also are seen, including splicing into either the intracellular or extracellular loops. There are even examples where splice variants generate a soluble binding site unable to generate a signal.

2. A Case-Study in Alternative Splicing: The *OPRM1* Gene

The *mu* opioid receptor is one of the most intensively spliced GPCRs (Fig. 2). After its initial isolation (26–29), a wide range of MOR-1 variants have been described using all the splicing mechanisms, with similar patterns in mice, rats, and humans. The primary series of variants involve C-terminal splicing, resulting in loss of the 12 amino acids in MOR-1 and their replacement by alternative, unique sequences (30–43). There are also a second set of variants generated by a second promoter associated with exon 11, located approximately 30 kb upstream of exon 1 and its promoter (32, 39, 44). The mouse exon 11-associated variants generate nine different splice variants (Fig. 3). Three of these also contain exon 1 and predict the same protein as in MOR-1 itself. However, the others predict truncated versions containing only six transmembrane domains (Fig. 3), lacking the first TM seen in the full length variants. However, there are also several variants encoding only a single transmembrane domain encoded by exon 1 (45). What makes this interesting is that this is the same transmembrane domain missing in the exon 11-associated 6TM variants. The only splicing pattern not yet identified in the *mu* opioid receptor involves splicing in either the intracellular or extracellular loops. Similar splicing with both 6 and 1TM variants has been observed in humans (44).

2.1. Biochemical Assessment of MOR-1 Splice Variants

The C-terminus full length variants were the first to be cloned and characterized. As shown in Fig. 4, they all are identical, except for the tip of the C-terminus. Since they have identical binding pockets, it is not surprising that all are highly selective for *mu* ligands and bind them with similar affinities, the only exception being mouse MOR-1B4 (39). This variant is quite unusual in many respects and needs further characterization.

Although there were few differences in binding affinity among the full length variants, significant differences were noted functionally, as assessed by opioid-stimulated ^{35}S-GTPγS binding (30, 32–35, 38, 44, 46). A number of *mu* opioids showed different potencies among the variants. More interesting, however, were the changes from variant to variant in the rank-order of the potencies and efficacies of the drugs, which also varied independently of each other.

Another interesting difference among the variants was their regional distributions. The regional distribution of *mu* opioid receptors has been well described, from early autoradiographic

a ***The mouse Oprm gene structure and alternative splicing***

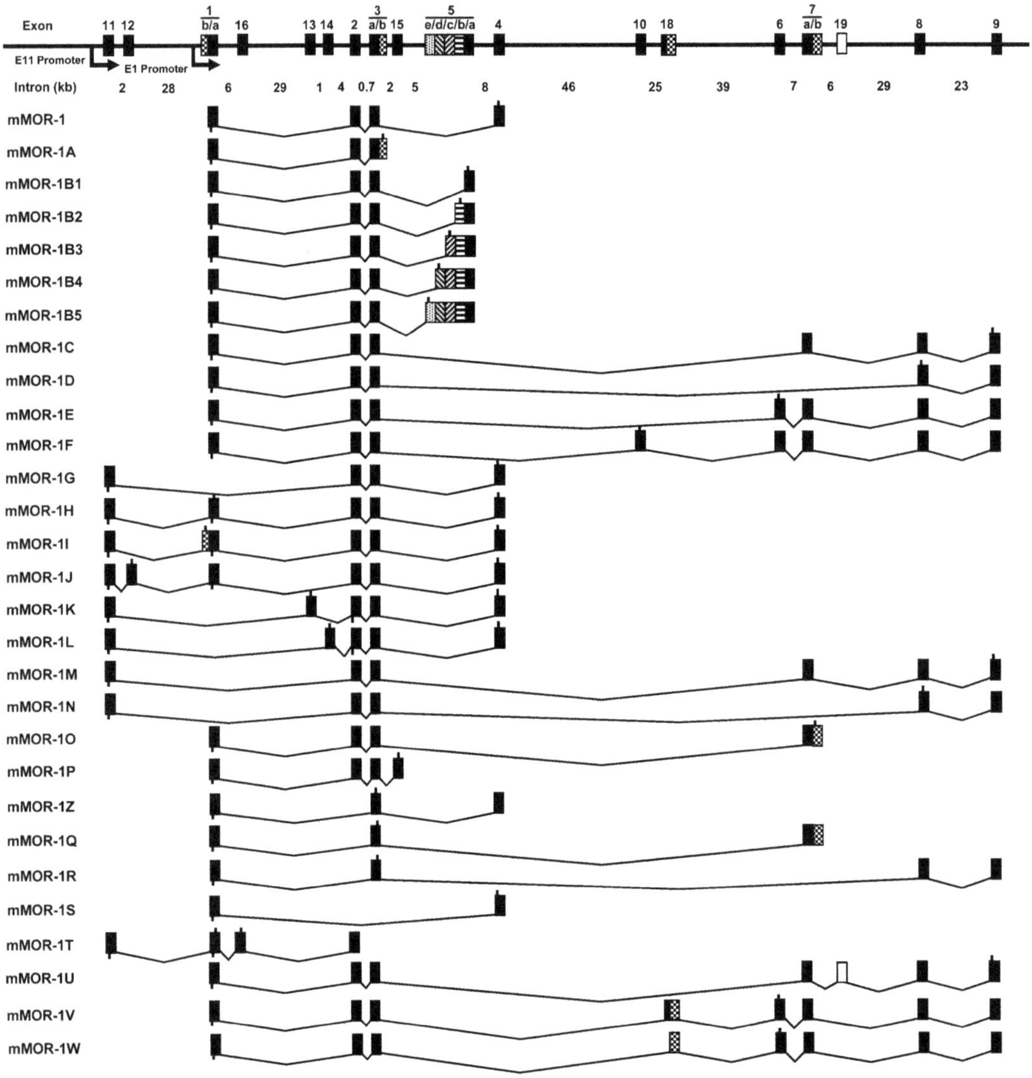

Fig. 2. Schematic of gene structure and alternative splicing of *OPRM1* genes. (**a**) The mouse *OPRM1* gene structure and alternative splicing. (**b**) The rat *OPRM1* gene structure and alternative splicing. (**c**) The human *OPRM1*gene structure and alternative splicing. Exons and introns are shown by *boxes* and *horizontal lines*, respectively. Promoters are indicated by *arrows*. Exons are numbered in the order in which they were identified. Translation start and stop points are shown by bars below and above exon boxes, respectively.

b ***The rat Oprm gene structure and alternative splicing***

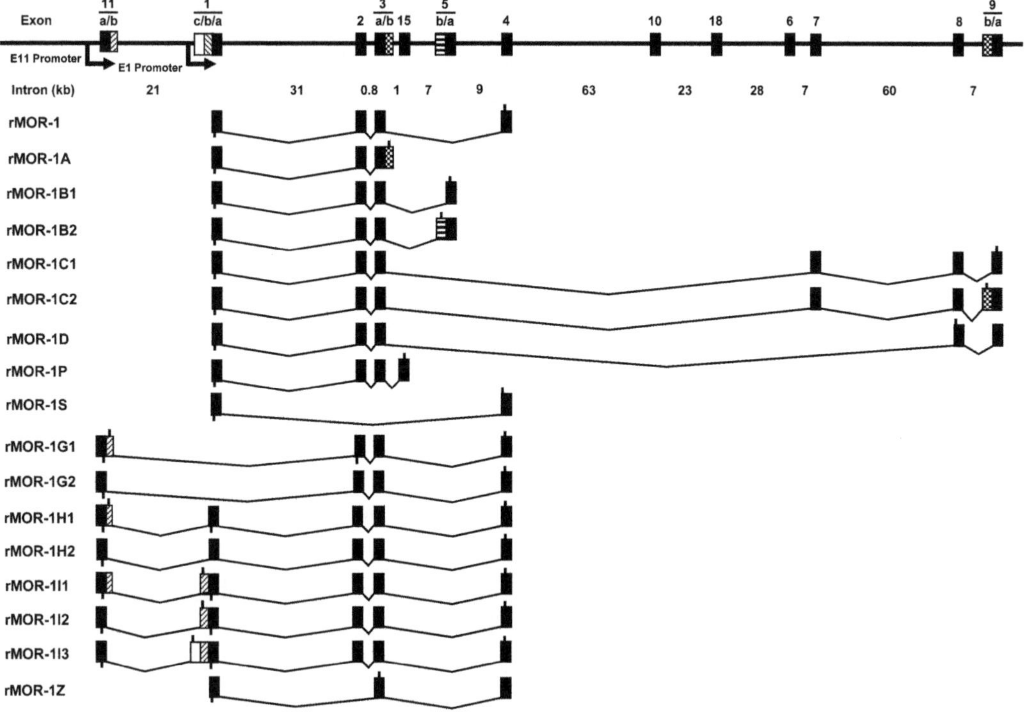

c ***The human Oprm gene structure and alternative splicing***

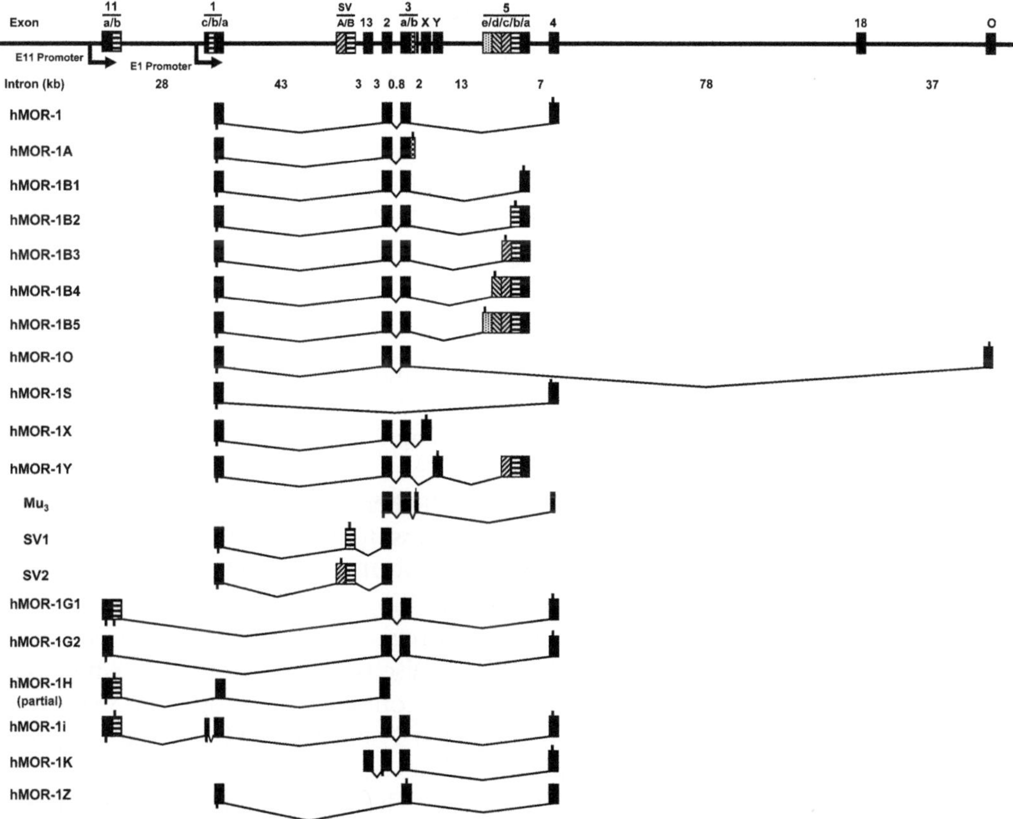

Fig. 2. (continued)

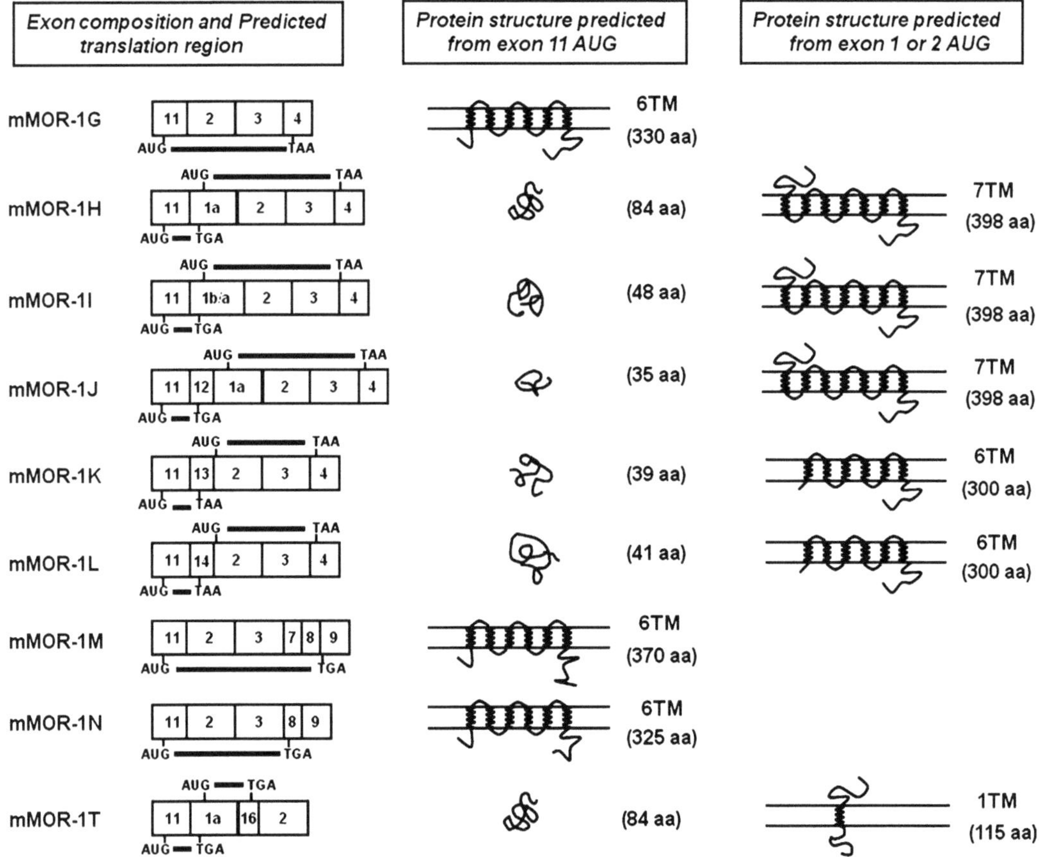

Fig. 3. Schematic of exon composition and predicted protein structure of exon 11-containing mouse MOR (mMOR) variants. The *left* panel shows the exon composition of the nine mMOR variants. The *middle* panel shows the predicted protein structures from exon 11 AUG start site. The *right* panel shows the predicted protein structures from the exon 1 or 2 AUG start sites.

approaches (47–50) to immunohistochemistry and *in situ* hybridization of the primary variant, MOR-1 (51–53). However, only recently have we obtained insights into the distributions of the variants.

A major question is whether a similar splicing pattern for the gene exists in all cells expressing it or whether splicing patterns differ from cell to cell. With MOR-1, the evidence for cell/region-specific splicing is strong. Using RT-PCR to amplify the various variants in different brain regions, their patterns differed markedly (Figs. 5 and 6). There also is evidence at the protein level for cell-specific splicing. Using confocal microscopy, epitopes associated with MOR-1 (i.e., exon 4) and MOR-1C (i.e., exons 7/8/9) labeled different, oftentimes adjacent, cells in the dorsal horn of the spinal cord (54, 55). The region- and cell-specific alternative splicing provides important insights of understanding the

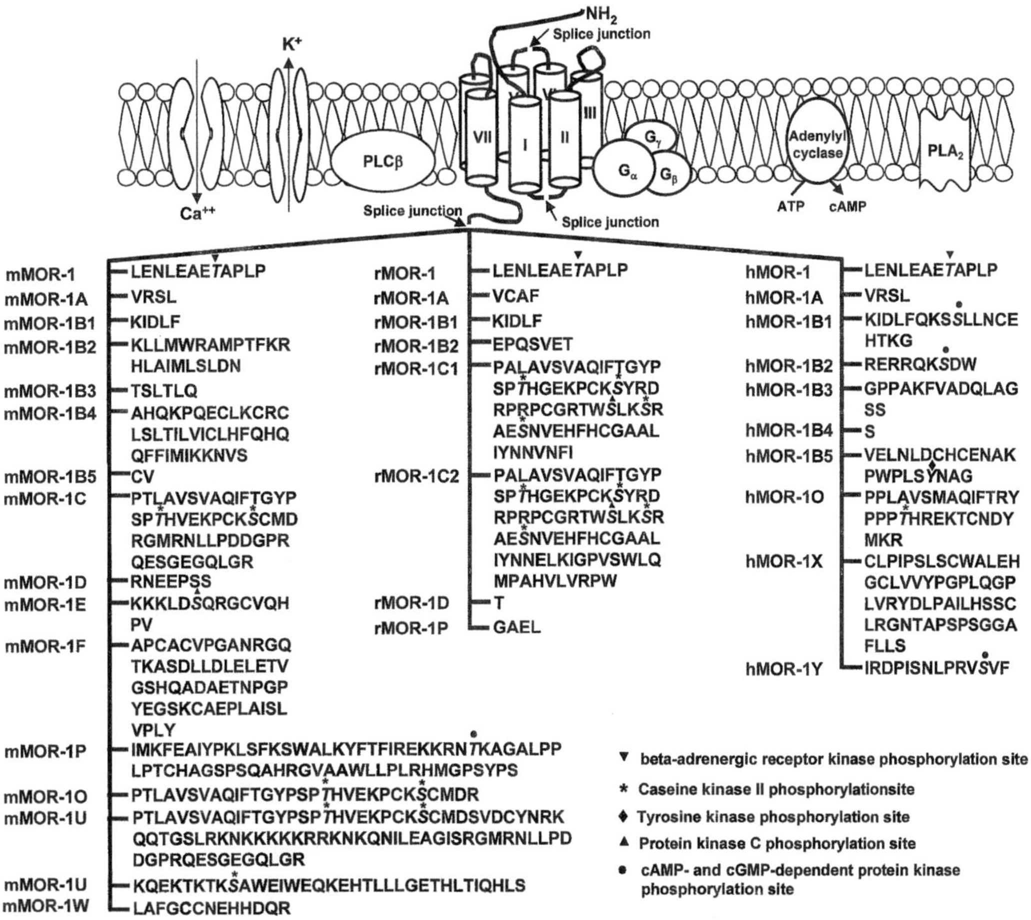

NH2
Splice junction
K+
Ca++
PLCβ
VII
I
II
III
Gγ
Gα
Gβ
Adenylyl cyclase
PLA2
ATP
cAMP
Splice junction
Splice junction
mMOR-1 LENLEAETAPLP
mMOR-1A VRSL
mMOR-1B1 KIDLF
mMOR-1B2 KLLMWRAMPTFKR HLAIMLSLDN
mMOR-1B3 TSLTLQ
mMOR-1B4 AHQKPQECLKCRC LSLTILVICLHFQHQ QFFIMIKKNVS
mMOR-1B5 CV
mMOR-1C PTLAVSVAQIFTGYP SPTHVEKPCKSCMD RGMRNLLPDDGPR QESGEGQLGR
mMOR-1D RNEEPSS
mMOR-1E KKKLDSQRGCVQH PV
mMOR-1F APCACVPGANRGQ TKASDLLDLELETV GSHQADAETNPGP YEGSKCAEPLAISL VPLY
mMOR-1P IMKFEAIYPKLSFKSWALKYFTFIREKKRNTKAGALPP LPTCHAGSPSQAHRGVAAWLLPLRHMGPSYPS
mMOR-1O PTLAVSVAQIFTGYPSPTHVEKPCKSCMDR
mMOR-1U PTLAVSVAQIFTGYPSPTHVEKPCKSCMDSVDCYNRK QQTGSLRKNKKKKRRKNKQNILEAGISRGMRNLLPD DGPRQESGEGQLGR
mMOR-1U KQEKTKTKSAWEIWEQKEHTLLLGETHLTIQHLS
mMOR-1W LAFGCCNEHHDQR
rMOR-1 LENLEAETAPLP
rMOR-1A VCAF
rMOR-1B1 KIDLF
rMOR-1B2 EPQSVET
rMOR-1C1 PALAVSVAQIFTGYP SPTHGEKPCKSYRD RPRPCGRTWSLKSR AESNVEHFHCGAAL IYNNVNFI
rMOR-1C2 PALAVSVAQIFTGYP SPTHGEKPCKSYRD RPRPCGRTWSLKSR AESNVEHFHCGAAL IYNNELKIGPVSWLQ MPAHVLVRPW
rMOR-1D T
rMOR-1P GAEL
hMOR-1 LENLEAETAPLP
hMOR-1A VRSL
hMOR-1B1 KIDLFQKSSLLNCE HTKG
hMOR-1B2 RERRQKSDW
hMOR-1B3 GPPAKFVADQLAG SS
hMOR-1B4 S
hMOR-1B5 VELNLDCHCENAK PWPLSYNAG
hMOR-1O PPLAVSMAQIFTRY PPPTHREKTCNDY MKR
hMOR-1X CLPIPSLSCWALEH GCLVVYPGPLQGP LVRYDLPAILHSSC LRGNTAPSPSGGA FLLS
hMOR-1Y IRDPISNLPRVSVF
▼ beta-adrenergic receptor kinase phosphorylation site
* Caseine kinase II phosphorylationsite
♦ Tyrosine kinase phosphorylation site
▲ Protein kinase C phosphorylation site
● cAMP- and cGMP-dependent protein kinase phosphorylation site

Fig. 4. Structure of full length C-terminal variants of mouse, rat, and human MOR.

a
Cortex
Cerebellum
Striatum
Thalamus
Hypothalamus
Brain Stem
Spinal Cord
PAG
Whole Brain
MOR-1E
MOR-1C
MOR-1D
β2MG
b
MOR-1
β2MG

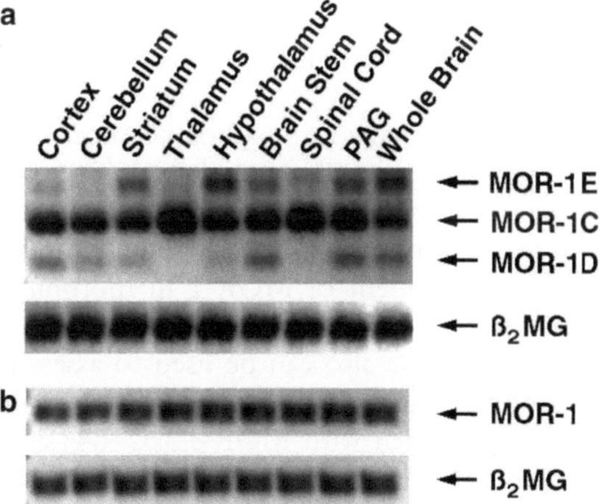

Fig. 5. Regional expression of MOR-1 splice variants. (**a**) Regional distribution of the mMOR-1C, mMOR-1D, mMOR-1E, and (**b**) mMOR-1 mRNAs determined by RT-PCR. β-microglobulin (MG) was used as RNA loading control.

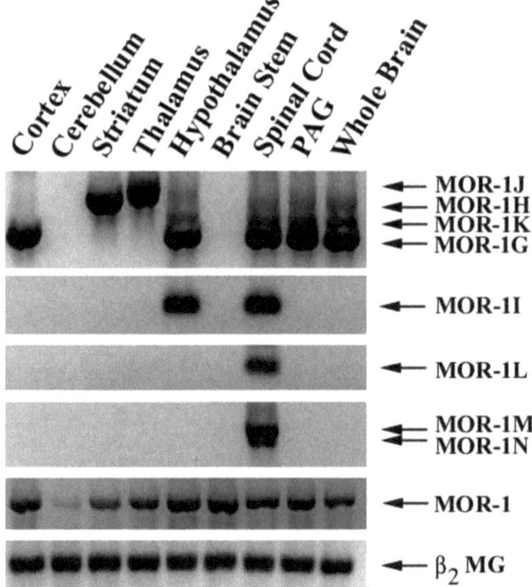

Fig. 6. Regional distribution of the exon 11-associated variants and mMOR-1 mRNAs determined by RT-PCR. β-microglobulin (MG) was used as RNA loading control.

functions of the splice variants. Dissecting the splicing mechanisms leading to the preferred generation of one, as opposed to another variant, will be quite interesting.

2.2. Pharmacological Assessment of OPRM1 Splice Variants

The isolation of such a vast array of MOR-1 splice variants immediately raises the question of why they are needed. Thus, their functional evaluation *in vivo* is important. One advantage of the opioid system is its well-established pharmacology. The most important action remains analgesia, although inhibition of gastrointestinal transit and respiratory depression also is easily measurable. Functionally significant differences among the splice variants *in vivo* have been suggested by both antisense approaches and by knockout mice.

2.2.1. Antisense

Antisense involves the administration of short oligodeoxynucleotides complementary to the sequence of the mRNA being targeted. Annealing of the antisense to the mRNA leads to its degradation. Following the initial cloning of MOR-1 in 1993, the involvement of MOR-1 in morphine analgesia was quickly established using an antisense targeting exon 1 (56–60).

Antisense also can be used to assess potential roles of splice variants, initially shown with the *delta* opioid receptor DOR-1 (61) and then with MOR-1 (62–64) and neuronal nitric oxide synthase (nNOS) (65). DOR-1 has three exons. In this study, antisense probes targeting each of the three exons within the mRNA effectively downregulated DOR-1 mRNA and protein (61). Thus, it was not necessary to target the antisense to the 5′

end of the mRNA to see an effective downregulation of the mRNA and protein. This ability to target individual exons within the mRNA opened the possibility of assessing splice variants by targeting sequences in one, but not the other, variants. Variants with exon skipping also can be assessed by targeting the region of the splice site, as demonstrated with nNOS (65). This approach has been termed antisense mapping.

Antisense mapping MOR-1 revealed some very interesting findings (62–64). Probes targeting exon 1 and 4 clearly impaired morphine analgesia. However, little effect was seen with probes targeting exon 2 or 3. Conversely, the exon 2 and 3 probes downregulated morphine-6β-glucuronide (M6G), a potent *mu* compound, while the same exon 1 and 4 probes that downregulated morphine analgesia were ineffectual against M6G. These findings implied different MOR-1 mechanisms for morphine analgesia as opposed to M6G analgesia at the molecular level, although many questions remained.

2.2.2. Knockout Mice

Antisense has many advantages, including the ease of its use and its ability to assess splice variants. However, it does not eliminate the targeted mRNA, but rather downregulates it, often by only 50–70%. Other targeting approaches, such as siRNA, have similar advantages and disadvantages (66). Disrupting the gene itself offers a way to eliminate the protein of interest completely. However, it has its own limitations. When dealing with behavior, it is not uncommon to see compensatory mechanisms develop, making the assessment of the disrupted gene difficult. However, a more intriguing issue arose with studies on nNOS in which disruption of specific exon led to the loss of only splice variants containing that exon (67). This is important, since a similar issue has arisen with MOR-1.

Several knockout mice targeting MOR-1 have been reported (68–72). Morphine lost its activity in all of them. However, the mouse generated by Pintar and colleagues provided an interesting model (72). Although it targeted exon 1, a number of splice variants associated with exon 11 were still expressed. Like the other MOR-1 knockout mice, morphine was inactive. However, both heroin and M6G retained full activity in these Pintar mice, although with a slight decrease in potency (72). Thus, as in the earlier antisense studies, these results implied that morphine's receptor mechanisms differed from those of both heroin and M6G.

Recently, another MOR-1 knockout mouse was reported with disruption of exon 11 (36). While this disrupted interfered with exon 11-associated variants, it did not significantly impact the traditional exon 1-associated variants. Thus, it was not surprising that morphine and methadone analgesia in these exon 11 knockout mice did not significantly differ from control wild-type mice. However, the analgesic activity of heroin and M6G was lowered in the animals. Together, these findings suggest that both exon 1 and exon 11-associated variants are involved with *mu* opioid receptor-mediated analgesia, but the exon 1 variants mediate morphine actions while the exon 11-associated variants are involved with heroin and M6G. The concept of two different sets of variants responsible for the actions of different groups of *mu* opioids is quite similar to that seen in the antisense studies.

3. Methods for Discovery of GPCR Splice Variants

3.1. RT-PCR

Since one or more exons are shared by different splice variants and their expression levels vary significantly, traditional cDNA library screening often fails to identify splice variants, particularly those with low abundance. RT-PCR has provided a valuable alternative of isolating splice variants with great sensitivity and specificity. In RT-PCR, rapid amplification of cDNA 5′-end (5′RACE) and 3′RACE approaches is commonly used for identifying the variants with new exons at either 5′-end or 3′-end. To identify the variants with alternative 5′ or 3′ splicing or with exon skipping or with new exons within introns, an internal exon scanning approach with primers from known exon sequences is a useful tool. Multiple variants from the *OPRM1* genes have been obtained with these approaches.

3.2. Gene Targeting

The idea of gene targeting (also referred to as targeted gene disruption or gene knockout) was developed by Mario Capecchi and Oliver Smithies in the early 1980s (73, 74). The establishment of totipotent, germ-line competent murine embryonic stem (ES) cells and the ability to engineer specific genes by homologous recombination in ES cells have essentially revolutionized the field (75, 76). Gene targeting technologies have quickly developed to a point where it is possible to selectively disrupt any gene or any specific exon. Basically, gene targeting involves constructing a targeting vector which includes gene sequences for homologous recombination and a positively selectable marker such as neomycin resistance gene (neo) and/or a negatively selectable marker such as thymidine kinase gene, transfecting the ES cells with the vector to generate targeted ES cells and using these mutated ES cells to produce chimeras and genetically defined offspring.

4. Future Directions

Our understanding of splice variants is continually growing. Within the GPCR field, we are seeing an increasing number of receptors that generate more than one gene product. The real challenge is to understand how these variants differ functionally. As shown with *OPRM1* gene, evidence is accumulating that the variants differ pharmacologically. How this occurs remains to be elucidated. Do the variants have an intrinsic difference in transduction? For example, the inserted sequence in the dopamine receptor variants, D_{2L} and D_{2S}, may impact coupling to G proteins. A number of GPCR have splicing at the C-terminus, which may also impact transduction. In some situations, the changes may not influence transduction directly, but may change the localization of the receptor, either in the cell or in different organs or brain regions may be important in understanding their pharmacology. The role of mutations in modulating the splicing and possibly even altering levels of specific variants also needs to be explored.

The story has become even more complex with the demonstration of receptor dimerization. Numerous examples exist in which the pharmacology of heterodimers differs markedly from either component alone (77, 78). How different splice variants may impact with dimers remains to be documented. Thus, it seems likely that alternative splicing will continue to increase in importance, both with GPCRs and all proteins in general.

5. Conclusions

Like most proteins, GPCRs undergo alternative splicing, with estimates of approximately 50% of them involved. Alternative splicing offers the ability of a single gene to generate dozens, or more, different proteins. In GPCRs, alternative splicing has typically been associated with both the N- and C-terminals of the receptors, with less extensive splicing in the interior. Alternative splicing of GPCRs is region- and cell-specific and can vary markedly among organs and even brain regions. The *mu* opioid receptor system provides insights into how these splice variants may mediate different pharmacological effects. Thus, alternative splicing offers an important way to expand the repertoire of GPCRs.

References

1. Markovic D, Challiss RA (2009) Alternative splicing of G protein-coupled receptors: physiology and pathophysiology. Cell Mol Life Sci **66**:3337–3352.
2. Sharp PA (1994) Split genes and RNA splicing. Cell 77:805–815.
3. Padgett RA, Grabowski PJ, Konarska MM et al (1986) Splicing of messenger RNA precursors. Annu Rev Biochem **55**:1119–1150.
4. Black DL (2003) Mechanisms of alternative pre-messenger RNA splicing. Annu Rev Biochem 72:291–336.
5. Black DL (1992) Activation of c-src neuron-specific splicing by an unusual RNA element in vivo and in vitro. Cell **69**:795–807.
6. Chan RC, Black DL (1997) The polypyrimidine tract binding protein binds upstream of neural cell-specific c-src exon N1 to repress the

splicing of the intron downstream. Mol Cell Biol **17**:4667–4676.

7. Chan RC, Black DL (1997) Conserved intron elements repress splicing of a neuron-specific c-src exon in vitro. Mol Cell Biol **17**:2970.
8. Chan RC, Black DL (1995) Conserved intron elements repress splicing of a neuron-specific c-src exon in vitro. Mol Cell Biol **15**:6377–6385.
9. Xu R, Teng J, Cooper TA (1993) The cardiac troponin T alternative exon contains a novel purine-rich positive splicing element. Mol Cell Biol **13**:3660–3674.
10. Ryan KJ, Cooper TA (1996) Muscle-specific splicing enhancers regulate inclusion of the cardiac troponin T alternative exon in embryonic skeletal muscle. Mol Cell Biol **16**:4014–4023.
11. Cooper TA (1998) Muscle-specific splicing of a heterologous exon mediated by a single muscle-specific splicing enhancer from the cardiac troponin T gene. Mol Cell Biol **18**:4519–4525.
12. Carstens RP, Wagner EJ, Garcia-Blanco MA (2000) An intronic splicing silencer causes skipping of the IIIb exon of fibroblast growth factor receptor 2 through involvement of polypyrimidine tract binding protein. Mol Cell Biol **20**:7388–7400.
13. Carstens RP, McKeehan WL, Garcia-Blanco MA (1998) An intronic sequence element mediates both activation and repression of rat fibroblast growth factor receptor 2 pre-mRNA splicing. Mol Cell Biol **18**:2205–2217.
14. Carstens RP, Eaton JV, Krigman HR et al (1997) Alternative splicing of fibroblast growth factor receptor 2 (FGF-R2) in human prostate cancer. Oncogene **15**:3059–3065.
15. Fu D, Skryabin BV, Brosius J et al (1995) Molecular cloning and characterization of the mouse dopamine D_3 receptor gene: An additional intron and an mRNA variant. DNA Cell Biol **14**:485–492.
16. Graveley BR (2000) Sorting out the complexity of SR protein functions. RNA **6**:1197–1211.
17. Hastings ML, Krainer AR (2001) Pre-mRNA splicing in the new millennium. Curr Opin Cell Biol **13**:302–309.
18. Ladd AN, Nguyen NH, Malhotra K et al (2004) CELF6, a member of the CELF family of RNA-binding proteins, regulates muscle-specific splicing enhancer-dependent alternative splicing. J Biol Chem **279**: 17756–17764.
19. Ladd AN, Charlet N, Cooper TA (2001) The CELF family of RNA binding proteins is implicated in cell-specific and developmentally regulated alternative splicing. Mol Cell Biol **21**:1285–1296.
20. Ladd AN, Taffet G, Hartley C et al (2005) Cardiac tissue-specific repression of CELF activity disrupts alternative splicing and causes cardiomyopathy. Mol Cell Biol **25**:6267–6278.
21. Jensen KB, Dredge BK, Stefani G et al (2000) Nova-1 regulates neuron-specific alternative splicing and is essential for neuronal viability. Neuron **25**:359–371.
22. Ule J, Stefani G, Mele A et al (2006) An RNA map predicting Nova-dependent splicing regulation. Nature **444**:580–586.
23. Ule J, Ule A, Spencer J et al (2005) Nova regulates brain-specific splicing to shape the synapse. Nat Genet **37**:844–852.
24. Giros B, Sokoloff P, Martres MP et al (1989) Alternative splicing directs the expression of two D_2 dopamine receptor isoforms. Nature **342**:923–929.
25. Monsma FJ, McVittie LD, Gerfen CR et al (1989) Multiple D_2 dopamine receptors produced by alternative RNA splicing. Nature **342**:926–929.
26. Chen Y, Mestek A, Liu J et al (1993) Molecular cloning and functional expression of a *mu*-opioid receptor from rat brain. Mol Pharmacol **44**:8–12.
27. Eppler CM, Hulmes JD, Wang J-B et al (1993) Purification and partial amino acid sequence of a *mu* opioid receptor from rat brain. J Biol Chem **268**:26447–26451.
28. Thompson RC, Mansour A, Akil H et al (1993) Cloning and pharmacological characterization of a rat *mu* opioid receptor. Neuron **11**:903–913.
29. Wang JB, Imai Y, Eppler CM et al (1993) *Mu* opiate receptor: cDNA cloning and expression. Proc Natl Acad Sci USA **90**:10230–10234.
30. Pan L, Xu J, Yu R et al (2005) Identification and characterization of six new alternatively spliced variants of the human *mu* opioid receptor gene, *OPRM1*. Neuroscience **133**:209–220.
31. Pan Y-X, Xu J, Rossi GC et al (1998) Cloning and expression of a novel splice variant of the mouse *mu*-opioid receptor (MOR-1) gene. Soc Neurosci Abstr **24**:524.
32. Pan Y-X, Xu J, Mahurter L et al (2001) Generation of the *mu* opioid receptor (MOR-1) protein by three new splice variants of the *OPRM1* gene. Proc Natl Acad Sci USA **98**:14084–14089.
33. Pan YX, Xu J, Bolan EA et al (1999) Identification and characterization of three

new alternatively spliced *mu* opioid receptor isoforms. Mol Pharmacol **56**:396–403.

34. Pan YX, Xu J, Mahurter L et al (2003) Identification and characterization of two new human *mu* opioid receptor splice variants, hMOR-1O and hMOR-1X. Biochem Biophys Res Commun **301**:1057–1061.
35. Pan YX, Xu J, Bolan E et al (2005) Identification of four novel exon 5 splice variants of the mouse *mu*-opioid receptor gene: functional consequences of C-terminal splicing. Mol Pharmacol **68**:866–875.
36. Pan YX, Xu J, Xu M et al (2009) Involvement of exon 11-associated variants of the *mu* opioid receptor MOR-1 in heroin, but not morphine, actions. Proc Natl Acad Sci USA **106**:4917-4922.
37. Pan YX, Xu J, Bolan E et al (2000) Isolation and expression of a novel alternatively spliced *mu* opioid receptor isoform, MOR-1F. FEBS Lett **466**:337–340.
38. Pasternak DA, Pan L, Xu J et al (2004) Identification of three new alternatively spliced variants of the rat *mu* opioid receptor gene: dissociation of affinity and efficacy. J Neurochem **91**:881–890.
39. Zhang Y, Pan YX, Kolesnikov Y et al (2006) Immunohistochemical labeling of the *mu* opioid receptor carboxy terminal splice variant mMOR-1B4 in the mouse central nervous system. Brain Res **1099**:33–43.
40. Doyle GA, Sheng XR, Lin SS et al (2007) Identification of five mouse *mu*-opioid receptor (MOR) gene (*OPRM1*) splice variants containing a newly identified alternatively spliced exon. Gene **395**:98–107.
41. Choi HS, Kim CS, Hwang CK et al (2006) The opioid ligand binding of human *mu*-opioid receptor is modulated by novel splice variants of the receptor. Biochem Biophys Res Commun **343**:1132–1140.
42. Kvam TM, Baar C, Rakvag TT et al (2004) Genetic analysis of the murine *mu* opioid receptor: increased complexity of *OPRM1* gene splicing. J Mol Med **82**:250–255.
43. Gris P, Gauthier J, Cheng P et al (2010) A novel alternatively spliced isoform of the *mu*-opioid receptor: functional antagonism. Mol Pain **6**:33.
44. Xu J, Xu M, Hurd YL et al (2009) Isolation and characterization of new exon 11-associated N-terminal splice variants of the human *mu* opioid receptor gene. J Neurochem **108**:962–972.
45. Du Y-L, Elliot K, Pan Y-X et al (1997) A splice variant of the *mu* opioid receptor is present in human SHSY-5Y cells. Soc Neurosci Abstr **23**:1206.
46. Bolan EA, Pasternak GW, Pan Y-X (2004) Functional analysis of MOR-1 splice variants of the *mu* opioid receptor gene, *OPRM1*. Synapse **51**:11–18.
47. Atweh SF, Kuhar MJ (1983) Distribution and physiological significance of opioid receptors in the brain. Br Med Bull **39**:47–52.
48. Atweh SF, Kuhar MJ (1977) Autoradiographic localization of opiate receptors in rat brain. III. The telencephalon. Brain Res **134**:393–405.
49. Atweh SF, Kuhar MJ (1977) Autoradiographic localization of opiate receptors in rat brain. I. Spinal cord and lower medulla. Brain Res **124**:53–67.
50. Atweh SF, Kuhar MJ (1977) Autoradiographic localization of opiate receptors in rat brain. II. The brain stem. Brain Res **129**:1–12.
51. Arvidsson U, Riedl M, Chakrabarti S et al (1995) Distribution and targeting of a *mu*-opioid receptor (MOR1) in brain and spinal cord. J Neurosci **15**:3328–3341.
52. Mansour A, Fox CA, Burke S et al (1994) *Mu*, *delta*, and *kappa* opioid receptor mRNA expression in the rat CNS: An in situ hybridization study. J Comp Neurol **350**:412–438.
53. Mansour A, Fox CA, Burke S et al (1994) Immunohistochemical localization of the *mu* opioid receptors. Regul Pept **54**:179–180.
54. Abbadie C, Pan Y-X, Drake CT et al (2000) Comparative immunhistochemical distributions of carboxy terminus epitopes from the *mu* opioid receptor splice variants MOR-1D, MOR-1 and MOR-1C in the mouse and rat central nervous systems. Neuroscience **100**:141–153.
55. Abbadie C, Pan Y-X, Pasternak GW (2000) Differential distribution in rat brain of *mu* opioid receptor carboxy terminal splice variants MOR-1C and MOR-1-like immunoreactivity: Evidence for region-specific processing. J Comp Neurol **419**:244–256.
56. Uhl GR, Childers S, Pasternak GW (1994) An opiate-receptor gene family reunion. Trends Neurosci **17**:89–93.
57. Rossi G, Pan YX, Cheng J et al (1994) Blockade of morphine analgesia by an antisense oligodeoxynucleotide against the *mu* receptor. Life Sci **54**:L375–L379.
58. Chen XH, Adams JU, Geller EB et al (1995) An antisense oligodeoxynucleotide to *mu*-opioid receptors inhibits *mu*-opioid receptor agonist-induced analgesia in rats. Eur J Pharmacol **275**:105–108.
59. Khasar SG, Gold MS, Dastmalchi S et al (1996) Selective attenuation of *mu*-opioid receptor-mediated effects in rat sensory neurons by intrathecal administration of antisense oligodeoxynucleotides. Neurosci Lett **218**:17–20.

60. Leventhal L, Cole JL, Rossi GC et al (1996) Antisense oligodeoxynucleotides against the MOR-1 clone alter weight and ingestive responses in rats. Brain Res **719**:78–84.
61. Standifer KM, Chien C-C, Wahlestedt C et al (1994) Selective loss of *delta* opioid analgesia and binding by antisense oligodeoxynucleotides to a *delta* opioid receptor. Neuron **12**:805–810.
62. Rossi GC, Brown GP, Leventhal L et al (1996) Novel receptor mechanisms for heroin and morphine-6 *beta*-glucuronide analgesia. Neurosci Lett **216**:1–4.
63. Leventhal L, Stevens LB, Rossi GC et al (1997) Antisense mapping of the MOR-1 opioid receptor clone: modulation of hyperphagia induced by DAMGO. J Pharmacol Exp Ther **282**:1402–1407.
64. Rossi GC, Leventhal L, Pan YX et al (1997) Antisense mapping of MOR-1 in rats: distinguishing between morphine and morphine-6*beta*-glucuronide antinociception. J Pharmacol Exp Ther **281**:109–114.
65. Kolesnikov YA, Pan YX, Babey AM et al (1997) Functionally differentiating two neuronal nitric oxide synthase isoforms through antisense mapping: Evidence for opposing NO actions on morphine analgesia and tolerance. Proc Natl Acad Sci USA **94**:8220–8225.
66. Philippe Sarret, Louis Doré-Savard, Pascal Tétreault, Valérie Bégin-Lavallée, Marc-André Dansereau, and Nicolas Beaudet (2011) Using RNA Interference to Downregulate G Protein-Coupled Receptors. In: Stevens CW (ed) Methods for the Discovery and Characterization of G Protein-Coupled Receptors. Springer, New York
67. Huang PL, Dawson TM, Bredt DS et al (1993) Targeted disruption of the neuronal nitric oxide synthase gene. Cell **75**:1273–1286.
68. Matthes HWD, Maldonado R, Simonin F et al (1996) Loss of morphine-induced analgesia, reward effect and withdrawal symptoms in mice lacking the *mu*-opioid-receptor gene. Nature **383**:819–823.
69. Sora I, Funada M, Uhl GR (1997) The *mu*-opioid receptor is necessary for [D-Pen^2,D-Pen^5]enkephalin-induced analgesia. Eur J Pharmacol **324**: R1-R2.
70. Sora I, Takahashi N, Funada M et al (1997) Opiate receptor knockout mice define *mu* receptor roles in endogenous nociceptive responses and morphine-induced analgesia. Proc Natl Acad Sci USA **94**:1544–1549.
71. Loh HH, Liu HC, Cavalli A et al (1998) *Mu* opioid receptor knockout in mice: effects on ligand-induced analgesia and morphine lethality. Mol Brain Res **54**:321–326.
72. Schuller AG, King MA, Zhang J et al (1999) Retention of heroin and morphine-6*beta*-glucuronide analgesia in a new line of mice lacking exon 1 of MOR-1. Nat Neurosci **2**:151–156.
73. Folger KR, Wong EA, Wahl G et al (1982) Patterns of integration of DNA microinjected into cultured mammalian cells: evidence for homologous recombination between injected plasmid DNA molecules. Mol Cell Biol **2**:1372–1387.
74. Hsiung N, Roginski RS, Henthorn P et al (1982) Introduction and expression of a fetal human globin gene in mouse fibroblasts. Mol Cell Biol **2**:401–411.
75. Evans CJ, Weber E, Barchas JD (1981) Isolation and characterization of a-N-acetyl *beta*-endorphin (1–26) from the rat posterior/intermediate pituitary lobe. Biochem Biophys Res Commun **102**:897–904.
76. Martin GR (1981) Isolation of a pluripotent cell line from early mouse embryos cultured in medium conditioned by teratocarcinoma stem cells. Proc Natl Acad Sci USA **78**:7634–7638.
77. Jordan BA, Devi LA (1999) G-protein-coupled receptor heterodimerization modulates receptor function. Nature **399**:697–700.
78. Pan Y-X, Bolan E, Pasternak GW (2002) Dimerization of morphine and orphanin FQ/nociceptin receptors: generation of a novel opioid receptor subtype. Biochem Biophys Res Commun **297**:659–663.

Chapter 3

Detecting Polymorphisms in G Protein-Coupled Receptor Genes

Dmitri Proudnikov, Vadim Yuferov, and Mary Jeanne Kreek

Abstract

The genes for G protein-coupled receptors (GPCRs) including those encoding the classical *mu*, *delta*, and *kappa* opioid receptors (MOR, DOR, and KOR); cannabinoid receptors (CB1); ACTH receptor (melanocortin receptor type 2, MC2R); and serotonin receptors (5HT1B) have been a focus of the studies of our group for a number of years since these receptors are involved in specific addictions. Genetic variants of GPCR genes have been associated with vulnerability to stress, anxiety, depression, and predisposition to develop drug addiction. To study these variants including single nucleotide polymorphisms (SNPs) and their allocation on alleles (haplotypes), our group developed special techniques (genotyping assays using polyacrylamide gel pad technology, molecular haplotyping assays based on the use of fluorescent PCR) and also used commercially available techniques and methodologies. Although these novel technologies allow rapid and reliable high-throughput analysis, in order to use them, the precise position of the polymorphic site should be known in advance. The contemporary genetic databases contain copious information on genetic variants. However, we found that some important functional variants are still unreported. Therefore, resequencing of the genes studied in specific populations is necessary. Each technology that we use has specific advantages that we will discuss below.

Key words: Genotyping, SNP, Molecular haplotyping, Microarrays, Single nucleotide extension, Polymorphism

1. Introduction

Inherited genetic variations affecting gene expression may play an important role in susceptibility to complex disorders (1, 2). Recent studies linked genetic variants of G protein-coupled receptors (GPCRs) to a number of neuropsychiatric conditions including stress, anxiety, depression, and predisposition to drug addiction. Our group has extensive experience in genetic studies of a number of genes encoding GPCRs including the *mu* and

Craig W. Stevens (ed.), *Methods for the Discovery and Characterization of G Protein-Coupled Receptors*, Neuromethods, vol. 60, DOI 10.1007/978-1-61779-179-6_3, © Springer Science+Business Media, LLC 2011

kappa opioid receptors (MOR and KOR), nociceptin/orphanin FQ receptor (ORL), adrenocorticotropic hormone receptor (ACTHR or melanocortin receptor type 2 receptor, MCR2), cannabinoid receptors (CB1), and serotonin receptors (5HT1B) (3–10). In our earlier studies of genotyping of GPCRs, we used the Sanger sequencing method and later used microarrays made in our laboratory based on the use of polyacrylamide gel pad technology. Advances of methodology in molecular biology and genetics during the last decade led to the emergence of a large number of novel techniques to study genetic variants. Nevertheless, regular resequencing using the Sanger method allows the discovery of novel, previously unreported functional genetic variants. Introduction of microarray technology including custom arrays makes it possible to genotype hundreds of thousands of individual polymorphisms at once in a single DNA sample. On the other hand, the fluorogenic exonuclease assay (TaqMan) makes it possible to genotype a single polymorphism of interest in hundreds of DNA samples simultaneously. Currently, we use both the microarray and TaqMan approaches depending on the particular goal of each study. Based on joint use of TaqMan and allele-specific amplification, we developed a new inexpensive technique for molecular haplotyping of SNPs (assignment of polymorphisms on a single DNA strand). The method was applied for high-throughput haplotyping of SNPs of *OPRK1* and *5HTR1B* (7, 11). Our improved methodology makes possible haplotyping of polymorphisms separated from each other by several thousand nucleotides. Here we will provide an overview of methods developed in our laboratory as well as commercially available methods that we routinely use for the study of genetic variants of GPCRs.

2. Methods to Detect GPCR Polymorphisms

2.1. Application of the Sanger Sequencing Technique for Analysis of Variants of GPCR

Since the introduction of the direct sequencing method for analysis of DNA by Sanger (12) and, as an alternative, Maxam and Gilbert sequencing (13) techniques, study of genetic polymorphisms has become a routine part of genetic research. The Maxam and Gilbert method is based on random fragmentation of nucleic acid by chemical means using four separate reactions, each of which is specific for only one type of nucleic base. This technique has become the basis for many methods that are currently in use in molecular biology, e.g., immobilization of probes on microarrays or fluorescent labeling of nucleic acids (14, 15).

The Sanger method, based on selective termination of the PCR reaction using dideoxynucleotide triphosphates, has become the most popular method for analysis of DNA sequences and has been in use without essential modification for nearly 3 decades. Although a number of techniques for high-throughput screening of genetic variants including TaqMan, pyrosequencing, microarrays, etc. have been developed, Sanger technique remains the essential tool for verification of the results of such screenings and also for discovery of the novel variants of the genes that are not reported in genetic databases.

Most of the genetic databases originally were built on the results of genotyping of primarily Caucasians, although they are now expanding the data to other ethnicities (e.g., www.1000genomes.org/page.php, hapmap.ncbi.nlm.nih.gov/hapmappopulations.html.en). Our work with variants of GPCR genes, including *OPRM1* (3–5), *OPRK1* (6), *5HTR1B* (7), *MC2R* (8), *CNR1* (9), and *OPRL1* (10) in three ethnic groups (African Americans, Caucasians, and Hispanics), has shown that the majority of the polymorphisms in these receptors are present in different frequencies in different ethnic groups. For example, we found that the potentially functional 17C>T polymorphism (rs1799972) of *OPRM1* is in high allelic frequency in African Americans (about 20%), but not in Caucasians (3, 16). We also found that the frequency of the variant -184G>A located in the promoter region of the gene for the other GPCR, *MC2R*, also varies dramatically in different ethnicities: 17% was found in Hispanics and 0.4% in Caucasians in a US cohort (8). Therefore, in studies of functionality of the -179A>G polymorphism in the promoter region of *MC2R* in a European cohort, the variant -184G>A was not reported (17). In our laboratory, we regularly perform Sanger resequencing of the coding and also promoter regions of genes being studied.

2.2. Use of Custom-Made Polyacrylamide Gel Pad-Based Microarrays for the Analysis of Polymorphisms

Genotyping of polymorphisms or gene expression profiling using microarrays has become an indispensable tool in contemporary molecular biology and genetics. Currently, the researcher has access to a wide variety of arrays, including 2- and 3-dimensional (2D and 3D) arrays. In our laboratory, to study GPCR variants, we applied disposable 2D arrays based on Affymetrix technology (18, 19) and reusable 3D arrays developed in the Engelhardt Institute of Molecular Biology (Moscow, Russia) and the Argonne National Laboratory (Argonne, IL) in the late 1980s (20). These 3D arrays consist of elements of polyacrylamide (PAA) gel $100 \times 100 \times 30$ μm separated from each other with 200 μm spacers. Each gel element contains a specific oligonucleotide probe chemically bound to polyacrylamide. The 3D structure of the PAA gel-based elements allows immobilization of much greater amounts of the oligonucleotide probe within each element compared to 2D arrays, therefore increasing the intensity of the signal from an element. Hybridization and enzymatic reactions within the acrylamide gel more closely resemble liquid-phase reactions, rather than solid-phase reactions on 2D support. This feature of these arrays is valuable for basic research studies, e.g., determining the thermodynamic parameters of the nucleic acid duplexes (21) and *de novo* sequencing of short nucleic acids (22). These microarrays were routinely reused up to 10 times without significant loss of hybridization properties or increase of the background (23). Such arrays were found to be useful for field applications that require rapid testing of a limited number of genetic variants in a small number of samples, e.g., identification of pathological bacteria in soil (24), quantification of the mutant component in polio vaccine (23), among other uses. They also might be useful for small laboratories or medical offices.

In our studies in the late 1990s, we developed 3D PAA arrays for genotyping the polymorphisms 17C>T (rs1799972) and 118A>G (rs1799971) of *OPRM1*. The array was then applied to genotype 36 human DNA samples (25). The results of this genotyping were in excellent concordance with the results of Sanger resequencing of the same samples.

2.3. Use of Single Nucleotide Extension (Minisequencing) on PAA-Based Microarrays for Analysis of Polymorphisms in GPCRs

We also used PAA gel pad technology for genotyping of SNPs 17C>T and 118A>G of *OPRM1* by single nucleotide extension. This technique can be used in two separate modes. In the first, an oligonucleotide probe is designed so that the last 3' nucleotide of the probe precedes the analyzed base. During enzymatic reaction, DNA polymerase incorporates only dideoxynucleotide triphosphate that is complementary to the analyzed base. If labeled with different fluorophores, several dideoxynucleotide triphosphates might be used simultaneously in the same reaction, or, if carrying the same fluorophore, might be used in separate reactions. An alternative approach is to design not one probe, but a set of four oligonucleotide probes for each polymorphism, so that the last 3' nucleotide of the probe is complementary to the analyzed base. This last approach we utilized to detect the polymorphism 17C>T ((25), Fig. 1). Later we (Proudnikov, LaForge, Mirzabekov, and Kreek, unpublished) developed an assay for genotyping 118A>G, 151G>A (rs1042753), and the unconfirmed 183C>T of *OPRM1* (Fig. 2). All microarrays were made manually. The fluorescent

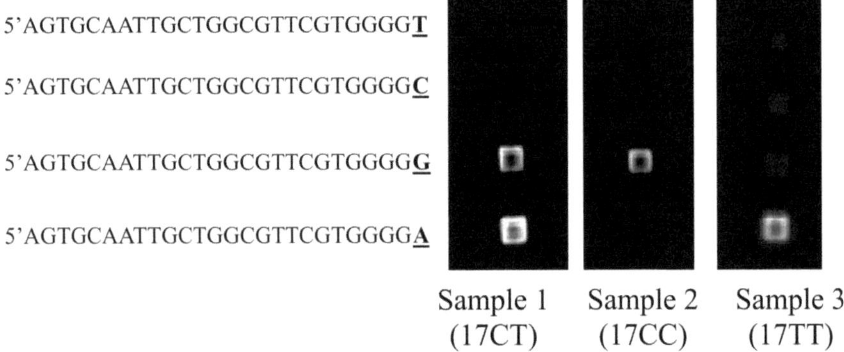

Fig. 1. Detection of the 17C>T polymorphism of *OPRM1* on PAA microarray (adapted from (25)). Sequences of the immobilized oligonucleotides and distribution of fluorescent signals after single nucleotide extension using different DNA templates are shown.

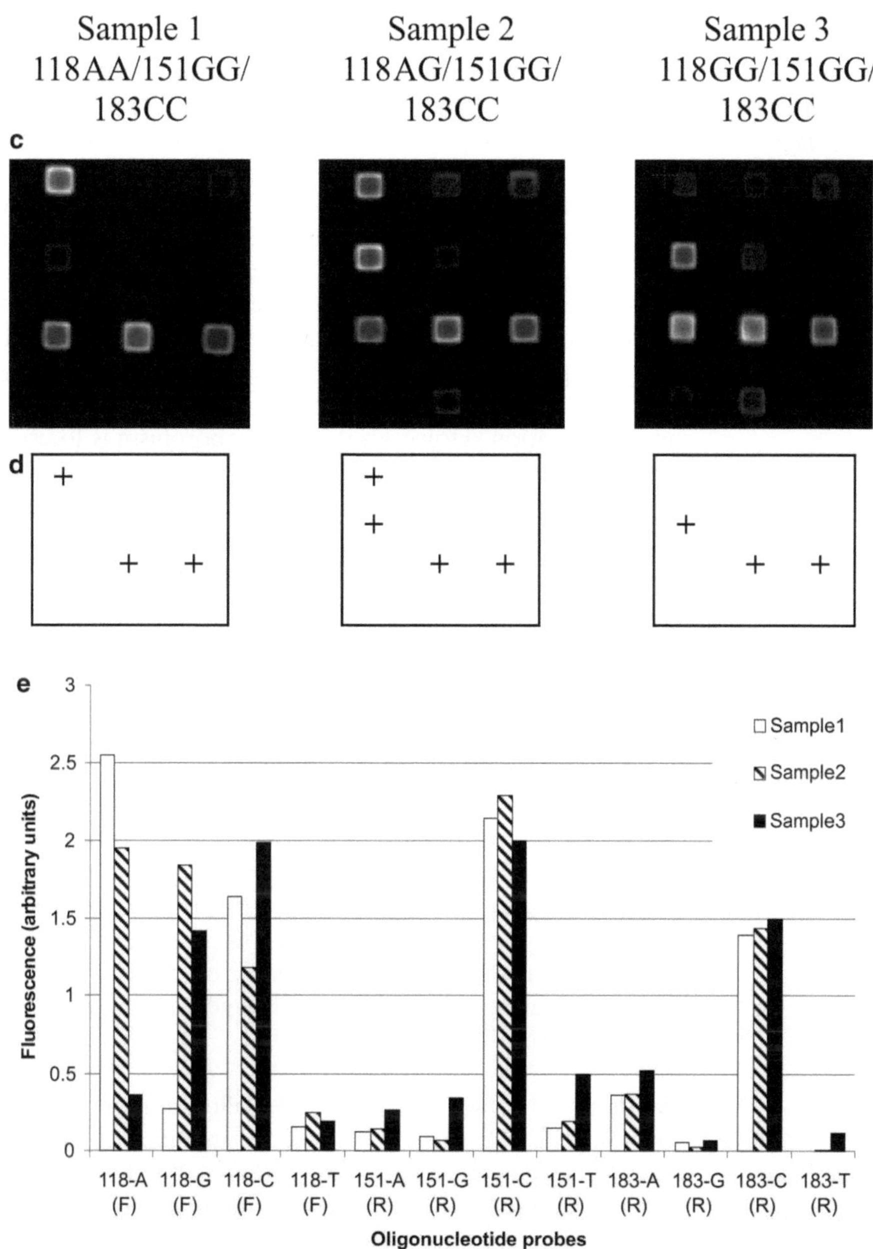

Fig. 2. Detection of the polymorphisms 118A>G, 151G>A, and 183C>T of *OPRM1* on PAA microarray. (**a**) The location of oligonucleotide probes on microarrays; (**b**) sequences of the immobilized oligonucleotides, 3'X is A, G, C or T nucleotide; (**c**) images of the microarrays after extension reaction; (**d**) the predicted pattern of single nucleotide extension of oligonucleotide probes using different DNA samples as the template; (**e**) distribution of fluorescent signals on microarrays after wash by electrophoresis.

signal pattern was as expected, with the exception of a false-positive signal from oligonucleotide 118-F extended with dC. The appearance of this signal was likely caused by the presence of a stable GC-rich palindrome GGCC at the 3' end of the extended oligonucleotide. The protocol for this novel assay technique is given in Sect. 3 of this chapter.

2.4. Application of the 5' Fluorogenic Exonuclease Assay (TaqMan) for High-Throughput Screening of GPCR Variants

Since early reports in the 1990s (26), the TaqMan assay has become a powerful tool for gene expression studies and for high-throughput genotyping of genetic variants. We routinely use this technique for both applications. Although it is common to use the shortest possible amplicon (50–80 bp), we found that this length is not optimal. Such short amplicons are very difficult to verify by Sanger sequencing. In both gene expression and genotyping assays we use amplicons with a minimal length of 150 bp that allows verification of the sequence amplified. Commercial introduction of oligonucleotides linked to the minor groove binder (MGB) as TaqMan probes increased the sensitivity and accuracy of the detection, since shorter oligonucleotide probes have higher stringency in hybridization. However, if the polymorphism is located in the middle of a palindrome or another region not well suited for hybridization, shortening the probe might reduce the specificity of hybridization, resulting in loss of signal. We found that in such cases use of regular non-MGB probes can be beneficial (27).

During our studies of the polymorphisms of the GPCRs, we found that some variants are located so close to each other that the oligonucleotide probes designed for one polymorphism would interfere with another one. For genotyping such polymorphisms, we successfully used degenerate probes that contain a specific nucleotide for the position studied and a mix of the nucleotide bases for the second position (11). Our experiments showed that in such assays, because of the use of degenerate probes, each main cluster of genotypes for the main polymorphism studied is often subdivided into subclusters in accord with the genotypes of the second polymorphism (11), which was confirmed by Sanger resequencing. Further development based on this phenomenon may lead to an assay that makes possible genotyping several genetic variants in one TaqMan assay.

2.5. Molecular Haplotyping of GPCRs: A Joint Application of the 5' Fluorogenic Exonuclease Assay (TaqMan) and Allele-Specific Amplification

Studies have shown that haplotypes (allocation of polymorphisms on a single DNA strand) in some cases are more relevant in regulation of cell mechanisms than individual polymorphisms, but commonly used SNP identification methods do not allow allelic assignment of a combination of SNPs (the "molecular haplotype"). Most haplotype analyses are performed using statistical estimation of the actual molecular haplotype. Our study showed that two commonly used programs for statistical haplotyping (SNPHAP and PHASE) provide similar results for some haplotype estimations, but

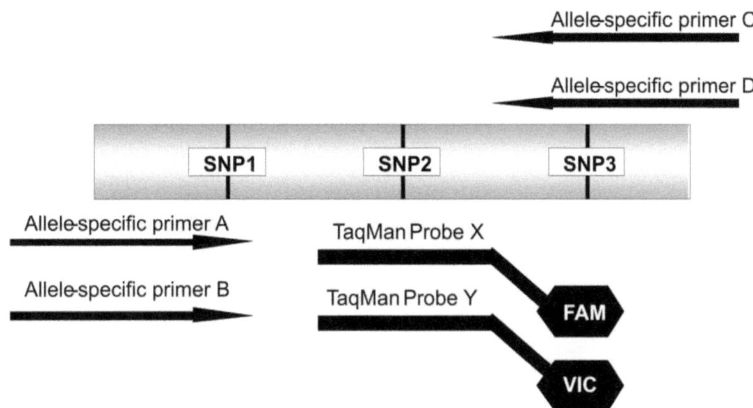

Fig. 3. General scheme of the method of molecular haplotype analysis of single nucleotide polymorphisms using allele-specific primers. Amplification was performed in four reactions using both TaqMan probes specific for different alleles of SNP2 and also different combinations of the allele-specific primers (A + C; A + D; B + C; B + D) complementary to the terminal polymorphisms SNP1 and SNP2, respectively.

for other haplotypes the results differ (27). Therefore, the question regarding the limitation of each of these programs should be addressed. This may be done by confirming results produced by these programs with an independent method for haplotype assignment, molecular haplotyping. We have developed a novel approach (Fig. 3) to molecular haplotyping of SNPs using fluorescent PCR (7, 11). We successfully tested this method for molecular haplotyping polymorphisms -261T>G (rs11568817), -161A>T (rs130058), 129C>T (rs6298) of *HTR1B* (7) and polymorphisms 843A>G (rs702764) and 846C>T (rs16918875) of *OPRK1* (11). We found that this method is simple, reliable, and less expensive than other available methods for molecular haplotyping and that it can be easily adopted for high-throughput use.

2.6. Genotyping Polymorphisms Using Affymetrix Arrays

The technology developed by Affymetrix for GeneChip® Arrays is based on the direct synthesis of oligonucleotide probes on a solid support that is then used as a part of the hybridization chamber. The use of a specially developed combination of photoremovable protective groups and masks in oligonucleotide synthesis requires a similar time for the synthesis of one or hundreds of thousands of oligonucleotide probes in one array.

In our laboratory, we used GeneChip® 10K and 100K Arrays consisting of 10,000 and 100,000 oligonucleotide probes designed by Affymetrix (18, 19). We found that such universal arrays containing a large number of genetic markers were suited for preliminary genetic studies. The limitation is the absence of many functional polymorphisms of moderate frequency that are of interest for particular areas of research. The considerable cost of arrays that are not reusable requires special approaches to minimize experimental costs, e.g., pooling of several samples from the same study group. Results of pooling experiments are acceptable in some situations but are often difficult to interpret (17, 28).

Using GeneChip® 10K arrays in our association studies, we identified a genotype pattern AG-TT-GG consisting of rs1714984, rs965972, and rs1867898, which is significantly associated with addiction to heroin (18). The genotype pattern GG-CT-GG of these variants was found to be associated with protection from addiction, and lacking this genotype pattern explained 83% of the population-attributable risk for developing heroin addiction. Using 100K arrays, we found a strong association of polymorphisms of many genes with heroin addiction, including genes not previously considered to be involved in addiction mechanisms (19).

2.7. Genotyping of Polymorphisms of the GPCRs Using Illumina Arrays

As an alternative to GeneChip® Affymetrix arrays, our group used the 1536-plex GoldenGate Custom Panel (Illumina, San Diego, CA) designed by Dr Goldman's group (29) and contained 1,350 SNPs from 130 genes implicated in heroin addiction (18) to analyze an association of the variants of the hypothesis-driven genes (30, 31). The use of 186 additional SNPs as ancestry informative markers (AIMs) allowed a population stratification analysis to confirm self-reported ethnic origins of the subjects. Using these arrays, we found an association with heroin addiction of a number of SNPs in genes for several receptors including adrenergic receptor α-A1 (*ADRA1A*), arginine vasopressin 1A receptor (*AVPR1A*), cholinergic receptor muscarinic 2 (*CHRM2*), dopamine receptor D1 (*DRD1*), gamma aminobutyric acid A receptor subunit β-3 (*GABRB3*), and glutamate receptor 2A (*GRIN2A*).

Currently, both Affymetrix and Illumina produce whole-genome arrays with over one million probes for genetic variations, as well as many specialized arrays for different applications.

2.8. Functional Analysis of Gene Polymorphisms

To study functionality of the genetic variants, a "genetical genomics" strategy that integrates DNA variation, gene expression, and disease phenotype has been developed (32, 33). Studies of the influence of genetic variations associated with specific phenotypes on the gene expression pattern were not always conclusive. The results of high-throughput methods for direct measurement of differences in allelic expression require validation using independent techniques (34).

We have applied the SNaPshot assay (Applied Biosystems, Foster City, CA) to study allele-specific gene expression and to identify *cis*-acting SNPs in the prodynorphin (*PDYN)* gene of human post-mortem brain (35). This method is based on a primer extension reaction and compares the relative level of each variant of mRNA transcript in a tissue from individuals who are heterozygous for an expressed polymorphism, so that each allele serves as an internal control for expression of the other allele. Our study of the caudate and nucleus accumbens, principal brain regions in the rewarding effects of drugs of abuse, provided the first evidence that the SNP rs910079 in the *PDYN* gene is a *cis*-acting polymorphism, related to differential *PDYN* gene expression in an allele-specific manner. The measurements of the total *PDYN* mRNA levels using our modified RNase protection assay (36) demonstrated a strong effect of the TTC and CCT haplotypes consisting of SNPs rs910080, rs910079, and rs2235749. The subjects with homozygous "protective" TTC haplotypes had significantly higher levels of *PDYN* mRNA compared to the subjects with homozygous "risk" haplotypes.

3. Protocols to Detect Polymorphisms in G Protein-Coupled Receptor Genes

3.1. DNA Isolation from the Blood

The Rockefeller University Hospital Institutional Review Board-approved informed consent for genetic studies was obtained from all subjects in our studies. DNA was isolated from blood using Gentra Puregene Blood Kit from Qiagen (Valencia, CA). DNA concentration was measured using Molecular Probes Quant-iT PicoGreen dsDNA (Invitrogen, Carlsbad, CA). The arrays have been manually produced as described (37). Oligonucleotide primers were custom-synthesized by the Midland Certified Company (Midland, TX); TaqMan probes were custom-synthesized by Applied Biosystems.

3.2. Genotyping of OPRM1 Polymorphisms Using Enzymatic Reactions on Polyacrylamide-Based Microarrays

Using 5–10 ng of purified genomic DNA, exon 1 of the *OPRM1* was preamplified using the following set of primers: 5'-CTGGCT ACCTCGCACAGC (For), 5'-GAGGGCCATGATCGTGAT (Rev), platinum DNA polymerase (Invitrogen), and the following program: 95°C for 2 min and then 30 cycles of 94°C for 30 s, 54°C for 30 s, 72°C for 30 s; then 72°C for 6 min. PCR products were purified using PCR QIAquick PCR Purification Kit (Qiagen). These products were then partially fragmented in 80% formic acid at 20°C for 10 min, and after precipitation with 2% $LiClO_4$ in acetone, treated with 10% piperidine in water for 1 h at 95°C, followed by precipitation with acetone.

Partially-fragmented amplified DNA (0.5 pmole) was diluted with 50 μl of 1× Thermosequenase® buffer (Amersham Biosciences, Piscataway, NJ) to 10 nM concentration and, after denaturation for 2 min at 95°C, was loaded onto a preheated (72°C) gel pad microarray. The array cluster was then covered with AmpliCover® discs and clips (Applied Biosystems) and incubated at 37°C for 24 h in an Isotemp® temperature incubator (Fisher, Pittsburgh, PA). The excess hybridization solution was removed and the array was dried in air for 30 min. A solution of 10 μM fluorescein-labeled dideoxynucleotide triphosphates and Thermosequenase® (50 U per 50 μl of reaction mix) in 1× Thermosequenase® buffer was loaded onto the microarray cluster with hybridized DNA and covered with two drops of mineral oil.

The microchip was sealed with AmpliCover® discs and clips and placed in the GeneAmp® in situ PCR system 1000 (Applied Biosystems) preheated to 95°C and incubated at this temperature for 10 s. Then the temperature was reduced to 67°C, and the single nucleotide extension reaction was carried out for 60 min. Arrays were washed with methylene chloride, alcohol, and water. Remaining unincorporated dideoxynucleotide triphosphates were removed by electrophoresis in 0.5× TBE with 1% TWEEN-100. The microchips were analyzed using a fluorescence microscope.

3.3. Molecular Haplotyping of the Polymorphisms of 5HT1B Using Joint Application of Allele-Specific Amplification and Fluorescent Exonuclease Assay (TaqMan)

TaqMan® MGB probes (CACCC**A**TGACCTCT, FAM labeled, and CACCC**T**TGACCTCTA, VIC labeled) for the serotonin 1B receptor gene (*5HT1B*, Gene Bank accession # M75128) were designed using Primer Express software (Applied Biosystems). For molecular haplotype analysis, first, the efficiency of discrimination of polymorphisms by different allele-specific primer sets was determined using qPCR™ Mastermix Plus for SYBR® Green 1 (VWR International, Bridgeport, NJ). Primers that showed the best discrimination were then used for haplotype analysis as shown in Fig. 3. The concentration of DNA samples used was in the range of 1–10 ng per μl and 1 μl of this solution was used per 5 μl of amplification reaction. Each DNA sample was amplified in four different PCR reactions using MGB TaqMAn probes and the combination of the following primers: CCAGCTCTTAGCAACCCAGtT (Primer A) + CTTTCCAGGGTAGGGAGAcG (Primer C), CCAGCTCTTAGCAACCCAGtG (Primer B) + ACTTTCCAGGGTAGGGAGAcA (Primer D), Primer A + Primer D, Primer B + Primer C. The first combination of the primers allows haplotyping of alleles T/A/C and T/T/C (the first base corresponds to the position −261, the second to position −161, the third to position 129), the second combination for the alleles G/A/T and G/T/T, the third combination for the alleles T/A/T and T/T/T, and the fourth combination for the alleles G/A/C and G/T/C. PCR cycling was performed using Platinum® quantitative PCR SuperMix-UDG (Invitrogen) on an ABI Prizm 7900 sequence detection system (Applied Biosystems) with monitoring of fluorescence in real-time mode. The fluorescent signal change (ΔR) vs. number of cycles was plotted and the threshold line at the level of 10–15% of the maximum signal was established. These threshold cycle values were imported into Excel (Microsoft, Santa Rosa, CA) for analyses.

4. Future Directions

Considering the importance of analysis of individual genomes for studies of complex diseases, further development of whole genome sequencing techniques should be pursued. Currently, the "next generation sequencing" technology able to read sequences of up to hundreds of millions of DNA (or cDNA) fragments is commercially available in three platforms: 454 System from Roche (Branford, CT), SOLiD System from Applied Biosystems, and HiSeq System from Illumina. Assembly of the resulting short sequences is a statistically challenging task that may be overcome by increasing the length of the sequence fragments. For example, in 2010, Roche plans to use a novel development in the 454 System to read fragments as long as 1,000 nucleotides. The selection of specific regions of genomic material may be accomplished by multiplex PCR amplification or by enrichment on microarrays (e.g., NimbleGene Sequence Capture Array by Roche).

5. Conclusions

The development of novel technologies including microarrays for analysis of genetic variations and building of genetic databases provided new opportunities for the genetics researcher. The use of the Human SNP Array 5.0 from Affymetrix containing on a single array one million SNPs and an additional 420,000 non-polymorphic probes including copy number variants to study genetic differences, or the Infinium BeadChips from Illumina containing 1.2 million genetic markers including up to 60,800 custom markers, makes feasible the whole-genome genotyping of known SNPs. The TaqMan assay remains a valuable tool for high-throughput analysis of the specific variants in a large number of subjects. The discovery of novel polymorphisms that might be associated with a particular disease still requires resequencing of a specific region of the gene of interest. The methods for analysis of combinations of genetic variations on a single allele, e.g., molecular haplotyping of more than five polymorphisms located within several thousand nucleotides from each other at high-throughput, require further development.

Acknowledgments

We thank Drs. Ann Ho and Orna Levran for a critical reading of this chapter and Susan Russo for editorial assistance. This work was supported by the NIH Grant P60-DA05130 (MJK) and Clinical and Translational Science Award Grant UL1 RR024143 (NCRR, Rockefeller University).

References

1. Sadée W, Dai Z (2005) Pharmacogenetics/genomics and personalized medicine. Hum Mol Genet **14**:R207–214
2. Knight JC (2005) Regulatory polymorphisms underlying complex disease traits. J Mol Med **83**:97–109
3. Bond C, LaForge KS, Tian M et al (1998) Single nucleotide polymorphism in the human *mu* opioid receptor gene alters *beta*-endorphin binding and activity: Possible implications for opiate addiction. Proc Natl Acad Sci USA **95**:9608–9613
4. Bart G, LaForge KS, Borg L et al (2006) Altered levels of basal cortisol in healthy subjects with a 118G allele in exon 1 of the *mu* opioid receptor gene. Neuropsychopharmacology **31**:2313–2317
5. Bart G, Heilig M, LaForge KS et al (2004) Substantial attributable risk related to a functional *mu*-opioid receptor gene polymorphism in association with heroin addiction in central Sweden. Mol Psychiatry **9**:547–549
6. Yuferov V, Fussell D, LaForge KS et al (2004) Redefinition of the human *kappa* opioid receptor gene (*OPRK1*) structure and association of haplotypes with opiate addiction. Pharmacogenetics **14**:793–804
7. Proudnikov D, LaForge KS, Hofflich H et al (2006) Association analysis of polymorphisms in serotonin 1B receptor (HTR1B) gene with heroin addiction: a comparison of molecular and statistically estimated haplotypes. Pharma cogenet Genomics **16**:25–36
8. Proudnikov D, Hamon S, Ott J et al (2008) Association of polymorphisms in the melanocortin receptor type 2 (MC2R, ACTH receptor) gene with heroin addiction. Neurosci Lett **435**:234–239
9. Proudnikov D, Kroslak T, Sipe JC et al (2010). Association of polymorphisms of the cannabinoid receptor (*CNR1*) and fatty acid amide hydrolase (*FAAH*) genes with heroin addiction: impact of long repeats of *CNR1*. Pharmacogenomics J **10**:232–242
10. Briant JA, Nielsen DA, Proudnikov D et al (2010) Evidence for association of two variants of the nociceptin/orphanin FQ receptor gene *OPRL1* with vulnerability to develop opiate addiction in Caucasians. Psychiatr Genet **20**:65–72
11. Proudnikov D, LaForge KS, Kreek MJ (2004) High-throughput molecular haplotype analysis (allelic assignment) of single-nucleotide polymorphisms by fluorescent polymerase chain reaction. Anal Biochem **335**:165–167
12. Sanger F, Nicklen S, Coulson AR (1977) DNA sequencing with chain-terminating inhibitors. Proc Natl Acad Sci USA **74**:5463–5467
13. Maxam AM, Gilbert W (1980) Sequencing end-labeled DNA with base-specific chemical cleavages. Methods Enzymol **65**:499–560
14. Proudnikov D, Mirzabekov A (1996) Chemical methods of DNA and RNA fluorescent labeling. Nucleic Acids Res **24**:4535–4542
15. Proudnikov D, Timofeev E, Mirzabekov A (1998) Immobilization of DNA in polyacrylamide gel for the manufacture of DNA and DNA-oligonucleotide microchips. Anal Biochem **259**:34–41
16. Bart G, Kreek MJ, Ott J et al (2005) Increased attributable risk related to a functional *mu* opioid receptor gene polymorphism in association with alcohol dependence in central Sweden. Neuropsychopharmacology **30**:417–422
17. Slawik M, Reisch N, Zwermann O et al (2004) Characterization of an adrenocorticotropin (ACTH) receptor promoter polymorphism leading to decreased adrenal responsiveness to ACTH. J Clin Endocrinol Metab **89**:3131–3137
18. Nielsen DA, Ji F, Yuferov V et al (2008) Genotype patterns that contribute to increased risk for or protection from developing heroin addiction. Mol Psychiatry **13**:417–428
19. Nielsen DA, Ji F, Yuferov V et al (2010) Genome-wide association study identifies genes that may contribute to risk for developing heroin addiction. Psychiatr Genet 20:207–214
20. Khrapko KR, Lysov YP, Khorlyn AA et al (1989) An oligonucleotide hybridization approach to DNA sequencing. FEBS Lett **256**:118–122
21. Fotin AV, Drobyshev AL, Proudnikov DY et al (1998) Parallel thermodynamic analysis of duplexes on oligodeoxyribonucleotide microchips. Nucleic Acids Res **26**:1515–1521
22. Chechetkin VR, Turygin AY, Proudnikov DY et al (2000) Sequencing by hybridization with the generic 6-mer oligonucleotide microarray: an advanced scheme for data processing. J Biomol Struct Dyn **18**:83–101
23. Proudnikov D, Kirillov E, Chumakov K et al (2000) Analysis of mutations in oral poliovirus vaccine by hybridization with generic oligonucleotide microchips. Biologicals **28**:57–66
24. Guschin DY, Mobarry BK, Proudnikov D et al (1997) Oligonucleotide microchips as genosensors for determinative and environmental studies in microbiology. Appl Environ Microbiol **63**:2397–2402

25. LaForge KS, Shick V, Spangler R et al (2000) Detection of single nucleotide polymorphisms of the human *mu* opioid receptor gene by hybridization or single nucleotide extension on custom oligonucleotide gelpad microchips: potential in studies of addiction. Am J Med Genet **96**:604–615

26. Witcombe D, Brownie J, Gillard HL et al (1998) A homogeneous fluorescence assay for PCR amplification to real-time single-tube genotyping. Clin Chem **44**:918–923

27. Nielsen DA, Barral S, Proudnikov D et al (2008) *TPH2* and *TPH1*: association of variants and interactions with heroin addiction. Behav Genet **38**:133–150

28. Drgon T, Zhang PW, Johnson C et al (2010) Genome wide association for addiction: replicated results and comparisons of two analytic approaches. PLoS One **5**:e8832

29. Hodgkinson CA, Yuan Q, Xu K et al (2008) Addictions biology: haplotype-based analysis for 130 candidate genes on a single array. Alcohol Alcohol **43**:505–515

30. Levran O, Londono D, O'Hara K et al (2008) Genetic susceptibility to heroin addiction: a candidate gene association study. Genes Brain Behav 7:720–729

31. Levran O, Londono D, O'Hara K et al (2009) Heroin addiction in African Americans: a hypothesis-driven association study. Genes Brain Behav **8**:531–540

32. Li J, Burmeister M (2005) Genetical genomics: combining genetics with gene expression analysis. Hum Mol Genet **14**:R163-169

33. Jansen RC, Nap JP (2001) Genetical genomics: the added value from segregation. Trends Genet **17**:388–391

34. Serre D, Gurd S, Ge B et al (2008) Differential allelic expression in the human genome: a robust approach to identify genetic and epigenetic *cis*-acting mechanisms regulating gene expression. PLoS Genet **4**:e1000006

35. Yuferov V, Ji F, Nielsen DA et al (2009) A functional haplotype implicated in vulnerability to develop cocaine dependence is associated with reduced PDYN expression in human brain. Neuropsychopharmacology **34**:1185–1197

36. Branch AD, Unterwald EM, Lee SE et al (1992) Quantitation of preproenkephalin mRNA levels in brain regions from male Fischer rats following chronic cocaine treatment using a recently developed solution hybridization assay. Brain Res Mol Brain Res **14**:231–238

37. Guschin D, Yershov G, Zaslavsky A et al (1997) Manual manufacturing of oligonucleotide, DNA, and protein microchips. Anal Biochem **250**:203–211

Chapter 4

Regulation of the Transcription of G Protein-Coupled Receptor Genes

Cheol Kyu Hwang, Ping-Yee Law, Li-Na Wei, and Horace H. Loh

Abstract

G protein-coupled receptors (GPCRs) participate in a variety of physiological functions and are major targets of pharmaceutical drugs. More than 600 GPCRs have been identified in the human genome. Although GPCRs are expressed in multiple tissues and individual tissues express multiple GPCRs, many have exclusive or increased expression within the central nervous system (CNS). These unique and diverse expression patterns raise fundamental questions as to the molecular mechanisms underlying the tissue/cell-specific distribution of GPCRs as well as the means by which their expression is altered in response to stimuli. Gene expression in mammals involves both transcriptional and posttranscriptional mechanisms. In this chapter, we provide an overview of the transcriptional regulation of GPCRs and discuss both established and emerging techniques to study transcriptional regulation.

Key words: ChIP-on-chip, Chromatin immunoprecipitation, Electrophoretic mobility shift assay, Epigenetic, G protein-coupled receptor, Gene expression, Histone

Abbreviations

ChIP	Chromatin immunoprecipitation
CpG	Cytosine–phosphate–guanine
DMS	Dimethyl sulfate
DNMT	DNA methyltransferase
DRRF	Dopamine receptor regulating factor
dsRNA	Double-stranded RNA
EMSA	Electrophoretic mobility shift assay
GPCRs	G protein-coupled receptors
LM-PCR	Ligation-mediated polymerase chain reaction
MNase	Micrococcal nuclease
MOR	μ Opioid receptor
qPCR	Quantitative PCR
ReChIP	Reciprocal ChIP

Craig W. Stevens (ed.), *Methods for the Discovery and Characterization of G Protein-Coupled Receptors*, Neuromethods, vol. 60, DOI 10.1007/978-1-61779-179-6_4, © Springer Science+Business Media, LLC 2011

RNAi	RNA interference
RT-PCR	Reverse transcription polymerase chain reaction
siRNAs	Small interfering RNAs

1. Introduction

G protein-coupled receptors (GPCRs), also known as seven-transmembrane receptors, are one of the largest superfamilies of proteins in the mammalian genome (1). The majority of neurotransmitters, hormones, and other signaling molecules in mammals exert their effects via GPCRs. GPCRs participate in a broad range of physiological functions by mediating extracellular signaling in cells, and more than half of all marketed drugs act on GPCRs (2). Molecular cloning studies and genomic data analyses have revealed at least 600 members of the GPCR superfamily in humans, comprising over 1% of the entire human genome. These GPCRs are classified by the GRAFS system into five main families: Glutamate, Rhodopsin, Adhesion, Frizzled/taste2, and Secretin (3). Based on the chromosomal position of the genes and "fingerprint" motifs, GPCRs within a specific GRAFS family have been shown to evolve from the same ancestral gene through gene duplication and exon shuffling (3). For example, comparative analyses of vertebrate opioid receptors using bioinformatics and data from human genome studies indicate that the four types of opioid receptors found in vertebrate species (MOR, DOR, KOR, and ORL) may have arose from an ancestral opioid unireceptor after two rounds of whole genome duplication (4). Divergent evolution of GPCR genes results in each having unique functions relating to the binding of a specific ligand (or no known ligand – the so-called "orphan receptors"), and unique patterns of transcriptional or translational expression in cells or tissues. Studying the transcriptional regulation of GPCR is a more direct means than studying translational regulation to determine the mechanisms underlying their unique distribution. However, when using older, established *in vitro* techniques, several factors (e.g., epigenetic regulation) must be considered before one can draw conclusions regarding the transcriptional mechanisms regulating GPCR expression.

Transcripts are primarily altered by transcriptional and post-transcriptional processes. Transcriptional regulation itself can be divided into three modes of influence: genetic (direct interaction of a control factor with the gene), modulation (interaction of a control factor with the transcription machinery), and epigenetic (nonsequence changes in DNA structure that influence transcription). These means of transcriptional regulation are discussed in greater detail in this chapter.

2. An Overview of GPCR Gene Transcription

Direct interaction with DNA is the simplest method by which a protein can alter transcription levels, and genes often have several protein binding sites upstream from the coding region that specifically regulate transcription. There are many classes of regulatory DNA binding sites in GPCR genes: enhancers, insulators, repressors, and silencers (see Table 1). The mechanisms for regulating transcription vary, from blocking key RNA polymerase binding sites on the DNA to direct activation or to promoting transcription by assisting polymerase binding.

Transcriptional mechanisms in mammals were long thought to be similar to those seen in prokaryotes, governed by specific positive or negative regulatory elements residing upstream of promoters. Activator or repressor proteins would bind to these regulatory elements, which then influenced the promoter's basic transcriptional machinery. The actual situation is however quite different, because mammalian DNA is wrapped with histones. The histone–DNA complex (i.e., chromatin) prevents binding of proteins to promoters unless assisted by specific regulatory molecules. That is, the *in vivo* template for transcription is the chromatin complex, rather than naked DNA. Any protein involved with transcriptional regulation must therefore interact with and penetrate the repressive chromatin structure. There are two steps of this process. First, the chromatin structure must be

Table 1
Proposed regulatory elements and transcription factors of GPCR genes

Receptor	Regulatory elements/Transcription factors (ref)
β1 adrenergic	AP2 (54), GRU (55), Sp1/CREB (56), Egr1 (57)
β2 adrenergic	GRE/Sp1 (58)
μ opioid (MOR)	PCBPs (59), Sox (60), NF-κB (61), PU.1 (62), NRSF (63, 64)
δ opioid (DOR)	Sp1 (39), USF (65), AP1/2 (66), Ik-2 (67)
κ opioid (KOR)	Sp1 (68), c-Myc (69), Ik (70)
D1A dopamine	AP2 (71), Sp1/Zic2 (72), ERE (73)
D2 dopamine	Sp1 (74), DRRF (47), NF-κB (75), AP1 (76)
M1 muscarinic	Sp1/NZF1/AP1/AP2/Ebox/NF-κB/Oct1 (77)
M2 muscarinic	GATA (78), LIF/CNTF (79)
M4 muscarinic	NRSF (80)
A1 adenosine	GATA4/Nkx2.5 (81), AGBP (82), NF-κB (83), GR (84)

converted from a closed to an open state by chromatin-remodeling complexes and/or histone modifications. Second, as a result of this remodeling, interaction of activators with the DNA is increased. Bound activators can then interact with the RNA polymerase complex to stabilize formation of the preinitiation complex. Interference with or modulation of any of these steps can activate or repress transcription (5). This view of transcription is supported by the increasing number of transcription factors now known to interact with histones by recruiting histone-modifying enzymes and/or engaging in chromatin remodeling activities (6, 7). The involvement of chromatin changes the traditional view of transcription such that different questions must be asked (and different experiments performed to address them) if we are to elucidate fully the cell- and tissue-specific expression of GPCR genes.

For several decades extensive research efforts on GPCRs focused on receptor signaling. These studies often ignored the necessity of GPCR genes being expressed in the right cells or at the right time, or the fact that their expression levels change dynamically under different conditions or in response to various ligands. As mentioned earlier, each GPCR gene is distributed uniquely in various tissues at both the mRNA and protein levels. For examples, the μ opioid receptor (MOR), a major target receptor of morphine and opioid peptides, is mainly expressed in the CNS, with densities varying greatly in different brain regions and displaying different functional roles (8). This unique spatial expression of the MOR gene is closely associated with epigenetic programming through chromatin modifications and DNA methylation (9). The MOR is also regulated temporally (10, 11) and is regulated by opioid ligands (12–14) and other nonopioid drugs (e.g., cocaine and haloperidol) in discrete brain regions (15, 16). However, the molecular mechanisms of MOR gene regulation like those of most other GPCR genes are still not completely understood.

2.1. GPCRs and Chromatin

The interaction between proteins and DNA is essential for many cellular functions, especially transcription. Transcription is controlled by the dynamic association of transcription factors and chromatin modifiers with target DNA sequences. These associations are modulated by modifications of DNA such as methylation of CpG dinucleotides (17), posttranslational modifications of histones (18), and incorporation of histone variants (19). These alterations that modify the composition of DNA and chromatin without changing DNA sequence are referred to as epigenetic modifications and they are themselves inheritable. The technique of chromatin immunoprecipitation (ChIP) can be used to investigate the location of posttranslationally modified histones and histone variants in the gene promoter and to map DNA target sites of transcription factors and other chromatin-associated proteins. Some of these histone modifications can serve as genetic hallmarks of whether transcription is active or inactive. For instance, hyperacetylation of the lysine residues of H3 and H4 histones is generally associated with the promoters of actively transcribed genes, whereas hypoacetylated histones correlate with gene silencing (20). Intriguingly, the lysine residues on histones can either be acetylated or methylated. For example, histone H3 lysine 4 methylation has been correlated with active gene expression (21), whereas H3 lysine 9 methylation has been linked to gene silencing and the assembly of heterochromatin (22). DNA methylation pathways and the histone code are functionally interactive. Through the binding of MeCP2 to 5-methyl-CpG, dinucleotides can recruit transcriptional corepressors with histone deacetylase activity, providing a link between DNA methylation and histone deacetylation. MeCP2 also associates with histone H3 lysine 9 methyltransferase activity, providing a mechanism for targeting repressive histone methylation to DNA-methylated promoters (11, 23).

More recently, the RNA interference (RNAi) pathway was included as an epigenetic regulatory feature, mainly silencing its target gene. Small interfering RNAs (siRNAs) are chemically synthesized 21-nucleotide sequences of double stranded RNA that match perfectly to the 21 nucleotides of their target sequence (24, 25) and act through RNAi pathways to silence gene expression either at the transcriptional or posttranscriptional level. When transfected into mammalian cells, siRNAs mediate silencing by forming heterochromatin, downregulating genes pretranscriptionally (26). Besides siRNAs, endogenous microRNAs can also act through RNAi pathways. However, microRNAs appear to modulate the translation of mRNA rather than knocking down RNA levels at the transcriptional level.

2.2. In Vitro Assays for Transcriptional Regulation

Over the last decade, numerous studies examined the regulatory domains of GPCR promoters. Although a full assessment of each of these studies is beyond the scope of this chapter, key references are noted in Table 1. Here we focus on selected examples that illustrate general principles and discuss problems and solutions relating to several methods used to study the role of transcription in regulating endogenous GPCR expression. Much of the information presented here is illustrated by work carried out over the last decade by our laboratory and those of others on the regulation of the opioid receptor genes.

2.2.1. Reporter Gene Assays

A typical series of experiments to understand the basis of a particular GPCR's expression pattern (this applies to other types of genes as well) might include isolating a genomic DNA fragment containing the GPCR gene, and then designing and constructing a series of reporter genes. These are then transiently transfected into expressing and nonexpressing cell lines, followed by measuring reporter gene activity to localize regulatory domains.

Individual regulatory elements can be identified using DNA–protein interaction assays such as electrophoretic mobility shift assays (EMSA) and DNase footprinting. Finally, these DNA elements can be used to identify DNA binding factors (e.g., transcription factors) using EMSA, screening expression libraries, or one-hybrid assays. Despite the general utility of this experimental sequence, there are a number of potential pitfalls that must be considered: (1) Transient transfection can generate high-copy episomes of several thousands (although only a small fraction of these are transiently transcribed), whereas only two copies of an endogenous GPCR gene are found in a diploid cell (27); (2) most cell lines used in such experiments do not provide an accurate representation of the *in vivo* condition; (3) chromatin organization on episomal DNA differs from that of chromosomal genes. Therefore, rather than considering such experiments as definitive, it is better to consider them as enabling: They provide useful data of candidate elements, the detailed roles of which must be verified in a more physiological setting.

2.2.2. Genomic Footprinting Assays

Various procedures exist to fine-map the sites at which transcription factors interact with their chromatin template (28). DNase I and dimethyl sulfate (DMS) genomic footprinting both rely on the blockage of a cleavage or modification site on genomic DNA by a bound transcription factor and its subsequent detection by ligation-mediated polymerase chain reaction (LM-PCR). The banding patterns of digested DNA are compared with those obtained using purified DNA, and the footprint is indicated by the differences in the digestion pattern. DMS footprinting has the distinct advantage of greater resolution than that obtained with DNase I, largely as a consequence of the much smaller size of DMS compared with DNase I. However, both procedures are limited by their inability to detect all protein interactions with genomic DNA. By itself, genomic footprinting cannot identify the protein(s) that are responsible for the footprint. Nevertheless, numerous studies have used footprinting to identify putative sites of interaction of transcription factors with their regulatory sites in neuronal genes.

2.3. In Vivo Assays for Transcriptional Regulation

Many transcriptional assays performed in the last 2 decades have relied mainly on EMSAs to demonstrate physical interactions between transcription factors and DNA *in vitro* and transient reporter gene assays to show functional interaction between recruited DNA and the transcriptional apparatus; the latter is usually carried out ex vitro but transgenic animals can be used to study reporter gene expression *in vivo*, e.g., using the κ opioid receptor (KOR) promoter (29). Both methods are useful in describing potential interactions between a transcription factor and its cognate *cis* regulatory elements but do not necessarily deliver information on interactions that occur *in vivo* between a transcription factor and endogenous genes.

2.4. Chromatin Immunoprecipitation

ChIP has become the method of choice for detecting the presence of proteins recruited to a specific genomic locus (Fig. 1). As such, it is the only procedure that can readily examine transcription factor occupancy, cofactor recruitment, and histone modification

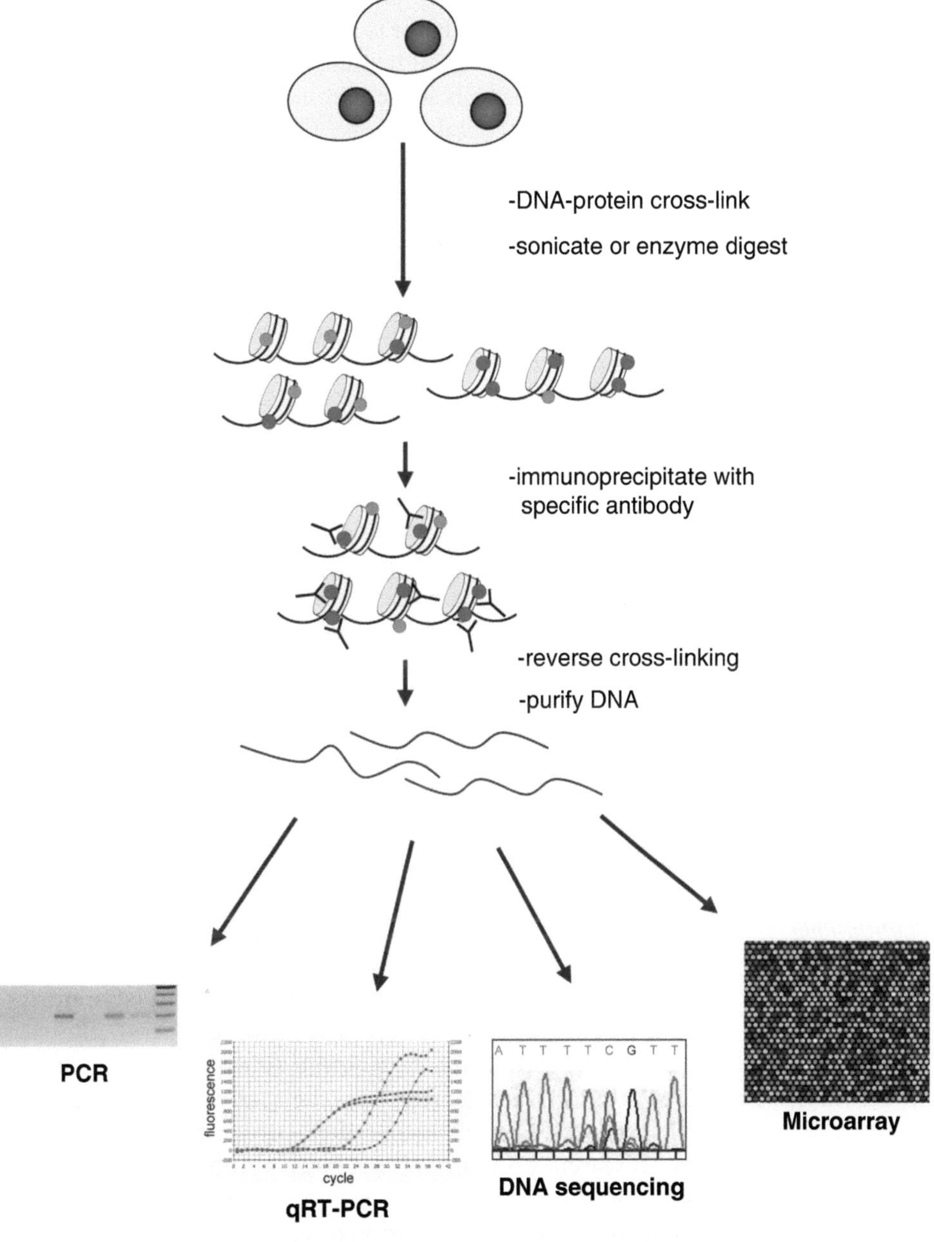

Fig. 1. Summary of the chromatin immunoprecipitation (ChIP) assay and various analytical methods.

status *in vivo*. There are numerous ChIP protocols, all involving immunoprecipitation of chromatin together with bound transcription factors and cofactors. The resulting "pull-down" is then analyzed by PCR (or by real-time quantitative PCR [qPCR]) or array analysis to assess the relative amount of DNA corresponding to specific genomic loci. Using this method, the relative enrichment (i.e., occupancy) of a region of chromatin by a transcription factor, cofactor, or specific histone can be found. The method can either use formaldehyde to crosslink chromatin and associated proteins (X-ChIP) or omit crosslinking (i.e., native ChIP [N-ChIP]) (30); the latter is more suitable for the analysis of histones because of their high affinity for DNA. In both X-ChIP and N-ChIP, chromatin is subsequently fragmented, either enzymatically with micrococcal nuclease (MNase, which digests DNA at the level of the linker, leaving the nucleosome intact) or by sonication to shear the chromatin into fragments of 200–1,000 bp (with an average of 500 bp). In many cases, reproducibility of shearing chromatin by sonication from multiple samples causes problems and requires proper optimization for X-ChIP. If this is the major obstacle in one lab, use of advanced bath-based sonicator commercially available (e.g., Bioruptor Sonicator, http://www.diagenode.com) may provide reproducible processing of multiple samples with a minimum of operator error. The type of detergent in lysis buffer can have a profound impact on shearing size of chromatin. For example, when preparing chromatin from neural stem cells, NP40 and SDS produce much smaller chromatin fragments than do Triton X-100-based buffers. The sheared chromatin is then precipitated with an antibody against a protein of interest. Crosslinks are reversed and the ChIP-enriched DNA is purified. The immunoprecipitated DNA sequences associated with the specific protein can be identified by PCR, quantitative PCR (qPCR) (9), DNA microarrays (i.e., ChIP-on-chip, described below) (31), molecular cloning and sequencing (32), or direct high-throughput sequencing (ChIP-seq) (33) (Fig. 1). ChIP techniques have been used frequently to study the transcriptional regulation of GPCR genes (11, 34, 35).

2.5. ChIP-On-Chip

Much effort has been devoted in the past several years to develop microarray technologies to understand the genome-wide localization of DNA binding proteins, and different laboratories have adopted a number of different technologies (36). Global target site profiling over whole-genome tiling arrays has shown that transcription factors regulate their target genes mostly from distant DNA regions rather than proximal promoters. Binding motif searches revealed that a large number of the *in vivo* target sites did not contain the expected or known *cis*-acting sequence, indicating that transcriptional regulation frequently involves crosstalk with other factors. "ChIP-on-chip" combines ChIP and microarray technology. The technique starts with conventional

X-ChIP: DNA and protein are crosslinked, followed by extraction, shearing, and immunoprecipitation of chromatin. The immunoprecipitated DNA is then labeled with a fluorescent dye and hybridized to microarrays containing genomic target sequences. Generally, target sequences have a higher ratio of ChIP signal/input signal, although there are several alternative means of normalizing the results of ChIP-on-chip studies. Some laboratories construct microarrays with PCR products or oligonucleotides in-house; oligonucleotide-bound chips are also available commercially (e.g., Affymetrix, Agilent, Nimblegen). Compared to conventional ChIP, the quantity of available material becomes even more of an issue with ChIP-on-chip when dealing with limited materials such as *in vivo* samples. In these instances, DNA amplification to prepare probe either before or during probe synthesis may also need to be considered. Amplification can be achieved using LM-PCR, which involves ligation of linkers to the ends of immunoprecipitated DNA fragments followed by PCR amplification. Because production of appropriate probes is crucial to successful ChIP-on-chip, the number of PCR cycles should be limited to minimize bias, i.e., <1,000-fold amplification (37). Once the data have been obtained, a number of software programs (e.g., MAT [http://liulab.dfci.harvard.edu/MAT/]; BAC [http://www.bioconductor.org/packages/bioc/html/BAC.html]) are available for the statistical analysis and normalization of ChIP-on-chip data.

3. Methods to Analyze the Transcriptional Regulation of GPCR Expression

Many general molecular biological techniques are available for analyzing transcriptional regulation (28, 38). In this section, established techniques are described in terms of their purposes and expected results; more detailed descriptions are given of emerging technologies.

3.1. Measuring and Detecting Transcription

Transcription can be measured and detected in a variety of ways.

3.1.1. Northern Blot

Northern blotting gives size and sequence information about specific mRNA molecules and can be used to measure mRNA levels quantitatively. In addition to its general utility, the size information allows the discrimination of alternately spliced transcripts: Several spliced isoforms of three opioid receptors were confirmed using this method (39–41). The identification of these isoforms could provide insight into the pharmacological complexity seen in the clinical and behavioral observations of opioid GPCRs.

3.1.2. Nuclear Run-On Assays

Nuclear run-on assays are designed to examine the genes being transcribed in a cell nucleus at a specific time (42). In addition to measuring the relative abundance of newly formed transcripts, this method allows changes in transcription rates to be measured, which often differ from steady-state mRNA levels used in northern blots and microarrays.

3.1.3. RT-PCR and Real-Time qPCR

Reverse transcription polymerase chain reaction (RT-PCR), as its name implies, first generates a DNA template from mRNA by reverse transcription. It is used primarily to measure mRNA abundance (28, 38). Real-time quantitative RT-PCR (qRT-PCR) uses nucleic acid detection dyes and a real-time detection system (43). For example, SYBR Green I has been used to detect transcripts of opioid receptors in many cells and tissue types, even microdissected brain regions (9). Briefly, 5 μg of total RNA is treated with DNase I and reverse-transcribed using primers combined with oligo (dT) and random hexamer. One-fortieth of this reaction is used for real-time qRT-PCR analysis of gene expression using SYBR Green I dye (Quantitect SYBR Green PCR kit; Qiagen) performed in an iCycler IQ (Bio-Rad). To calculate relative mRNA gene expression, amplification curves of a test sample and standard samples that contained 10^1–10^8 molecules of the gene of interest (e.g., the MOR expression plasmid pmMUEG) (44) are monitored, and the number of target molecules in the test sample is analyzed using qCalculator ver. 1.0 software (45) (http://www.gene-quantification.de/download.html#qcalculator) based on the mathematical model of Pfaffl (46). The number of target molecules is normalized against that obtained for β-actin, used as an internal control. The specificity of RT-PCR primers is determined using a melt curve after the amplification to show that only a single species of qPCR product results from the reaction. Single PCR products also can be sequenced or verified on an agarose gel.

3.1.4. DNA Microarray

One limitation of Northern blots and RT-qPCR methods is that they are good for detecting whether a single gene is being expressed, but they quickly become impractical if many genes within a sample are being studied. Using DNA microarrays, transcript levels for many genes (i.e., expression profiling) can be measured simultaneously. Recent advances in microarray technology allow for the quantification of transcript levels for every known gene in several organisms' genomes (including humans) on a single array. Alternatively, "tag-based" technologies (e.g., serial analysis of gene expression, which can provide a relative measure of the cellular concentration of different messenger RNAs) can be used. The great advantage of tag-based methods is the "open architecture," allowing for the exact measurement of any transcript whether the sequence is known or not.

3.1.5. In Situ Hybridization and Immunofluorescence

Analyses of gene expression can establish more than its quantification. Gene expression can also be localized at the RNA level by *in situ* hybridization with a suitably tagged complementary mRNA strand, and at the protein level by immunofluorescence with labeled antibodies, followed by microscopic examination. Numerous studies have used these detection methods *in vivo*; e.g., Hwang et al. (47) used both procedures to detect dopamine receptor regulating factor (DRRF).

3.2. Analysis of cis-Regulatory Sequences

3.2.1. Gel-Shift Assays

The EMSA (also known as the gel shift or band shift assay) is a technique for studying gene regulation and determining protein–DNA interactions. The assay is based on the observation that complexes of protein and DNA migrate through a nondenaturing polyacrylamide gel more slowly than free DNA fragments or double-stranded oligonucleotides. The EMSA is carried out by first incubating a protein or mix of proteins (such as nuclear or cell extract) with a ^{32}P end-labeled DNA fragment containing the putative protein binding site. The reaction products are then analyzed on a nondenaturing polyacrylamide gel. The specificity of the DNA-binding protein for the putative binding site is established by competition experiments using DNA fragments or oligonucleotides containing a binding site for the protein of interest or other unrelated DNA sequences.

3.2.2. Promoter Reporter Gene Assays

Reporter genes are generally used in cultured cells to determine if a DNA fragment has promoter activity itself or regulatory activity for a promoter. The promoter DNA fragment is fused to the reporter gene in a single DNA construct and transfected into the cell or organism. Commonly used reporter genes usually involving fluorescence (e.g., green fluorescent protein) or enzymes (e.g., chloramphenicol acetyltransferase, luciferase) produce a visually identifiable reaction product.

3.3. Determination of Trans-Acting Proteins

3.3.1. Transcriptional Factor Binding Site Searches

There are a number of tools and databases that allow searches for transcriptional factors or analysis of a given DNA sequence of interest to predict transcriptional factor binding sites: Match™ (www.gene-regulation.com/pub/programs.html#match), sponsored by BIOBASE GmbH, the public version of this program (1.0) is designed to search nucleotide sequences for potential transcription factor binding sites; MatInspector (www.genomatix.de/products/MatInspector/index), Genomatix's tool for transcription factor binding site analysis; Tfbind (tfbind.hgc.jp/), software for searching transcription factor binding sites (including TATA boxes, GC boxes, CCAAT boxes, transcription start sites, etc.) using weighted matrices from the transcription factor database Transfac R.3.4 (developed by Wingender et al.); and Tfsearch (www.cbrc.jp/research/db/TFSEARCH.html), which given a DNA sequence, this site returns potential transcriptional factor binding sites on the sequence.

3.3.2. Supershift Assays

Supershift assays are used to detect DNA–protein interactions. They are similar in theory and execution to EMSAs, except they rely on antibody binding to the epitope of interest to alter migration through the gel.

3.3.3. Affinity Purification

Affinity chromatography purification is a means of separating DNA-binding proteins based on a highly specific biological interaction, such as that between regulatory DNA and protein. The method has been used by Choi et al. to isolate MOR promoter binding proteins (48).

3.3.4. siRNA-Mediated Gene Silencing

Briefly, this method knocks down the target transcription factor to determine its endogenous role. RNAi using double-stranded RNA (dsRNA) molecules of approximately 20–25 nucleotides (i.e., siRNAs) is a powerful method for preventing the expression of a particular gene. The dsRNA dominantly silences gene expression in a sequence-specific manner by causing the corresponding endogenous mRNA to be degraded. The technique was first developed in *Caenorhabditis elegans*, and was rapidly applied to a wide range of organisms. Methods for expressing siRNAs in cells in culture and *in vivo* using viral vectors, and for transfecting cells with synthetic siRNAs, have been developed and are being used to establish the functions of specific proteins in various cell types and organisms. For example, chemically synthesized or *in vitro* transcribed siRNAs can be transfected into cells, injected into mice, or introduced into plants. siRNAs can also be expressed endogenously from siRNA expression vectors or PCR products in cells or in transgenic animals.

3.3.5. ChIP Assays

Because of its utility as a starting point for a number of detailed analytical systems (Fig. 1), the basic X-ChIP assay is described in detail below. For mammalian studies, the number of cells required depends on the cell type, but usually $2–10 \times 10^6$ cells per ChIP. Therefore, it is important to verify the number of individual cells cultured in each laboratory. The following protocol is based on the use of mouse P19 embryonal carcinoma cells. The procedure is designed for cells in 100-mm dishes, so the volumes listed below should be modified as necessary.

3.3.5.1. Methods

1. Replace the medium with 10 ml of fresh culture medium.
2. Crosslink DNA–protein complexes by adding 270 μl of 37% formaldehyde to the culture medium (i.e., to a final concentration of 1%). Incubate at room temperature for 10 min (see Note 1).
3. Add 1.47 ml of 1 M glycine per 10 ml of culture medium to stop the crosslinking.
4. Place the plate on ice. Wash twice with 10 ml of ice-cold phosphate-buffered saline.

5. Add 1 ml of cell suspension buffer (25 mM HEPES [pH 7.8], 1.5 mM $MgCl_2$, 10 mM KCl, 0.1% NP-40, 1 mM DTT, 0.5 mM PMSF, protease inhibitor cocktail [Roche]) to the cells. Harvest using a cell scraper.
6. Incubate the cells on ice for 10 min. Homogenize using 10–20 strokes in a Dounce homogenizer. Check the nuclei in microscope by mixing an aliquot with equal volume of 0.4% trypan blue.
7. Centrifuge the suspension at 2,000 rpm for 5 min.
8. Resuspend the pelleted nuclei in 5–10 ml of lysis buffer (50 mM HEPES [pH 7.9], 140 mM NaCl, 1 mM EDTA, 1% Triton X-100, 0.1% sodium deoxycholate, 0.1% SDS, 0.5 mM PMSF, protease inhibitor cocktail).
9. Sonicate 10–15 times using a Heat Systems Sonicator at a setting of 40%, for 10–20 s each burst. Keep the sample in ice and allow cooling for 1 min between each sonication. The fragment size produced should be 200–1,000 nucleotides (see Note 2).
10. Centrifuge at 14,000 rpm for 15 min. Measure the protein concentration of the supernatant (using, e.g., the Bio-Rad assay). The supernatant contains crude soluble chromatin. Use 1–5 mg protein per immunoprecipitation.
11. Add sonicated salmon sperm DNA (to a final concentration of 1 μg/ml) and BSA (to a final concentration of 1 mg/ml) to the supernatant.
12. Preclear the lysate by incubating with constant rotation with 40–50 μl of Protein-A or -G agarose per ml of lysate for 2 h in a cold room (see Note 3).
13. Centrifuge the samples at 2,000 rpm for 5 min. Reserve the supernatant containing the precleared soluble chromatin (see Note 4).
14. Reserve a 50-μl aliquot (i.e., one-twentieth of the amount used per immunoprecipitation) at −20°C for preparation of input DNA.
15. Divide the sample into 1-ml aliquots in Eppendorf tubes for immunoprecipitation.
16. Add 2–5 μg antibody. Rotate for 2 h in the cold room (see Note 5).
17. Add 40 μl of Protein-A/G agarose per immunoprecipitation (equilibrated as above). Incubate overnight by constant rotation in the cold room.
18. Centrifuge the beads at 6,000 rpm for 3 min.
19. Wash twice with 1 ml of lysis buffer. Each wash includes 10 min of constant rotation in the cold room.

20. Wash twice with 1 ml of wash buffer A (i.e., lysis buffer containing 500 mM NaCl).
21. Wash twice with 1 ml of wash buffer B (20 mM Tris [pH 8.0], 1 mM EDTA, 250 mM LiCl, 0.5% NP-40, 0.5% sodium deoxycholate, 0.5 mM PMSF, protease inhibitor cocktail).
22. Wash twice with 1 ml of TE buffer (10 mM Tris [pH 8.0], 1 mM EDTA).
23. Add 200 μl of elution buffer (50 mM Tris [pH 8.0], 1 mM EDTA, 1% SDS, 50 mM $NaHCO_3$) to the beads. Incubate for 10 min at 65°C.
24. Centrifuge at 14,000 rpm for 1 min. Transfer the supernatant to a new tube. Elute the beads again, as above. Combine the eluates to a final volume of 400 μl (adjust with elution buffer, if necessary).
25. Thaw the 50-μl input sample and supplement with 350 μl of elution buffer. Add 21 μl of 4 M NaCl to both the input and immunoprecipitated samples.
26. Incubate the samples for at least 5 h at 65°C (see Note 6).
27. Add 1 μl of RNase A (from a 10-mg/ml, DNase-free stock). Incubate for 1 h at 37°C.
28. Add 4 μl of 0.5 M EDTA and 2 μl of 10 mg/ml proteinase K. Incubate for 2 h at 42°C.
29. Recover the DNA by phenol/chloroform extraction.
30. Add 1 μl of 20 mg/ml glycogen, 40 μl of 3 M sodium acetate and 1 ml of ethanol. Vortex and allow to precipitate overnight at −20°C.
31. Centrifuge at 14,000 rpm for 30 min. Wash once with 70% ethanol.
32. Resuspend the immunoprecipitate and input samples in 100 μl of 10 mM Tris (pH 7.5). Analyze by PCR.

3.3.5.2. Notes

1. Crosslinking time influences ChIP efficiency. A 10-min crosslinking period is suitable for ChIP analyses of histone modifications, but longer crosslinking times (up to 30 min) can be used for ChIP studies on transcription factors. An incubation time of 15 min is standard in our laboratory.
2. An Ultrasonic Processor XL (Heat Systems Sonicator) works efficiently and almost independently of cell concentration. We routinely sonicate 12 times (10 s each) with intervals of 30 s on ice. It is very important that sample stays cold during sonication. Typically, short sonications result in high recoveries (% ChIP/input) but low resolution, whereas longer sonication times result in lower recovery but higher resolution. Nevertheless, gel images can be inconclusive, and the procedure may need to be adjusted to provide an optimal signal/background ratio.

3. Before preclearing, wash the resins once with lysis buffer.
4. At this point, the samples can be frozen at –80°C. However, the use of fresh chromatin clearly increases recovery. Storage for longer than 1–2 months is not recommended. Chromatin prepared in lysis buffer containing 1% SDS can be stored at 4°C for 1–2 days until use.
5. The optimal ratio for transcriptional factors is ~2–4 μg of antibody per $1–2 \times 10^6$ cells, although a different ratio was observed for antihistone antibodies. Generally, the concentration of the antibody should be determined empirically.
6. Although this de-crosslinking step can be performed overnight, such prolonged incubation can be problematic.

3.3.6. Reciprocal ChIP (ReChIP)

Generally, gene transcription is controlled by several transcription factors interacting with DNA regulatory elements in the nucleus in conjunction with ubiquitous and/or tissue-specific cofactors. Therefore, it is useful to determine if multiple proteins "cohabit" in a particular cell type or at a specific developmental stage. ReChIP uses two independent rounds of immunoprecipitation using high-quality antibodies to determine the *in vivo* colocalization of two different proteins interacting with, or in close contact to, the same genomic locus. Basically, a round of ChIP is performed as described above. The immunocomplexes are eluted by rotating at 37°C with two changes (30 min each) of 30 μl of 10 mM DTT. One-sixth of the eluent volume is reserved as the "first ChIP assay." The supernatant is diluted 1:40 in ChIP dilution buffer, antibody against the second protein of interest is added, and the new immunocomplexes are allowed to form by incubating for 1 h at room temperature on a rocking platform. The immunocomplexes are collected by incubating on a rocking platform with 40 μl of a Protein-A or -G agarose slurry (Upstate) for 1 h at room temperature and washing as indicated above. Both the first and second ChIP samples are eluted twice by adding 250 μl of elution buffer (1% SDS in 100 mM $NaHCO_3$) and incubating with rotation for 10 min at 65°C. After centrifugation, the supernatants are collected and the crosslinking is reversed by adding NaCl to final concentration of 200 mM and incubating for 4 h at 65°C. The remaining proteins are digested by adding proteinase K to a final concentration of 40 μg/ml and incubation for 1 h at 45°C. The DNA is recovered by phenol/chloroform/isoamyl alcohol extraction and precipitation with 0.1 volumes of 5 M lithium chloride and 2 volumes of ethanol, using glycogen as a carrier. This method was used to successfully colocalize neuron-restrictive silencer factor (NRSF) and Sp3 in the neuron-restrictive silencer element of the MOR promoter (49) and to determine the coordination of chromatin modifications by MeCP2 and DNA methyltransferase (DNMT) in specific brain regions (9).

3.4. Epigenetic Regulation of GPCRs

3.4.1. DNA Methylation

DNA methylation is a postreplication event involving the addition of a methyl group to the 5 positions of a CpG dinucleotide, catalyzed by DNMT. Procedures for DNA methylation analysis can be divided roughly into two types: global and gene-specific. For global methylation analyses, chromatographic methods and methyl-accepting capacity assays measure the overall level of methyl cytosines in the genome. By contrast, numerous techniques have been developed for gene-specific methylation analyses. Early studies used methylation-sensitive restriction enzymes to digest DNA, followed by detection on Southern blots or PCR amplification. Recently, to identify unknown methylation hotspots or methylated CpG islands in the genome, several genome-wide detection methods have been developed (e.g., restriction landmark genomic scanning for methylation and CpG island microarrays) (50). Additionally, bisulfite-reaction-based methods such as methylation-specific PCR and bisulfite genomic sequencing PCR have become very popular (9, 11). Briefly, genomic DNA from cell lines (e.g., P19 cells, neuroblastoma NS20Y cells, or brain tissue samples) is isolated using TRI reagent and linearized with the restriction enzyme EcoRV. Bisulfite treatment of DNA is performed using an EZ DNA Methylation-Gold Kit™ (Zymo Research) according to the manufacturer's instructions. The resulting bisulfite-modified DNA is amplified by PCR using methylation-specific primers. The PCR products are cloned into a pCRII-TOPO vector (Invitrogen) and clones from all the cells containing a correctly sized insert are analyzed by DNA sequencing. The PCR products can also be used for restriction enzyme-assisted methylation analyses. After finding suitable methyl-CpG-sensitive restriction enzymes based on DNA sequence information for the gene of interest (e.g., BstBI, ClaI, Hpy188I, and HpyCH4IV for the MOR gene), the samples are incubated for 1 h at the recommended temperature. The input of each PCR product should be adjusted for equivalent amounts in the restriction enzyme reactions. The resulting DNA is loaded onto a 2.5% agarose gel and quantified by Kodak molecular imaging software.

3.4.2. Micrococcal Nuclease–Southern Blot Assay

MNase is distinct from other nucleases in that it is able to cut double strands within nucleosome linker regions, but only produce single-strand nicks within the nucleosome itself. Consequently, MNase is widely used to determine the nucleosomal complex formation of a DNA fragment of interest (28). In addition, if the nucleosomes are consistently positioned, MNase can be used to determine their approximate positions in a region of DNA, as in a recent study of the κ opioid receptor (KOR) promoter (34, 51). In this procedure, normal (i.e., undifferentiated) and differentiated cells (e.g., P19 cells) are crosslinked as in the ChIP procedure described above. Nuclei are isolated from ~10^8 cells, suspended in

1 ml of wash buffer (10 mM Tris–HCl [pH 7.4], 15 mM NaCl, 50 mM KCl, 0.15 mM spermine, 0.5 mM spermidine, 8.5% sucrose) and digested with 20–120 U of MNase (Worthington) for 6 min at 37°C. Reactions are stopped by adding 100 μl of 5 M NaCl and 100 μl of stop solution (10% SDS, 125 mM EDTA, 1 mg/ml proteinase K) and incubating for 3 h at 50°C. The temperature is increased to 65°C for at least 4 h to reverse crosslinking. Purified genomic DNA is analyzed by Southern blot hybridization using DNA probes for MNase digestion labeled with [α-^{32}P] dCTP by the random priming method.

3.4.3. MNase-Mediated LM-PCR

This procedure is described in detail elsewhere (28, 34); here we present some minor modifications. For nucleosome mapping, 1 μg of MNase-digested DNA is phosphorylated and ligated with annealed linker DNA (linker 1: 1, 5′-GCGGTGACCCGGG AGATCTGAATTC-3′; linker 2: 5′-GAATTCAGATC-3′) overnight at 16°C. The ligated DNAs are amplified by PCR using a gene-specific primer and linker 1, as follows: 25 cycles of 1 min (4 min for the first cycle) at 95°C, 2 min at 63°C, 3 min at 76°C, and an additional 5 min extension at 76°C. Labeling is performed by adding a second gene-specific primer 3′end-labeled with [γ-^{32}P]ATP and continuing PCR for an additional three cycles: 1 min (4 min for the first cycle) at 95°C, 2 min at 68°C, and 10 min at 76°C. Amplified and labeled DNA fragments are resolved on a 6% polyacrylamide/urea gel, followed by exposure to a PhosphorImager screen.

4. Future Directions

Two decades of research have revealed that during the course of cellular communication, transcriptional regulation of GPCR genes changes continually via their ligand–receptor interactions. Furthermore, such changes do not just affect individual GPCR genes but rather entire groups of the genes, collectively representing a change in the GPCR transcriptional program. A series of transcription factors operate, particularly during cellular differentiation, such that many indirect changes occur in response to individual transcription factors. Our future goal is to represent the transcriptional machinery of a cell as an interactive network in a manner analogous to a circuit diagram. The first step is to identify all the potential targets of all the transcription factors expressed by a particular cell. This is a highly redundant process and presumes that the activity of a cell or a group of cells can be demonstrated as the sum of its components. In yeast, this approach has been performed successfully (52), and even computationally (53). However, there is a huge difference in the genomic complexity between yeast and mammalian cells. Yeast cells contain only a quarter of the genes and a fraction of the noncoding RNA of a mammalian cell. Further, yeast can easily be synchronized in cultures consisting of a single cell type whereas there are hundreds or more different neuronal cells in the mammalian brain. More importantly, yeast have a simpler response system to signals whereas a neuron must communicate with other cells after receiving signals. Nevertheless, yeast genetics have provided a wealth of information regarding GPCR gene regulation that is also relevant to mammalian neurobiology, and despite the differences in complexity, these limitations might well be overcome in the future.

5. Conclusions

Several techniques have been developed to demonstrate the regulation of GPCR gene transcription. A single experimental method cannot elucidate the complexity of mammalian GPCR transcription, but additional techniques can compensate for or confirm preliminary results obtained from a single method. Although chromatin-related experiments are emphasized in this chapter, *in vitro* methods such as EMSA and reporter assays still provide valuable information. Indeed, it is often useful to perform *in vitro* studies on GPCR transcription first before proceeding to more physiologically relevant experiments.

Acknowledgments

This work was supported by NIH Grants DA000564, DA001583, DA011806, K05-DA070554 (HHL), DA011190, DA013926 (LW), and by the A&F Stark Fund of the Minnesota Medical Foundation. We thank Dr. Martin Winer and Mr. Bradley J. Stish for editorial assistance with the manuscript.

References

1. Vassilatis DK, Hohmann JG, Zeng H et al (2003) The G protein-coupled receptor repertoires of human and mouse. Proc Natl Acad Sci USA **100**:4903–4908
2. Lundstrom K (2009) An overview on GPCRs and drug discovery: structure-based drug design and structural biology on GPCRs. Methods Mol Biol **552**:51–66
3. Fredriksson R, Lagerstrom MC, Lundin LG et al (2003) The G-protein-coupled receptors in the human genome form five main families. Phylogenetic analysis, paralogon groups, and fingerprints. Mol Pharmacol **63**:1256–1272
4. Stevens CW (2009) The evolution of vertebrate opioid receptors. Front Biosci **14**:1247–1269
5. Lemon B, Tjian R (2000) Orchestrated response: a symphony of transcription factors for gene control. Genes Dev **14**:2551–2569
6. Hsieh J, Gage FH (2005) Chromatin remodeling in neural development and plasticity. Curr Opin Cell Biol **17**:664–671

7. Lomvardas S, Thanos D (2002) Opening chromatin. Mol Cell **9**:209–211
8. Mansour A, Fox CA, Akil H et al (1995) Opioid-receptor mRNA expression in the rat CNS: anatomical and functional implications. Trends Neurosci **18**:22–29
9. Hwang CK, Song KY, Kim CS et al (2009) Epigenetic programming of *mu* opioid receptor gene in mouse brain is regulated by MeCP2 and Brg1 chromatin remodeling factor. J Cell Mol Med **13**:3591–3615.
10. Chen HC, Wei LN, Loh HH (1999) Expression of *mu-*, *kappa-* and *delta*-opioid receptors in P19 mouse embryonal carcinoma cells. Neuroscience **92**:1143–1155
11. Hwang CK, Song KY, Kim CS et al (2007) Evidence of endogenous *mu* opioid receptor regulation by epigenetic control of the promoters. Mol Cell Biol **27**:4720–4736
12. Suzuki S, Miyagi T, Chuang TK et al (2000) Morphine upregulates *mu* opioid receptors of human and monkey lymphocytes. Biochem Biophys Res Commun **279**:621–628
13. Ruzicka BB, Akil H (1997) The interleukin-1*beta*-mediated regulation of proenkephalin and opioid receptor messenger RNA in primary astrocyte-enriched cultures. Neuroscience **79**:517–524
14. Chang SL, Felix B, Jiang Y et al (2001) Actions of endotoxin and morphine. Adv Exp Med Biol **493**:187–196
15. Azaryan AV, Clock BJ, Rosenberger JG et al (1998) Transient upregulation of *mu* opioid receptor mRNA levels in nucleus accumbens during chronic cocaine administration. Can J Physiol Pharmacol **76**:278–283
16. Delfs JM, Yu L, Ellison GD et al (1994) Regulation of *mu*-opioid receptor mRNA in rat globus pallidus: effects of enkephalin increases induced by short- and long-term haloperidol administration. J Neurochem **63**:777–780
17. Antequera F (2003) Structure, function and evolution of CpG island promoters. Cell Mol Life Sci **60**:1647–1658
18. Kouzarides T (2007) Chromatin modifications and their function. Cell **128**:693–705
19. Mito Y, Henikoff JG, Henikoff S (2005) Genome-scale profiling of histone H3.3 replacement patterns. Nat Genet **37**:1090–1097
20. Struhl K (1998) Histone acetylation and transcriptional regulatory mechanisms. Genes Dev **12**:599–606
21. Zegerman P, Canas B, Pappin D et al (2002) Histone H3 lysine 4 methylation disrupts binding of nucleosome remodeling and deacetylase (NuRD) repressor complex. J Biol Chem **277**:11621–11624
22. Lachner M, O'Carroll D, Rea S et al (2001) Methylation of histone H3 lysine 9 creates a binding site for HP1 proteins. Nature **410**:116–120
23. Fuks F, Hurd PJ, Wolf D et al (2003) The methyl-CpG-binding protein MeCP2 links DNA methylation to histone methylation. J Biol Chem **278**:4035–4040
24. Fire A, Xu S, Montgomery MK et al (1998) Potent and specific genetic interference by double-stranded RNA in *Caenorhabditis elegans*. Nature **391**:806–811
25. Mello CC, Conte D, Jr. (2004) Revealing the world of RNA interference. Nature **431**:338–342
26. Holmquist GP, Ashley T (2006) Chromosome organization and chromatin modification: influence on genome function and evolution. Cytogenet Genome Res **114**:96–125
27. Smith CL, Hager GL (1997) Transcriptional regulation of mammalian genes *in vivo*. A tale of two templates. J Biol Chem **272**: 27493–27496
28. Carey M, Peterson CL, Smale ST (2009) Transcriptional regulation in eukaryotes : concepts, strategies, and techniques, 2nd ed., Cold Spring Harbor Laboratory Press, Cold Spring Harbor, New York
29. Hu X, Cao S, Loh HH et al (1999) Promoter activity of mouse *kappa* opioid receptor gene in transgenic mouse. Brain Res Mol Brain Res **69**:35–43
30. O'Neill LP, Turner BM (1996) Immunoprecipitation of chromatin. Methods Enzymol **274**:189–197
31. Lee TI, Johnstone SE, Young RA (2006) Chromatin immunoprecipitation and microarray-based analysis of protein location. Nat Protoc **1**:729–748
32. Wei CL, Wu Q, Vega VB et al (2006) A global map of p53 transcription-factor binding sites in the human genome. Cell **124**:207–219
33. Barski A, Cuddapah S, Cui K et al (2007) High-resolution profiling of histone methylations in the human genome. Cell **129**:823–837
34. Park SW, Huq MD, Loh HH et al (2005) Retinoic acid-induced chromatin remodeling of mouse *kappa* opioid receptor gene. J Neurosci **25**:3350–3357
35. Wei LN (2008) Epigenetic control of the expression of opioid receptor genes. Epigenetics **3**:119–121
36. Collas P (2009) Chromatin immunoprecipitation assays : methods and protocols, Humana Press, Totowa, New Jersey.
37. Ren B, Robert F, Wyrick JJ et al (2000) Genome-wide location and function of DNA binding proteins. Science **290**:2306–2309

38. Sambrook J, Russell DW (2001) Molecular cloning : a laboratory manual, 3rd ed., Cold Spring Harbor Laboratory Press, Cold Spring Harbor, New York
39. Smirnov D, Im HJ, Loh HH (2001) *Delta*-opioid receptor gene: effect of Sp1 factor on transcriptional regulation *in vivo*. Mol Pharmacol **60**:331–340
40. Bi J, Hu X, Loh HH et al (2001) Regulation of mouse *kappa* opioid receptor gene expression by retinoids. J Neurosci **21**:1590–1599
41. Pan YX, Xu J, Bolan E et al (2005) Identification of four novel exon 5 splice variants of the mouse *mu*-opioid receptor gene: functional consequences of C-terminal splicing. Mol Pharmacol **68**:866–875
42. Kim SS, Pandey KK, Choi HS et al (2005) Poly(C) binding protein family is a transcription factor in *mu*-opioid receptor gene expression. Mol Pharmacol **68**:729–736
43. Dorak MT (2006) Real-time PCR, Taylor & Francis, New York.
44. Song KY, Hwang CK, Kim CS et al (2007) Translational repression of mouse *mu* opioid receptor expression via leaky scanning. Nucleic Acids Res **35**:1501–1513
45. Gilsbach R, Kouta M, Bonisch H et al (2006) Comparison of *in vitro* and *in vivo* reference genes for internal standardization of real-time PCR data. Biotechniques **40**:173–177
46. Pfaffl MW (2001) A new mathematical model for relative quantification in real-time RT-PCR. Nucleic Acids Res **29**:e45
47. Hwang CK, D'Souza UM, Eisch AJ et al (2001) Dopamine receptor regulating factor, DRRF: a zinc finger transcription factor. Proc Natl Acad Sci USA **98**:7558–7563
48. Choi HS, Song KY, Hwang CK et al (2008) A proteomics approach for identification of single strand DNA-binding proteins involved in transcriptional regulation of mouse *mu* opioid receptor gene. Mol Cell Proteomics 7:1517–1529
49. Kim CS, Choi HS, Hwang CK et al (2006) Evidence of the neuron-restrictive silencer factor (NRSF) interaction with Sp3 and its synergic repression to the *mu* opioid receptor (MOR) gene. Nucleic Acids Res **34**:6392–6403
50. Tost J (2009) DNA methylation : methods and protocols, 2nd ed., Humana Press, Totowa, New Jersey.
51. Ausubel FM. (2003) Current protocols in molecular biology. Wiley, New York.
52. Lee TI, Rinaldi NJ, Robert F et al (2002) Transcriptional regulatory networks in *Saccharomyces cerevisiae*. Science **298**:799–804
53. Bar-Joseph Z, Gerber GK, Lee TI et al (2003) Computational discovery of gene modules and regulatory networks. Nat Biotechnol **21**:1337–1342
54. Kirigiti P, Yang YF, Li X et al (2000) Rat *beta*1-adrenergic receptor regulatory region containing consensus AP-2 elements recognizes novel transactivator proteins. Mol Cell Biol Res Commun **3**:181–192
55. Tseng YT, Stabila JP, Nguyen TT et al (2001) A novel glucocorticoid regulatory unit mediates the hormone responsiveness of the *beta*1-adrenergic receptor gene. Mol Cell Endocrinol **181**:165–178
56. Lee K, Richardson CD, Razik MA et al (1998) Multiple potential regulatory elements in the 5' flanking region of the human *alpha*1a-adrenergic receptor. DNA Seq **8**:271–276
57. Bahouth SW, Beauchamp MJ, Vu KN (2002) Reciprocal regulation of *beta*(1)-adrenergic receptor gene transcription by Sp1 and early growth response gene 1: induction of EGR-1 inhibits the expression of the *beta*(1)-adrenergic receptor gene. Mol Pharmacol **61**:379–390
58. Malbon CC, Hadcock JR (1988) Evidence that glucocorticoid response elements in the 5'-noncoding region of the hamster *beta* 2-adrenergic receptor gene are obligate for glucocorticoid regulation of receptor mRNA levels. Biochem Biophys Res Commun **154**:676–681
59. Ko JL, Loh HH (2005) Poly C binding protein, a single-stranded DNA binding protein, regulates mouse *mu*-opioid receptor gene expression. J Neurochem **93**:749–761
60. Hwang CK, Wu X, Wang G et al (2003) Mouse *mu* opioid receptor distal promoter transcriptional regulation by SOX proteins. J Biol Chem **278**:3742–3750
61. Kraus J, Borner C, Giannini E et al (2003) The role of nuclear factor *kappa*B in tumor necrosis factor-regulated transcription of the human *mu*-opioid receptor gene. Mol Pharmacol **64**:876–884
62. Hwang CK, Kim CS, Choi HS et al (2004) Transcriptional regulation of mouse *mu* opioid receptor gene by PU.1. J Biol Chem **279**:19764–19774
63. Kim CS, Hwang CK, Choi HS et al (2004) Neuron-restrictive silencer factor (NRSF) functions as a repressor in neuronal cells to regulate the *mu* opioid receptor gene. J Biol Chem **279**:46464–46473
64. Andria ML, Simon EJ (2001) Identification of a neurorestrictive suppressor element (NRSE) in the human *mu*-opioid receptor gene. Brain Res Mol Brain Res **91**:73–80

65. Liu HC, Shen JT, Augustin LB et al (1999) Transcriptional regulation of mouse *delta*-opioid receptor gene. J Biol Chem **274**:23617–23626
66. Woltje M, Kraus J, Hollt V (2000) Regulation of mouse *delta*-opioid receptor gene transcription: involvement of the transcription factors AP-1 and AP-2. J Neurochem **74**:1355–1362
67. Sun P, Loh HH (2003) Transcriptional regulation of mouse *delta*-opioid receptor gene. Ikaros-2 and upstream stimulatory factor synergize in trans-activating mouse *delta*-opioid receptor gene in T cells. J Biol Chem **278**:2304–2308
68. Li J, Park SW, Loh HH et al (2002) Induction of the mouse *kappa*-opioid receptor gene by retinoic acid in P19 cells. J Biol Chem **277**:39967–39972
69. Park SW, Li J, Loh HH et al (2002) A novel signaling pathway of nitric oxide on transcriptional regulation of mouse *kappa* opioid receptor gene. J Neurosci **22**:7941–7947
70. Hu X, Bi J, Loh HH et al (2001) An intronic Ikaros-binding element mediates retinoic acid suppression of the *kappa* opioid receptor gene, accompanied by histone deacetylation on the promoters. J Biol Chem **276**:4597–4603
71. Minowa MT, Minowa T, Mouradian MM (1993) Activator region analysis of the human D1A dopamine receptor gene. J Biol Chem **268**:23544–23551
72. Yang Y, Hwang CK, Junn E et al (2000) ZIC2 and Sp3 repress Sp1-induced activation of the human D1A dopamine receptor gene. J Biol Chem **275**:38863–38869
73. Lee SH, Mouradian MM (1999) Up-regulation of D1A dopamine receptor gene transcription by estrogen. Mol Cell Endocrinol **156**:151–157
74. Yajima S, Lee SH, Minowa T et al (1998) Sp family transcription factors regulate expression of rat D2 dopamine receptor gene. DNA Cell Biol **17**:471–479
75. Bontempi S, Fiorentini C, Busi C et al (2007) Identification and characterization of two nuclear factor-*kappa*B sites in the regulatory region of the dopamine D2 receptor. Endocrinology **148**:2563–2570
76. Wang J, Miller JC, Friedhoff AJ (1997) Differential regulation of D2 receptor gene expression by transcription factor AP-1 in cultured cells. J Neurosci Res **50**:23–31
77. Pepitoni S, Wood IC, Buckley NJ (1997) Structure of the M1 muscarinic acetylcholine receptor gene and its promoter. J Biol Chem **272**:17112–17117
78. Rosoff ML, Nathanson NM (1998) GATA factor-dependent regulation of cardiac M2 muscarinic acetylcholine gene transcription. J Biol Chem **273**:9124–9129
79. Laszlo GS, Rosoff ML, Amieux PS et al (2006) Multiple promoter elements required for leukemia inhibitory factor-stimulated M2 muscarinic acetylcholine receptor promoter activity. J Neurochem **98**:1302–1315
80. Saffen D, Mieda M, Okamura M et al (1999) Control elements of muscarinic receptor gene expression. Life Sci **64**:479–486
81. Rivkees SA, Chen M, Kulkarni J et al (1999) Characterization of the murine A1 adenosine receptor promoter, potent regulation by GATA-4 and Nkx2.5. J Biol Chem **274**:14204–14209
82. Ren H, Stiles GL (1998) A single-stranded DNA binding site in the human A1 adenosine receptor gene promoter. Mol Pharmacol **53**:43–51
83. Nie Z, Mei Y, Ford M et al (1998) Oxidative stress increases A1 adenosine receptor expression by activating nuclear factor *kappa* B. Mol Pharmacol **53**:663–669
84. Ren H, Stiles GL (1999) Dexamethasone stimulates human A1 adenosine receptor (A1AR) gene expression through multiple regulatory sites in promoter B. Mol Pharmacol **55**:309–316

Chapter 5

Deciphering the Evolution of G Protein-Coupled Receptors in Vertebrates

Craig W. Stevens

Abstract

G protein-coupled receptors (GPCRs) are ancestrally related membrane proteins on cells that mediate the pharmacological effect of most drugs and neurotransmitters. GPCRs are the largest group of membrane receptor proteins encoded in the human genome. Using the case study of vertebrate opioid receptors, this chapter introduces an evolutionary approach to understanding pharmacological selectivity, predicted from sequence analysis, and confirmed by experimental studies. The same approach can be used to examine receptor function and applies to other families of GPCRs besides the opioid receptor family. Opioid receptors consist of a family of four closely related proteins expressed in all vertebrates examined. The three types of opioid receptors shown unequivocally to mediate analgesia in animal models and in humans are the *mu* (MOR), *delta* (DOR), and *kappa* (KOR) opioid receptor proteins. The role of the fourth member of the opioid receptor family, the nociceptin or orphanin FQ receptor (ORL), in producing analgesia is not as clear. There are now cDNA sequences for all four types of opioid receptors that are expressed in the brain of six species from three different classes of vertebrates. This chapter presents a comparative analysis of vertebrate opioid receptors using bioinformatics and data from recent human genome studies. Results indicate that opioid receptor genes most likely arose by gene duplication, that there appears to be an evolutionary vector of opioid receptor type divergence in sequence and function, and that the *hMOR* gene shows evidence of positive selection or adaptive evolution in *Homo sapiens*. Additionally, unlike many typical reviews, this paper highlights the methods used to come to these conclusions.

Key words: Opioid receptors, Molecular evolution, Bioinformatics, Positive selection, Gene duplication

1. Introduction

G protein-coupled receptors (GPCRs) represent the largest group of cell membrane proteins encoded in the human genome. There are at least 2,000 GPCRs within the rhodopsin-like superfamily of proteins recognized in the genome, with more than a third of

Craig W. Stevens (ed.), *Methods for the Discovery and Characterization of G Protein-Coupled Receptors*, Neuromethods, vol. 60, DOI 10.1007/978-1-61779-179-6_5, © Springer Science+Business Media, LLC 2011

these receptors classified as "orphan GPCRs" in that their endogenous ligand is not known (1–3). Genes that encode GPCRs represent about 5% of the known coding sequences, and their receptors are acted on by more than 60% of prescription drugs and the targets for new drug development in the pharmaceutical industry (4). GPCRs are classified by their sequence homology which in most part parallels earlier receptor classification schemes based solely on receptor function and ligand selectivity (5, 6).

Given the numerous members of the GPCR superfamily present in all vertebrate animals, it is of great interest to understand the molecular evolution of these receptor proteins. Molecular evolution is the process driving the changes that occur in the amino acid sequence of the receptor protein over evolutionary time and among the different classes of vertebrate species. Ultimately we would like to be able to describe the "vector" of this molecular evolution such that trends might be detected in the selectivity of receptor proteins or in evolving receptor efficacy or other aspects of receptor function. One such vector of molecular evolution we recently discovered is the increase in type-selectivity among opioid receptor proteins throughout vertebrate evolution, as discussed further below.

The pursuit to understand GPCR evolution is not just cerebral onanism. The correlation between receptor amino acid sequence and receptor function (while obvious from first principles: sequence determines structure determines function) is at the crux of fully understanding receptor selectivity and ensuing pharmacological action. Additionally, a detailed understanding of receptor sequence impacting type-selectivity and the direction of vertebrate receptor evolution is needed for upcoming gene therapy which will undoubtedly include expression of exogenous GPCR proteins. One can imagine an engineered GPCR gene encoding a novel GPCR protein that is more efficacious in the presence of the endogenous ligand. It may also be beneficial to have an engineered GPCR that is more selective to existing medications compared to the wild-type GPCR. These are just a few examples which portend enormous clinical applications and pharmaceutical industry profit that will eventually be feasible with increased understanding of the molecular evolution of GPCRs.

This chapter will eschew a formal bioinformatics approach and describe how cloning nonmammalian opioid receptors by our research group led to novel hypotheses of opioid receptor evolution using only the simplest of bioinformatic methods. The effort here is to elucidate the methods and results that fueled novel hypotheses on the evolution of the opioid receptors. Of course, there is no reason to suspect that the methods and hypotheses generated from the study of vertebrate opioid receptors are not applicable to other families of GPCRs. For those interested in reading hard-core bioinformatics papers or possessing a deeper background of bioinformatics relating to receptor evolution, recent papers take a much broader genomic analysis of GPCR families or trace chromosomal synteny of opioid receptor genes (7–9).

1.1. The Beginning of G Protein-Coupled Receptors

Before examining the evolution of vertebrate GPCRs, the origin of an ancestral GPCR protein is of interest. Do earlier-evolved life forms also contain genes for GPCR proteins? All life on earth is classified into three domains or superkingdoms: Archea, Bacteria, and Eucaryotes, although there are other schemes with up to six kingdoms (10). This classification does not include viruses; some taxonomists seem to ignore them and unable to reach a conclusion whether they meet the definition of a life form or not. Viruses are simply self-replicating bits of DNA or RNA, and one wonders if self-replication is sufficient by itself to serve as the sole criterion for life. Prions self-replicate also, but do not use RNA or DNA to do so; maybe heritable information transfer is the key to life. Of course, this leads us to think that more complex behaving species are somehow more alive, but at the end of the day, self-replication is really the only thing that even *Homo sapiens* accomplishes.

The origin of viruses is not known. Two main theories exist that viruses are precellular and arose before the time of the most primitive cells, or that viruses are the results of a massively reduced cellular parasite to its core replication machinery (11, 12). These scenarios are not mutually incompatible as some viruses may have arisen by either route. In any event, there are viruses that encode GPCR-like proteins. The transcription of these genes produces receptors not destined for use by the virus but to be embedded in the host cell's membrane. Most notably, human herpesviruses encode for various chemokine receptors that may help the virus gain a foothold in the host organism (13, 14). Thus, if viruses are precellular ancestors of cells, it is possible that genes for GPCR proteins in all life forms may have arisen from initial viral transfection, akin to life itself on Earth originating due to extraterrestrial panspermia.

A second way that GPCRs could have initially arisen is through the duplication and fusion of a gene that encoded a three helices transmembrane (TM) receptor protein (15). It is hypothesized that gene duplication occurred in the evolution of an ancestral GPCR gene, such that helices 5–7 originated as duplicates of helices 1–3. This gives rise to the amino acid similarities between helices 1–3 and 5–7 of many GPCRs (15). This suggests that ancestral GPCRs may not have been as structurally complete as extant GPCRs found today. Regardless of the ultimate origin of ancestral GPCR, the extensive duplication, expansion, and utilization of GPCRs by all vertebrates suggest that this superfamily has and is undergoing positive selection to generate the GPCRs expressed today. The next sections introduce the opioid receptors in general and studies on opioid systems in nonmammalian species in particular.

1.2. A Brief History of Opioid Receptors

The classical opioid receptors are membrane proteins of the Type A or rhodopsin-like GPCRs that mediate the analgesic effects of opioid agents like morphine. Opioid receptors also enable substance abuse with drugs like heroin as well as the burgeoning problem of prescription opioid abuse. The natural ligands for opioid receptors are endogenous opioid peptides; most notably *beta*-endorphin, met- and leu-enkephalin, dynorphin, and endomorphin. The role of endogenous opioid receptors in physiological processes is vast; nearly any PUBMED search with the Boolean parameters of "any *physiological process* and opioid" brings up pages of references. This is true for terms like reproduction, growth, development, respiration, blood pressure, renal function, thermoregulation, endocrinology, seizures, stress, immunology, pregnancy, and aging. It is also true when considering behaviors using the terms like eating, drinking, sex, learning, memory, and locomotion. While there appears to be no limit to the effects of opioids, both endogenous and exogenous, on any system in the brain and body, underlying all opioid effects is the initial event of an opioid molecule binding to an opioid receptor.

There was a time when only the structure–activity relationships of various ligands defined GPCRs. In this way, Martin et al. initially proposed multiple opioid receptors based on the differential action of morphine and ketocyclazocine in a spinal dog model (16). Martin called them *mu* (Greek "m" for morphine; MOR) and *kappa* (for ketocyclazocine; KOR) opioid receptors. Later, Hans Kosterlitz et al., working with the newly discovered enkephalin peptides in the mouse vas deferens preparation, proposed a *delta* (for deferens; DOR) opioid receptor type (17). Much work toward the end of the last century solidified the structure–activity relationship of these primary opioid receptor proteins, aided in large part by the development of highly selective (type-specific) opioid agonists and antagonists by Takemori and Portoghese (18–20). Common highly selective opioid antagonists used are *beta*-funaltrexamine (*beta*-FNA, for MOR) (18), naltrindole (NTI, for DOR) (19), and nor-binaltorphimine (nor-BNI, for KOR) (20). Pharmacological evidence of three separate opioid receptors mediating analgesia in mammalian models was confirmed using these antagonists. Evidence of stereospecific opioid binding sites from radioligand binding studies using rodent brain homogenates did not emerge until 1973 from the labs of Snyder, Simon, and Terenius (21–23). Ongoing refinements of such "grind and bind" studies clearly established three main opioid receptor binding profiles (24–26) and a number of pharmacological subtypes for each of the main receptors (27, 28).

The ultimate discovery and isolation of the cDNA for the mouse *delta* opioid receptor (mDOR[1]) from the labs of Evans in the USA and Kieffer in France in 1992 (29, 30) for the first time linked an opioid receptor sequence to a pharmacological type of opioid receptor. Both groups screened a cDNA library derived from NG108-15, a species-hybrid cell line derived from mouse neuroblastoma x rat glioma cells. Once the mDOR cDNA sequence was known, homology cloning led to the identification of receptor sequences for numerous rodent and monkey species and the human opioid receptors, hMOR (31), hDOR (32, 33), and hKOR (34–36). About this time, cloning studies revealed another "opioid receptor-like" (ORL) sequence from numerous labs that did not match the known MOR, DOR, and KOR sequences (37–43). This opioid receptor-like protein remained an "orphan" receptor until the endogenous ligand for ORL was identified as the neuropeptide named nociceptin by the lab of Meunier (44) or named orphanin FQ from the team assembled by Civelli (45). This makes the nociceptin (or orphanin FQ) receptor (ORL) the fourth member of the opioid receptor family expressed in the nervous system of mammals (5, 46). However, the results of nociceptin or orphanin FQ binding to the ORL protein are not clear as behavioral studies show hyperalgesia (the origin of the name *nociceptin*) or analgesia, or sometimes no effect at all (47). Additionally, the vast majority of opioids do not bind at all to ORL receptors, nor do the opioid antagonists, naloxone or naltrexone. For this reason, MOR, DOR, and KOR are considered the classical opioid receptor types of the family and ORL a first-cousin. *In vivo* studies using antisense oligodeoxynucleotides designed from MOR, DOR, and KOR cDNA sequences confirm that each one of the three opioid receptor proteins can independently produce analgesia in mammalian models (48–50).

There is evidence of opioid receptor subtypes in mammalian species (e.g., mu_1, mu_2, $delta_1$, $delta_2$, $kappa_1$, $kappa_2$, $kappa_3$) which may result from alternative splicing of receptor mRNA (51–54). Interestingly, it appears that many more splice variants are noted for MOR than for the other types of vertebrate opioid receptors. Driven by the prodigious efforts of Pan and Pasternak, there are now at least 20 alternative splice variants of mMOR (55) and at least six variants of hMOR (56). By contrast, only a few alternative splice variants are reported for KOR (57, 58), and even less for DOR (57) and ORL (59). This finding may reflect sampling bias as MOR as the most clinically relevant receptor type may have

[1]The convention for naming opioid receptor proteins here follows a hybrid system. For example, the *mu* opioid receptor in humans, mice and rats is hMOR, mMOR, and rMOR, but for all other species a representation of the Linnaean binomial taxonomy is used such that in the zebrafish, *Danio rerio*, the receptor is drMOR, and in the leopard frog, *Rana pipiens*, rpMOR.

been examined to a greater degree, but more likely supports the notion that MOR is very different from the other types of opioid receptors, an aspect of vertebrate MOR discussed further below. Pasternak and colleagues present the methods and their results for the discovery of GPCR splice variants in Chap. 2 of this book.

1.3. Evidence for Opioid Receptors in Nonmammalian Vertebrates

Early evidence for the existence of opioid receptors in nonmammalian vertebrates came from behavioral and radioligand binding studies using morphine and other opioid agents. The best studied nonmammalian vertebrates with regard to opioid systems are amphibians, so the focus is on these studies. Thus, studies using amphibians provide at least one comparison of opioid receptors in earlier-evolved species to compare with later-evolved mammals including humans.

Initial studies of the analgesic or antinociceptive effects of opioids in amphibians (*Rana pipiens*) were conducted using non-selective opioid agonists, endogenous opioid peptides, and antagonists (60–63). Tolerance to the analgesic effects of daily morphine administration was documented (64), and stress-induced release of endogenous opioids produced analgesia which was potentiated by enkephalinase inhibitors and blocked by naltrexone (65, 66). These studies showed that both exogenous opioid agonists and endogenous opioid peptides could raise the nociceptive threshold in amphibians by an action at opioid receptors. Other behavioral studies of opioid effects in amphibians include an investigation of the effects of opioids on noxious and non-noxious sensory modalities (67, 68) and reports of analgesia produced by remifentanil (69) and xendorphin (70).

The relative antinociceptive potency of *mu*, *delta*, or *kappa* opioid agonists after systemic, intraspinal or intracerebroventricular administration in amphibians was highly correlated to that observed in typical mammalian models and to the relative analgesic potency of opioid analgesics in human clinical studies (71–73). These data established the amphibian model as a robust and predictive alternative or adjunct model for the testing of opioid analgesics (74, 75). In the one study examining the behavioral effects of nociceptin in a nonmammalian model, nociceptin produced antinociceptive effects following spinal administration in *Rana pipiens*, which was blocked by ORL antagonists but not by opioid antagonists (76).

There has been only one study in amphibians examining opioid agonist effects with coadministration of highly selective MOR, DOR, and KOR antagonists. *Beta*-FNA, NTI, and nor-BNI blocked the antinociceptive effects produced by all three types of selective opioid agonists in amphibians (77). For example, *beta*-FNA, a highly selective MOR antagonist in mammalian models, blocked the effects of selective MOR, DOR, and KOR agonists. This provided an initial data for our initial hypothesis that opioid receptors are less type-selective in earlier-evolved vertebrates compared to humans and other mammals, a hypothesis further supported by additional data as described below.

Radioligand binding studies also supported the existence of functional opioid receptors in nonmammalian vertebrates, including fish (78–81), amphibians (82–86) and reptiles (87–89). Many reports examining opioid binding sites using the European water frog, *Rana esculenta*, were produced from the Hungarian research group headed by Wolleman, Simon, Borsodi, and Benyhe (90–97). The main difference observed between amphibians and mammals was a greater affinity of *mu* and *delta* selective opioids for the amphibian *kappa*-like site, and a lesser affinity for *kappa*-selective opioids, compared to mammalian *kappa* opioid receptors (84, 98). These results provided the first hint from radioligand studies that opioid receptors in nonmammalian species may be less selective than their mammalian counterparts.

In our laboratory, brain and spinal cord membranes from *Rana pipiens* showed a single high-affinity site which was displaced by *mu*, *delta*, and *kappa* selective opioid agonists with apparent affinities ranging from 1.86 nM to 31 μM. Surprisingly, the highly selective opioid antagonists (*beta*-FNA, NTI, and nor-BNI) displaced [^{3}H]-naloxone binding with equal affinity to opioid receptors in brain and spinal cord tissue, each with an apparent affinity of about 3.0 nM (99, 100). This finding was consistent with behavioral studies showing nonselectivity of type-selective antagonists and also supports the hypothesis that opioid receptors from earlier-evolved vertebrates are less selective than mammalian receptors (77).

However, using the selective opioid agonist radioligands, [^{3}H]-DAMGO (MOR), [^{3}H]-DPDPE (DOR), and [^{3}H]-U65953 (KOR), three distinct opioid binding sites were identified based on different binding densities and selective competitive displacement of agonist radioligand by *mu*, *delta*, and *kappa* opioid ligands (101). As opposed to their equipotent displacement of [^{3}H]-naloxone binding, the type-selective opioid antagonists (*beta*-FNA, nor-BNI, and NTI) were highly selective in displacing the binding of their respective *mu*, *delta*, and *kappa* opioid agonist radioligands (101).

The summary of the above studies is that brain tissue of nonmammalian vertebrates contains the three types of classical opioid binding sites, similar to those characterized in mammalian tissues, but with significant differences in the binding of selective opioid ligands. There is one study characterizing ORL sites in amphibians by the Hungarian group showing high-affinity, saturable binding of labeled nociceptin to brain homogenates (90).

2. The Vertebrate Opioid Receptor Dataset

Prior to the identification of any nonmammalian, full-length opioid receptor sequences, a study from Evans group utilized PCR and degenerate primers isolated opioid receptor-like fragments from genomic DNA obtained from each of the major vertebrate classes but not in any of the invertebrates tested (102, 103). The first full-length clone sequenced from a nonmammalian species was ccMOR from the brain of the white suckerfish, *Catostomus commersoni* (104). Other nonmammalian species with full-length opioid receptors cloned are the zebrafish, *Danio rerio,* by Rodriguez and colleagues (105–108) and the rough-skinned newt, *Taricha granulosa,* in the laboratory of Moore (109–111). Using the same sets of primers that Evans group used in the first phylogenetic study of opioid receptors (102), our group was able to clone all four opioid receptor types expressed in *Rana pipiens* brain tissue: rpMOR, rpDOR, rpKOR, and rpORL (112). These were the first opioid receptors cloned from *Class Amphibia*, and, on the amino acid level, showed 70–84% identity with their orthologous mammalian counterparts.

At present, the vertebrate dataset of opioid receptor cDNA sequences yields 24 nucleotide and 24 protein sequences for phylogenetic analysis. As shown in Table 1, at present there is a dataset of six vertebrates from three different classes that have all four opioid receptor sequences from brain tissue cDNA deposited in GENBANK. As the focus of this chapter is the relationship between amino acid sequence and pharmacological action throughout evolution, the analyses were done using the protein data.

Table 1
Existing vertebrate species with cloned cDNA sequences for *MOR, DOR, KOR, and ORL* receptors

			Access #s			
Vertebrate class	Common name	Genus species	MOR	DOR	KOR	ORL
Pisces	Zebrafish	*Danio rerio*	AAK01143	CAA04862	AAG60607	AAN46747
Amphibia	Leopard frog	*Rana pipiens*	AAQ09991	AAQ09992	ABY82593	AAR08905
Amphibia	Newt	*Taricha granulosa*	AAV28689	AAV28690	AAU15126	AAU26067
Mammalia	Rat	*Rattus norvigecus*	AAA41630	AAA19939	AAA18261	AAA50827
Mammalia	Mouse	*Mus muscularis*	AAA86878	AAA37522	AAA39363	CAA62922
Mammalia	Human	*Homo sapiens*	AAA20580	AAA83426	AAC50158	AAA84913

Access numbers are given for the NCBI protein database

2.1. Alignment of Opioid Receptor Protein Sequences

A ClustalW alignment of the present dataset of vertebrate opioid receptors shows a high degree of sequence similarity both within types of opioid receptors and among all 24 sequences (Fig. 1). Key residues thought to be important for the function of GPCRs are present, such as the DRY motif at alignment positions 189–191 (113, 114) and the conserved Cys–Cys bridge between EL1 and EL2 provided by residues at 165 and 245 (115, 116). Na^+ is known to regulate the agonist binding of GPCRs and is dependent on an allosteric binding site at the conserved Asp (134) in the TM2 domain (117, 118). There is high overall homology among existing vertebrate MOR, DOR, KOR, and ORL proteins, with nearly identical sequences among the seven transmembrane helical regions. Intracellular loop domains were also highly conserved and the N-terminus regions were most divergent, followed by the C-terminus and extracellular loop domains.

There are also obvious differences between mammalian and nonmammalian sequences. For example, mammalian DOR contains an extended C-terminus sequence compared to the nonmammalian species. More striking is the 11 amino acid C-terminus extension of mammalian MOR (C-terminus, alignment positions 419–429) compared to the shorter sequence in earlier-evolved vertebrates. Phosphorylation sites along this extended sequence include Thr (position 425) which is critical for internalization and desensitization of rat *mu* opioid receptors (119). Interestingly, this "add-on" piece of protein represents translation of exon 4, which is lacking in some hMOR and mMOR splice variants (55, 56). The differences observed in amphibians with regard to selective opioid ligands in behavioral and radioligand binding studies may result from other substitutions in the primary amino acid sequence. The *mu* opioid antagonist, *beta*-FNA, is dependent on Lys (234) in EL2 for covalent attachment and selective *mu* opioid blockade (120). The amphibian sequences show conservation of this residue in rpMOR, rpDOR, and rpKOR, which may account for the pan-antagonism of the type-selective antagonists cited above. The action of the *kappa*-selective opioid antagonist, nor-BNI, is dependent on the *kappa*-specific Glu (335) in EL3 (121), and the amphibian sequence contains a Val in this position which may account for diminished type-selectivity in amphibians.

2.2. Phylogenetic Analysis of Vertebrate Opioid Receptors

The analyses below are based on the sequence of the canonical MOR, DOR, KOR, and ORL proteins for each species listed in Table 1. Each type of opioid receptor (each set of orthologs) provided a pattern of vertebrate evolution consistent with established fossil evidence and phenotypic characteristics (Fig. 2). The radial tree was rooted with the sequences of rhodopsin (RHO, lower left) from each species with sequence available in Genbank. The "tree" was rooted with an outlier group of sequences, those of

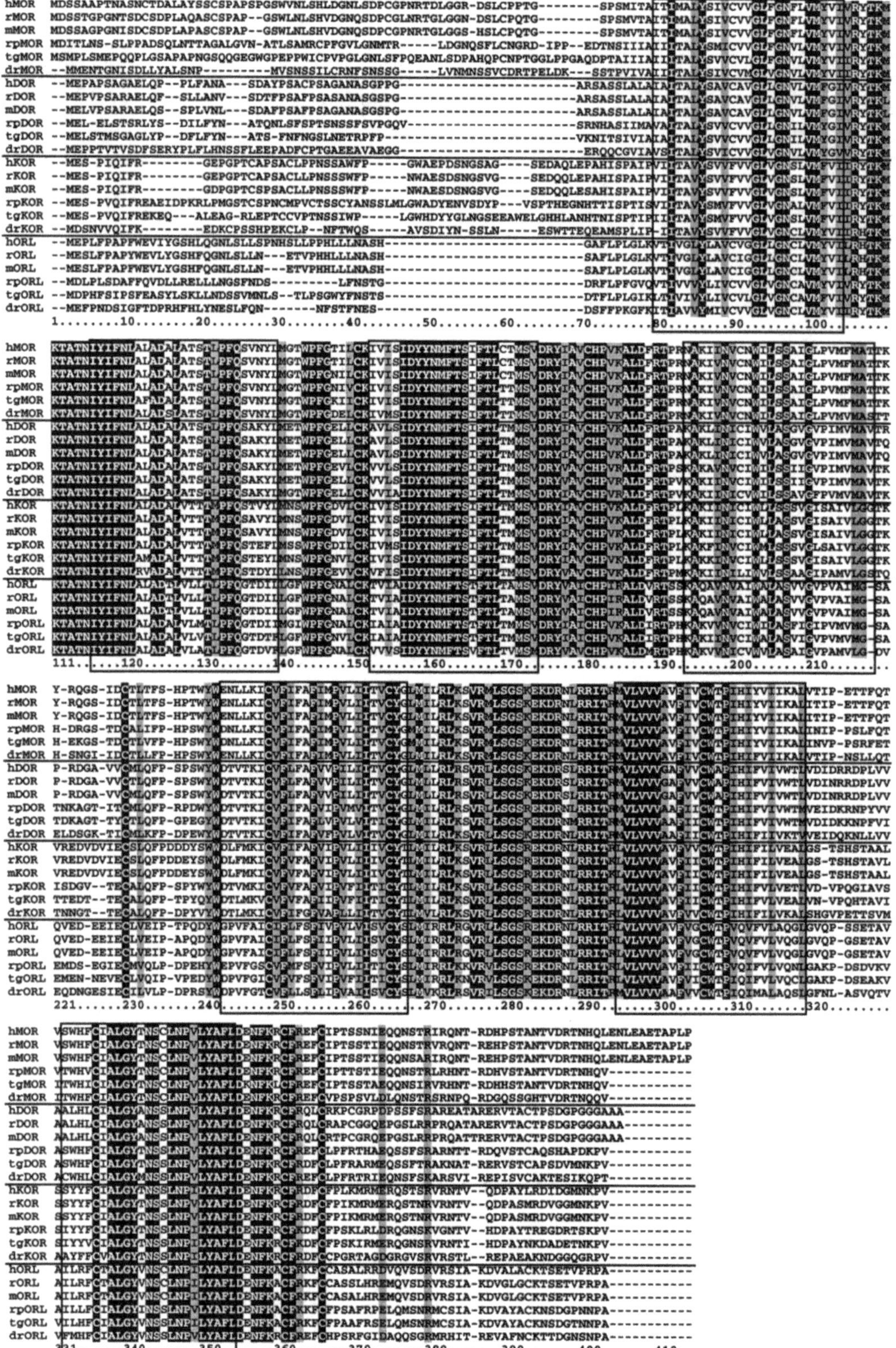

Fig. 1. Alignment of protein sequences of available sets of four opioid receptor types in six vertebrates. Protein alignment was done using ClustalW with default values. Identical sites are indicated by *white text on black background*, conservative substitutions noted by *gray background* by using Boxshade (see Table 4). Boxed-in domains indicate the seven transmembrane regions of the receptor proteins and numbers below the sequences are the alignment positions.

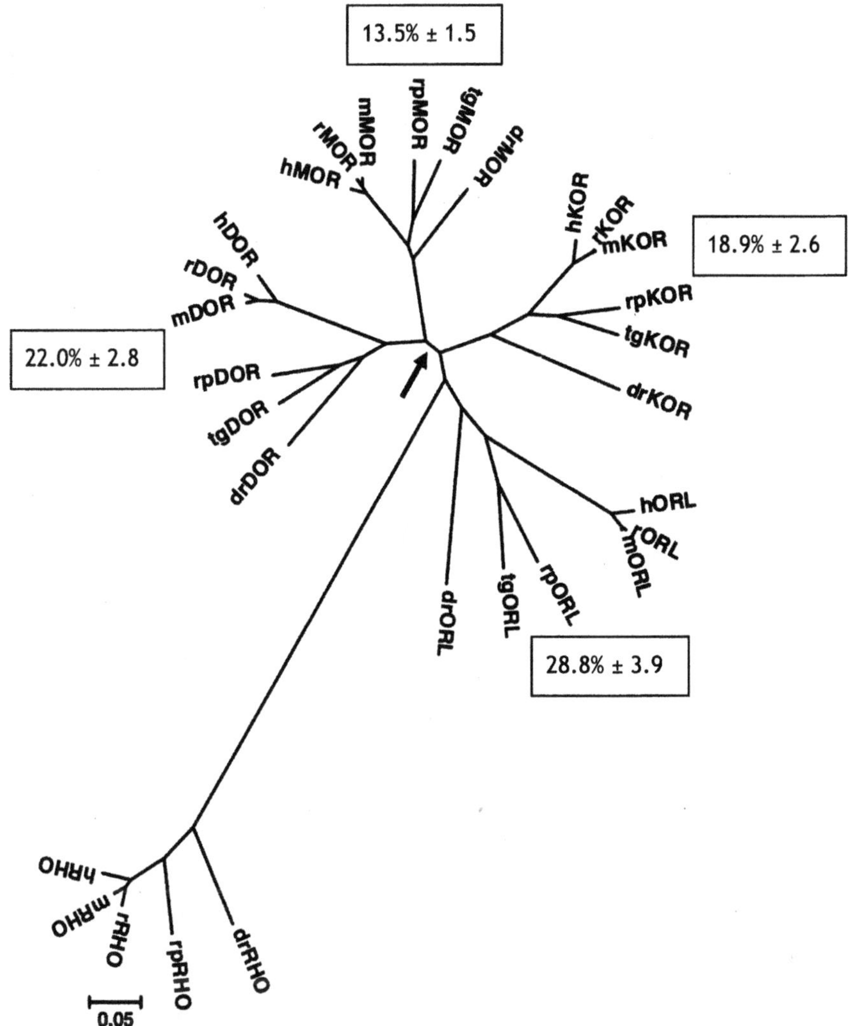

Fig. 2. Phylogenetic analysis of MOR, DOR, KOR, and ORL sequences in six vertebrates. Bioinformatics software (MEGA4) was used to generate a radial phylogenetic tree using the neighbor-joining method, rooted with the available matching sequences of rhodopsin (RHO). Protein sequences were from species listed in Table 1, where access numbers may also be found. Key: humans, *Homo sapiens* (h); mouse, *Mus musculus* (m); rat, *Rattus norvegicus* (r); leopard frog, *Rana pipiens* (rp); rough-skinned newt, *Taricha granulosa* (tg); and the zebrafish, *Danio rerio* (dr). The *arrow* shows the bifurcation of MOR and DOR sequences from KOR and ORL. Values in boxes by each opioid receptor type are the mean ± SEM of the pairwise distance (% divergence) among members of each type. Branch length is equal to the proportional difference among the sequences (scale bar = 0.05 or 5% difference in amino acid sequence).

the rhodopsin (RHO) protein for each species where available. Overall, the four groups of opioid receptor sequences formed a dyad, with MOR and DOR sequences sharing a common ancestor (node) and KOR and ORL sharing a different common node. A similar, but unrooted tree was generated after all four types of opioid receptors were cloned and sequenced in another amphibian, a newt (109).

The graphical view of MOR and DOR sharing one fork of the tree and KOR and ORL sharing the other is supported by an earlier study comparing mouse and rat opioid receptor protein sequences by Chaturvedi et al. (122). They noted that the pairings of MOR and DOR, and KOR and ORL, were the most closely related among the rodent sequences. Using the present vertebrate dataset and simple bioinformatics, these findings were confirmed and extended to all available vertebrates (123).

2.3. Divergence and Convergence of Opioid Receptor Types

As pharmacological selectivity is correlated to similarity of amino acid sequences at the GPCR family level (e.g., opioid receptors vs. muscarinic receptors), type-selectivity within family members (e.g., MOR, DOR, or KOR) is also correlated with percent identity or similarity. The results of pairwise BLAST analysis (124) for the three classical types of opioid receptor sequences within each species yielded a rank order of divergence such that MOR, DOR, and KOR proteins were more closely related to each other in earlier-evolved vertebrates than in humans and other mammals (see Table 2). Insofar as divergence of molecular sequence is related to the greater type selectivity of opioid receptors, this finding gave rise to the hypothesis that opioid receptors are more type-selective in mammals than in nonmammalian species.

To state in another way, the molecular evolution of opioid receptors has a "vector" of increased type-selectivity as reflected in the greatest divergence of MOR, DOR, and KOR proteins in mammalian species, especially in *Homo sapiens.* In other words, opioid receptors in earlier-evolved vertebrates are less type-selective. Given the balancing constraints of maintaining the common sequence for opioid receptor family membership and that of adaptive evolution for greater type-selectivity, it is likely that even minor differences in MOR, DOR, and KOR sequences between vertebrate species could be detected in the binding and efficacy of selective opioid ligands. This is supported by data from studies of the first nonmammalian opioid receptor, ccMOR, cloned from the white suckerfish, *Catostomus commersoni* (104). In this study, ccMOR expressed in HEK cells bound the nonselective opioid antagonist naloxone with high-affinity to HEK cell membranes; however, the *mu*-selective opioid agonist, DAMGO, displaced naloxone with surprisingly low affinity. Additionally, nonselective opioid ligands bound well to zebrafish, *Danio rerio delta* opioid receptor (drDOR in Fig. 2) expressed in HEK cells but low affinity was observed for selective *mu*, *delta*, or *kappa* opioid ligands (125). Studies of opioid receptors cloned from a second amphibian species, the rough-skinned newt, *Taricha granulosa*, led to the conclusion that the opioid type-selectivity of newt opioid receptors was less stringent than that in mammals (109). Recent *in vitro* data comparing amphibian (rpMOR) and human (hMOR) opioid receptors expressed in CHO cells showed significant less

Table 2
Comparison of vertebrate MOR, DOR, and KOR protein sequences within vertebrate species and by group

Group/species	MOR vs. DOR[a]	MOR vs. KOR	DOR vs. KOR	Species mean (sem)[b]	Group mean (sem)
Percent amino acid identity					
Nonmammals					
Danio rerio	70	62	65	65.7 (2.3)	
Rana pipiens	73	65	63	67.0 (3.1)	
Taricha granulosa	70	66	68	68.0 (1.2)	66.9 (1.2)†
Mammals					
Rattus norvegicus	66	61	61	62.7 (1.7)	
Mus musculus	61	60	61	60.7 (0.3)	
Homo sapiens	62	60	59	60.3 (0.9)	61.2 (0.7)
Percent amino acid similarity					
Nonmammals					
Danio rerio	82	79	79	80.0 (1.0)*	
Rana pipiens	85	81	77	81.0 (2.3)*	
Taricha granulosa	82	80	82	81.3 (0.7)*	80.8 (0.7)†
Mammals					
Rattus norvegicus	77	75	75	75.6 (0.7)	
Mus musculus	72	75	74	73.6 (0.9)	
Homo sapiens	73	74	72	73.0 (0.6)	74.1 (0.5)

[a]ver. BLASTP 2.2.14, settings: matrix = Blossum62, gap open = 11, gap extension = 1, x-drop-off = 50, expect = 10.00, wordsize = 3, and filter off
[b]Standard error of the mean
*Denotes significantly different% similarity than rat, mouse, and human mean values ($p<0.05$, one-way ANOVA followed by *post-hoc* Newman–Kuels test). † denotes group means ($N=9$) different for identity and similarity at $p<0.01$, Student's t-test

affinity of selective opioid ligands with rpMOR compared to hMOR (126).

Besides comparing the MOR, DOR, KOR sequences to each other within species, which led to the discovery that the opioid receptors are less like each other in mammals than nonmammals (divergence), analysis of receptor types as a group shows that vertebrate MOR proteins are more similar to each other than ORL proteins (convergence), likewise for the KOR proteins vs. ORL proteins. As shown in Fig. 2 (boxed values), the mean distance of the MOR group is 13.5% (which means that after aligning all six MOR primary sequences, 13.5% of the sites were not identical), 18.9% for the KOR group, and 22.0% for the DOR group. For the group of vertebrate ORL sequences, 28.8% of sites were not identical. Figure 3 graphs the values for each receptor type, showing that both MOR and KOR are significantly less divergent, thus more convergent, than the group of ORL proteins.

The conclusion at this point from simple bioinformatic analyses is that vertebrate opioid receptors are less divergent in earlier-evolved vertebrates; however, the groups of MOR and KOR proteins show signs of convergence throughout the vertebrate spectrum. Vertebrate MOR proteins are more similar to DOR proteins, with KOR more similar to ORL. Vertebrate ORL is most similar to rhodopsin. Next, data from the human genome and other results from the world of evolutionary biology are applied to the story of the evolution of vertebrate opioid receptors.

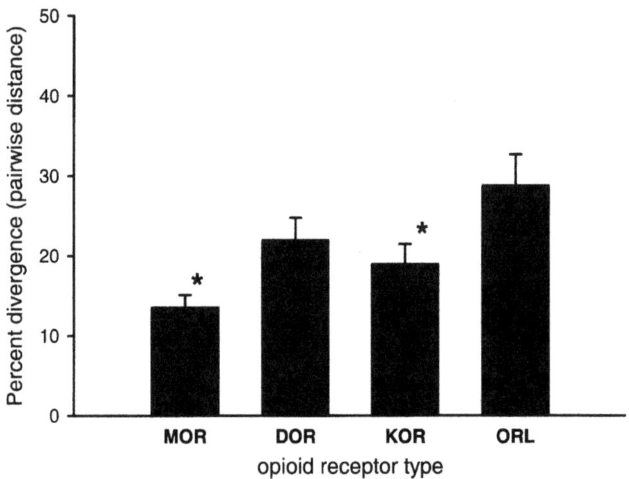

Fig. 3. Percent divergence plotted by opioid receptor type. Pairwise distance was calculated for each group of six amino acid sequences after alignment by ClustalW using Mega4 (Poisson correction, uniform rate, gaps deleted). Mean plus SEM are plotted for each group. *Asterisks* indicate that the MOR and the KOR group significantly less divergent than the ORL group (one-way ANOVA followed by a post-hoc Newman–Keuls test, $p=0.004$; SigmaStat v. 3.1).

3. The Human Genome and the Evolution of Opioid Receptors

This section reviews the recent data available since the completion of the human genome project, including opioid receptor gene location, the results on studies of receptor variation at single nucleotide positions (SNPs), and background information on the identification and characteristics of duplicate genes.

3.1. Opioid Receptor Genes in the Human Genome

The three classical opioid receptor genes and the fourth member of the opioid receptor gene family, the ORL or nociceptin receptor gene, are mapped to four different chromosomes in the human genome, as shown in Fig. 4 (5). The classical opioid and ORL receptor gene loci are part of a broader set of paralogs, or common suites of genes across chromosomes that may indicate gene duplication, which is outlined next.

3.2. Duplication of Opioid Receptor Genes

Although not widely known among other scientists, it is accepted by most evolutionary biologists that two rounds of genome-wide duplication (paleoploidization) occurred early in vertebrate evolution, originally stated by Ohno and called the 2R hypothesis (127, 128). Besides the initial explosion of quadrupled genes at the heart of vertebrate evolution, it is thought that duplicate genes (paralogs) have a gene-birth rate of one new pair of duplicates every million years (129). The human genome contains at least 5% of its protein genes as paralogs, and paralogous regions on chromosomes are known containing stretches of genes lined up in to related genes on stretches of sequence on another chromosome (130, 131). Most paralogs do not survive, becoming pseudogenes or otherwise nonfunctional DNA sequence (132, 133). The fate of the duplicate genes that do survive is characterized by asymmetrical divergence such that one member of the pair continues as before and the other is "free" to undergo adaptive evolution (134, 135).

The phylogenetic pattern of (AB) (CD) in the "tree" of vertebrate opioid receptor proteins (shown in Fig. 2) is expected from the application of the 2R hypothesis to a single ancestral gene (136). Applied to the family of opioid receptor genes, the 2R hypothesis supports the idea that the genes encoding MOR, DOR, KOR, and ORL within a species are paralogs, i.e., genes related by duplication. The evidence for gene duplication for opioid receptors is also supported by the location of MOR, DOR, KOR, and ORL genes on paralogous regions of human chromosomes (5). Human MOR and DOR genes are mapped to chromosome 6 and 1, respectively, while KOR and ORL genes are mapped to chromosomes 8 and 20, as shown in Fig. 4. The specific pairing of MOR with DOR, and KOR with ORL on separate branches of the sequence tree is supported by a genomic study

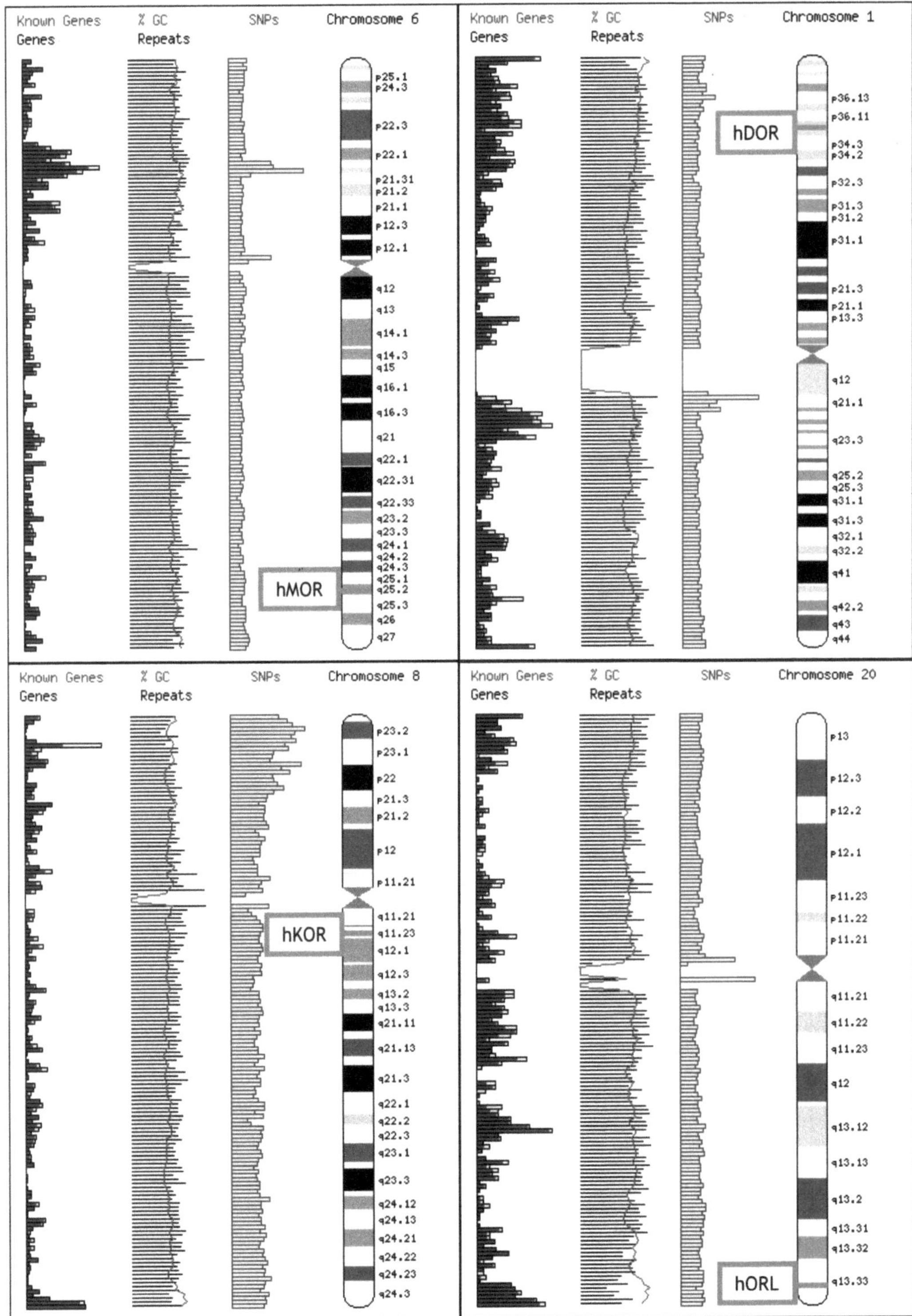

Fig. 4. Chromosomal mapping of human opioid receptor genes. Image adapted from screen captures of the Ensembl Human MapView (Table 4). Note that *hMOR*, *hDOR*, *hKOR*, and *hORL* map to chromosomes 6, 1, 8, and 20, respectively. Also shown in vertical columns are the density of known genes, the % GC content, and SNPs cataloged along the chromosomes.

whereby the greatest number of gene duplicates was found between paralogous regions on chromosomes 1 and 6, and between paralogous regions on chromosomes 8 and 20 (130).

3.3. Variation in Human Opioid Receptor Genes

The vast information available from human genomic data is beginning to allow studies of *intraspecific* variation due to differences in the nucleotide code from one human to the other. Generically called polymorphisms, variation data is obtained by sampling the genome of numerous individuals from different population groups and cataloging sites where single nucleotide bases differ; i.e., single nucleotide polymorphisms or SNPs. Analysis of SNPs, or sets of associated SNPs called haplotypes, provides an insight into the evolution of particular genes and is especially useful in determining genetic defects leading to disease states or medical conditions (137). Although the majority of SNPs are found among intronic, repeatable sequences, or other lengths of noncoding DNA, SNPs found in coding or exonic DNA will be classified as either *synonymous* (mutating the nucleotide base does not change the amino acid, due to degeneracy in the codon table) or *nonsynonymous* (changing the amino acid at that place on the protein).

Studies of polymorphism in human opioid receptor genes have focused on the association of particular SNPs with opioid-dependent and alcohol-dependent populations (138–141). Most studies have focused on the A118G allele of *hMOR* which changes the amino acid Asn to Asp (N40D) in the N-TERMINUS of the hMOR protein. Kreek and colleagues tell their story of GPCR human polymorphisms in Chap. 3 above. The location of numerous *hMOR* SNPs to the 5' untranslated region (5'UTR) and upstream promoter regions as well as 3' UTR regions of hMOR does not affect the receptor structure but alters the expression of hMOR and may explain individual differences in pain sensitivity, drug-dependency, and the clinical efficacy of morphine (142–145). Altering the expression of genes, by adaptive evolution in non-coding DNA, is a viable outlet for Darwinian or positive selection (146, 147). There are fewer studies of the *hKOR* polymorphism (139, 148, 149), *hDOR* (150), and only one on a single SNP in the promoter region of *hORL* (151).

As shown in Table 3 and graphically in Fig. 5, *hMOR* has the greatest number of SNPs (Total SNPs column) found along its gene sequence, with 356 SNPs in the SNP reference database accessible from the HapMap project. This contrasts with the lowest number of SNPs observed along the *hORL* gene, with only 13 total SNPs. This table was constructed from data available from the HapMap website, using Release #21a, from January 2007 (NCBI build 35) and includes at least 3.8 million genotyped SNPs. While the population-based approach taken by the contributors to HapMap does not allow for complete genome sequencing of all individuals, there does not appear to be a sampling or

Table 3
Number and types of single-nucleotide polymorphisms (SNPs) found in human opioid receptor genes

Gene	Total SNPs	Intron SNPs	Exon SNPs	Nonsyn SNPs	Syn SNPs	nonsynSNP/ synSNP	% nonsynSNP
hMOR	356	335	21	15	6	2.50	71.43
hDOR	351	248	4	2	2	0.50	50.00
hKOR	81	74	7	2	4	0.50	28.57
hORL	13	9	4	1	3	0.33	25.00

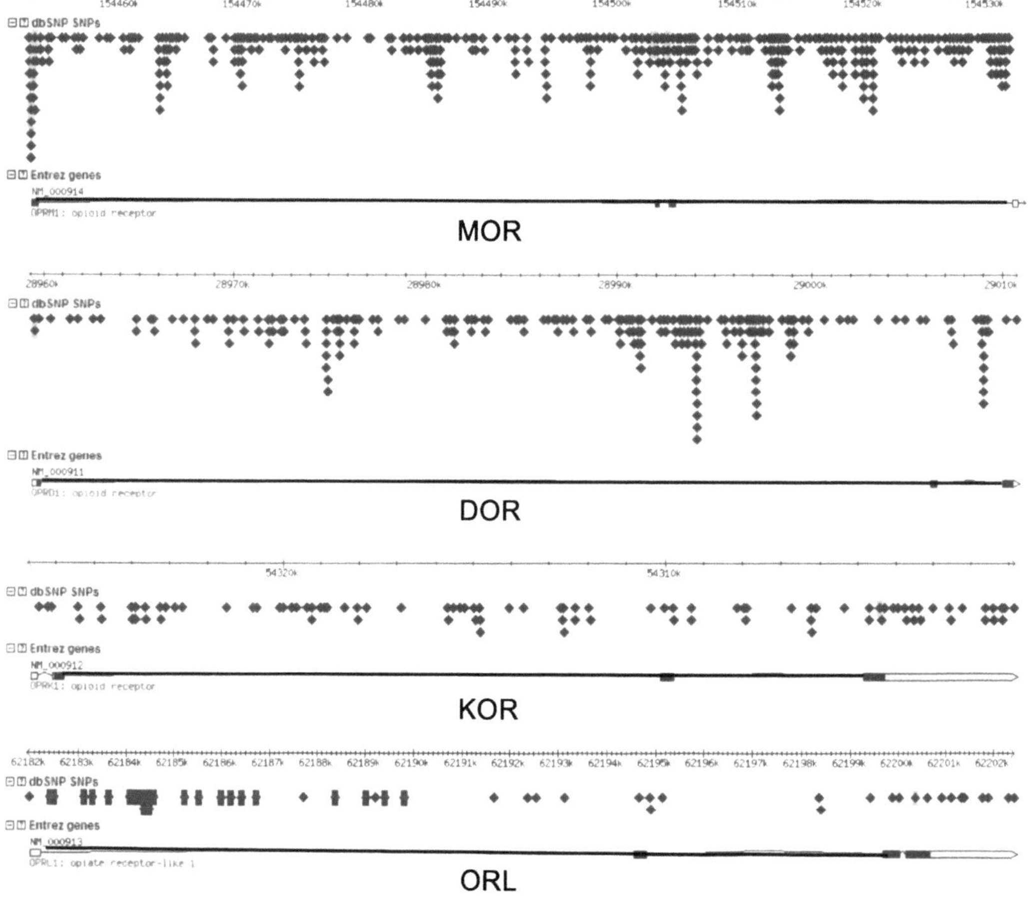

Fig. 5. Chart of single-nucleotide polymorphisms for each human opioid receptor gene. Adapted from screen captures from the over three million database of SNPs (dbSNP) available at the HapMap website (Table 4). Each *triangular* symbol represents a single SNP; the opioid receptor genes are noted by their database acronyms (*OPRM1, OPRD1, OPRK1, and OPRL1* for *hMOR, hDOR, hKOR,* and *hORL*, respectively). Exons are designated by *boxed* regions along the line representing the genes; *grayed boxes* represent untranslated regions of the gene (5' to the left, 3' on the right) found on exon regions.

ascertainment bias that could account for the results shown in Table 3 to explain the fact that *hMOR* has more identified SNPs (152). The rank order of total SNPs in the opioid receptor genes is *hMOR* > *hDOR* >> *hKOR* > *hORL*. Of particular interest for the molecular evolution of opioid receptors with regard to primary structure are SNPs found within exonic regions of the receptor genes. There were 21 exonic SNPs for *hMOR*, three times higher than any other type of opioid receptor gene. Of these SNPs in exonic regions (coding SNP), 15 were found to be nonsynonymous (non-synSNP) and six synonymous (synSNP). These values yielded a very high ratio of non-synSNP/synSNP for *hMOR* of 2.5; i.e., 71% of exonic SNPs in *hMOR* result in an amino acid substitution. The ratio of non-synSNP/synSNP for the other three opioid receptor genes ranged from 0.33 to 0.50. Figure 6 shows the total number of nonsynonymous SNPs per human opioid receptor gene type.

The high degree of polymorphism in the *hMOR* gene resulting in amino acid changes compared to the other types of opioid receptor genes suggests that *hMOR* is under greater selection pressure and undergoing adaptive evolution. Under the nearly neutral theory of evolution, all four opioid receptor genes should show rates of nonsynonymous mutations much less than synonymous substitution yielding a non-synSNP/synSNP ration of << 1.0 (153–155). Relatively speaking, the alleles in *hMOR* have not been fixed by purifying selection to the same degree as they are in *hDOR*, *hKOR*, and *hORL*. Comparing the ***interspecific*** differences (divergence) in vertebrate MOR, DOR, KOR, and ORL with the ***intraspecific*** variation (polymorphism) of *Homo*

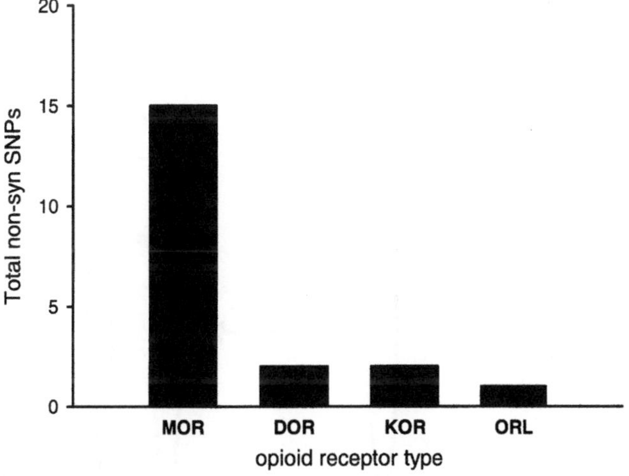

Fig. 6. Plot of total nonsynonymous single nucleotide polymorphisms (non-synSNPs) by type of human opioid receptor gene. See Table 3 for additional SNP data for human opioid receptor genes. There are a total of 15 non-syn SNPs in the exonic regions of hMOR, with 1 or 2 for the other types of opioid receptor genes.

sapiens hMOR, *hDOR*, *hKOR*, and *hORL* is hampered by the lack of comprehensive SNP data from the five other vertebrate species. It is interesting that the most studied nonsynonymous SNP in hMOR (A118G) is also found in the same alignment position as a nonsynonymous SNP in the MOR gene of the rhesus monkey (156). This suggests that SNP data from the population studies of nonhuman and nonmammalian genomes will be a fruitful approach for comparative analysis of intraspecific variation. It can be stated that in the most recently evolved species examined (humans), *hMOR* is the most polymorphic opioid receptor gene, followed by *hDOR*, *hKOR*, and *hORL*. This finding suggests greater forces of adaptive evolution operating on MOR genes since the time of its birth by duplication, as outlined in the next section.

4. The Molecular Evolution of Vertebrate Opioid Receptors

A synthesis of the results from the above analysis comparing the sequences of vertebrate opioid receptors, by species and by type of opioid receptor, along with behavioral and binding data of opioids in nonmammalian species and recent data gleaned from human genome studies suggests the following scenario for the evolution of vertebrate opioid receptors (Fig. 7). Early in vertebrate evolution, there may have existed a single opioid unireceptor gene. The unireceptor gene is likely a duplicate of a proto-unireceptor gene, perhaps from the time of the arthropod and chordate split, long before the posited 2R whole genome duplication at the root of the vertebrate evolution. The first round of genome duplication early in chordate evolution produced the

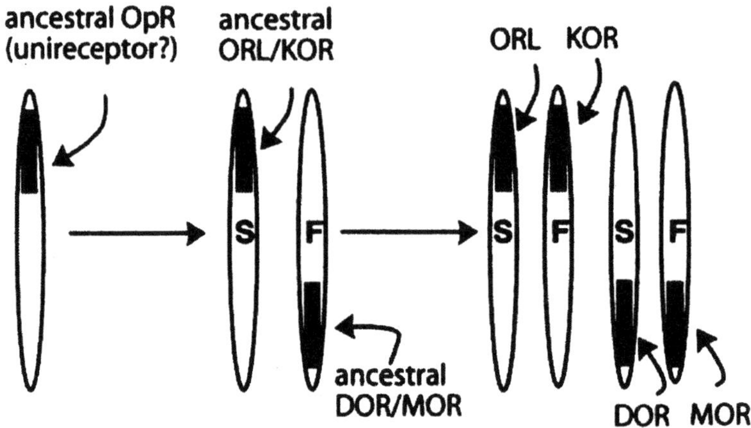

Fig. 7. The molecular evolution of vertebrate opioid receptors. For simplicity, the genes are referred to by the same acronym as the opioid receptor proteins they encode. "S" denotes slow and "F" fast rate of adaptive evolution. See text for further details.

ancestral *DOR/MOR* and *ORL/KOR* genes. The second round of genome duplication led to the four opioid receptors present in all extant vertebrates today.

That the initial duplication led to an ancestral *DOR/MOR* gene is supported by the finding that the sequences of MOR and DOR are most identical to each other, as shown above by pairwise comparison and graphically by the bifurcating pattern of the phylogenetic analysis shown in Fig. 2. Likewise, the idea that KOR and ORL shared a common ancestral gene (*KOR/ORL* in Fig. 2) is supported by the above analysis as well as other data from ligand binding that suggest KOR and ORL maintain close functional ties. As mentioned above, it is known that duplicate genes undergo asymmetrical divergence such that one gene is under relaxed constraint, showing an increased rate of adaptive evolution (positive selection), while the other gene maintains ancestral structure and function (133, 134, 157). This asymmetry for the opioid receptor gene duplicates is noted by an “F” for faster rate of adaptive evolution or positive selection (letter on the inside of the chromosome) and conversely by an “S” for the gene duplicate with a slower rate of evolution.

The gene encoding *hMOR*, and not any of the other opioid receptor types, was one of only nine genes controlling brain size or behavior that showed a significantly increased rate of protein evolution in the *Homo sapiens* genome compared to primate and rodent genomes (158). Thus, *MOR* is assigned an “F” and conversely *DOR* is the “S” member of the pair (see Fig. 7). Likewise, because ORL protein is most closely related to the rhodopsin (RHO) sequences in the dendrogram, it is assigned an “S” (maintaining more ancestral characteristics) and *KOR* is the faster evolving member of the pair. This is also supported by the finding that the vertebrate MOR, as a group, are more closely related to each other than vertebrate ORL (boxed values in Fig. 2). Likewise, vertebrate KOR proteins share more common sequence as a group than do the vertebrate ORL proteins. Stated another way, the least divergence among all vertebrates is seen in the MOR proteins, followed by the KOR group, suggesting that adaptive evolution is greatest in the *MOR* genes (confirmed with the human genome studies above), less so for the *KOR* gene, and the least for *DOR* and *ORL*. Going back to the first duplication event, the finding that vertebrate ORL maintained most ancestral characteristics (most closely related to RHO) supports the assignment of the *ORL/KOR* duplicate gene as the “slow” ancestor gene and *DOR/MOR* as the “fast” duplicate.

Although many GPCR and other gene families in the human genome appear deficient in an even number of duplicate genes due to gene deletion or mutation into pseudogenes (159), it is possible that the opioid receptor gene family avoided this fate due to a gene dosage effect (increased signaling by doubling opioid receptor expression) that provided a selective advantage, as noted for other gene families (134). It is of greater interest to note that the target of most clinically used opioids is the *mu* opioid receptor (MOR), and selective *mu* opioid agonists are the most efficacious type of opioid analgesics in the clinic and in animal models. The rank order of opioid receptor efficacy in a number of animal models (MOR >> KOR ≥ DOR >> ORL) approximates the rank order of the protein sequence identity of opioid receptor types among vertebrates, suggesting adaptive evolution of MOR (and to a lesser extent, KOR) to be a "better" opioid receptor. Thus, the molecular evolution of vertebrate opioid receptors may provide a striking example of Darwinian or positive selection at the receptor level with convergent evolution in all vertebrate species leading to the most efficacious opioid receptor in the shape of the MOR protein.

5. Methods to Decipher the Evolution of Vertebrate GPCRs

There is a dizzying array of digital tools and websites that can be used to investigate the evolution of GPCRs by bioinformatic analyses. Table 4 lists the tools and websites that were used for the simple bioinformatic analyses presented in this chapter. Brief notes on their use in also given in Table 4. Further comments for some of the methods are given below.

Multiple alignments of vertebrate opioid receptor proteins (Fig. 1) were made using the CLUSTAL-W program (160) as implemented on the EBI website (see Table 4). Default programs settings were used to obtain a complete multiple alignment of the total vertebrate dataset. The alignment file from the CLUSTAL-W program was used as the input file to the MEGA4 program. For construction of the phylogenetic trees (dendrograms) and pairwise differences among opioid receptors, the software MEGA4 was used (161). Using the alignment editor program bundled with MEGA4, gapped columns were removed and the multiple alignment file saved and opened in the phylogeny subprogram of MEGA4.

The neighbor-joining method (NJ) was selected in this program as the algorithm used to determine the topology of the dendrograms. The NJ method has a high degree of accuracy (162) and was used for previous studies examining receptor phylogeny (7, 153, 163). The reliability of branching points (nodes) was determined by the bootstrap method. The number of bootstrap repetitions was set at 1,000 within the program. The branch length represents the proportion of different amino acids from each node, and the value is denoted below the branch (scale bar given on dendrogram, see Fig. 2). Branch lengths were generated by pairwise distance using the MEGA4 software package.

Table 4
Bioinformatic resources used in deciphering the evolution of opioid receptors

Name	Internet location[a]	Function	Comments
BLAST	www.ncbi.nlm.nih.gov/BLAST	Identifies homologous sequences in GenBank to query sequence	Used to test homology of novel *OpRs* and other vertebrate *OpRs* already banked
BLAST-P	www.ncbi.nlm.nih.gov/BLAST (available link)	Pairwise BLAST analysis on two submitted sequences	Used to generate pairwise homology between novel *OpRs* and other vertebrate *OpRs*
GPCR SUB-FAMILY CLASSIFIER	www.soe.ucsc.edu/research/compbio/gpcr	Uses machine intelligence to determine GPCR classification	Additional test of novel *OpRs* as members of GPCR-Opioid family
CLUSTAL-W	www.ebi.ac.uk/clustalw	Multiple alignment program on submitted sequences	Used to align novel *OpRs* to existing vertebrate *OpRs*
BOXSHADE	mobyle.pasteur.fr/cgi-bin/MobylePortal/portal.py?form=boxshade	Provides tools to display aligned sequences	Used to display the *OpRs* vertebrate dataset
MEGA (V 4.0) MOLECULAR EVOLUTIONARY GENETICS ANALYSIS	www.megasoftware.net/ (free download, runs locally)	A suite of programs including phylogenetic analysis, distance matrix, and dendrogram generation and editing	Used for phylogenetic analysis and tree topology of novel *OpRs* and all vertebrate *OpRs*, used for partition analysis using receptor domains
ENSEMBL HAPMAP	www.hapmap.org	Identifies and catalogs SNPs and haplotypes	Used to compare SNPs of *OpRs*
ENSEMBL HUMAN MAPVIEW	www.ensembl.org/Homo_sapiens/index.html	Displays genes on human chromosomes	Used to map *OpRs* and paralogs
RATE SHIFT ANALYSIS	www.daimi.au.dk/~compbio/rateshift/	Identifies site-specific rate-shifts in evolutionarily related proteins	Used to determine potential a.a. sites that are crucial to type-selectivity

[a]URL is valid at time of writing (web addresses often change) and multiple URLs exist for many programs

Initially, a series of 2-way BLAST-P analyses was made for each comparison of opioid receptor family proteins in each species (*MOR* vs. *DOR, MOR* vs. *KOR,* and *DOR* vs. *KOR*). These values were obtained for both percent identity and percent similarity; these results are shown in Table 2. To calculate the percent divergence shown in Fig. 3, the distance matrix available in MEGA4 is used to directly calculate the mean distance of all four types of opioid receptors proteins for each species. Data collected for the other figures and tables came from the tools listed in Table 4 as described in the "Notes" portion of the table.

6. Future Directions

The results presented here are a first approximation as the opioid receptor sequence dataset is limited. Complete cDNA sequences and the conceptual translation of the protein amino acids for MOR, DOR, KOR, and ORL are only available from six species representing three vertebrate classes. Specifically, opioid receptor sequences cloned from brain tissue cDNA of key species representing *Class Reptilia* or *Class Aves* are not available, not to mention *Class Agnatha* (hagfish and lamprey). While new animal genomes are appearing online with increasing rapidity, these results are not of high-fidelity nor confirmed by expression in brain tissue of that species.

Understanding the evolution of vertebrate opioid receptor proteins contributes to the fundamental model of molecular attraction between opioid drugs and their receptor proteins. The primary structure of the protein sequence must be deterministic in each aspect of measured opioid function including ligand binding, conformational change, and signal transduction. The conceit of this evolutionary approach is not only to explain differences observed in the present opioid receptor proteins expressed in various vertebrate species but also to seek the pattern of opioid receptor evolution to come. This will allow the design of engineered opioid receptors, most likely a variant of the MOR protein (super-MOR?) that could be usurped to provide unsurpassed analgesia, perhaps solely from endogenous opioid activation, once the inevitable arrival of gene therapy is secured.

Experiments can be performed to support the hypothesis of opioid receptor by gene duplication by using ancestral sequence analysis, as was done for steroid receptors (164). Using bioinformatics and available sequences, the most likely DOR/MOR protein sequence can be determined, a synthetic *DOR/MOR* gene made, transfected in CHO cells, and the ancestral receptor protein characterized. Predictions would include the binding of both

MOR and DOR opioid agonists (but not KOR or ORL agonists) but perhaps less robust binding to the DOR/MOR receptor than to either MOR or DOR protein alone.

Many questions and analyses remain. Is there a relationship between number of splice variants and SNPs? The MOR protein has more of both compared to the other types. What specific amino acids determine the divergence in type-selectivity? It is still not known how *mu*, *kappa*, and *delta* opioid selectivity correlates to the primary amino acid structure of individual opioid receptor proteins. Specifically, it is not known what exact residues or domains of the primary amino acid sequence differ in each of the three proteins (MOR, DOR, and KOR) to produce *mu*, *delta*, or *kappa* opioid agonist type-selectivity. Continued analysis of the vertebrate dataset coupled with experimental data should provide some answers soon. What type of adaptive evolution is occurring? Certainly not purifying selection for all opioid receptor genes, but does the presence of a large number of SNPs in MOR signal positive or Darwinian selection, and if so, by balancing selection or diversifying selection? What are the selection pressures driving the convergent evolution of opioid receptors in different species, especially MOR and to a lesser extent KOR, and how are they balanced against the evolutionary vector of divergence of opioid receptor types? Do the different types of opioid receptor genes evolve at different rates? These and other questions must wait for now.

Future bioinformatic analyses include evolutionary trace studies, whereby single amino acid sites are compared among different species and significance levels applied would give critical input to specific domain hypothesis and ligand selectivity (165). Single site analysis also contributes to the *in silico* modeling and structural outcome. For example, does selectivity arise from a more restricted general binding pocket as shown by substitution with bulkier amino acids? Is there an alteration of charge in selective areas consistent with altered ligand selectivity? Are some sites of the receptor sequence undergoing more rapid molecular evolution than other sites and, if so, what is the selection pressure?

Finally, through the use of cell culture studies, hypotheses concerning the molecular evolution of opioid receptors should be tested. Significant functional differences in transfected amphibian and human MOR are seen when tested for. For example, the amphibian *mu* opioid receptor (rpMOR) did show decreased affinity to selective MOR opioid agonists compared to hMOR in transfected cell lines (126). Such studies need to be carried out with a number of cloned opioid receptors from different vertebrate species and correlated with sequence analysis and bioinformatics. While this chapter reviewed the analyses and results using the case example of opioid receptors, efforts using other GPCR families of receptors should be equally as fruitful.

7. Conclusions

In summary, the degree of opioid receptor sequence divergence within species was correlated with vertebrate evolution. As the primary amino acid sequence (structure) is a determinant of the selective binding (function) of opioid receptors, the correlation of sequence divergence with vertebrate evolution shows that opioid receptor proteins exhibit an evolutionary vector of increased type-selectivity. Additionally, the rapid rate of adaptive evolution of vertebrate MOR suggests an evolutionary vector of increased opioid receptor function. The selective pressure underlying the rapid adaptation of MOR is similar in all vertebrate species, an example of convergent evolution. It is of greater interest to note that the target of most clinically used opioids is the *mu* opioid receptor, and selective *mu* opioids are the most potent type of opioid drug in the clinic and in animal models. This provides a striking example of Darwinian fitness at the molecular level should the selection pressure turn out to be a "better" opioid receptor. If confirmed, these findings provide a unique understanding of the pharmacology of vertebrate opioid receptors and the many other families of GPCR proteins encoded by duplicate genes.

Acknowledgments

The author gratefully acknowledges the past and continued support of the National Institutes of Health-NIDA through research grants DA R15-12448 and DA R29-07326, as well as the state of Oklahoma through the Oklahoma Center for the Advancement of Science and Technology (OCAST) Health Research Contracts. Many thanks also to my colleagues at the International Narcotics Research Conference (www.inrcworld.org) for their encouragement and inspiration. This chapter is dedicated to my children in celebration of their entry into adulthood.

References

1. Attwood TK (2001) A compendium of specific motifs for diagnosing GPCR subtypes. Trends Pharmacol Sci **22:**162–165.
2. Perez DM (2003) The evolutionarily triumphant G-protein-coupled receptor. Mol Pharmacol **63:**1202–1205.
3. Perez DM (2005) From plants to man: the GPCR "tree of life". Mol Pharmacol **67:**1383–1384.
4. Bockaert J, Pin JP (1999) Molecular tinkering of G protein-coupled receptors: an evolutionary success. The EMBO Journal **18:**1723–1729.
5. Fredriksson R, Lagerstrom MC, Lundin LG et al (2003) The G-protein-coupled receptors in the human genome form five main families: Phylogenetic analysis, paralogon groups, and fingerprints. Mol Pharmacol **63:**1256–1272.

6. Fredriksson R, Schiotch H.B. (2005) The repertoire of G-protein-coupled receptors in fully sequenced genomes. Mol Pharmacol **67:**1414–1425.
7. Josefsson L-G (2000) Evidence for kinship between diverse G-protein coupled receptors. Gene **239:**333–340.
8. Sundstrom G, Dreborg S, Larhammar D (2010) Concomitant duplications of opioid peptide and receptor genes before the origin of jawed vertebrates. PLoS One **5:**e10512.
9. Dreborg S, Sundstrom G, Larsson TA et al (2008) Evolution of vertebrate opioid receptors. Proc Natl Acad Sci USA **105:**15487–15492.
10. Cavalier-Smith T (1998) A revised six-kingdom system of life. Biol Rev Camb Philos Soc **73:**203–266.
11. Iyer LM, Balaji S, Koonin EV et al (2006) Evolutionary genomics of nucleo-cytoplasmic large DNA viruses. Virus Res **117:**156–184.
12. Bamford DH (2003) Do viruses form lineages across different domains of life? Res Microbiol **154:**231–236.
13. Maussang D, Vischer HF, Schreiber A et al (2009) Pharmacological and biochemical characterization of human cytomegalovirus-encoded G protein-coupled receptors. Methods Enzymol **460:**151–171.
14. Maussang D, Vischer HF, Leurs R et al (2009) Herpesvirus-encoded G protein-coupled receptors as modulators of cellular function. Mol Pharmacol **76:**692–701.
15. Taylor EW, Agarwal A (1993) Sequence homology between bacteriorhodopsin and G-protein coupled receptors: exon shuffling or evolution by duplication? FEBS Lett **325:**161–166.
16. Martin WR, Eades CG, Thompson JA et al (1976) The effects of morphine-and nalorphine-like drugs in the nondependent and morphine-dependent chronic spinal dog. J Pharmacol Exp Ther **197:**517–532.
17. Lord JAH, Waterfield AA, Hughes J et al (1977) Endogenous opioid peptides: multiple agonists and receptors. Nature **267:**495–499.
18. Portoghese PS, Sultana M, Takemori AE (1988) Naltrindole, a highly selective and potent non-peptide *delta* opioid receptor antagonist. Eur J Pharm **146:**185–186.
19. Takemori AE, Ho BY, Naeseth JS et al (1988) Nor-binaltorphimine, a highly selective *kappa*-opioid antagonist in analgesic and receptor binding assays. J Pharmacol Exp Ther **246:**255–258.
20. Ward SJ, Portoghese PS, Takemori AE (1982) Pharmacological profiles of *beta*-funaltrexamine (β-FNA) and *beta*-chlornaltrexamine (β-CNA) on the mouse vas deferens preparation. Eur J Pharmacol **80:**377–384.
21. Pert CB, Snyder SH (1973) Opiate receptor: demonstration in nervous tissue. Science **179:**1011–1014.
22. Simon EJ, Hiller JM, Edelman I (1973) Stereospecific binding of the potent narcotic analgesic [^{3}H] etorphine to rat-brain homogenate. Proc Natl Acad Sci USA **70:**1947–1949.
23. Terenius L (1973) Stereospecific interaction between narcotic analgesics and a synaptic plasma membrane fraction of rat cerebral cortex. Acta Pharmacol Toxicol **32:**317–320.
24. Loh H, Smith A (1990) Molecular characterization of opioid receptors. Ann Rev Pharmacol Toxicol **30:**123–147.
25. Simon EJ (1986) Recent studies on opioid receptors: heterogeneity and purification. Annal NY Acad Sci **463:**31–45.
26. Simon EJ (1981) Opiate receptors: some recent developments. In: Lamble JW (ed) Towards Understanding Receptors, Elsevier/North-Holland Biomedical Press, New York.
27. Zaki PA, Bilsky EJ, Vanderah TW et al (1996) Opioid receptor types and subtypes: The *delta* receptor as a model. Ann Rev Pharmacol Toxicol **36:**379–401.
28. Wollemann M, Benyhe S, Simon J (1993) The *kappa* opioid receptor: evidence for the different subtypes. Life Sci **52:**599–611.
29. Kieffer BL, Befort K, Gaveriaux-Ruff C et al (1992) The *delta*-opioid receptor: Isolation of a cDNA by expression cloning and pharmacological characterization. Proc Natl Acad Sci USA **89:**12048–12052.
30. Evans CJ, Keith DE, Morrison H et al (1992) Cloning of a *delta* opioid receptor by functional expression. Science **258:**1952–1955.
31. Wang JB, Johnson PS, Persico AM et al (1994) Human *mu* opiate receptor. cDNA and genomic clones, pharmacologic characterization and chromosomal assignment. FEBS Lett **338:**217–222.
32. Knapp RJ, Malatynska E, Fang L et al (1994) Identification of a human *delta* opioid receptor: cloning and expression. Life Sci **54:**L463–L469.
33. Simonin F, Befort K, Gaveriaux-Ruff C et al (1994) The human *delta*-opioid receptor: genomic organization, cDNA cloning, functional expression, and distribution in human brain. Mol Pharmacol **46:**1015–1021.

34. Mansson E, Bare L, Yang D (1994) Isolation of a human *kappa* opioid receptor cDNA from placenta. Biochem Biophys Res Commun **202:**1431–1437.
35. Zhu J, Chen C, Xue JC et al (1995) Cloning of a human *kappa* opioid receptor from the brain. Life Sci **56:**L201–L207.
36. Simonin F, Gaveriaux-Ruff C, Befort K et al (1995) *Kappa*-opioid receptor in humans: cDNA and genomic cloning, chromosomal assignment, functional expression, pharmacology, and expression pattern in the central nervous system. Proc Natl Acad Sci USA **92:**7006–7010.
37. Mollereau C, Parmentier M, Mailleux P et al (1994) ORL1, a novel member of the opioid receptor family: cloning, functional expression and localization. FEBS Lett **341:**33–38.
38. Fukuda K, Kato S, Mori K et al (1994) cDNA cloning and regional distribution of a novel member of the opioid receptor family. FEBS **343:**42–46.
39. Chen Y, Fan Y, Liu J et al (1994) Molecular cloning, tissue distribution and chromosomal localization of a novel member of the opioid receptor gene family. FEBS Lett **347:**279–283.
40. Bunzow JR, Saez C, Mortrud M et al (1994) Molecular cloning and tissue distribution of a putative member of the rat opioid receptor gene family that is not a *mu*, *delta*, or *kappa* opioid receptor type. FEBS Letters **347:**284–288.
41. Wang JB, Johnson PS, Imai Y et al (1994) cDNA cloning of an orphan opiate receptor gene family member and its splice variant. FEBS Lett **348:**75–79.
42. Lachowicz JE, Shen Y, Monsma FJ Jr, et al (1995) Molecular cloning of a novel G protein-coupled receptor related to the opiate receptor family. J Neurochem **64:**34–40.
43. Wick MJ, Minnerath SR, Lin X et al (1994) Isolation of a novel cDNA encoding a putative membrane receptor with high homology to the cloned *mu*, *delta*, and *kappa* opioid receptors. Brain Res **27:**37–44.
44. Meunier J-C, Mollereau C, Toll L et al (1995) Isolation and structure of the endogenous agonist of opioid receptor-like ORL1 receptor. Nature **377:**532–535.
45. Reinscheid RK, Nothacker HP, Bourson A et al (1995) Orphanin FQ: a neuropeptide that activates an opioid-like G protein-coupled receptor. Science **270:**792–794.
46. Meunier J-C (1997) Nociceptin/orphanin FQ and the opioid receptor-like ORL1 receptor. Eur J Pharmacol **340:**1–15.
47. Mogil JS, Pasternak G (2001) The molecular and behavioral pharmacology of the orphanin FQ/nociceptin peptide and receptor family. Pharmacol Rev **53:**381–415.
48. Chen X, Liu-Chen LY, Tallarida RJ et al (1996) Use of a *mu*-antisense oligodeoxynucleotide as a *mu* opioid receptor noncompetitive antagonist *in vivo*. Neurochem Res **21:**1363–1368.
49. Wang HQ, Kampine JP, Tseng L-F (1996) Antisense oligodeoxynucleotide to a *delta*-opioid receptor messenger RNA selectively blocks the antinociception induced by intracerebroventricularly administered *delta*-, but not *mu*-, or *kappa*-opioid receptor agonists in the mouse. Neurosci **75:**445–452.
50. Pasternak GW, Standifer KM (1995) Mapping of opioid receptors using antisense oligodeoxynucleotides: correlating their molecular biology and pharmacology. Trends Pharmacol Sci **16:**344–350.
51. Uhl GR, Childers S, Pasternak G (1994) An opiate-receptor gene family reunion. Trends Neurosci **17:**89–93.
52. Pan Y-X, Xu J, Bolan E et al (1999) Identification and characterization of three new alternatively spliced *mu*-opioid receptor isoforms. Mol Pharmacol **56:**396–403.
53. Pan Y-X, Cheng J, Xu J et al (1995) Cloning and functional characterization through antisense mapping of a *kappa*$_3$-related opioid receptor. Mol Pharmacol **47:**1180–1188.
54. Pasternak GW (2005) Molecular biology of opioid analgesia. J Pain Symptom Manag **29:**S2–S9.
55. Pan YX, Xu J, Bolan E et al (2005) Identification of four novel exon 5 splice variants of the mouse *mu*-opioid receptor gene: functional consequences of C-terminal splicing. Mol Pharmacol **68:**866–875.
56. Pan YX, Xu J, Mahurter L et al (2003) Identification and characterization of two new human *mu* opioid receptor splice variants, hMOR-1O and hMOR-1X. Biochem Biophys Res Commun **301:**1057–1061.
57. Gaveriaux-Ruff C, Peluso J, Befort K et al (1997) Detection of opioid receptor mRNA by RT-PCR reveals alternative splicing for the *delta*- and *kappa*-opioid receptors. Mol Brain Res **48:**298–304.
58. Wei LN, Law PY, Loh HH (2004) Post-transcriptional regulation of opioid receptors in the nervous system. Front Biosci **9:**1665–1679.
59. Pan YX, Xu J, Wan BL et al (1998) Identification and differential regional expression of KOR-3/ORL-1 gene splice variants in mouse brain. FEBS Lett **435:**65–68.
60. Stevens CW, Pezalla PD (1983) A spinal site mediates opiate analgesia in frogs. Life Sci **33:**2097–2103.

61. Stevens CW, Pezalla PD (1984) Naloxone blocks the analgesic action of levorphanol but not of dextrorphan in the leopard frog. Brain Res **301**:171–174.
62. Pezalla PD, Stevens CW (1984) Behavioral effects of morphine, levorphanol, dextrorphan and naloxone in the frog *Rana pipiens*. Pharmacol Biochem Behav **21**:213–217.
63. Stevens CW, Pezalla PD, Yaksh TL (1987) Spinal antinociceptive action of three representative opioid peptides in frogs. Brain Res **402**:201–203.
64. Stevens CW, Kirkendall K (1993) Time course and magnitude of tolerance to the analgesic effects of systemic morphine in amphibians. Life Sci **52**:PL111–116.
65. Pezalla PD, Dicig M (1984) Stress-induced analgesia in frogs: evidence for the involvement of an opioid system. Brain Res **296**:356–360.
66. Stevens CW, Sangha S, Ogg BG (1995) Analgesia produced by immobilization stress and an enkephalinase-inhibitor in amphibians. Pharmacol Biochem Behav **51**:675–680.
67. Willenbring S, Stevens CW (1996) Thermal, mechanical and chemical peripheral sensation in amphibians: opioid and adrenergic effects. Life Sci **58**:125–133.
68. Willenbring S, Stevens CW (1997) Spinal *mu*, *delta* and *kappa* opioids alter chemical, mechanical and thermal sensitivities in amphibians. Life Sci **61**:2167–2176.
69. Mohan S, Stevens CW (2006) Systemic and spinal administration of the *mu* opioid, remifentanil, produces antinociception in amphibians. Eur J Pharmacol **534**:89–94.
70. Stevens CW, Toth G, Borsodi A et al (2007) Xendorphin B1, a novel opioid-like peptide determined from a *Xenopus laevis* brain cDNA library, produces opioid antinociception after spinal administration in amphibians. Brain Res Bull **71**:628–632.
71. Stevens CW, Klopp AJ, Facello JA (1994) Analgesic potency of *mu* and *kappa* opioids after systemic administration in amphibians. J Pharmacol Exp Ther **269**:1086–1093.
72. Stevens CW (1996) Relative analgesic potency of *mu*, *delta* and *kappa* opioids after spinal administration in amphibians. J Pharmacol Exp Ther **276**:440–448.
73. Stevens CW, Rothe KS (1997) Supraspinal administration of opioids with selectivity for *mu*, *delta* and *kappa* -opioid receptors produces analgesia in amphibians. Eur J Pharmacol **331**:15–21.
74. Stevens CW (2004) Opioid research in amphibians: an alternative pain model yielding insights on the evolution of opioid receptors. Brain Res Rev **46**:204–215.
75. Stevens, C. W. (2008) Non-mammalian models for the study of pain. In: Conn PM (ed) Sourcebook of Models for Biomedical Research, Humana Press, Totowa.
76. Stevens CW, Martin KK, Stahlheber BW (2008) Nociceptin produces antinociception after spinal administration in amphibians. Pharmacol Biochem Behav **91**:436–440.
77. Stevens CW, Newman LC (1999) Spinal administration of selective opioid antagonists in amphibians: evidence for an opioid unireceptor. Life Sci **64**:PL125–PL130.
78. Gonzalez-Nunez V, Barrallo A, Traynor JR et al (2006) Characterization of opioid binding sites in zebrafish brain. J Pharmacol Exp Ther **316**:900–904.
79. Brooks AI, Standifer KM, Cheng J et al (1994) Opioid binding in giant toad and goldfish brain. Receptor **4**:55–62.
80. Bird DJ, Jackson M, Baker BI et al (1988) Opioid binding sites in the fish brain: An autoradiographic study. Gen Comp Endocrinol **70**:49–62.
81. Buatti MC, Pasternak GW (1981) Multiple opiate receptors: phylogenetic differences. Brain Res **218**:400–405.
82. Simon EJ, Hiller JM, Groth J et al (1982) The nature of opiate receptors in toad brain. Life Sci **31**:1367–1370.
83. Zawilska J, Lajtha A, Borsodi A (1988) Selective protection of benzomorphan binding sites against inactivation by N-ethylmaleimide: Evidence for *kappa* opioid receptors in frog brain. J Neurochem **51**:736–739.
84. Benyhe S, Varga E, Hepp J et al (1990) Characterization of *kappa*1 and *kappa*2 opioid binding sites in frog (*Rana esculenta*) brain membrane. Neurochem Res **15**:899–904.
85. Simon J, Benyhe S, Hepp J et al (1987) Purification of *kappa*-opioid receptor subtype from frog brain. Neuropeptides **10**:19–28.
86. Simon J, Benyhe S, Borsodi A et al (1985) Separation of *kappa*-opioid receptor subtype from frog brain. FEBS Letters **183**:395–397.
87. Pert CB, Aposhian D, Snyder SH (1974) Phylogenetic distribution of opiate binding. Brain Res **75**:356–361.
88. Xia Y, Haddad GG (2001) Major difference in the expression of *delta*- and *mu*-opioid receptors between turtle and rat brain. J Comp Neurol **436**:202–210.
89. Bakalkin GY, Pivovarov AS, Kobylyansky AG et al (1989) Lateralization of opioid receptors in turtle visual cortex. Brain Res **480**:268–276.

90. Benyhe S, Monory K, Farkas J et al (1999) Nociceptin binding sites in frog (*Rana esculenta*) brain membranes. Biochem Biophys Res Commun **260**:592–596.
91. Wollemann M, Farkas J, Toth G et al (1999) Comparison of the endogenous heptapeptide met-enkephalin-arg^{6}-phe^{7} binding in amphibian and mammalian brain. Acta Biologica Hungarica **50**:297–307.
92. Benyhe S, Ketevan A, Simon J et al (1997) Affinity labelling of frog brain opioid receptors by dynorphin$_{(1-10)}$ chloromethyl ketone. Neuropeptides **31**:52–59.
93. Moitra J, Öktem HA, Borsodi A (1995) Thermodynamic parameters of frog brain *kappa*-opioid receptors. J Neurochem **65**: 798–801.
94. Benyhe S, Simon J, Borsodi A et al (1994) [^{3}H]Dynorphin 1–8 binding sites in frog (*Rana esculenta*) brain membranes. Neuropeptides **26**:359–364.
95. Wollemann M, Farkas J, Toth G et al (1994) Characterization of [^{3}H] met-enkephalin-arg6-phe7 binding to opioid receptors in frog brain membrane preparations. J Neurochem **63**:1460–1465.
96. Benyhe S, Szucs M, Borsodi A et al (1992) Species differences in the stereoselectivity of *kappa* opioid binding sites for [^{3}H]U-69593 and [^{3}H] ethylketocyclazocine. Life Sci **51**:1647–1655.
97. Simon J, Benyhe S, Hepp J et al (1990) Method for isolation of *kappa*-opioid binding sites by dynorphin affinity chromatography. J Neurosci Res **25**:549–555.
98. Mollereau C, Pascaud A, Baillat G et al (1988) Evidence for a new type of opioid binding site in the brain of the frog *Rana ridibunda*. Eur J Pharmacol **150**:75–84.
99. Newman LC, Wallace DR, Stevens CW (2000) Selective opioid agonist and antagonist displacement of [^{3}H]-naloxone binding in amphibian brain. Eur J Pharmacol **397**:255–262.
100. Newman LC, Wallace DR, Stevens CW (2000) Selective opioid receptor agonist and antagonist displacement of [^{3}H]-naloxone binding in amphibian spinal cord. Brain Res **884**:184–191.
101. Newman LC, Sands SS, Wallace DR et al (2002) Characterization of *mu*, *kappa*, and *delta* opioid binding in amphibian whole brain tissue homogenates. J Pharmacol Exp Ther **301**:364–370.
102. Li X, Keith DEJr, Evans CJ (1996) *Mu* opioid receptor-like sequences are present throughout vertebrate evolution. J Mol Evol **43**:179–184.
103. Li X, Keith DE, Jr., Evans CJ (1996) Multiple opioid receptor-like genes are identified in diverse vertebrate phyla. FEBS Lett **397**:25–29.
104. Darlison MG, Greten FR, Harvey RJ et al (1997) Opioid receptors from a lower vertebrate (*Catostomus commersoni*): Sequence, pharmacology, coupling to a G-protein-gated inward-rectifying potassium channel (GIRK1), and evolution. Proc Natl Acad Sci USA **94**:8214–8219.
105. Barrallo A, González-Sarmiento R, Porteros A et al (1998) Cloning, molecular characterization, and distribution of a gene homologous to d opioid receptor from zebrafish (*Danio rerio*). Biochem Biophys Res Commun **245**:544–548.
106. Barrallo A, González-Sarmiento R, Alvar F et al (2000) ZFOR2, a new opioid receptor-like gene from the teleost zebrafish (*Danio rerio*). Mol Brain Res **84**:1–6.
107. Alvarez FA, Rodriguez-Martin I, Gonzalez-Nunez V et al (2006) New *kappa* opioid receptor from zebrafish *Danio rerio*. Neurosci Lett **405**:94–99.
108. Porteros A, Garcia-Isodoro M, Barrallo A et al (1999) Expression of ZFOR1, a *delta*-opioid receptor, in the central nervous system of the zebrafish (*Danio rerio*). J Comp Neurol **412**:429–438.
109. Bradford CS, Walthers EA, Stanley DJ et al (2006) *Delta* and *mu* opioid receptors from the brain of a urodele amphibian, the rough-skinned newt, *Taricha granulosa*: Cloning, heterologous expression, and pharmacological characterization. Gen Comp Endocrinol **146**:275–290.
110. Bradford CS, Walthers EA, Searcy BT et al (2005) Cloning, heterologous expression and pharmacological characterization of a *kappa* opioid receptor from the brain of the rough-skinned newt, *Taricha granulosa*. J Mol Endocrinol **34**:809–823.
111. Walthers EA, Bradford CS, Moore FL (2005) Cloning, pharmacological characterization and tissue distribution of an ORL1 opioid receptor from an amphibian, the rough-skinned newt, *Taricha granulosa*. J Mol Endocrinol **34**:247–256.
112. Stevens CW, Brasel CM, Mohan S (2007) Cloning and bioinformatics of amphibian *mu, delta, kappa,* and nociceptin opioid receptors expressed in brain tissue: Evidence for opioid receptor divergence in mammals. Neurosci Lett **419**:189–194.
113. Minami M, Satoh M (1995) Molecular biology of the opioid receptors: structures,

functions and distributions. Neurosci Res **23:**121–145.

114. Gether U (2002) Uncovering molecular mechanisms involved in activation of G protein-coupled receptors. Endo Rev **21:**90–113.
115. Shahrestanifar MS, Howells RD (1996) Sensitivity of opioid receptor binding to N-substituted maleimides and methanethiosulfonate derivatives. Neurochem Res **21:**1295–1299.
116. Palczewski K, Kumasaka T, Hori T et al (2000) Crystal structure of rhodopsin: a G protein-coupled receptor. Science **289:**739–746.
117. Lomize AL, Pogozheva ID, Mosberg HI (1999) Structural organization of G-protein-coupled receptors. J Comput Aided Mol Des **13:**352–353.
118. Christopoulos A, Kenakin T (2002) G protein-coupled receptor allosterism and complexing. Pharmacol Rev **54:**323–374.
119. Wang HL, Chang WT, Hsu CY et al (2002) Identification of two C-terminal amino acids, Ser(355) and Thr(357), required for short-term homologous desensitization of *mu*-opioid receptors. Biochem Pharmacol **15:**257–266.
120. Law P-Y, Wong YH, Loh HH (1999) Mutational analysis of the structure and function of opioid receptors. Biopoly **51:**440–455.
121. Ferguson DM, Kramer S, Metzger TG et al (2000) Isosteric replacement of acidic with neutral residues in extracellular loop-2 of the *kappa*-opioid receptor does not affect dynorphin A(1–13) affinity and function. J Biol Chem **43:**1251–1252.
122. Chaturvedi K, Christoffers KH, Singh K et al (2000) Structure and regulation of opioid receptors. Biopolymers **55:**334–346.
123. Stevens, C. W. (2005) Molecular evolution of vertebrate opioid receptor proteins: A preview. In Capasso A (ed) Recent Developments in Pain Research, 2005. Research Signpost, Trivandrum.
124. Altschul SF, Gish W, Miller W et al (1990) Basic local alignment search tool. J Mol Biol **215:**403–410.
125. Rodríguez RE, Barrallo A, Garcia-Malvar F et al (2000) Characterization of ZFOR1, a putative *delta*-opioid receptor from the teleost zebrafish. Neurosci Lett **288:**207–210.
126. Brasel CM, Sawyer GW, Stevens CW (2008) A pharmacological comparison of the cloned frog and human *mu* opioid receptors reveals differences in opioid affinity and function. Eur J Pharmacol **599:**36–43.
127. Ohno, S (1970) Evolution by Gene Duplication. George Allen and Unwin, London.
128. Lundin LG, Larhammar D, Hallbook F (2003) Numerous groups of chromosomal regional paralogies strongly indicate two genome doublings at the root of the vertebrates. J Struct Funct Genomics **3:**53–63.
129. Zhang J (2003) Evolution by gene duplication: an update. Trends Ecol Evol **18:**292–298.
130. McLysaght A, Hokamp K, Wolfe KH (2002) Extensive genomic duplication during early chordate evolution. Nat Genet **31:**200–204.
131. Hokamp K, McLysaght A, Wolfe KH (2003) The 2R hypothesis and the human genome sequence. J Struct Funct Genomics **3:**95–110.
132. Krylov DM, Wolf YI, Rogozin IB et al (2003) Gene loss, protein sequence divergence, gene dispensability, expression level, and interactivity are correlated in eukaryotic evolution. Genome Res **13:**2229–2235.
133. Prince VE, Pickett FB (2002) Splitting pairs: the diverging fates of duplicated genes. Nat Rev Genet **3:**827–837.
134. Kondrashhov FA, Rogozin IB, Wolf YI et al (2002) Selection in the evolution of gene duplications. Genome Biol **3:**0008.1–0008.9.
135. Jordan IK, Wolf YI, Koonin EV (2004) Duplicated genes evolve slower than singletons despite the initial rate increase. BMC Evol Biol **4:**22.
136. Makalowski W (2001) Are we polyploids? A brief history of one hypothesis. Genome Res **11:**667–670.
137. Hoehe MR (2003) Haplotypes and the systematic analysis of genetic variation in genes and genomes. Pharmacogenomics **4:**547–570.
138. Rommelspacher H, Smolka M, Schmidt LG et al (2001) Genetic analysis of the *mu*-opioid receptor in alcohol-dependent individuals. Alcohol **24:**129–135.
139. LaForge KS, Yuferov V, Kreek MJ (2000) Opioid receptor and peptide gene polymorphisms: potential implications for addictions. Eur J Pharmacol **410:**249–268.
140. Bond C, LaForge KS, Tian M et al (1998) Single-nucleotide polymorphism in the human *mu* opioid receptor gene alters *beta*-endorphin binding and activity: possible implications for opiate addiction. Proc Natl Acad Sci USA **95:**9608–9613.
141. Hoehe MR, Kopke K, Wendel B et al (2000) Sequence variability and candidate gene analysis in complex disease: association of *mu* opioid receptor gene variation with substance dependence. Hum Mol Genet **9:**2895–2908.

142. Mayer P, Hollt V (2006) Pharmacogenetics of opioid receptors and addiction. Pharm Genomics **16:**1–7.
143. Klepstad P, Dale O, Skorpen F et al (2005) Genetic variability and clinical efficacy of morphine. Acta Anaesthesiol Scand **49:**902–908.
144. Bayerer B, Stamer U, Hoeft A et al (2007) Genomic variations and transcriptional regulation of the human *mu*-opioid receptor gene. Eur J Pain **11:**421–427.
145. Uhl GR, Sora I, Wang Z (1999) The *mu* opiate receptor as a candidate gene for pain: polymorphisms, variations in expression, nociception, and opiate responses. Proc Natl Acad Sci USA **96:**7752–7755.
146. Gu J, Gu X (2003) Induced gene expression in human brain after the split from chimpanzee. Trends Genet **19:**63–65.
147. Jordan IK, Marino-Ramirez L, Koonin EV (2005) Evolutionary significance of gene expression divergence. Gene **345:**119–126.
148. Yuferov V, Fussell D, LaForge KS et al (2004) Redefinition of the human *kappa* opioid receptor gene (OPRK1) structure and association of haplotypes with opiate addiction. Pharmacogenetics **14:**793–804.
149. Saito M, Ehringer MA, Toth R et al (2003) Variants of *kappa*-opioid receptor gene and mRNA in alcohol-preferring and alcohol-avoiding mice. Alcohol **29:**39–49.
150. Kobayashi H, Hata H, Ujike H et al (2006) Association analysis of *delta*-opioid receptor gene polymorphisms in methamphetamine dependence/psychosis. Am J Med Genet B Neuropsychiatr Genet **141:**482–486.
151. Ito E, Xie GX, Maruyama K et al (2000) A core-promoter region functions bi-directionally for human opioid-receptor-like gene ORL1 and its 5'-adjacent gene GAIP. J Mol Biol **304:**259–270.
152. Clark AG, Hubisz MJ, Bustamante CD et al (2005) Ascertainment bias in studies of human genome-wide polymorphism. Genome Research **15:**1496–1502.
153. Iwama H, Gojobori T (2002) Identification of neurotransmitter receptor genes under significantly relaxed selective constraint by orthologous gene comparisons between humans and rodents. Mol Biol Evol **19:**1891–1901.
154. Rana BK, Hewett-Emmett D, Jin L et al (1999) High polymorphism at the human melanocortin 1 receptor locus. Genetics **151:**1547–1557.
155. McPartland JM, Norris RW, Kilpatrick CW (2007) Tempo and mode in the endocannaboinoid system. J Mol Evol **65:**267–276.
156. Ikeda K, Ide S, Han W et al (2005) How individual sensitivity to opiates can be predicted by gene analyses. Trends Pharmacol Sci **26:**311–317.
157. Merritt TJ, Quattro JM (2001) Evidence for a period of directional selection following gene duplication in a neurally expressed locus of triosephosphate isomerase. Genetics **159:**689–697.
158. Dorus S, Vallender EJ, Evans PD et al (2004) Accelerated evolution of nervous system genes in the origin of *Homo sapiens.* Cell **119:**1027–1040.
159. Wolfe KH, Li WH (2003) Molecular evolution meets the genomics revolution. Nat Genet **33:**255–265.
160. Thompson JD, Higgins DG, Gibson TJ (1994) CLUSTAL W: improving the sensitivity of progressive multiple sequence alignment through sequence weighting, position-specific gap penalties and weight matrix choice. Nucleic Acids Res **22:**4673–4680.
161. Kumar S, Tamura K, Nei M (2004) MEGA3: Integrated software for Molecular Evolutionary Genetics Analysis and sequence alignment. Brief Bioinform **5:**150–163.
162. Zhang J, Nei M (1997) Accuracies of ancestral amino acid sequences inferred by the parsimony, likelihood, and distance methods. J Mol Evol **44:**S139–S146.
163. Madsen O, Willemsen D, Ursing B et al (2002) Molecular evolution of the mammalian *alpha* 2B adrenergic receptor. Mol Biol Evol **19:**2150–2160.
164. Ortlund EA, Bridgham JT, Redinbo MR et al (2007) Crystal structure of an ancient protein: evolution by conformational epistasis. Science **317:**1544–1548.
165. Madabushi S, Gross AK, Philippi A et al (2004) Evolutionary trace of G protein-coupled receptors reveals clusters of residues that determine global and class-specific functions. J Bio Chem **279:**8126–8132.

Part II

The Trafficking of G Protein-Coupled Receptors

Chapter 6

A New Approach to Visualize Endogenously Expressed G Protein-Coupled Receptors in Tissues and Living Cells

Grégory Scherrer and Brigitte L. Kieffer

Abstract

G protein-coupled receptors (GPCRs) represent the largest family of membrane receptors. These proteins respond to a broad diversity of environmental stimuli and ligands, modulate most physiological processes, and represent prime therapeutic targets. Detecting GPCRs *in vivo*, however, remains a challenge and this limitation hampers our knowledge of receptor physiology. Autoradiographic ligand binding procedures provide low-resolution information, and the development of specific antibodies for immunohistochemistry has proven difficult. Tagged GPCRs have mainly been used in heterologous overexpression systems and cellular models. Here we describe an innovative approach where a fluorescent protein is fused to a GPCR *in vivo*. Using a knockin methodology, one can produce mutant mice that express a functional fluorescent receptor in place of the native receptor, and at physiological levels. We have pioneered this approach with the *delta* opioid receptor, implicated in both pain and emotional disorders. Here we describe these unique knockin reporter mice, and address potential pitfalls of the strategy. We report our first observations using this tool, and exemplify its usefulness at the level of receptor anatomy, function, and adaptations to drugs, with a particular focus on pain processes. This approach is potentially applicable to any GPCR, using an increasing choice among fluorescent reporter proteins, and offers unprecedented perspectives toward understanding GPCR biology and developing novel drugs of therapeutic interest.

Key words: Green fluorescent protein, *Delta* opioid receptor, Knockin mouse, G protein-coupled receptors, Pain, Anxiety, Depression, Trafficking

1. Introduction

Localizing a protein in tissue sections is often critical to understanding its function. This is perhaps best exemplified by proteins that are specifically expressed in a given cell type or organ (i.e., taste receptors expressed in taste buds), or by proteins that have a particular subcellular distribution (i.e., nuclear localization

Craig W. Stevens (ed.), *Methods for the Discovery and Characterization of G Protein-Coupled Receptors*, Neuromethods, vol. 60, DOI 10.1007/978-1-61779-179-6_6, © Springer Science+Business Media, LLC 2011

for transcription factors, synapses for vesicular transporters). G protein-coupled receptors (GPCRs) are no exception to this rule (1). Unfortunately, localizing GPCRs in tissues is not an easy task. Two main approaches are either binding of selective ligands to tissue sections or tissue labeling with antibodies. Both methods have proven unquestionably useful, but also have significant limitations such that alternative approaches need to be developed.

Regarding ligand binding, radioligands are commonly used in autoradiographic procedures for detecting endogenously expressed GPCRs. Although this approach provides a helpful view on the general receptor distribution, resolution is low and does not reach the cellular level. Most ligands cannot be cross-linked to the GPCR by aldehyde fixation; therefore, ligand autoradiography is difficult to combine with any other histological techniques (immunohistochemistry, in situ hybridization) to characterize labeled cells. Also the generation of labeled ligands with high specific activity, and retaining high affinity and selectivity for their targets, may represent a true challenge.

The second approach consists in generating anti-GPCR antibodies. Immunodetection of a protein in tissues is highly sensitive, provides a good spatial resolution, and allows the characterization of positive cells. However, for undetermined reasons about which one can only speculate, it has proven difficult to generate specific, high-affinity, antibodies to detect GPCRs in tissue sections. The apparent low antigenicity of GPCRs may result from their intrinsic transmembrane nature or their multiple conformations. Importantly, the classical preadsorption control using the immunogenic peptide indicates that the antibody indeed recognizes the peptide, but not that immunoreactivity in a given tissue arises from the targeted GPCR. This problem became particularly obvious when knockout mice became available. For a number of anti-GPCR antibodies, immunoreactivity turned out to remain intact in tissues from knockout mice even though the signal was lost after preadsorption with the immunogenic peptide; however, these data are rarely published (see for example (2)). Altogether it appears that many anti-GPCRs antibodies may in fact recognize a different molecule and that their labeling patterns misrepresent the true distribution of the receptor. In the absence of appropriate control, i.e., knockout mice in which translation of the epitope-containing exon is disrupted, it is therefore risky to rely solely on antibodies for anatomical studies of GPCR expression.

Finally, regardless of issues of sensitivity, resolution, or specificity, neither of the two above approaches allows real-time monitoring of endogenous GPCR trafficking during their life cycle, from synthesis to degradation. In this chapter, we present a novel approach combining mouse genetics and fluorescence imaging that not only circumvents all these problems but also offers new possibilities for studying GPCRs in their native environment. The method consists in generating a mutant mouse, in which a detectable genetically encoded molecule, the reporter, is fused to a GPCR using homologous recombination in embryonic stem (ES) cells. In the resulting knockin mice, the labeled receptor is functional and replaces the endogenous receptor. Because this receptor is expressed at native sites and endogenous levels, the knockin approach allows one to explore receptor biology and function under physiological conditions. This methodology avoids confounds associated with ectopic receptor overexpression approaches, which are used otherwise to study GPCRs *in vivo*. The best reporter candidates are fluorescent proteins, whose archetypal member is the green fluorescent protein (GFP) (3).We used the *delta* opioid receptor (DOR) as a model to develop this innovative approach. The DOR is a class A GPCR expressed predominantly in the nervous system, which regulates numerous physiological processes (4), including pain (5–10), shows anxiolytic and antidepressant activity (11–21), and modulates motor impulsivity (22).

Here we describe the construction and characterization of a knockin mouse line (thereafter referred as to DOR-eGFP mice), where enhanced GFP (eGFP) is fused to the C-terminus of the DOR. We highlight critical aspects of the methodology used for creating the DOR-eGFP mouse line, which are generally applicable for creating knockin reporter lines for other GPCRs. We present a number of important findings that have resulted from the analysis of DOR-eGFP mice in the opioid receptor field, and more generally in the context of GPCR biology and drug design. Finally, we discuss the wide array of possibilities offered by this unique tool throughout the article.

2. Methodology: Details and Pitfalls in the Creation of the Reporter DOR-eGFP Knockin Mouse Line

2.1. The Tagged Receptor

A first step toward the generation of knockin mice expressing a fluorescently labeled GPCR is to determine whether the fusion protein can be produced *in vitro* as a functional receptor. This is not as simple as it seems, since any tag may disrupt protein expression and function. For many GPCRs, numerous constructs have been generated and evaluated in transfected cells. Basically, three parameters can be modified: the nature of the tag (fluorescent protein, FLAG-tag, HA-tag, etc.), its position within the fusion protein (N-terminus, loops, C-terminus), and the composition of the linker region between the GPCR and the tag (amino acid identity and length). Several studies have indicated that C-terminal fusions allow correct receptor folding, insertion, and export to the cell surface, and minimally modify ligand binding and receptor signaling (23). We found that fusing the enhanced version of the GFP to the DOR C-terminus using a six amino-acid linker (GSIAT) leads to expression of a fully functional fluorescent receptor *in vitro* (Fig. 1a and (24)). In brief, we established two HEK293 cell lines that stably express either the native DOR or the DOR-eGFP fusion protein, and compared receptor properties in the two cell lines (24). Radioligand binding experiments showed that opioid ligands display similar affinities for native and DOR-eGFP receptors, regardless of their nature and structure

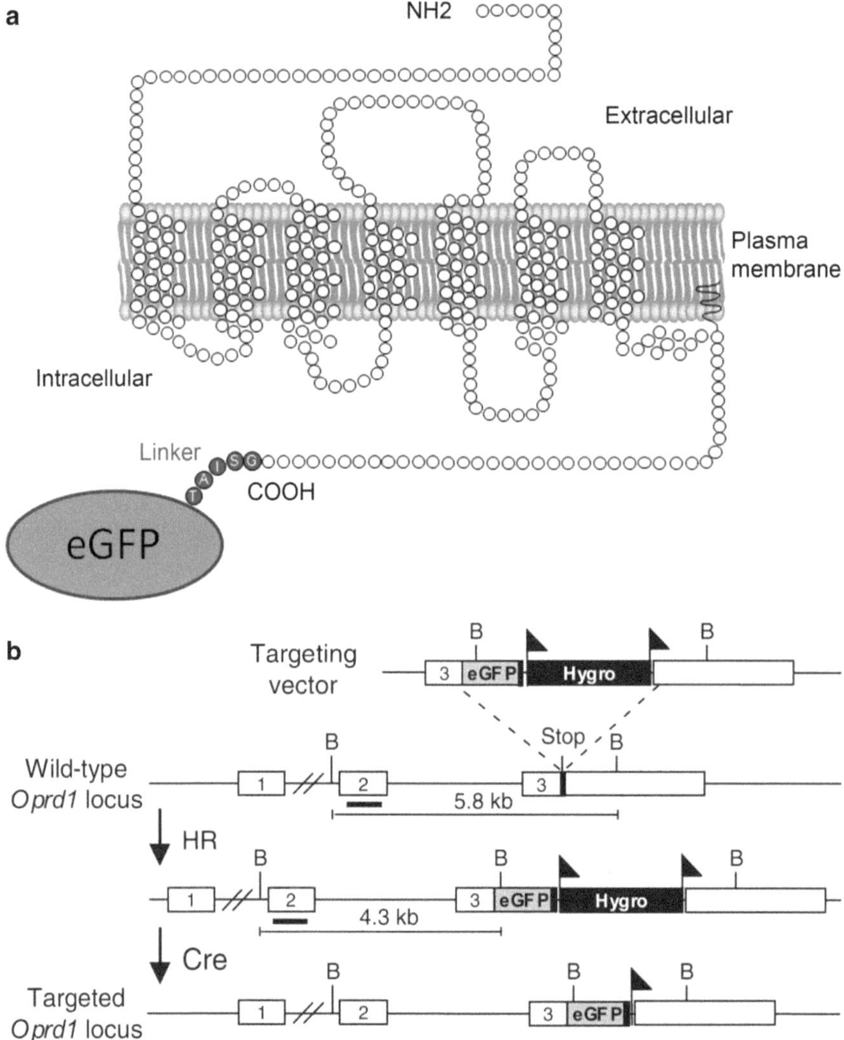

Fig. 1. Tagging *delta* opioid receptors (DORs) *in vivo* with eGFP by homologous recombination (HR) in mice. (**a**) Schematic structure of the DOR-eGFP receptor. The eGFP tag is fused to the C-terminus of the receptor by means of a five amino acid linker (GSIAT). (**b**) Strategy used to insert the eGFP cDNA in place of the stop codon within the *Oprd1* gene. *Oprd1* exons, eGFP cDNA, and the floxed hygromycin cassette are displayed as *empty box*, gray box, and *black triangles*, respectively. HR was followed by Cre recombinase treatment (Cre) in ES cells. Adapted from (24).

(endogenous opioid peptides, synthetic alkaloid ligands, agonists, antagonists), suggesting that DOR-eGFP folding in the membrane is intact (24). We also examined G protein activation following incubation of membrane preparations from either native or DOR-eGFP receptor with the agonist deltorphin II. Potency and maximal effect of deltorphin II were identical for the two receptors, indicating that receptor signaling is not impaired by a potential steric hindrance, at least at the level of Go/i protein coupling (24). Following the same line of reasoning, previous studies in transfected cells showed that DORs are internalized following activation, and that internalized receptors subsequently undergo lysosomal or proteasomal degradation (25–28). These events require the binding of multiple molecules to intracellular domains of the receptor, including kinases, *beta*-arrestin, and ubiquitin (29). Here, also, the eGFP tag may hamper receptor interactions with adaptor or regulatory proteins, and it was therefore important to test whether DOR-eGFP internalizes and desensitizes. In the stable DOR-eGFP HEK293 cell line, we found the fluorescent signal mainly located at the cell surface under basal conditions, and observed that application of DOR agonists (Met-enkephalin, deltorphin 2, or SNC80) resulted in DOR-eGFP internalization, as reported previously (26). Further, DOR-eGFP was downregulated following prolonged exposure to deltorphin 2, as was the native receptor in a control experiment and concordant with earlier data (30, 31). Altogether, these data suggested that DOR-eGFP is fully functional *in vitro* (24). Hence, despite its large size, the GSIAT-eGFP tag seemed to provide enough flexibility for interacting proteins to access the receptor. This construct was therefore selected to create the knockin mouse line.

Importantly, one can never exclude that the fused reporter modifies an aspect of receptor function that has not been tested, or is difficult to detect or even is unknown. In the case of DOR-eGFP, all the experiments that we have so far performed using the knockin mouse line confirm our pilot analysis in transfected cells, in that we have not detected alterations of DOR function at the cellular level or *in vivo* in any experiment comparing mutant DOR-eGFP mice and their wild-type controls (see below and detailed data in (24, 32)).

2.2. The Targeting Strategy

We are far from understanding all the processes that control gene transcription and RNA stability. As a consequence, we are unable to predict with certainty how insertion of an exogenous DNA sequence into a GPCR gene will impact on receptor expression.

Therefore, the outcome of a gene knockin experiment is generally difficult to predict. Minimizing modifications within the genome, however, should optimize chances of success. To create the DOR-eGFP mouse, we constructed a targeting vector where a sequence encoding the GSIAT-eGFP tag is inserted in frame to replace the STOP codon in exon 3 of the DOR gene (*oprd1*) (Fig. 1b). In this construct, all the *Oprd1* gene sequences regulating gene transcription, mRNA stability, and protein translation are unchanged, and will therefore be intact in the final recombined allele. The targeting construct also must contain a gene conferring resistance to an antibiotic, to select for recombined alleles in ES cells. Notably, this resistance gene should be located as close as possible to the modifying eGFP linker-tag sequence (less than 300 base pairs) so that selection leads to ES clones recombining the entire targeting vector. In the course of creating DOR-eGFP mice, we used a first construct, where this particular distance was higher (700 base pairs), and obtained ES clones with integration of the neomycin resistance gene but not the linker-eGFP fused gene. Finally, it is also critical to remove the resistance gene after selection of the positive ES cells, to reduce the risk of modifying gene transcription. To this aim, we used a neomycin resistance gene flanked by loxP sites. Once a correct positive ES cell was obtained, we removed the selection gene by transfecting ES cells with a Cre recombinase-expressing vector and further selecting an appropriate ES clone in the absence of neomycin. Overall, the final mutant DOR-eGFP allele contained the desired targeted GSIAT-eGFP insertion and one remaining lox*P* sequence (34 base pairs).

In spite of these precautions, there was a slight modification of *Oprd1* gene processing in DOR-eGFP mice. We found an approximately twofold increase in DOR-eGFP mRNA levels compared to wild type in brain extracts. One possible explanation is that the exogenous GSIAT-eGFP sequence increases mRNA stability. Receptor density was increased accordingly (320 vs. 160 femtomoles/mg brain membrane protein), as was the maximal level of G protein activation (Emax) for deltorphin 2 and SNC80, but not Met-enkephalin. Altogether these modifications are subtle. In these mice, receptor expression remains in the femtomolar range and is by no means comparable to overexpression levels achieved in transfected systems (picomolar range). In addition, the slightly higher receptor number does not seem to impact on receptor distribution throughout the nervous system (see section below). Further, all behavioral responses to agonists that we have tested are identical to controls (see section below), suggesting the existence of a receptor reserve in wild-type animals. Altogether, all the data that we have collected so far suggest that DOR levels in DOR-eGFP mice indeed reflect physiological conditions.

3. Receptor Expression and Function in the DOR-eGFP Knockin Line

Once the DOR-eGFP knockin mouse line was established, we first tested whether the DOR-eGFP fusion protein is functional in mutant animals. We performed a series of biochemical and pharmacological assays on membrane protein preparations from brains of both DOR-eGFP mice and their control littermates. Western blot analysis showed an 80-kDa eGFP-immunoreactive protein, corresponding to the predicted size of the fusion protein, that was detected in both heterozygous and homozygous mutants but not wild-type controls (24). Ligand-binding experiments indicated that regardless of their nature (partial or full agonists, antagonists) or structure (peptides or alkaloids) all the tested opioid ligands display identical affinity for DOR-eGFP and wild-type receptors (24). We also performed a G protein activation assay using prototypical *delta* opioid agonists. We found that SNC80, deltorphin 2, and Met-enkephalin potencies (EC50) were similar in knockin and wild-type mice (Fig. 2a and (24))

3.1. Biochemical and Pharmacological Assays Reveal Intact Receptor Expression and Signaling

Together, these results indicate that DOR-eGFP is expressed and functional in tissues from DOR-eGFP knockin mice.

3.2. DOR-eGFP Distribution in Brain and Trafficking in Live Neurons Are Consistent with Previous Binding and Cellular Data

Initially, we investigated DOR-eGFP distribution in the brains of mutant mice by examining brain slices under a fluorescence microscope. The eGFP fluorescence pattern was identical to the previously reported DOR distribution based on radioligand autoradiographic binding, the specificity of which was demonstrated using DOR knockout mice (Fig. 2b and (33)). For example, intense fluorescence was observed throughout the striatum and in the basolateral nucleus of the amygdala (24). The high level of correspondence between DOR-eGFP fluorescence and radioligand binding data indicates that the eGFP fusion protein is expressed in neurons that normally express the native receptor. Additionally, the presence of eGFP-labeled and native receptors at identical sites throughout the brain strongly suggests that the DOR-eGFP fusion protein is transported to the correct subcellular neuronal compartments, such as dendrites or axon terminals. Finally, in accordance with previous data using transfected cells (25, 27, 28, 34), DOR-eGFP was internalized upon exposure to an agonist either in striatal or hippocampal primary cultures from knockin mice or *in vivo* (24, 32). Together, therefore, these observations strongly support the notion that DOR-eGFP is expressed normally and traffics as expected in the mutant line.

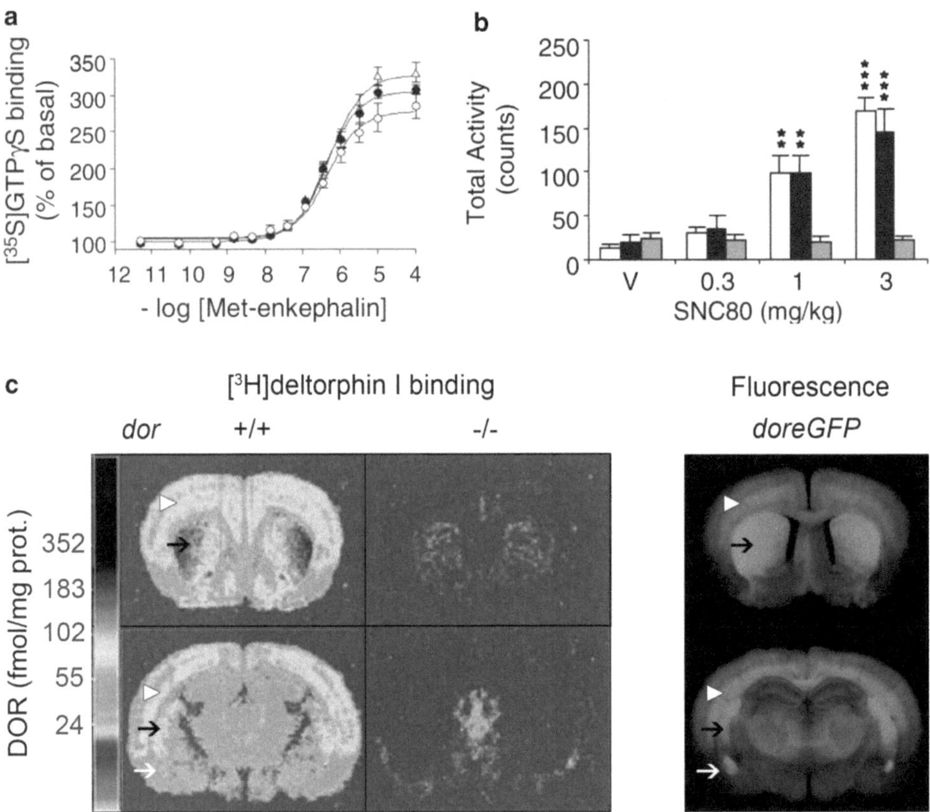

Fig. 2. DOR-eGFP distribution and function are similar to that of the native DOR. (**a**) The endogenous DOR agonist Met-enkephalin activates G proteins with similar half-maximal effective concentration in brain membranes from wild-type (*white circles*), heterozygous (*black circles*) and homozygous (*white triangles*) DOR-eGFP mice. (**b**) The DOR agonist SNC80 dose-dependently increases locomotor activity similarly in control and DOR-eGFP mice. The stimulating effect of SNC80 is identical in DOR-eGFP knockin mice (*black bars*) and wild-type littermates (*white bars*) and absent in DOR knockout mice (*gray bars*), demonstrating intact function of the DOR-eGFP receptor *in vivo*. Two and three *asterisks* correspond to *P* values for treatment effect <0.01 and 0.001, respectively. (**c**) The distribution pattern of DOR-eGFP in the brain is identical to that of the native DOR in wild-type mice (dor +/+) as revealed by binding of the DOR agonist [^{3}H] deltorphin I (computer-enhanced autoradiograms) (33). Absence of [^{3}H]deltorphin I in brain slices from DOR knockout mice (dor –/–) demonstrates binding specificity. DORs are expressed at high levels in the cortex (*white arrowheads*), striatum (*black arrows*), and amygdala (*white arrows*). Adapted from (24) and (33).

3.3. DOR-eGFP Knockin Mice Show Behavioral Responses to Agonist That Are Identical to Wild-Type Mice

Next, we tested whether DOR-eGFP mice respond normally to DOR agonists. We used an actimetry test, and examined the well-documented stimulant effect of SNC80 (35–40). We found a dose-dependent increase of locomotor activity that was comparable in DOR-eGFP mice and their wild-type littermates, and absent in DOR knockout mice (Fig. 2c and (24)). Furthermore, we tested the analgesic properties of DOR agonists whose efficacy has been demonstrated in animal models of chronic pain (5–10, 41–56). In the Complete Freund's Adjuvant model of inflammatory pain, we found that analgesic effects of both SNC80 (the prototypical nonpeptidic DOR agonist (57)) and AR-M100390 (a recently developed SNC80 analog, see (58)) are identical in DOR-eGFP mutant mice and wild-type controls (32). Hence, despite a slightly higher receptor number in DOR-eGFP mice (see above), DOR-mediated responses *in vivo* are unchanged in the knockin mouse line.

In summary, we used biochemical, pharmacological, histological, and behavioral approaches to assess the functionality of DOR-eGFP receptors. Together our results indicate that knockin mice produce a fully functional DOR-eGFP fusion protein. The eGFP tag does not seem to modify receptor distribution and biology, at least for all the classical responses that we have tested, and all assays that we have used. This mouse line therefore represents a unique tool to gain insights into several aspects of DOR function, via imaging approaches and with a resolution power that was previously inaccessible.

4. Refining *Delta* Opioid Receptor Distribution in the Nervous System

4.1. Characterization of DOR-eGFP-Expressing Neurons in the Brain

The tagged-GPCR knockin mouse may be used to study receptor distribution in tissues and characterize cells in which the receptor is expressed. First, the receptor may be directly visualized by confocal microscopy in the case that the fluorescent signal is strong enough, or labeled with an anti-eGFP antibody for signal amplification. Second, detection of the tagged receptor and other proteins can be combined in immunohistochemical colocalization experiments.

One set of experiments using DOR-eGFP mice aimed at confirming previously published results on the neurochemical nature of DOR-expressing neurons in the brain. As an example, in situ hybridization and radioligand binding studies have shown that the DOR is expressed at high levels in the striatum (33, 59), a forebrain region involved in motor control and motivational behaviors. The distribution of *Oprd1* and choline acetyl-transferase mRNAs was investigated in adjacent striatal sections by in situ hybridization, and a large overlap was found between DOR-expressing neurons and cholinergic neurons (60, 61). We immunostained striatal sections from DOR-eGFP knockin mice with the same cholinergic marker, and found that 76% of cholinergic neurons indeed express the DOR, in accordance with the previous study (Fig. 3a and (24)). This experiment confirmed the likely importance of DORs in the modulation of cholinergic interneurons of the striatum.

Further colocalization experiments in many brain areas are underway to provide a complete description of DOR anatomy in neuronal circuits where these receptors operate.

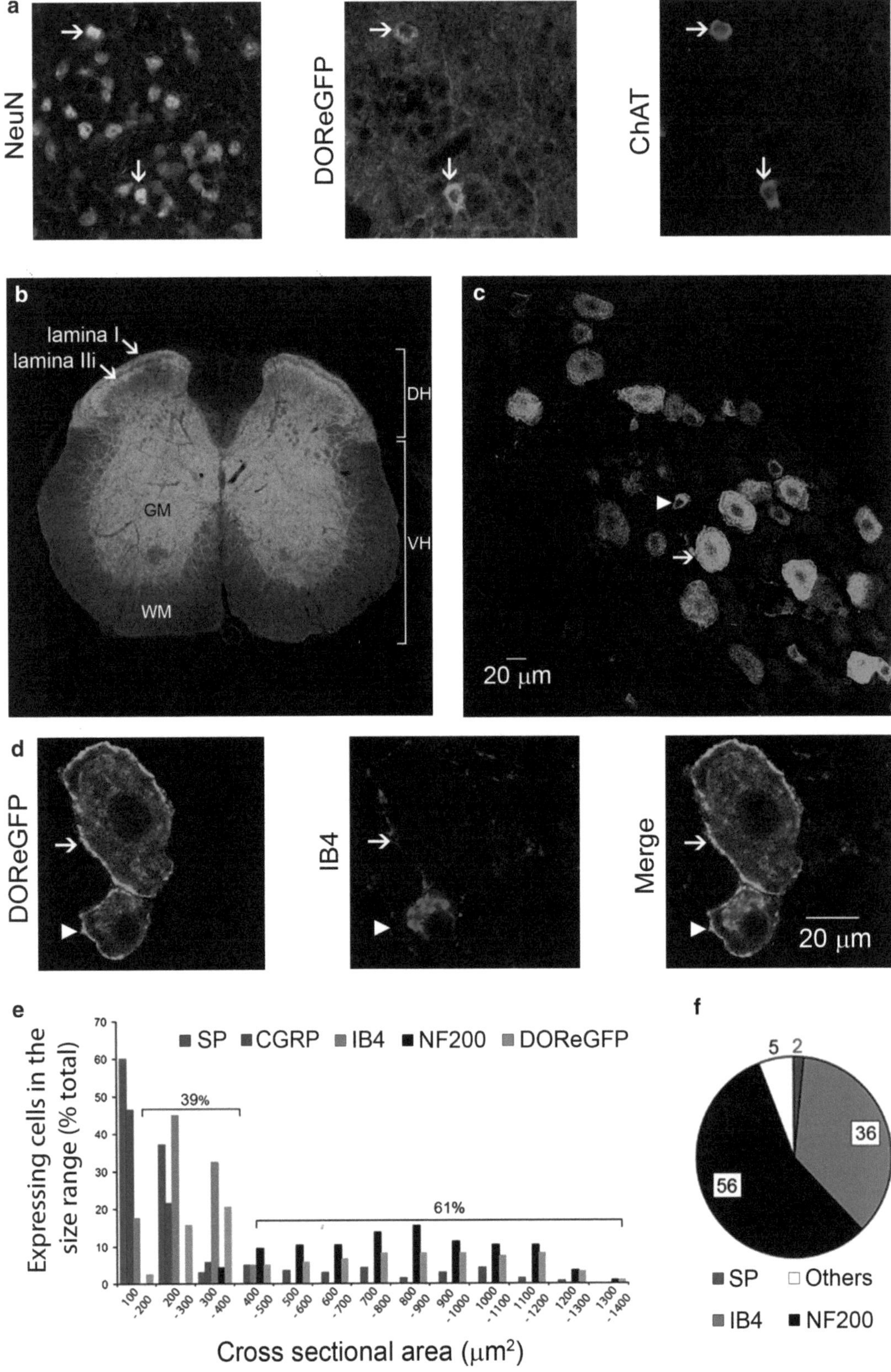
a
NeuN
DOReGFP
ChAT
b
lamina I
lamina IIi
DH
VH
GM
WM
c
20 μm
d
DOReGFP
IB4
Merge
20 μm
e
SP
CGRP
IB4
NF200
DOReGFP
Expressing cells in the size range (% total)
39%
61%
Cross sectional area (μm²)
f
5
2
36
56
SP
Others
IB4
NF200

4.2. DOR-eGFP Distribution in the Dorsal Root Ganglia and the Spinal Cord

One interesting issue is the exact localization of DORs within the distinct types of sensory neurons in the dorsal root ganglia (DRG), as well as in the distinct laminae of the spinal cord where these neurons project. These areas are main neuroanatomical substrates for pain processing, and DOR expression at these sites remained unclear.

On one hand, *in situ* hybridization studies in rodents have shown that only about 3% of substance P-positive DRG neurons express the DOR at a high level (62). DOR mRNA was otherwise abundant in large DRG cells with myelinated axons as well as restricted populations of small-diameter DRG neurons (63). In the rodent spinal cord, DOR mRNA was found expressed by a large number of neurons distributed throughout the ventral and dorsal horns (63, 64). Accordingly, (^{125}I)-deltorphin-labeled *delta* opioid binding sites were detected throughout the gray matter of the spinal cord (laminae I–X) (63). On the other hand, numerous studies using an antibody raised against amino acids 3–17 of the N-terminus of the DOR (65) reported a different staining pattern in DRG and spinal cord. Some studies indicated that, in the DRG, most immunoreactive neurons are of small diameter and express the neuropeptide substance P (66–68). Further, immunoreactivity in the spinal cord was concentrated in the superficial dorsal horn (laminae I and II), colocalized with substance P and was significantly reduced after deafferentation, indicating that immunoreactivity arises from staining of central terminals of peptidergic (substance P-expressing) DRG neurons (65, 66, 69–71).

Hence the localization of DORs at this level of pain pathways was uncertain, limiting our understanding of pain-reducing effects of DOR agonists (9) or cooperation between *mu* opioid receptors and DORs in pain control (72). We used the DOR-eGFP mice to study DOR distribution in DRG and spinal cord (see Fig. 1 and Supplementary Table 1 in (73)). We found that 61% DOR-eGFP-expressing cells are large diameter DRG neurons

Fig. 3. DOR-eGFP distribution at main sites of DOR expression in the nervous system. (**a**) Cholinergic (ChAT) neurons (NeuN) in the striatum express the *delta* opioid receptor (DOR-eGFP). (**b**) In the spinal cord, DOR-eGFP is present throughout the gray matter (GM) and enriched in lamina I and in the ventral portion of inner lamina II (IIi) of the dorsal horn (DH). *WM* white matter; *VH* ventral horn. (**c**) In the dorsal root ganglia, DOR-eGFP is expressed in subsets of small (*arrowhead*) and medium-to-large (*arrow*) diameter neurons. (**d**) Labeling of dorsal root ganglion (DRG) neurons with the marker isolectin B4 (IB4) reveals that small-diameter neurons expressing DOR-eGFP belong mainly to the nonpeptidergic subset of DRG neurons (IB4 positive). *Arrow*, large-diameter DRG neuron expressing DOR-eGFP. (**e**) Size distribution of DOR-eGFP-expressing dorsal root ganglion neurons compared to those labeled for substance P, IB4 or neurofilament 200 (NF200). DOR-eGFP is expressed both in small diameter and medium-to-large diameter cells. (**f**) Many DOR-eGFP+ cells coexpress the marker of myelinated neurons NF200 (56%). The next most abundant population of DOR-eGFP+ cells are small-diameter, unmyelinated cells that express IB4+ (36%). Few DOR-eGFP+ cells express substance P (SP, 2%), corresponding to the peptidergic subpopulation. Adapted from (73).

and have myelinated axons, and that the DOR-eGFP signal is present throughout the gray matter of the spinal cord (Fig. 3b, c, e and (73)). In addition about 2% DOR-eGFP DRG neurons coexpress substance P (Fig. 3f), and DOR-eGFP is present in less than 1% of all substance P-positive neurons, indicating that colocalization between the DOR and substance P is limited (73).

In the course of these studies, we observed that immunostaining obtained with several anti-DOR antibodies was intact in tissues from two independently generated strains of *Oprd1* knockout mice (73), suggesting that immunohistochemistry was not the best approach to detect DORs in tissues.

We subsequently took advantage of the possibility to label DOR-eGFP and specific neuronal populations simultaneously, and characterized DOR expression in the periphery (skin, viscera), at the level of the DRG, and in the spinal cord. We found that DOR-eGFP is preferentially expressed in putative mechanoresponsive DRG neurons innervating the skin, but not viscera (73). Thus, DOR-eGFP is found in the nonpeptidergic population of small-diameter DRG neurons (IB4-positive) that detect noxious mechanical stimuli (74, 75). Additionally, DOR-eGFP is expressed in myelinated large-diameter DRG neurons, known as A fibers, some of which respond to touch and play a role in the development of chronic mechanical hypersensitivity after tissue or nerve injury (inflammatory or neuropathic pain, respectively) (76). Finally, DOR-eGFP is expressed in the most ventral part of lamina 2 of the spinal cord, a region containing PKCγ-expressing interneurons critical for touch-evoked neuropathic pain (77–79). By contrast, DOR-eGFP is rarely expressed in peptidergic small-diameter DRG neurons that express the heat sensor transient receptor potential vanilloid 1 (TRPV1) (80) and are essential for normal responsiveness to noxious heat stimuli (74, 81). Thus, approximately 5% of all DOR-eGFP-positive DRG neurons express TRPV1, and DOR-eGFP is expressed in about 1% of TRPV1-positive neurons (Fig. 3d–f). These observations led us to design a series of functional experiments addressing the role of DORs in mechanosensation. We found that DOR agonists, indeed, more efficiently reduce mechanical pain than heat pain. The data also indicated that *delta* and *mu* opioid receptors play very distinct roles in pain processing at DRG and spinal cord levels, based on both histological and behavioral evidence (73). Together, these studies using the DOR-eGFP mice have provided an anatomical substrate and mechanism of action for the analgesic properties of DOR agonists in situations of chronic, mechanically evoked pain.

5. Subcellular Localization and Trafficking of *Delta* Opioid Receptors *In Vivo*

5.1. DOR-eGFP Is Available at the Cell Surface

GPCRs are membrane proteins present at the cell surface where they respond to extracellular stimuli. GPCRs traffic in and out the cell surface, and both receptor endocytosis and externalization regulate receptor activity (82–84). Several lines of evidence have suggested that the cell surface pool of DORs increases after various stimuli. Thus, chronic morphine treatment (85–88), deletion of the *mu* opioid receptor (89), peripheral injury (86, 88), activation of purinergic receptors (67), administration of proalgesic compounds such as capsaicin (67, 88) and bradykinin (90), and forced swim stress (91) have been suggested to promote the translocation of DORs to the cell surface (for a review, see (9)). Additionally, a few studies have shown that some of these pretreatments induce a gain of function of DOR agonists, as measured by electrophysiological or behavioral methods. This was interpreted as a result of receptor redistribution and increase of cell surface receptors (85, 89, 92–96). Together, these observations have led to the proposal that, in some cases, delivery of a prestimulus is necessary to increase the efficacy of DOR agonists.

However, this hypothesis is difficult to reconcile with a large number of other studies using *delta* opioid agonists. For example, these compounds can exert their locomotor (24, 35–40), antidepressant (12–20, 97), anxiolytic (12, 14, 21), analgesic (8, 73, 98–103), or proconvulsive (17, 104, 105) effects independently of any prestimulus that would induce translocation of DORs to the cell surface. These pharmacological data rather indicate that DORs are available and can be activated under basal conditions.

Using DOR-eGFP knockin mice, we examined fluorescence distribution in brain neurons, at the subcellular level. We performed a series of live-imaging experiments using primary striatal and hippocampal neurons cultured from mutant mice. We found that DOR-eGFP fluorescence was present at the cell surface of neurons under basal conditions, and that application of the agonists Met-enkephalin, deltorphin 2, or SNC80 to the media triggered a rapid (minutes) and massive (up to 100%) internalization of the fluorescent signal (Fig. 4a and (24)). We also investigated the subcellular localization of DOR-eGFP in tissues of knockin mice. In striatal, hippocampal, cortical, and DRG neurons, as well as in neurons in the olfactory bulb, DOR-eGFP fluorescence was clearly predominant at the cell surface (Fig. 4b–d and (24)). Further, as observed in primary culture, systemic administration of SNC80 induced a dramatic reduction of cell-surface fluorescence concomitant with the appearance of brightly fluorescent intracellular vesicles in DOR-eGFP-expressing neurons throughout the nervous system (Fig. 4b–d). This response was observed

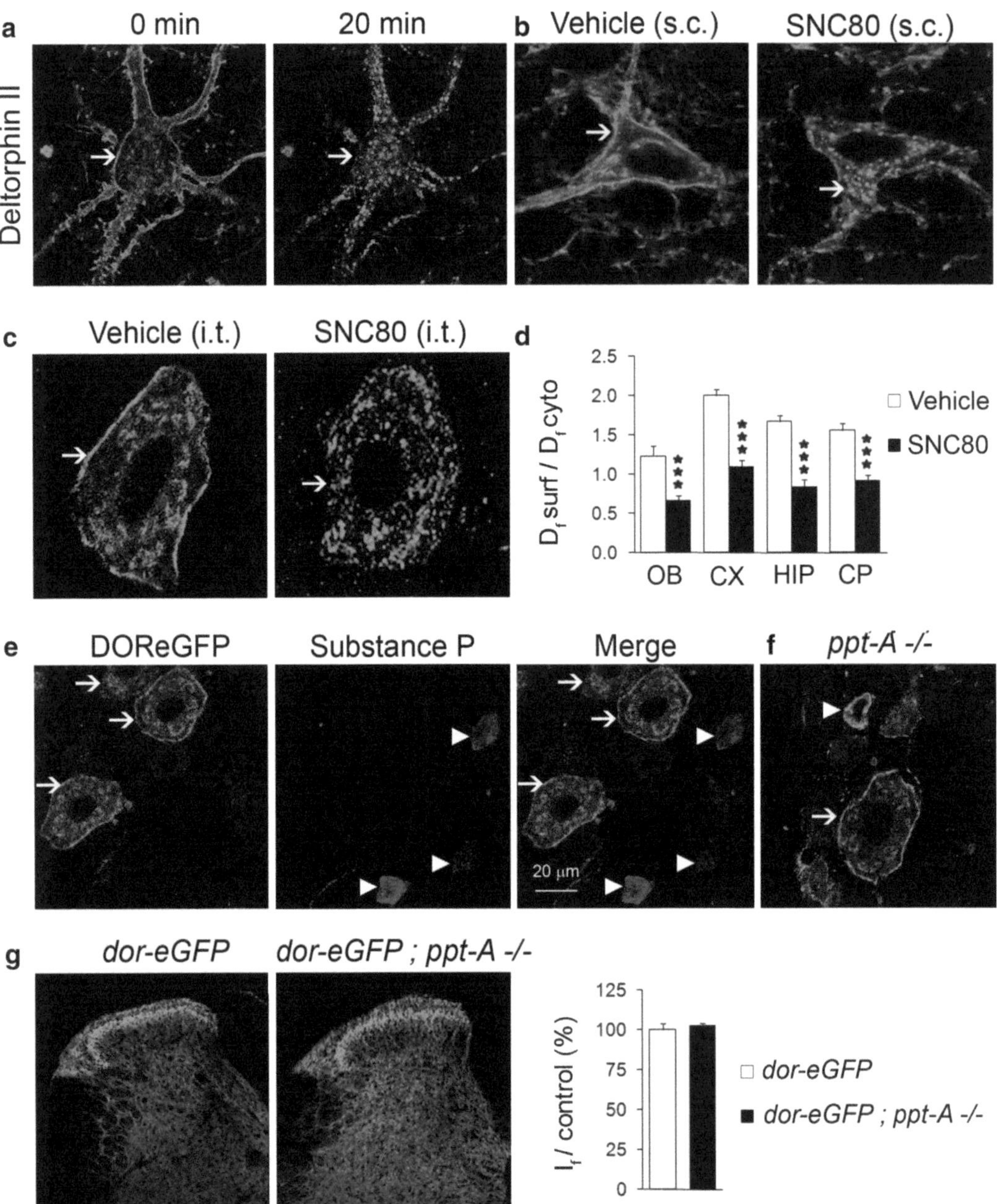

Fig. 4. DOR-eGFP mice report for DOR trafficking *in vivo*. (**a**) Real-time confocal imaging shows internalization of DOR-eGFP in primary striatal neurons (*arrow*) after exposure to the DOR agonist deltorphin 2. (**b**) Subcutaneous injection of SNC80 to DOR-eGFP mice triggers receptor internalization throughout the central nervous system. Shown here are images of DOR-eGFP-expressing neurons in the hippocampus (HIP) 2 h after vehicle (*left panel*) or SNC80 (*right panel*) administration. (**c**) Intrathecal injection of SNC80 induces activation and internalization of DOR-eGFP in dorsal root ganglion neurons. Note receptor localization at the cell surface following vehicle injection (*arrow*), as observed under basal condition. (**d**) Quantification of DOR-eGFP internalization in neurons of the olfactory bulb (OB), cerebral cortex (CX), HIP, and caudate putamen (CP) 2 h following subcutaneous injection of vehicle or SNC80. Fluorescence density at the cell surface (Df surf) vs. cytoplasmic fluorescence in the cytoplasm (Df cyto) is lower after SNC80 (*black bars*) compared to saline (*white bars*) injection. Three *asterisks* correspond to *P* value for treatment effect <0.001. (**e**) Colabeling of dorsal root ganglion (DRG) sections with specific antibodies demonstrates that substance P (*arrowhead*) and DOR-eGFP (*arrow*)

within a time window where the agonist produces hyperlocomotion (24) or analgesia (32).

These findings suggest that a functional receptor pool is available at the cell surface and responsive to agonists under basal conditions, without prior need for externalization. This does not exclude the possibility that intracellular pools of receptors are recruited to the cell surface, while surface receptors internalize. The putative outward receptor movement may have been preferentially detected in some of the previous studies, but was not detected under our experimental conditions. Furthermore, our observation of easily traceable receptor internalization *in vivo* upon agonist stimulation has an interesting additional implication. DOR-eGFP knockin mice may be used to report agonist bioavailability and diffusion in tissues following drug administration, an approach which may be extremely useful for understanding drug effects *in vivo* or when novel agonists are being developed for clinical use.

5.2. DOR Transport in Dorsal Root Ganglia Neurons

Regarding receptor localization within neuronal compartments, the case of DRG neurons is particularly intriguing. In these cells, DOR immunoreactivity was reported to be concentrated in neuropeptide (substance P and calcitonin gene-related peptide)-containing large dense core vesicles (LDCVs) of peptidergic small-diameter neurons (66–68). Thus, instead of, or in addition to, being continuously trafficked to the cell surface via a constitutive exocytosis pathway, a large part of surface DORs in DRG neurons may arise from the regulated secretory pathway. Hence, surface recruitment of this particular receptor pool would require a stimulus that, in turn, triggers the fusion of DOR-containing LDCVs with the plasma membrane (67). This proposal provided a mechanism for the prestimulus externalization hypothesis. A further proposal from these studies was that DORs colocalize and possibly interact with substance P within LCDVs of DRG neurons, a hypothesis based on the observation that DOR immunoreactivity is lost in the spinal cord of knockout mice lacking substance P (68). Hence receptors with analgesic activity (DORs) and neuropeptides with proalgesic properties (substance P) would be cotransported via the secretory pathway to axon terminals, and regulate pain in a coordinated manner at the level of the spinal cord (106).

Fig. 4. (continued) are expressed by distinct neuronal populations. Note the size difference between unmyelinated substance P-expressing small DRG neurons, and myelinated, DOR-eGFP expressing, large-diameter DRG neurons. (**f**) DOR-eGFP is intact at the cell surface of dorsal root ganglion neurons in mice lacking substance P (*ppt-A* knockout mice), indicating that substance P is not necessary for receptor transport to the plasma membrane. (**g**) DOR-eGFP is present at axon terminals of dorsal root ganglion neurons in the spinal cord of *ppt-A* knockout mice, demonstrating that DOR-eGFP transport is independent from substance P. Quantification of DOR-eGFP fluorescence intensity in the superficial DH following anti-eGFP immunostaining of spinal cord sections is shown. Adapted from (24) and (73).

We addressed this interesting hypothesis using DOR-eGFP mice. We compared the DOR-eGFP fluorescent signal in DRG neurons from DOR-eGFP mice, as well as DOR-eGFP mice crossed with substance P knockout mice (*ppt-A* knockout mice). We found that receptor localization was identical in the two mouse lines (Fig. 4e, f) (73). Thus, in both the presence and absence of substance P, the DOR-eGFP signal is predominantly localized at the cell surface of DRG neurons, and present in the spinal cord. We further examined DOR radioligand binding sites in the spinal cord using autoradiography, and found similar binding patterns in *ppt-A* knockout mice and wild-type littermates (Fig. 4g and (73)), indicating that DORs are transported from the DRG to the spinal cord independently from substance P. Together with our data on antibody specificity and DOR-eGFP distribution in DRG (see Sect. 6.4.2), these results call into question the previously proposed direct interaction between the DOR and substance P, and suggest that mechanisms whereby these two important players modulate pain information at the level of the spinal cord are likely independent.

6. Implications of *Delta* Opioid Receptor Internalization *In Vivo*

6.1. DOR-eGFP Signaling and Internalization Co-occur In Vivo

Stimulation of a GPCR by an extracellular stimulus, either physiological or synthetic, triggers intracellular receptor signaling via heterotrimeric G proteins. This process is highly regulated and receptor activation is typically accompanied by desensitization of receptor signaling, a complex, feedback-regulated process whereby receptor responsiveness decreases upon continued agonist stimulation (82–84, 107). Among the many events contributing to the desensitization process, receptor trafficking is considered to be a key process in the regulation of receptor signaling. At present, most receptor trafficking studies are performed using receptor overexpression in heterologous systems and their physiological relevance is limited. These approaches may not reflect *in vivo* conditions in terms of receptor density and sites of receptor expression, including cell type and correct subcellular compartments, particularly for neurons. Additionally, most studies are conducted in cellular models, and provide no understanding of how receptor trafficking influences integrated responses in the living organism. DOR-eGFP mice represent a unique tool circumventing some of these limitations, since the receptor is expressed in native neurons, at physiological levels, and behavioral measurements may be combined with cellular imaging. Until now, our attempts to conduct behavioral testing and imaging simultaneously have not succeeded. Fluorescent imaging techniques in freely moving animals are developing rapidly (108–110) and may provide in the future sufficient sensitivity and resolution to visualize receptor trafficking in living DOR-eGFP mice. At present, our experimental design is based on behavioral observation first, followed by animal sacrifice and ex vivo imaging in tissue sections.

Our first study combining *in vivo* behavioral testing and ex vivo imaging showed that increasing doses of the *delta* opioid agonist SNC80 produce a dose-dependent locomotor activation that correlates with a dose-dependent decrease of the cell-surface fluorescent signal (24). Our second study (32) showed that, in the CFA model of inflammatory pain, SNC80 concomitantly produces pain-reducing effects in DOR-eGFP mice and internalizes the receptor. A parallel biochemical analysis further demonstrated that internalized receptors are unable to activate G proteins in brain membrane preparations. Collectively, the data indicate that receptor activation and internalization are coupled processes *in vivo*. Agonist binding to DOR-eGFP triggers receptor signaling that translates into locomotor and analgesic responses. At the same time, receptor internalization and uncoupling from G proteins occur, and the consequences of this second set of events are likely critical to desensitize the receptor response.

6.2. DOR-eGFP Internalization Is Responsible for Acute Behavioral Desensitization

To specifically address the role of DOR-eGFP in behavioral desensitization, we administered a second dose of SNC80 in both the locomotor (24) and analgesia (32) studies. The second injection was performed at a time when receptors are fully internalized and the behavioral effect of the first injection has terminated (locomotion: 2 h after first injection, analgesia: 4 h after first injection). Data from the two experimental set-ups showed a complete loss of SNC80 effects, indicating that behavioral desensitization had developed.

To further examine whether receptor internalization indeed is a key event in the *in vivo* desensitization process, we compared the analgesic efficacy of SNC80 with that of AR-M100390, a DOR agonist known to poorly internalize DORs (111). We found that the two agonists bound to the DOR-eGFP receptor and triggered G protein activation with similar potency and efficacy, and that a first treatment with the two agonists equally reduced CFA-induced inflammatory pain. The two agonists, therefore, show highly similar pharmacology in DOR-eGFP mice, both *in vitro* and *in vivo*. We observed, however, that AR-M100390 does not internalize DOR-eGFP receptors *in vivo*, even at higher doses, confirming the low internalizing efficacy of this agonist in the DOR-eGFP reporter line.

We then measured the pain-reducing effects of a second agonist injection, comparing the high (SNC80)- and the low (AR-M100390)-internalizing agonists. Animals initially treated with SNC80 showed no analgesic response to a second dose of either agonists. By contrast, treatment with AR-M100390 showed an intact analgesic response to a subsequent injection of either agonists (Fig. 5a). We examined receptor internalization

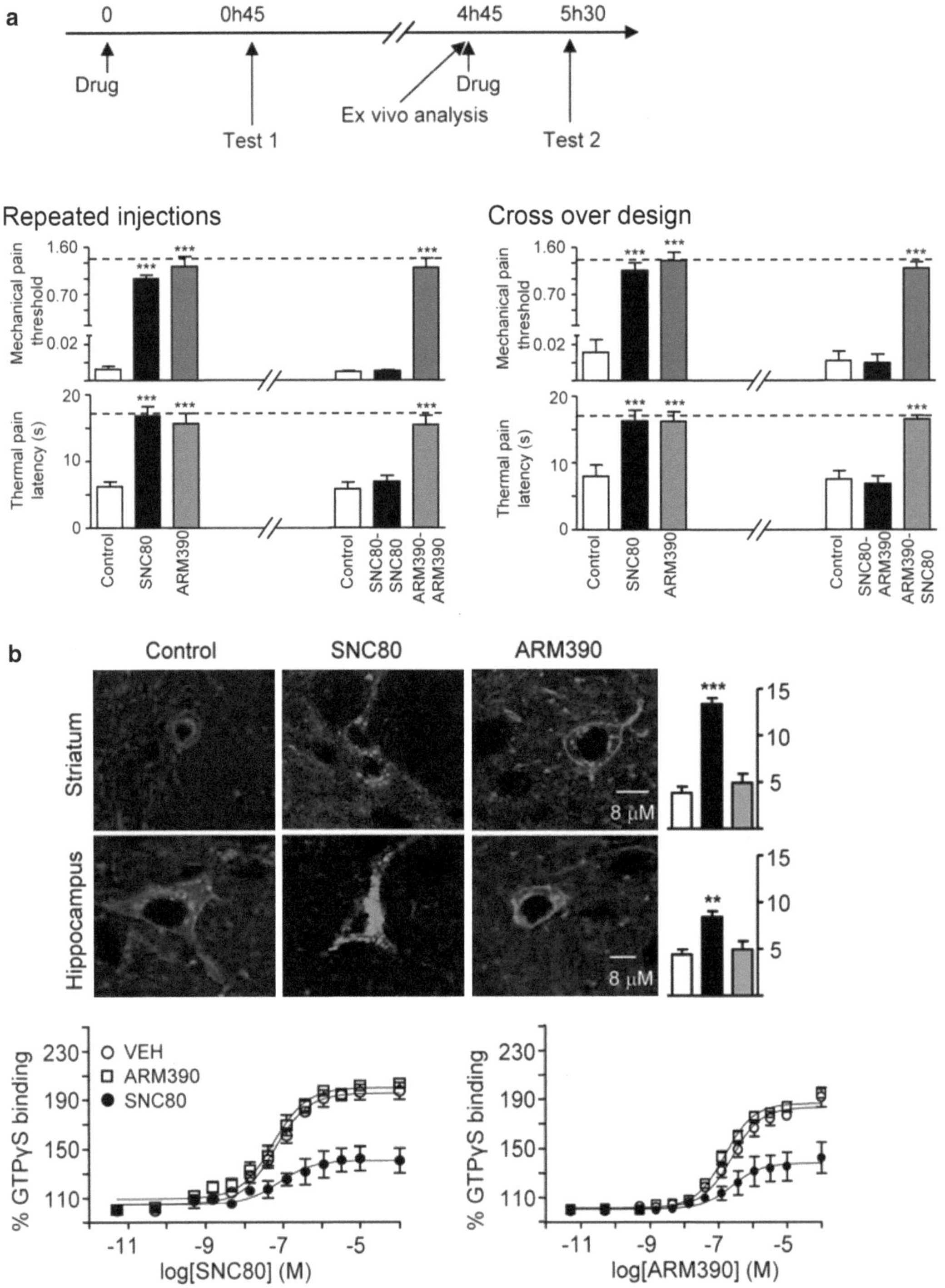

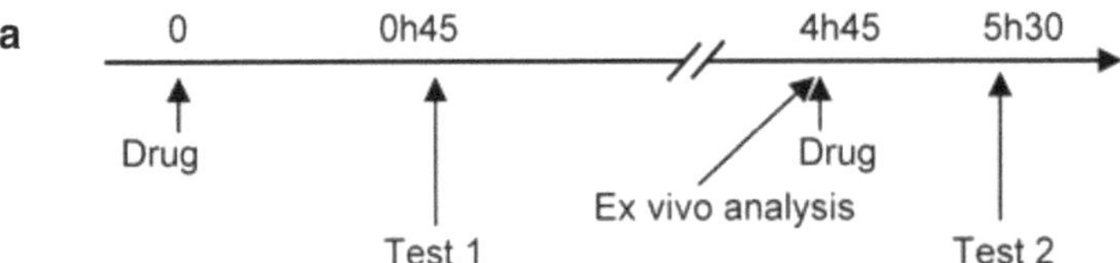

Fig. 5. The high-(SNC80) but not the low-(AR-M100390)-internalizing agonist triggers behavioral desensitization. (**a**) *Top*, time line of the experiments. *Low panels*: mechanical (CFA paw) and thermal (CFA tail) responses in animals treated with vehicle (control, *white bars*), SNC80 (*black bars*), or AR-M100390 (*gray bars*). *Left bars*: test 1, first injection. *Right bars*, test 2: animals rechallenged 4 h later with the same drug (*left*) or the other drug (*right*). *Dashed lines* represent baseline mechanical or thermal responses pre-CFA. Mice treated with SNC80, but not AR-M100390, do not respond to a

and G protein uncoupling at the time when animals received the second injection. SNC80 triggered complete receptor internalization and uncoupling in brain and spinal cord, whereas AR-M100390 modified neither receptor localization nor coupling (Fig. 5b). Overall, the data establish that receptor localization fully controls receptor function *in vivo*. Importantly, we observed similar SNC80- but not AR-M100390-induced behavioral desensitization using wild-type mice (C57Bl/6 mice). This observation demonstrates that the importance of internalization in drug efficacy and receptor function is not specific to the DOR-eGFP fusion receptor, and can be generalized to native DORs *in vivo*.

Finally, receptor internalization and loss of function were temporary, since full DOR-eGFP receptor responses seemed to be restored 24 h after SNC80 administration. The question arises of whether repeated agonist administration at 24-h intervals results in tolerance, and should this be the case, which mechanisms either distinct from or associated with receptor internalization may be involved. Tolerance to some behavioral effects of SNC80 was reported earlier (112), and our current studies address DOR-eGFP subcellular localization and function in several chronic treatment paradigms. Future studies will determine whether the use of DOR-eGFP mice as a screening tool to identify novel noninternalizing agonists is a valuable approach to develop drugs with increased efficacy. This has important implications for drug discovery, particularly in the field of neurological and psychiatric (e.g., chronic pain and depression) disorders where long-term treatments are required and DORs represent a valid therapeutic target.

7. Future Directions

The DOR-eGFP mouse line has proven to be an extraordinary tool to gain insights into DOR biology, and many more data are to come. Anatomical studies of DOR-eGFP mice have confirmed previous knowledge, and have allowed revisiting controversial issues, particularly concerning the distribution of the receptor in pain pathways. In the future, extensive mapping of DORs in

Fig. 5. (continued) second injection of any drug. (**b**) Mice were treated as in (**a**), but instead of the second drug administration, tissue was harvested for ex vivo analysis. *Top*, DOR-eGFP internalization. Brain regions were analyzed by confocal microscopy and representative images are shown. Mean intracellular DOR-eGFP fluorescence was quantified (*white bars*, control group; *black bars*, SNC80 group; *gray bars*, ARM390 group). *Bottom*, DOR-eGFP coupling to G proteins in brain membranes. [^{35}S]GTPγS concentration–response curves to SNC80 and ARM390 are shown. Mice treated with SNC80, but not AR-M100390, show receptor internalization and desensitization *ex vivo*. Adapted from (32).

the central nervous system combined with colocalization with neuronal markers will provide a strong neuroanatomical basis toward understanding receptor function.

Further colocalization and energy transfer approaches with other opioid receptors, and other GPCRs or critical signaling proteins, will provide a basis to study postulated protein–protein interactions that regulate neurotransmission. For example, we have addressed the issue of *mu*-DOR colocalization in DRGs. To this aim we have used combined immunohistochemistry with an anti-GFP antibody and an anti-*mu* opioid receptor antibody with demonstrated specificity (Evans laboratory, UCLA, USA, see supplementary Fig. 8 in (73)) in tissues from DOR-eGFP mice. Our data show limited (<5%) but detectable colocalization of the two receptors, indicating that potential opioid receptor interactions may occur at the cellular level, but in a highly restricted neuronal population only. These data will help designing future experiments addressing the function of DRGs neurons in which the two receptors coexist. More broadly, *mu-delta* receptor colocalization studies will help clarifying the long-standing and highly debated issue of functional *mu-delta* interactions *in vivo* (discussed for example in (5, 113)). Beyond colocalization, immunoprecipitation experiments using anti-GFP antibodies will be used to characterize interacting protein partners, using candidate or proteomic approaches.

In terms of receptor anatomy and physiology, an interesting approach will be the use of DOR-eGFP mice as reporters for endogenous DOR activation. At present, DOR-eGFP imaging has been performed solely in naïve mice, or mice receiving drug injections. Systemic administration of the agonist SNC80 produces massive receptor internalization throughout the nervous system, which reflects the dramatic and widespread action of drugs used in pharmacological treatments. The receptor, however, is normally activated endogenously under certain conditions, and this activation process is at the basis of receptor physiology. Because endogenous opioid peptide release is difficult to detect and map throughout the nervous system, DOR-eGFP mice may offer an interesting opportunity to report endogenous DOR activation under naturalistic conditions that recruit the opioid system (e.g., reward, stress, and pain). Along these lines, labeling other opioid receptors, which are also targets of opioid peptides, would be extremely useful, since DOR-eGFP internalization will only detect part of the opioid system activation.

More generally, the same knockin approach is applicable to other GPCRs. Technical feasibility to produce these lines does not represent the major limitation. Potential problems rather lie in the imaging aspect, since the physiological expression level and natural distribution of the receptor may not allow clear and unambiguous detection of the fluorescent signal at cellular or subcellular levels. Fluorescent protein engineering and discovery have progressed tremendously, and novel genetically encoded reporters with improved fluorescent properties are now available (114, 115). Fluorescent proteins of many colors may therefore be coupled to GPCRs, and this appealing possibility opens the way to produce distinct GPCR reporter mouse lines using diverse, and possibly complementary, fluorophores. Knockin mice could then be crossed to study GPCR interactions and dimerization *in vivo*, a phenomenon with major biological and therapeutic implications, whose physiological relevance is difficult to establish (116).

8. Conclusions

Receptor imaging in living animals (117) represents a main goal in neuroscience. Positron emission tomography and magnetic resonance imaging offer unique access to receptor anatomy and occupancy in the brain, and are particularly suited for human studies. The spatial and temporal resolution of these approaches, however, remains limited. Because receptor subcellular localization and trafficking in neurons are critical to understanding receptor function, there is tremendous interest in developing noninvasive imaging approaches in living animals, which achieve subcellular resolution and allow direct visualization of receptor movements in neurons. In this context, it is expected that strategies similar to the DOR-eGFP *in vivo* labeling approach described here, and have that been tested in the past for rhodopsin (118), will be developed and evolve toward noninvasive applications for the detection of GPCRs. The combination of mouse genetic engineering with high-resolution optical imaging should definitely clarify how single receptor molecules operate in highly organized cellular networks, and orchestrate physiological or pathological processes *in vivo.*

Acknowledgments

We would like to thank Allan Basbaum for initiating the project with us, and for his support and mentoring (GS) in revisiting some aspects of DOR anatomy and function in pain processing. We would like to acknowledge the essential implication of Amynah Pradhan in the behavioral desensitization aspect of these studies, and thank her for her contribution. We thank Petra Tryoen-Toth for her contribution in the initial characterization of mutant mice. We thank Claire Gavériaux-Ruff and Shannon Shields for critical reading and comments. We thank JL Vonesch and the IGBMC Imaging Platform. Research (BK) was supported by the CNRS,

INSERM, the Université de Strasbourg, the ANR grant IMOP, the NIH NIDA grant #DA05010, and the Shirley and Stefan Hatos Neuroscience Research Foundation. AP was supported by INSERM-FRSQ. GS was recipient of doctoral grants from French Research Ministry and Fondation pour la Recherche Médicale and was supported by Fondation pour la Recherche Médicale and NIH grants NS14627 and NS48499 during his postdoc in Allan Basbaum laboratory.

References

1. Pierce KL, Premont RT, Lefkowitz RJ (2002) Seven-transmembrane receptors. Nat Rev Mol Cell Biol **3**:639–650
2. Moser N, Mechawar N, Jones I, Gochberg-Sarver A et al (2007) Evaluating the suitability of nicotinic acetylcholine receptor antibodies for standard immunodetection procedures. J Neurochem **102**:479–492
3. Tsien RY (1998) The green fluorescent protein. Ann Rev Biochem **67**:509–544
4. Chang, K-J, Woods, JH (2004) The *Delta* Receptor. Marcel Dekker, New York
5. Kieffer BL, Gaveriaux-Ruff C (2002) Exploring the opioid system by gene knockout. Prog Neurobiol **66**:285–306
6. Nadal X, Banos JE, Kieffer BL et al (2006) Neuropathic pain is enhanced in *delta*-opioid receptor knockout mice. Eur J Neurosci **23**:830–834
7. Gaveriaux-Ruff C, Karchewski LA, Hever X et al (2008) Inflammatory pain is enhanced in *delta* opioid receptor-knockout mice. Eur J Neurosci **27**:2558–2567
8. Narita M, Suzuki T (2004) *Delta* opioid receptor-mediated antinociception/analgesia. In The *Delta* Receptor (Chang K-J, Woods JH, Ed), pp 331–354, Marcel Dekker, New York
9. Cahill CM, Holdridge SV, Morinville A (2007) Trafficking of *delta*-opioid receptors and other G-protein-coupled receptors: implications for pain and analgesia. Trends Pharmacol Sci **28**:23-31
10. Vanderah TW (2010) *Delta* and *kappa* opioid receptors as suitable drug targets for pain. Clin J Pain **26 Suppl 10**:S10–15
11. Filliol D, Ghozland S, Chluba J et al (2000) Mice deficient for *delta*- and *mu*-opioid receptors exhibit opposing alterations of emotional responses. Nat Genet **25**:195–200
12. Vergura R, Balboni G, Spagnolo B et al (2008) Anxiolytic- and antidepressant-like activities of H-Dmt-Tic-NH-CH(CH2-COOH)-Bid (UFP-512), a novel selective *delta* opioid receptor agonist. Peptides **29**:93–103
13. Tejedor-Real P, Mico JA, Smadja C et al (1998) Involvement of *delta*-opioid receptors in the effects induced by endogenous enkephalins on learned helplessness model. Eur J Pharmacol **354**:1–7
14. Saitoh A, Kimura Y, Suzuki T et al (2004) Potential anxiolytic and antidepressant-like activities of SNC80, a selective *delta*-opioid agonist, in behavioral models in rodents. J Pharmacol Sci **95**:374–380
15. Saitoh A, Yamada M, Takahashi K et al (2008) Antidepressant-like effects of the *delta*-opioid receptor agonist SNC80 ([(+)-4-[(alphaR)-alpha-[(2S,5R)-2,5-dimethyl-4-(2-propenyl)-1-piperazinyl]-(3-methoxyphenyl) methyl]-N,N-diethylbenzamide) in an olfactory bulbectomized rat model. Brain Res **1208**:160–169
16. Torregrossa MM, Jutkiewicz EM, Mosberg HI et al (2006) Peptidic *delta* opioid receptor agonists produce antidepressant-like effects in the forced swim test and regulate BDNF mRNA expression in rats. Brain Res **1069**: 172–181
17. Broom DC, Jutkiewicz EM, Folk JE et al (2002) Convulsant activity of a non-peptidic *delta*-opioid receptor agonist is not required for its antidepressant-like effects in Sprague-Dawley rats. Psychopharmacology (Berl) **164**:42–48
18. Broom DC, Jutkiewicz EM, Folk JE et al (2002) Nonpeptidic *delta*-opioid receptor agonists reduce immobility in the forced swim assay in rats. Neuropsychopharm **26**:744–755
19. Broom DC, Jutkiewicz EM, Rice KC et al (2002) Behavioral effects of *delta*-opioid receptor agonists: potential antidepressants? Jpn J Pharmacol **90**:1–6
20. Jutkiewicz EM, Rice KC, Woods JH et al (2003) Effects of the *delta*-opioid receptor

agonist SNC80 on learning relative to its antidepressant-like effects in rats. Behav Pharmacol **14**:509–516

21. Hirata H, Sonoda S, Agui S et al (2007) Rubiscolin-6, a *delta* opioid peptide derived from spinach Rubisco, has anxiolytic effect via activating *sigma*1 and dopamine D1 receptors, Peptides **28**:1998–2003
22. Olmstead MC, Ouagazzal AM, Kieffer BL (2009) *Mu* and *delta* opioid receptors oppositely regulate motor impulsivity in the signaled nose poke task. PLoS One **4**:e4410
23. Kallal L, Benovic JL (2000) Using green fluorescent proteins to study G-protein-coupled receptor localization and trafficking. Trends Pharmacol Sci **21**:175–180
24. Scherrer G, Tryoen-Toth P, Filliol D et al (2006) Knockin mice expressing fluorescent *delta*-opioid receptors uncover G protein-coupled receptor dynamics in vivo. Proc Natl Acad Sci USA **103**:9691–9696
25. Ko JL, Arvidsson U, Williams FG et al (1999) Visualization of time-dependent redistribution of *delta*-opioid receptors in neuronal cells during prolonged agonist exposure. Brain Res Mol Brain Res **69**:171–185
26. Law PY, Kouhen OM, Solberg J et al (2000) Deltorphin II-induced rapid desensitization of *delta*-opioid receptor requires both phosphorylation and internalization of the receptor. J Biol Chem **275**:32057–32065
27. Whistler JL, Tsao P, von Zastrow M (2001) A phosphorylation-regulated brake mechanism controls the initial endocytosis of opioid receptors but is not required for post-endocytic sorting to lysosomes. J Biol Chem **276**:34331–34338
28. Lecoq I, Marie N, Jauzac P et al (2004) Different regulation of human *delta*-opioid receptors by SNC-80 [(+)-4-[(*alpha*R)-*alpha*-((2S,5R)-4-allyl-2,5-dimethyl-1-piperazinyl)-3-meth oxybenzyl]-N,N-diethylbenzamide] and endogenous enkephalins. J Pharmacol Exp Ther **310**:666–677
29. Eisinger DA, Schulz R (2005) Mechanism and consequences of *delta*-opioid receptor internalization. Crit Rev Neurobiol **17**:1–26
30. Afify EA, Law PY, Riedl M et al (1998) Role of carboxyl terminus of *mu*-and *delta*-opioid receptor in agonist-induced down-regulation. Brain Res Mol Brain Res **54**:24–34
31. Cvejic S, Trapaidze N, Cyr C et al (1996) Thr353, located within the COOH-terminal tail of the *delta* opiate receptor, is involved in receptor down-regulation. J Biol Chem **271**:4073–4076
32. Pradhan AA, Becker JA, Scherrer G et al (2009) In vivo *delta* opioid receptor internalization controls behavioral effects of agonists. PLoS One **4**:e5425
33. Goody RJ, Oakley SM, Filliol D et al (2002) Quantitative autoradiographic mapping of opioid receptors in the brain of *delta*-opioid receptor gene knockout mice. Brain Res **945**:9–19
34. Law PY, Hom DS, Loh HH (1984) Down-regulation of opiate receptor in neuroblastoma x glioma NG108-15 hybrid cells: Chloroquine promotes accumulation of tritiated enkephalin in the lysosomes. J Biol Chem **259**:4096–4104
35. Negri L, Noviello L, Noviello V (1996) Antinociceptive and behavioral effects of synthetic deltorphin analogs. Eur J Pharmacol **296**:9–16
36. Spina L, Longoni R, Mulas A et al (1998) Dopamine-dependent behavioural stimulation by non-peptide *delta* opioids BW373U86 and SNC 80: Locomotion, rearing and stereotypies in intact rats. Behav Pharmacol **9**:1–8
37. Ito S, Mori T, Sawaguchi T (2006) Differential effects of *mu*-opioid, *delta*-opioid and *kappa*-opioid receptor agonists on dopamine receptor agonist-induced climbing behavior in mice. Behav Pharmacol **17**:691–701
38. Jutkiewicz EM, Eller EB, Folk JE et al (2004) *Delta*-opioid agonists: differential efficacy and potency of SNC80, its 3-OH (SNC86) and 3-desoxy (SNC162) derivatives in Sprague-Dawley rats. J Pharmacol Exp Ther **309**:173–181
39. Longoni R, Spina L, Mulas A et al (1991) (D-Ala2)deltorphin II: D1-dependent stereotypies and stimulation of dopamine release in the nucleus accumbens. J Neurosci **11**:1565–1576
40. Fraser GL, Parenteau H, Tu TM et al (2000) The effects of *delta* agonists on locomotor activity in habituated and non-habituated rats. Life Sci **67**:913–922
41. Gallantine EL, Meert TF (2005) A comparison of the antinociceptive and adverse effects of the *mu*-opioid agonist morphine and the *delta*-opioid agonist SNC80. Basic Clin Pharmacol Toxicol **97**:39–51
42. Pacheco DF, Reis GM, Francischi JN et al (2005) *Delta*-opioid receptor agonist SNC80 elicits peripheral antinociception via *delta*(1) and *delta*(2) receptors and activation of the l-arginine/nitric oxide/cyclic GMP pathway. Life Sci **78**:54–60
43. Cao CQ, Hong Y, Dray A et al (2001) Spinal *delta*-opioid receptors mediate suppression of

systemic SNC80 on excitability of the flexor reflex in normal and inflamed rat. Eur J Pharmacol **418**:79–87

44. Brainin-Mattos J, Smith ND, Malkmus S et al (2006) Cancer-related bone pain is attenuated by a systemically available *delta*-opioid receptor agonist. Pain **122**:174–181
45. Aceto MD, May EL, Harris LS et al (2007) Pharmacological studies with a nonpeptidic, *delta*-opioid (-)-(1R,5R,9R)-5,9-dimethyl-2'-hydroxy-2-(6-hydroxyhexyl)-6,7-benzomorphan hydrochloride ((-)-NIH 11082). Eur J Pharmacol **566**:88–93
46. Le Bourdonnec B, Windh RT, Ajello CW et al (2008) Potent, orally bioavailable *delta* opioid receptor agonists for the treatment of pain: discovery of N,N-diethyl-4-(5-hydroxyspiro[chromene-2,4'-piperidine]-4-yl)benzamide (ADL5859). J Med Chem **51**:5893–5896
47. Le Bourdonnec B, Windh RT, Leister LK et al (2009) Spirocyclic *delta* opioid receptor agonists for the treatment of pain: discovery of N,N-diethyl-3-hydroxy-4-(spiro[chromene-2,4'-piperidine]-4-yl) benzamide (ADL5747). J Med Chem **52**:5685–5702
48. Jones P, Griffin AM, Gawell L et al (2009) N,N-Diethyl-4-[(3-hydroxyphenyl)(piperidin-4-yl)amino] benzamide derivatives: the development of diaryl amino piperidines as potent *delta* opioid receptor agonists with in vivo antinociceptive activity in rodent models. Bioorg Med Chem Lett **19**:5994–5998
49. Mika J, Przewlocki R, Przewlocka B (2001) The role of *delta*-opioid receptor subtypes in neuropathic pain. Eur J Pharmacol **415**:31–37
50. Cahill CM, Morinville A, Hoffert C et al (2003) Up-regulation and trafficking of *delta* opioid receptor in a model of chronic inflammation: implications for pain control. Pain **101**:199–208
51. Holdridge SV, Cahill CM (2007) Spinal administration of a *delta* opioid receptor agonist attenuates hyperalgesia and allodynia in a rat model of neuropathic pain. Eur J Pain **11**:685–693
52. Kabli N, Cahill CM (2007) Anti-allodynic effects of peripheral *delta* opioid receptors in neuropathic pain. Pain **127**:84–93
53. Hervera A, Leanez S, Negrete R et al (2009) The peripheral administration of a nitric oxide donor potentiates the local antinociceptive effects of a DOR agonist during chronic inflammatory pain in mice. Naunyn Schmiedebergs Arch Pharmacol **380**:345–352
54. Hurley RW, Hammond DL (2000) The analgesic effects of supraspinal *mu* and *delta* opioid receptor agonists are potentiated during persistent inflammation. J Neurosci **20**:1249–1259
55. Fraser GL, Gaudreau GA, Clarke PB et al (2000) Antihyperalgesic effects of *delta* opioid agonists in a rat model of chronic inflammation. Br J Pharmacol **129**:1668–1672
56. Gendron L, Pintar JE, Chavkin C (2007) Essential role of *mu* opioid receptor in the regulation of *delta* opioid receptor-mediated antihyperalgesia. Neuroscience **150**:807–817
57. Knapp RJ, Santoro G, De Leon IA et al (1996) Structure-activity relationships for SNC80 and related compounds at cloned human *delta* and *mu* opioid receptors, J Pharmacol Exp Ther **277**:1284–1291
58. Wei ZY, Brown W, Takasaki B et al (2000) N,N-Diethyl-4-(phenylpiperidin-4-ylidenemethyl) benzamide: a novel, exceptionally selective, potent *delta* opioid receptor agonist with oral bioavailability and its analogues. J Med Chem **43**:3895–3905
59. Mansour A, Fox CA, Burke S et al (1994) *Mu*, *delta*, and *kappa* opioid receptor mRNA expression in the rat CNS: an in situ hybridization study. J Comp Neurol **350**:412–438
60. Georges F, Normand E, Bloch B et al (1998) Opioid receptor gene expression in the rat brain during ontogeny, with special reference to the mesostriatal system: an in situ hybridization study. Brain Res Dev Brain Res **109**:187–199
61. Le Moine C, Kieffer B, Gaveriaux-Ruff C et al (1994) *Delta*-opioid receptor gene expression in the mouse forebrain: localization in cholinergic neurons of the striatum. Neuroscience **62**:635–640
62. Minami M, Maekawa K, Yabuuchi K et al (1995) Double in situ hybridization study on coexistence of *mu*-, *delta*- and *kappa*-opioid receptor mRNAs with preprotachykinin A mRNA in the rat dorsal root ganglia. Brain Res Mol Brain Res **30**:203–210
63. Mennicken F, Zhang J, Hoffert C et al (2003) Phylogenetic changes in the expression of *delta* opioid receptors in spinal cord and dorsal root ganglia. J Comp Neurol **465**:349–360
64. Cahill CM, McClellan KA, Morinville A et al (2001) Immunohistochemical distribution of *delta* opioid receptors in the rat central nervous system: evidence for somatodendritic labeling and antigen-specific cellular compartmentalization. J Comp Neurol **440**:65–84
65. Dado RJ, Law PY, Loh HH et al (1993) Immunofluorescent identification of a

delta-opioid receptor on primary afferent nerve terminals. Neuroreport **5**:341–344

66. Zhang X, Bao L, Arvidsson U et al (1998) Localization and regulation of the *delta*-opioid receptor in dorsal root ganglia and spinal cord of the rat and monkey: evidence for association with the membrane of large dense-core vesicles. Neuroscience **82**:1225–1242
67. Bao L, Jin SX, Zhang C et al (2003) Activation of *delta* opioid receptors induces receptor insertion and neuropeptide secretion. Neuron **37**:121–133
68. Guan JS, Xu ZZ, Gao H et al (2005) Interaction with vesicle luminal protachykinin regulates surface expression of *delta*-opioid receptors and opioid analgesia. Cell **122**:619–631
69. Riedl MS, Schnell SA, Overland AC et al (2009) Coexpression of *alpha* 2A-adrenergic and *delta*-opioid receptors in substance P-containing terminals in rat dorsal horn. J Comp Neurol **513**:385–398
70. Stone LS, Vulchanova L, Riedl MS et al (2004) Effects of peripheral nerve injury on *delta* opioid receptor (DOR) immunoreactivity in the rat spinal cord. Neurosci Lett **361**:208–211
71. Robertson B, Schulte G, Elde R et al (1999) Effects of sciatic nerve injuries on *delta* -opioid receptor and substance P immunoreactivities in the superficial dorsal horn of the rat. Eur J Pain **3**:115–129
72. Rozenfeld R, Abul-Husn NS, Gomez I et al (2007) An emerging role for the *delta* opioid receptor in the regulation of *mu* opioid receptor function. ScientWorld J 7:64–73
73. Scherrer G, Imamachi N, Cao YQ et al (2009) Dissociation of the opioid receptor mechanisms that control mechanical and heat pain. Cell **137**:1148–1159
74. Cavanaugh DJ, Lee H, Lo L et al (2009) Distinct subsets of unmyelinated primary sensory fibers mediate behavioral responses to noxious thermal and mechanical stimuli. Proc Natl Acad Sci USA **106**:9075–9080
75. Dussor G, Zylka MJ, Anderson DJ et al (2008) Cutaneous sensory neurons expressing the Mrgprd receptor sense extracellular ATP and are putative nociceptors. J Neurophysiol **99**:1581–1589
76. Basbaum AI, Bautista DM, Scherrer G et al (2009) Cellular and molecular mechanisms of pain. Cell **139**:267–284
77. Malmberg AB, Chen C, Tonegawa S et al (1997) Preserved acute pain and reduced neuropathic pain in mice lacking PKC*gamma*. Science **278**:279–283
78. Miraucourt LS, Dallel R, Voisin DL (2007) Glycine inhibitory dysfunction turns touch into pain through PKC*gamma* interneurons. PLoS One **2**:e1116
79. Miraucourt LS, Moisset X, Dallel R et al (2009) Glycine inhibitory dysfunction induces a selectively dynamic, morphine-resistant, and neurokinin 1 receptor-independent mechanical allodynia. J Neurosci **29**:2519–2527
80. Caterina MJ, Schumacher MA, Tominaga M et al (1997) The capsaicin receptor: a heat-activated ion channel in the pain pathway. Nature **389**:816–824
81. Caterina MJ, Leffler A, Malmberg AB et al (2000) Impaired nociception and pain sensation in mice lacking the capsaicin receptor. Science **288**:306–313
82. Tan CM, Brady AE, Nickols HH et al (2004) Membrane trafficking of G protein-coupled receptors. Ann Rev Pharmacol Toxicol **44**:559–609
83. Hanyaloglu AC, von Zastrow M (2008) Regulation of GPCRs by endocytic membrane trafficking and its potential implications. Ann Rev Pharmacol Toxicol **48**:537–568
84. Duvernay MT, Filipeanu CM, Wu G (2005) The regulatory mechanisms of export trafficking of G protein-coupled receptors. Cell Signal **17**:1457–1465
85. Cahill CM, Morinville A, Lee MC et al (2001) Prolonged morphine treatment targets *delta* opioid receptors to neuronal plasma membranes and enhances *delta*-mediated antinociception. J Neurosci **21**:7598–7607
86. Morinville A, Cahill CM, Aibak H et al (2004) Morphine-induced changes in *delta* opioid receptor trafficking are linked to somatosensory processing in the rat spinal cord. J Neurosci **24**:5549–5559
87. Lucido AL, Morinville A, Gendron L et al (2005) Prolonged morphine treatment selectively increases membrane recruitment of *delta*-opioid receptors in mouse basal ganglia. J Mol Neurosci **25**:207–214
88. Gendron L, Lucido AL, Mennicken F et al (2006) Morphine and pain-related stimuli enhance cell surface availability of somatic *delta*-opioid receptors in rat dorsal root ganglia. J Neurosci **26**:953–962
89. Walwyn W, Maidment NT, Sanders M et al (2005) Induction of *delta* opioid receptor function by up-regulation of membrane

receptors in mouse primary afferent neurons. Mol Pharmacol **68**:1688–1698

90. Patwardhan AM, Berg KA, Akopain AN et al (2005) Bradykinin-induced functional competence and trafficking of the *delta*-opioid receptor in trigeminal nociceptors. J Neurosci **25**:8825–8832
91. Commons KG (2003) Translocation of presynaptic *delta* opioid receptors in the ventrolateral periaqueductal gray after swim stress. J Comp Neurol **464**:197–207
92. Chieng B, Christie MJ (2009) Chronic morphine treatment induces functional *delta*-opioid receptors in amygdala neurons that project to periaqueductal grey. Neuropharmacology **57**:430–437
93. Hack SP, Bagley EE, Chieng BC et al (2005) Induction of *delta*-opioid receptor function in the midbrain after chronic morphine treatment. J Neurosci **25**:3192–3198
94. Ma J, Zhang Y, Kalyuzhny AE et al (2006) Emergence of functional *delta*-opioid receptors induced by long-term treatment with morphine. Mol Pharmacol **69**:1137–1145
95. Morinville A, Cahill CM, Esdaile MJ et al (2003) Regulation of *delta*-opioid receptor trafficking via *mu*-opioid receptor stimulation: evidence from *mu*-opioid receptor knock-out mice. J Neurosci **23**:4888–4898
96. Pradhan AA, Siau C, Constantin A et al (2006) Chronic morphine administration results in tolerance to *delta* opioid receptor-mediated antinociception. Neuroscience **141**:947–954
97. Jutkiewicz EM (2006) The antidepressant-like effects of *delta*-opioid receptor agonists. Mol Interv **6**:162–169
98. Fraser GL, Pradhan AA, Clarke PB et al (2000) Supraspinal antinociceptive response to [D-Pen(2,5)]-enkephalin (DPDPE) is pharmacologically distinct from that to other *delta*-agonists in the rat. J Pharmacol Exp Ther **295**:1135–1141
99. Thorat SN, Hammond DL (1997) Modulation of nociception by microinjection of *delta*-1 and *delta*-2 opioid receptor ligands in the ventromedial medulla of the rat. J Pharmacol Exp Ther **283**:1185–1192
100. Ossipov MH, Kovelowski CJ, Nichols ML et al (1995) Characterization of supraspinal antinociceptive actions of opioid *delta* agonists in the rat. Pain **62**:287–293
101. Heyman JS, Mosberg HI, Porreca F (1986) Evidence for *delta* receptor mediation of [D-Pen2,D-Pen5]-enkephalin (DPDPE) analgesia in mice. NIDA Res Monogr **75**:442–445
102. Qi JA, Mosberg HI, Porreca F (1990) Antinociceptive effects of [D-Ala2]deltorphin II, a highly selective *delta* agonist in vivo. Life Sci **47**:PL43–47
103. Bilsky EJ, Calderon SN, Wang T et al (1995) SNC 80, a selective, nonpeptidic and systemically active opioid *delta* agonist. J Pharmacol Exp Ther **273**:359–366
104. Negus SS, Gatch MB, Mello NK et al (1998) Behavioral effects of the *delta*-selective opioid agonist SNC80 and related compounds in rhesus monkeys. J Pharmacol Exp Ther **286**:362–375
105. Danielsson I, Gasior M, Stevenson GW et al (2006) Electroencephalographic and convulsant effects of the *delta* opioid agonist SNC80 in rhesus monkeys. Pharmacol Biochem Behav **85**:428–434
106. Zhang X, Bao L, Guan JS (2006) Role of delivery and trafficking of *delta*-opioid peptide receptors in opioid analgesia and tolerance. Trends Pharmacol Sci **27**:324–329
107. Ferguson SS (2001) Evolving concepts in G protein-coupled receptor endocytosis: the role in receptor desensitization and signaling. Pharmacol Rev **53**:1–24
108. Holschneider DP, Maarek JM (2004) Mapping brain function in freely moving subjects. Neurosci Biobehav Rev **28**:449–461
109. Holschneider DP, Maarek JM (2008) Brain maps on the go: functional imaging during motor challenge in animals. Methods **45**:255–261
110. Miesenbock G, Kevrekidis IG (2005) Optical imaging and control of genetically designated neurons in functioning circuits. Ann Rev Neurosci **28**:533–563
111. Marie N, Landemore G, Debout C et al (2003) Pharmacological characterization of AR-M1000390 at human *delta* opioid receptors. Life Sci **73**:1691–1704
112. Jutkiewicz EM, Kaminsky ST, Rice KC et al (2005) Differential behavioral tolerance to the *delta*-opioidagonistSNC80([(+)-4-[(*alpha*R)-*alpha*-[(2S,5R)-2,5-dimethyl-4-(2-propenyl)-1-piperazinyl]-(3-methoxyphenyl) methyl]-N,N-diethylbenzamide) in Sprague-Dawley rats. J Pharmacol Exp Ther **315**:414–422
113. Rios CD, Jordan BA, Gomes I et al (2001) G-protein-coupled receptor dimerization: modulation of receptor function. Pharmacol Ther **92**:71–87
114. Hadjantonakis AK, Dickinson ME, Fraser SE et al (2003) Technicolour transgenics: imaging tools for functional genomics in the mouse. Nat Rev Genet **4**:613–625

115. Shaner NC, Campbell RE, Steinbach PA et al (2004) Improved monomeric red, orange and yellow fluorescent proteins derived from *Discosoma sp* red fluorescent protein. Nat Biotechnol **22**:1567–1572
116. Pin JP, Neubig R, Bouvier M et al (2007) International Union of Basic and Clinical Pharmacology LXVII Recommendations for the recognition and nomenclature of G protein-coupled receptor heteromultimers. Pharmacol Rev **59**:5–13
117. Baker, M (2010) Whole-animal imaging: Probe progress. Nature **463**:979
118. Chan F, Bradley A, Wensel TG, Wilson JH (2004) Knock-in human rhodopsin-GFP fusions as mouse models for human disease and targets for gene therapy. Proc Natl Acad Sci USA **101**:9109–9114

Chapter 7

Posttranslational Regulation of G Protein-Coupled Receptors

Yu Qiu and Ping-Yee Law

Abstract

The G protein-coupled receptors (GPCRs) are a superfamily of transmembrane receptors that structurally possess an extracellular amino terminus, seven transmembrane domains linked by extracellular and intracellular loops, and a cytoplasmic carboxyl terminus. They are synthesized by ribosomes and enter into the endoplasmic reticulum (ER), from which they are transported to Golgi apparatus and the trans-Golgi network (TGN) and finally move to the plasma membrane. At the plasma membrane, GPCRs receive environmental stimuli and relay the message to the cells. During these processes, GPCRs undergo posttranslational modifications that regulate their maturation, their function at the cell surface and even the ultimate fate of the internalized receptor after agonist treatment. There are four major types of posttranslational modifications – glycosylation, phosphorylation, palmitoylation, and ubiquitination, each of which has distinct roles in expression and function of GPCRs. In this chapter we discuss the methods to study these posttranslational modifications and the findings of post-translational modifications and their functional consequences on GPCRs, using opioid receptors as the main examples. Moreover, the detailed steps of the main methods are depicted and also our thoughts on future directions of this avenue of research.

Key words: G protein-coupled receptors, Posttranslational regulation, Opioid receptor, Glycosylation, Phosphorylation, Palmitoylation, Ubiquitination

1. Introduction

The G protein-coupled receptors (GPCRs) are a superfamily of transmembrane receptors that structurally possess an extracellular amino terminus, seven α-helical transmembrane domains linked by extracellular and intracellular loops and a cytoplasmic carboxyl terminus. Functionally, all these receptors must bind to a ligand and exert their agonist-stimulated signaling by coupling to the heterotrimeric G proteins (1). GPCRs are synthesized by

Craig W. Stevens (ed.), *Methods for the Discovery and Characterization of G Protein-Coupled Receptors*, Neuromethods, vol. 60, DOI 10.1007/978-1-61779-179-6_7, © Springer Science+Business Media, LLC 2011

ribosomes attached at the cytosolic face of the ER. Then they enter cotranslationally into ER lumen, where they are also folded and assembled. Properly folded receptors are then transported from the ER to the ER-Golgi intermediate complex (ERGIC), the Golgi apparatus, and the trans-Golgi network (TGN). Finally, mature receptors move from the TGN to the plasma membrane, their functional destination. During these transportation processes, receptors undergo posttranslational modifications to attain mature status (2). At the plasma membrane GPCRs receive environmental stimuli and thus relay the message to the cells. The stimulated GPCRs may undergo internalization. Internalized receptors are either recycled back to the plasma membrane or targeted to the lysosomes for degradation. The posttranslational modifications of the receptor can also occur after the activation of receptor and during the intracellular trafficking processes. Meanwhile, the posttranslational modifications can further influence the signaling and trafficking of the receptor.

Over the last several decades, a vast amount of information has been gathered regarding the posttranslational modifications of GPCRs that are comprised mostly of glycosylation, phosphorylation, palmitoylation, and ubiquitination. During the early phases of this research, various techniques were employed. To study glycosylation, methods from observing the different electrophoretic mobilities in sodium dodecyl sulfate polyacrylamide gel electrophoresis (SDS-PAGE), enzymatic deglycosylation to site-directed mutagenesis have been extensively used. [^{32}P] orthophosphate labeling, mutagenesis, phosphopeptide maps, and phosphor-specific antibody, mass spectrometry are usually employed to study receptor phosphorylation and its functional consequences. [^{3}H]Palmitate incorporation and mutagenesis are widely used in the study of palmitoylation. For ubiquitination, protease inhibitors are initially employed and later ubiquitin antibody combined with coimmunoprecipitation is commonly used. There is some limitation for each method. In certain situations, it is more comprehensive to interpret data based on several methods. Mutagenesis is widely used since the cloning of the receptors, but great care must be taken to interpret results from experiments with mutagenesis of specific residues, which may also be responsible for other important functions apart from the possible posttranslational modifications. In this chapter, we will review the posttranslational modifications of GPCRs, mostly using opioid receptors as examples, which serve as excellent models because they undergo glycosylation, palmitoylation, phosphorylation, and ubiquitination. Moreover, these processes affect the receptor expression, distribution on cell membrane, intracellular trafficking, and the interaction with other cellular proteins. In addition, the methods involved in these studies are also discussed.

2. Glycosylation of Opioid Receptors

Glycosylation has two forms: N-linked glycosylation and O-linked glycosylation. N-glycosylation is the most common posttranslational modification of GPCRs. N-glycosylation, which means oligosaccharide is linked to the nitrogen in the side-chain amid of asparagine residues, is initiated in the ER and completed during transport through the Golgi. The most common O-glycosylation involves the transfer of *N*-acetyl-galactosamine to a serine or threonine residue on the protein acceptor. There are also other types of O-glycosylation such as fucosyl and galactosyl types.

The study of GPCRs glycosylation is started with the employment of glycosidases to remove the sugar residues and glycosylation inhibitor tunicamycin to prevent the glycosylation. The absence of sugar residues increases the electrophoretic mobility of protein in SDS-PAGE, which can be detected by specific antibody or radioligand labeling of receptor. In the case of tunicamycin, the decrease of [^{3}H]-glucosamine incorporation to the protein can also be used to demonstrate the involvement of glycosylation. Combined with radioligand binding assay and assay of receptor signaling pathways such as adenylyl cyclase, glycosidases, and the glycosylation inhibitor tunicamycin can be used to study the impact of glycosylation on the receptor expression and function. By using the glycosidases neuraminidase, α-mannosidase, and tunicamycin, it was demonstrated that glycosylation is required for expression of β_2 adrenergic receptor (β_2AR) but not required for their function (3, 4). Using tunicamycin, it was also shown that glycosylation plays a role in expression of opioid receptors but not in function (5, 6). Glycosidases include neuraminidase that cleaves terminal *N*-aceylneuraminic acid, α-mannosidase that cleaves α form of mannose, and endoglycosidase H that cleaves asparagine-linked mannose rich oligosaccharides. There is another glycosidase called Peptide N-Glycosidase F (PNGase F) which removes all types of N-linked glycosylation. O-glycosydase that hydrolyses O-linked glycan chains is specific for the detection of O-glycosylation. The fact that these glycosidases cleaves the glycan chain differentially is utilized to identify the nature of the carbohydrate moieties linked to the receptor protein. It was shown that the major carbohydrate moieties of μ opioid receptor (MOR) are the complex type of N-linked glycans (7). Removal of N-linked glycans in MOR results in that the migration of MOR in SDS-PAGE moves from a molecular mass of 60–88 kDa to 43 kDa (7). Using endoglycosidase H, PNGase F and O-glycosydase, both N-glycosylation and O-glycosylation was demonstrated in the δ opioid receptor (DOR) (8) and κ opioid receptor (KOR) (9).

Lectin affinity chromatography is another method to display the glycosylation of receptors. Wheat germ lectin, which can specifically bind to both N-linked and O-linked glycosylated proteins (10), is coupled to agarose and the solubilized receptors can be applied to it and then washed and eluted. The amount and presence of receptor in which fraction provide the information of receptor glycosylation. Bowen and Kooper have used this method to show that the glycosylation of opioid receptors is different in cow striatum and rat whole brain (11). Moreover, lectin affinity chromatography can also be used to enrich the receptor for further analysis such as receptor phosphorylation assay.

After the receptor is cloned, the site-directed mutagenesis is employed to further study the glycosylation site(s) of receptor. It has been postulated that N-glycosylation occurs exclusively at the consensus sequence Asn-X-Ser/Thr, where X is any amino acid except Pro or Asp (12) which is usually located within the extracellular N-terminal region. By using site-directed mutagenesis, the two amino-terminal consensus glycosylation sites are shown both to be utilized in β_2AR (13). Also, by this method, human β_2AR was found to possess an additional glycosylation site in the second extracellular loop (14). Moreover, mutation of glycosylation sites was also shown to reduce ~50% of plasma membrane expression of β_2AR (13). For rat MOR, there are five putative glycosylation sites at the asparagine residues 9, 31, 38, 46, and 53 in the N-terminus. Mutation of all these five sites causes the receptor to be retained in the ER. Single mutation mutants of each of the five putative glycosylation sites of MOR exhibit a slightly faster mobility in SDS-PAGE than wild type MOR, suggesting that all five asparagines residues are glycosylated (15). Furthermore, all five glycosylation sites are found to contribute to the cell surface expression of receptor with asparagine residues 9, 31, 38, and 46 playing more crucial roles (15). The glycosylation of MOR also affects its interaction with Ribophorin I, which participates in transport of MOR from the ER to the cell surface (15). For KOR, the two putative glycosylation sites asparagine residues 25 and 39 are shown to be both glycosylated (9). Similarly, glycosylation of KOR plays important roles in stability and trafficking along the biosynthesis pathway. As mentioned above, the specific residues involved in glycosylation may also be responsible for other important functions. The results obtained by mutagenesis may also imply functions other than glycosylation. A natural occurred polymorphism of human MOR in which a potential N-glycosylation site is mutated (Asn^{40} to Asp) displays a higher affinity for β-endorphin as compared with the wild type (16). Whether this effect can be attributed to glycosylation is still not determined. However, heterologous expression of this polymorphism of MOR, unlike wild type MOR that is mainly localized in the lipid rafts domains of plasma membranes, is shown to be partitioned mainly to the nonraft domains (unpublished observations).

The development of receptor antibodies greatly eases the detection of different mobility of glycoproteins. The antibodies to MOR detect various molecular weight forms of receptor in which some are corresponding to the glycosylated form (17). Brain-region specific glycosylation of rat MOR is also demonstrated by a MOR C-tail specific antibody (18). Moreover, the MOR in the caudate putamen with higher median relative molecular masses is mainly associated with lipid rafts, while the MOR in the thalamus with lower median relative molecular masses distributes into lipid raft and nonraft domains evenly (18). Combined with glycosidases, pulse-chase labeling and biotinylation labeling, using antibody to flag-tagged DOR shows that N-glycosylation of DOR is initiated cotranslationally and is completed when the receptor reaches the TGN, and export from the ER is a limiting step in the overall maturation process of the receptor (8).

3. Phosphorylation of Opioid Receptors

Phosphorylation of opioid receptors has long been reported. Using *in vitro* ^{32}P incorporation method, morphine was found to enhance the phosphorylation of a 58 kDa protein in mouse brain membrane, which was believed to be MOR (19). But the concrete demonstration of opioid receptor phosphorylation was achieved after the cloning of opioid receptors. DOR was the first opioid receptor to display receptor phosphorylation (20). Then the phosphorylation of MOR and KOR was subsequently demonstrated (21, 22). The methods used in these studies are the metabolic labeling with [^{32}P]orthophosphate or the ^{32}P incorporation through [γ-^{32}P]ATP in the cultured cells expressing the interested receptors, followed by immunoprecipitation, SDS gel electrophoresis, and autoradiography. [^{32}P]orthophosphate labeling has been a main method to study receptor phosphorylation for several decades, while ^{32}P incorporation through [γ-^{32}P]ATP is mainly employed in *in vitro* phosphorylation studies. To increase the sensitivity of [^{32}P]orthophosphate labeling, the enrichment of receptors by lectin affinity chromatography can be incorporated. Moreover, the differential receptor phosphorylation induced by different agonists was also demonstrated. Morphine induces less phosphorylation of MOR while etorphine evokes robust receptor phosphorylation (23).

Using phosphopeptide mapping in which the ^{32}P labeled receptor is cleaved by cyanogen bromide to generate peptides, the phosphorylation of MOR was demonstrated to be located in the carboxyl terminus (24). Further localization of phosphorylation residues was performed by site-directed mutagenesis combined with ^{32}P labeling. In our laboratory, systematic mutations of Ser/Thr residues within the carboxyl tail sequence have identified Ser^{363}, Thr^{370}, and Ser^{375} as phosphorylation sites for MOR (24). However, Thr^{394} was identified as a phosphorylation site in other laboratory (25). The discrepancy may be due to the different mutation strategies and the sensitivity of ^{32}P labeling. Mutagenesis may also result in conformational changes of the carboxyl tail or a modification of a protein kinase recognition motif. A more comprehensive mutation strategy was employed in our laboratory with different combinations of mutated phosphorylation sites. In this way, Ser^{363}, Thr^{370} were recognized to be basally phosphorylated and Thr^{370}, Ser^{375} phosphorylated upon agonist activation (24). Using same methods, the phosphorylation sites of DOR were identified to be Thr^{358} and Ser^{363} in our laboratory with the phosphorylation of Ser^{363} occurring prior to the phosphorylation of Thr^{358} (26). One other laboratory also identified the same phosphorylation sites of DOR using mutagenesis and *in vitro* phosphorylation assay (27), however without defining the hierarchical pattern of phosphorylation. The discrepancy may also be due to the different mutation strategies, the methods employed, and the receptor or protein kinase levels. As for KOR, Ser^{369} was identified as a phosphorylation site upon agonist activation (28). The specific antibody to phosphorylated proteins also provides a way to study receptor phosphorylation and leads to the identification of more phosphorylated sites. The antibodies specific for the phosphorylated Ser^{375} of MOR and the phosphorylated Ser^{363} of DOR have been developed and successfully utilized to demonstrate the agonist-induced receptor phosphorylation of MOR and DOR (29, 30). Receptor phosphorylation at tyrosine residues of MOR (31) and KOR (32) has long been proposed by altered receptor function upon mutation of tyrosine residues facing intracellularly. But, it is with phosphor-tyrosine specific antibody, Tyr^{366} was identified as a phosphorylation site of MOR after chronic treatment of morphine and naloxone precipitation (33). Using an antibody that specifically recognizes the Tyr^{166}-phosphorylated MOR, Tyr^{166} was identified as another tyrosine phosphorylation site by concomitant activation of tyrosine kinase (34). Phospho-tyrosine specific antibody also contributes to the finding of phosphorylation of Tyr^{318} in DOR (35, 36). Mass spectrometry is another method to detect the phosphorylation sites of GPCRs. Bovine rhodopsin is the first GPCR to be examined by mass spectrometry which identified Ser^{338} and Ser^{343} as the major phosphorylation sites within its C-terminal region (37). The agonist-induced phosphorylation of β_2AR is also demonstrated to be localized exclusively in a proximal portion (between residues 339–369) of the carboxyl-terminal cytoplasmic domain (38).

It is widely accepted that GPCRs are usually phosphorylated by the G protein-coupled receptor kinase (GRK) family upon agonist activation. However, other kinases such as protein kinase

C (PKC), protein kinase A (PKA), casein kinases, and receptor tyrosine kinases are also involved in GPCR phosphorylation (39, 40). In our laboratory, using *in vitro* phosphorylation assay and purified GRK proteins, Ser^{375} but not the Thr^{370} residue of MOR was found to be phosphorylated by GRK2, whereas GRK5 did not phosphorylate the carboxyl tail of MOR (unpublished observations). Also using *in vitro* phosphorylation assay but with purified PKA, it was shown that PKA can phosphorylate MOR in the presence of morphine but not (D-Ala^2, N-$MePhe^4$,Gly-ol^5]-enkephalin (DAMGO) (41). When interpreting these results, one should keep in mind that *in vitro* analysis does not represent the cellular responses. The involved protein kinase(s) in cellular environment can be examined by overexpression of wild type or dominant-negative mutants of kinases. Using this method, it was suggested that DOR was phosphorylated by GRK2 or with other GRK subtypes (20, 27). By overexpression of dominant-negative mutant GRK2, DAMGO-induced phosphorylation of MOR was attenuated (42). Protein kinase activators and inhibitors are also utilized to study the kinases involved in receptor phosphorylation. Using PKC inhibitors demonstrated that PKC was not involved in the agonist-induced phosphorylation of MOR (42, 43). Using PKC activators, an additional phosphorylation site Ser^{344} was found for DOR and was proved to be phosphorylated by PKC in an agonist-independent manner (44). Regarding these data, it seems that the employment of the dominant-negative mutants and the inhibitors of protein kinases may reflect the inherent cellular responses better, but protein kinase inhibitors may have pleiotropic effects. Mutagenesis leads to identification of the involvement of calcium/calmodulin-dependent protein kinase II (CaM kinase II) and two putative CaM kinase II sites Ser^{261} and Ser^{266} in the third intracellular loop in the phosphorylation of MOR, however, the direct evidence of phosphorylation has not been provided (45).

Studies on the regulation of the β_2AR have established a widely accepted dogma that receptor phosphorylation induced by agonist activation promotes the association of arrestins to activated receptor, which not only uncouples the receptor from G protein to blunt signaling (desensitization) but also recruits and physically bridges the receptor to the endocytic machinery (internalization) (46, 47). The early observation that receptor phosphorylation of MOR and DOR correlates with receptor desensitization indicates that opioid receptor also follows this general rule (43, 48). More detailed studies are based on the information of phosphorylation sites and the involved protein kinases, therefore various methods mentioned above are broadly employed. Mutation of agonist-dependent phosphorylation site Ser^{375} in MOR reduced receptor internalization (24). Overexpression of GRK enhances desensitization while overexpression of its dominant-negative mutant attenuates desensitization of DOR (20). Mutation of phosphorylation sites attenuates desensitization and internalization of DOR (26). Mutation of Ser^{369} to Ala in KOR also results in blunting of the agonist-induced desensitization (28). Besides these methods, the animals with GRK or β-arrestin knockout are employed for *in vivo* studies. Desensitization of MOR does not occur after chronic morphine treatment in the β-arrestin 2 knockout mice. Moreover, these animals fail to develop antinociceptive tolerance (49). Another study with the GRK3 knockout mice demonstrates that these animals develop significantly less analgesic tolerance to KOR agonist (50). All these results support the model that the agonist-induced phosphorylation of opioid receptor plays an important role in β-arrestin recruitment and subsequent desensitization and internalization. However, the observation that arrestin recruitment is independent of receptor phosphorylation is also obtained from the studies with phosphorylation-deficient mutant of the receptor. MOR and DOR mutants with mutated phosphorylation sites still display agonist-induced internalization and desensitization (51, 52). This observation is further confirmed by showing that the overexpression of β-arrestins increases the phosphorylation-independent internalization, whereas internalization is blocked in β-arrestins-deficient cells or by overexpression of dominant-negative mutants of β-arrestins (51, 53). Further studies demonstrated that receptor phosphorylation does modulate the selectivity between β-arrestin 1 and 2 (51, 53).

By utilizing phosphorylation-deficient mutants, receptor phosphorylation is revealed to play a significant role in determining the ultimate fate of the internalized receptors. A mutant MOR with Thr^{394} replaced recycles more quickly than the wild type receptor (54), whereas a mutant MOR where agonist-induced phosphorylation sites after Ser^{363} are removed does not recycle after internalization (55). A study with mutation of three phosphorylation sites in MOR indicates that the nonphosphorylated MOR recycles through Rab11 at a slower rate than the wild type receptor which recycles through Rab4 (52). Mutation of phosphorylation sites of DOR leads mutant receptor to degradation (56). These data suggest that the phosphorylation status of the internalized receptor influence the cellular responsiveness or downstream signaling. This is further supported by the finding that receptor phosphorylation affects adenylyl cyclase superactivation, a hallmark for the cellular adaptive response after chronic opioid stimulation. Mutation of Thr^{394} in MOR abolishes the adenylyl cyclase superactivation induced by DAMGO (57). Tyr^{366} phosphorylation by Src kinase is responsible for the switch of MOR signal from initial inhibition to superactivation of adenylyl cyclase activity (33).

4. Palmitoylation of Opioid Receptors

Palmitoylation is one type of covalent lipid modification of proteins and has been broadly reported for GPCRs. Palmitoylation is mostly observed by the attachment of palmitate to cysteine residues through a thioester linkage (58). Metabolic labeling with [^{3}H]palmitate followed by immunoprecipitation, SDS-PAGE, and fluorography is the conventional approach to assessing the palmitoylation of proteins. However, [^{3}H]palmitate labeling cannot provide quantitative information regarding the extent of acylation and may not detect palmitoylation in proteins that have relatively low levels of expression or are not highly palmitoylated (59). Recently a new method, called fatty acyl exchange labeling, was developed. This method is highly sensitive and allows for quantitative estimates of palmitoylation (59). But some care must be taken in that both methods can yield false positives. The identification of palmitoylation sites is achieved by mutation of cysteine residues. The first GPCR found to be palmitoylated was rhodopsin with two adjacent cysteines in the C-terminal cytoplasmic tail found to be palmitoylated (60, 61). Soon after that study, the palmitoylation of β_2AR was demonstrated (62). Using metabolic labeling with (^{3}H)palmitate, the palmitoylation of MOR and DOR was also shown (63, 64). It was suggested that the palmitoylation usually occurs on cysteine residues in the cytoplasmic tail, positioned 10–14 amino acids downstream of the last transmembrane domain (65). However, mutation of two conserved cysteine residues in the C-terminus of MOR does not affect receptor palmitoylation (63). Thus, Cys^{170} in the second intracellular loop may be the palmitoylation site. This is confirmed by the finding that mutation of Cys^{170} abolishes receptor palmitoylation detected by fatty acyl exchange labeling (unpublished observations). Moreover, mass spectrometry can also be used to detect the palmitoylation of proteins. The palmitoylation of β_2AR at Cys^{341} (60) is verified by this method (38).

Palmitoylation of GPCRs can occur during or shortly after protein synthesis. This is the case for DOR of which palmitoylation is shown to occur initially during or shortly after export from the ER (64). Using 2-bromopalmitate to block the palmitoylation inhibits the cell surface expression of the receptor (64), indicating that palmitoylation is required for receptor maturation. However, this study does not preclude that the 2-bromopalmitate could inhibit the palmitoylation of proteins that are needed for the export of the receptor. Palmitoylation can also occur after the mature GPCRs have reached its specific location, where GPCRs may undergo cycles of palmitoylation and depalmitoylation and be particularly dependent on agonist stimulation (66). Agonist stimulation has been reported to mediate an increase in palmitate turnover in β_2AR (67). For opioid receptors, agonist-induced receptor palmitoylation was observed in DOR (64), however, its role in receptor function has not been elucidated.

Palmitic acid covalently bound to β_2AR interacts with cholesterol and is linked to its crystallized structure (68). Using a mutant of MOR with mutated palmitoylation site Cys[170] demonstrated that palmitic acid was responsible for the incorporation of cholesterol to the receptor and thus influencing the retention of receptor in membrane rafts and receptor coupling to G proteins (unpublished observations).

5. Ubiquitination of Opioid Receptors

Ubiquitination is a posttranslational modification involving the covalent addition of a small protein, ubiquitin, to the lysine side chains of substrate protein. Ubiquitination is catalyzed by three enzymes acting in tandem: an E1, a ubiquitin-activating enzyme, an E2, a ubiquitin-carrying enzyme; and an E3, a ubiquitin ligase (69). Ubiquitin is a highly conserved 76-amino acid polypeptide, which has 7 lysines at positions 6, 11, 27, 29, 33, 48, and 63. These lysine residues (generally at Lys[48]) can subsequently be self-conjugated to additional ubiquitin molecules, generating polyubiquitin chains. Lys[48]-linked polyubiquitin usually directs the degradation of cytosolic and nuclear proteins by 26S proteasomes (70). Polyubiquitin chain linked to other lysine residues such as Lys[63] is associated with nonproteasomal pathways (69). The monoubiquitination of protein appears to participate in the endosomal sorting of the receptor, directing proteins to the late endosomes and degradation in the lysosomes. The usage of protease inhibitors can provide a clue whether ubiquitination machinery is involved in receptor function or not. More detailed study with ubiquitin-specific antibody or epitope tagged ubiquitin combined with immunoprecipitation and SDS-PAGE can substantially prove the existence of receptor ubiquitination. By these methods, ubiquitination of β_2AR was shown (71). Again, mutation of putative ubiquitination sites provides an efficient way to study the effect of ubiquitination on receptor function. A study with a mutant β_2AR having all lysine residues changed suggested that receptor ubiquitination is essential for receptor degradation (71). Moreover, the ubiquitination of β-arrestin is also shown to participate in the internalization of β_2AR (71). Furthermore, the ubiquitination status of β-arrestin determines the stability of the receptor-β-arrestin complex and subsequent cellular trafficking of the receptor (72).

Protease inhibitors can attenuate agonist-induced downregulation of MOR and DOR (73). In addition, protease inhibitors can also cause accumulation of DOR in the cytosol in the absence of agonist (74). These data suggest that receptor ubiquitination of MOR and DOR may exist and be responsible for quality control of newly-synthesized receptor and trafficking of internalized receptor. Using ubiquitin-specific antibody, the ubiquitination of MOR and DOR was shown (73, 74). Overexpression of ubiquitin can facilitate the detection of ubiquitination of proteins. This strategy was employed in study of KOR where it was found that agonist-induced ubiquitination of KOR contains predominantly Lys^{63}-linked polyubiquitin chains and is involved in agonist-induced downregulation (75). However, overexpression of ubiquitin may cause artificial ubiquitination of receptor which may not be the inherent cellular response. Moreover, the proteasomal inhibitor used to enhance the detected signal of receptor ubiquitination in these studies also impairs the credibility of results. The ubiquitination of β-arrestin 2 upon agonist activation of opioid receptors is also observed. Activation of MOR that tends to recycle after internalization induces transient ubiquitination of β-arrestin 2, while activation of DOR that usually goes to degradation after internalization causes sustained ubiquitination of β-arrestin 2 (unpublished observations), which supports the hypothesis that the intracellular trafficking of internalized receptor is dictated by the ubiquitination status of β-arrestin.

6. Other Types of Posttranslational Modifications of GPCRs

Other covalent modifications such as isoprenylation and sumoylation were discovered for GPCRs, but have not been reported for opioid receptors. Here, we briefly summarize these two types of posttranslational modifications of GPCRs.

6.1. Isoprenylation of GPCRs

Isoprenylation, one of the lipid modifications of proteins, is characterized by the attachment of an isoprenoid moiety to the sulfhydryl group of a cysteine residue through a thioether bond. Isoprenylation of proteins can be demonstrated by the metabolic labeling of $[^{3}H]$mevalonolactone (76). Using this method and site-directed mutagenesis, isoprenylation of prostacyclin receptor was identified, which is the only GPCR so far shown to be isoprenylated. Isoprenylation of prostacyclin receptor does not influence ligand binding but is required for functional G protein coupling and efficient agonist-induced receptor internalization (76). Prostacyclin receptor is also palmitoylated. Isoprenylation and palmitoylation may collectively regulate the receptor signaling (77).

6.2. Sumoylation of GPCRs

Sumoylation, mechanistically analogous to ubiquitination, is a covalent linkage of an ubiquitin-like peptide – SUMO to the ε-amino group of lysine residues of target proteins (78). Sumoylation usually occurs at the consensus sumoylation motifs Φ-Lys-X-Asp/Glu, where Φ is an aliphatic amino acid and X is any amino acid (79). Sumoylation is involved in a number of biological processes such as promoting protein transport between the cytoplasm and the nucleus, blocking other lysine dependent modification like ubiquitination, transcription regulation, and maintaining genome integrity (80). Using yeast, two-hybrid screen with C-terminal domain of metabotropic glutamate receptor 8a and 8b as baits, Sumol and other proteins in sumoylation cascade were identified as interacting proteins. Using Sumo-specific antibody and mutagenesis of the consensus sumoylation motif, the sumoylation of metabotropic glutamate receptor was further confirmed in cellular system (81). Studies of the sumoylation of GPCRs are still at an early stage and the functional significance of GPCR sumoylation is not yet clear.

7. Methods to Decipher the Posttranslational Regulation of GPCRs

A number of methods for study of the posttranslational modifications of GPCRs and their functional consequences have been mentioned above. Among these methods, site-directed mutagenesis has been extensively used for the study of all four types of posttranslational modifications. However, the detailed protocol is not provided here because of the rudimentary use of commercially available mutagenesis kit and the different strategy of mutations for one or more amino acids. In this section, we will depict the details of some conventional and newly developed methods for the study of posttranslational regulation of GPCRs.

7.1. Detection of Receptor Phosphorylation by [^{32}P]Orthophosphate Labeling

1. Seed cells expressing the interested receptors in 100 mm dishes at 70–80% confluence. Twenty-four hours later, the cell confluence should reach 90–100% which is ready for receptor phosphorylation assay.
2. Wash cells twice with phosphate-free DMEM and incubate the cells in 4 mL of the same medium for 1 h at 37°C and 10% CO_2. Then add [^{32}P]orthophosphate to get the final concentration of 100 μCi/mL and incubate the cells for 2 h at 37°C and 10% CO_2.
3. At the end of incubation, add agonists and/or other compounds for designated times. Then put the dishes on ice to stop the reaction. Remove the labeling medium rapidly and wash cells twice with ice-cold phosphate-buffered saline (PBS).

4. Add 1 mL lysis buffer (25 mM HEPES, pH 7.4, 1% Triton X-100, 5 mM EDTA, with 100 μg/mL bacitracin, 10 μg/mL leupeptin, 0.1 mM phenylmethylsulfonyl fluoride, 100 μg/mL soybean trypsin inhibitor, 10 μg/mL pepstatin A, and 20 μg/mL benzamidine as protease inhibitors and 50 mM sodium fluoride, 10 mM sodium pyrophosphate, and 0.1 mM sodium vanadate as phosphatase inhibitors) and scrape cells into microcentrifuge tubes. Rotate the samples for 1 h at 4°C, and then centrifuge the samples at 14,000 × *g* for 15 min at 4°C. Transfer supernatants to new tubes. Go to step 7 if there is no need to enrich the glycoproteins in samples by wheat germ lectin affinity chromatography.
5. Add one volume of lysis buffer without Triton X-100 to supernatants to dilute the sample. Load samples to 1 mL wheat germ lectin affinity columns preequilibrated with Buffer A (25 mM HEPES, pH 7.4, 100 mM NaCl, and 0.1% Triton X-100). Incubate for 30 min at 4°C.
6. Wash columns with 10 mL of ice-cold Buffer A 3 times. Elute the bound glycoproteins with 3 mL ice-cold Buffer A containing 0.5 mM *N*-acetylglucosamine and the protease/phosphatase inhibitors as indicated in step 4.
7. Add appropriate amount of antibody of the interested receptor to the eluates or supernatants and 60 μL prewashed immunopure protein A-agarose (or protein G-agarose) beads. Incubate overnight at 4°C.
8. Wash beads twice with Buffer A and 3 times with Buffer A without NaCl. Add 50 μL of SDS-PAGE sample buffer (62.5 mM Tris buffer, pH 6.8, 2% SDS, 10% glycerol, 5% 2-mercaptoethanol, and 0.001% bromophenol blue) to elute proteins from beads. Heat the eluates at 42°C for 1 h.
9. Run 10% SDS-PAGE to separate proteins. After electrophoresis, dry gels and put it into cassette for 2 days (adjust the time according to the intensity of signal). Use the PhosphorImager Storm 840 system (Molecular Dynamics, Sunnyvale, CA) to visualize and quantify the phosphorylated proteins.

7.2. In Vitro Phosphorylation Assay

Using the phosphorylation assay of carboxyl tail of MOR by GRK2 kinase as an example, the general procedure to conduct the *in vitro* phosphorylation assay is provided.

1. Construct the plasmids of GST fusion proteins with carboxyl tail of MOR, express and purify the proteins.
2. Set the reaction tube as follows in total of 20 μL:

   GST-carboxyl tail of MOR fusion protein (0.5 μg/μL) 0.5 μL

   [γ-^{32}P] ATP (10 μCi/μL) 2 μL

   Purified GRK2 (0.09 μg/μL) 0.4 μL

Phosphorylation buffer (20 mM Tris buffer, pH 7.5, 2 mM EDTA, 7.5 mM $MgCl_2$) 17.1 μL

3. Incubate the reaction tubes at 37°C for 30 min.
4. Add 20 μL of SDS-PAGE sample buffer to stop the reaction, boil the samples for 3–5 min.
5. Run 10% SDS-PAGE to separate proteins. After electrophoresis, dry gels and put it into cassette overnight (adjust the time according to the intensity of signal). Use the PhosphorImager Storm 840 system (Molecular Dynamics, Sunnyvale, CA) to visualize and quantify the phosphorylated proteins.

7.3. Palmitoylation Assay by Immunoprecipitation and Fatty Acyl Exchange Labeling

1. Seed cells expressing the interested receptors in 100 mm dishes at 70–80% confluence. Twenty four hour later, the cell confluence should reach 90–100%.
2. Treat cells with agonists and/or other compounds for designated times. Then put the dishes on ice to stop the reaction. Wash cells twice with ice-cold PBS.
3. Add 500 μL of lysis buffer (50 mM Tris-HCl, pH 7.5, 150 mM NaCl, 0.25% sodium deoxycholate, 0.1% Nonidet P-40, 0.5% Triton X-100, 0.1% digitonin, 50 mM sodium fluoride, 1 mM dithiothreitol, 0.5 mM phenylmethylsulfonyl fluoride, 50 mM sodium pyrophosphate, 1 mM sodium vanadate, and 1X protease inhibitor cocktail; Roche, Indianapolis, IN) and scrape cells into microcentrifuge tubes. Rotate the samples for 20 min at 4°C, and then centrifuge the samples at 14,000×g for 15 min at 4°C. Transfer supernatants to new tubes.
4. Add appropriate amount of antibody of the interested receptor to the supernatants and 30 μL prewashed immunopure protein A-agarose (or protein G-agarose) beads. Incubate overnight at 4°C.
5. Wash beads with lysis buffer 5 times. Incubate the beads in 500 μL lysis buffer with 50 mM *N*-ethylmaleimide (NEM) for 2 h at room temperature to block free sulfhydryls.
6. Wash beads with lysis buffer 5 times. Incubate the beads in 500 μL lysis buffer with 1 M hydroxylamine for 2 h at room temperature to remove thioester linked palmitic acid.
7. Wash beads with lysis buffer 5 times. Incubate the beads in 500 μL lysis buffer with 40 μM biotin-conjugated 1-biotinamido-4-[4 -(maleimidomethyl)cyclohexanecarboxamide] butane (Biotin-BMCC) for 2 h at room temperature to label the free sulfhydryl generated by hydroxylamine.
8. Wash beads with lysis buffer 5 times. Add 30 μL of SDS-PAGE sample buffer. Heat the sample at 65°C for 30 min.
9. Run 10% SDS-PAGE to separate proteins. After electrophoresis, transfer to a polyvinylidene difluoride membrane.

10. Incubate the membrane in 10% nonfat milk in TTBS (0.1% Tween 20, 50 mM Tris-HCl, pH 7.4, and 150 mM NaCl) for 1 h at room temperature.
11. Incubate the membrane with alkaline phosphatase-conjugated streptavidin in 10% nonfat milk in TTBS for 1 h at room temperature.
12. Wash the membrane with TTBS for 10 min, repeat 3 times.
13. Develop the membrane with the ECF substrate. Use the PhosphorImager Storm 840 system (Molecular Dynamics, Sunnyvale, CA) to visualize and quantify the palmitoylated proteins.

7.4. Ubiquitination Assay

1. Seed cells expressing the interested receptors in 100 mm dishes at 70–80% confluence. Twenty four hour later, the cell confluence should reach 90–100%. If using epitope tagged ubiquitin (such as HA-ubiquitin) for labeling, transfect the cells with plasmids of epitope tagged ubiquitin 24–48 h before the assay.
2. Treat cells with agonists and/or other compounds for designated times. Then put the dishes on ice to stop the reaction. Wash cells twice with ice-cold PBS.
3. Add 500 μL of lysis buffer (50 mM Tris-HCl, pH 7.5, 150 mM NaCl, 1% Triton X-100, 1 mM EDTA, 20 μM NEM, 1 mM sodium fluoride, 1 mM phenylmethylsulfonyl fluoride, 1 mM sodium vanadate, and 1X protease inhibitor cocktail; Roche, Indianapolis, IN) and scrape cells into microcentrifuge tubes. Rotate the samples for 1 h at 4°C, and then centrifuge the samples at 14,000 × *g* for 15 min at 4°C. Transfer supernatants to new tubes.
4. Add appropriate amount of antibody of the interested receptor to the supernatants and 30 μL prewashed immunopure protein A-agarose (or protein G-agarose) beads. Incubate overnight at 4°C.
5. Wash beads with lysis buffer 5 times. Add 30 μL of SDS-PAGE sample buffer to elute proteins from beads. Heat the sample at 65°C for 30 min.
6. Run 10% SDS-PAGE to separate proteins. After electrophoresis, transfer to a polyvinylidene difluoride membrane.
7. Incubate the membrane in 10% nonfat milk in TTBS for 1 h at room temperature.
8. Incubate the membrane with anti-ubiquitin antibody (1:500) in 10% nonfat milk in TTBS for 1 h at room temperature. For samples done with epitope tagged ubiquitin, incubate the membrane with anti-epitope antibody (such as mouse anti-HA from Covance, 1:1,000) in 10% nonfat milk in TTBS for 1 h at room temperature.

9. Wash the membrane with TTBS for 10 min, repeat 3 times.
10. Incubate the membrane with alkaline phosphatase-conjugated secondary antibody in 10% nonfat milk in TTBS for 1 h at room temperature.
11. Wash the membrane with TTBS for 10 min, repeat 3 times.
12. Develop the membrane with the ECF substrate. The PhosphorImager Storm 840 system (Molecular Dynamics, Sunnyvale, CA) can be used to visualize and quantify the ubiquitinated proteins.

8. Future Directions

For study of posttranslational modifications of GPCRs, a method which could monitor the dynamic process of modification would provide more information for functional meanings. Bioluminescence resonance energy transfer (BRET) has been successfully developed for real-time monitoring of ubiquitination of β-arrestin 2 in living cells (82). If the method could be extended to the monitoring of ubiquitination of GPCRs, it should be more informative to understand the procedure and function of the ubiquitination of GPCRs. Site-directed mutagenesis is heavily used and plays key roles in study of functional consequences of the posttranslational modifications of GPCRs. However, this method has critical disadvantages – mutation of single or multiple amino acids may cause conformational changes of proteins; the substituted amino acid(s) may bear function(s) other than as acceptor for the linked groups – which could mislead the functional meanings of its related modifications. Other methods such as enzymatic deglycosylation, inhibitors to prevent glycosylation and palmitoylation or protein kinase inhibitors could extend their effect to all the related proteins including the proteins critical for GPCR functions such as G proteins and thus may also lead to the incorrect interpretation of the data. Thus, a new method which could specifically remove the linked groups from amino acid of a specific protein without perturbing the physiological environment of cells would provide more reliable data.

9. Conclusions

Using various methods, posttranslational modifications of GPCRs have been delineated and have been found to be complex processes. Moreover, posttranslational modifications of GPCRs dynamically affect a broad spectrum of biological processes, including receptor expression, G protein coupling, receptor internalization and desensitization, receptor intracellular trafficking, and downregulation. The study of posttranslational modifications of GPCRs provides valuable insight into the molecular mechanisms involved in receptor functions. With the development of new methods, our understanding of posttranslational regulation of GPCRs should show dramatic advancement in the near future.

References

1. Pierce KL, Premont RT and Lefkowitz RJ (2002) Seven-transmembrane receptors. Nat Rev Mol Cell Biol 3:639–650
2. Achour L, Labbe-Jullie C, Scott MG et al (2008) An escort for GPCRs: implications for regulation of receptor density at the cell surface. Trends Pharmacol Sci **29**:528–535
3. Stiles GL, Benovic JL, Caron MG et al (1984) Mammalian *beta*-adrenergic receptors. Distinct glycoprotein populations containing high mannose or complex type carbohydrate chains. J Biol Chem **259**:8655–8663
4. George ST, Ruoho AE and Malbon CC (1986) N-glycosylation in expression and function of *beta*-adrenergic receptors. J Biol Chem **261**:16559–16564
5. Law PY, Ungar HG, Hom DS et al (1985) Effects of cycloheximide and tunicamycin on opiate receptor activities in neuroblastoma X glioma NG108-15 hybrid cells. Biochem Pharmacol **34**:9–17.
6. McLawhon RW, Cermak D, Ellory JC et al (1983) Glycosylation-dependent regulation of opiate (enkephalin) receptors in neurotumor cells. J Neurochem **41**:1286–1296.
7. Liu-Chen LY, Chen C and Phillips CA (1993) *Beta*-(3H)funaltrexamine-labeled *mu*-opioid receptors: species variations in molecular mass and glycosylation by complex-type, N-linked oligosaccharides. Mol Pharmacol **44**:749–756
8. Petaja-Repo UE, Hogue M, Laperriere A et al (2000) Export from the endoplasmic reticulum represents the limiting step in the maturation and cell surface expression of the human *delta* opioid receptor. J Biol Chem **275**:13727–13736
9. Li JG, Chen C and Liu-Chen LY (2007) N-Glycosylation of the human *kappa* opioid receptor enhances its stability but slows its trafficking along the biosynthesis pathway. Biochemistry **46**: 10960–10970
10. Gallagher JT, Morris A and Dexter TM (1985) Identification of two binding sites for wheat-germ agglutinin on polylactosamine-type oligosaccharides. Biochem J **231**:115–122
11. Bowen WD and Kooper G (1986) Photoaffinity labeling of opiate receptors with 3H-etorphine: possible species differences in glycosylation. NIDA Res Monogr **75**:17–20
12. Kristiansen K (2004) Molecular mechanisms of ligand binding, signaling, and regulation within the superfamily of G-protein-coupled receptors: molecular modeling and mutagenesis approaches to receptor structure and function. Pharmacol Ther **103**:21–80
13. Rands E, Candelore MR, Cheung AH et al (1990) Mutational analysis of *beta*-adrenergic receptor glycosylation. J Biol Chem **265**:10759–10764
14. Mialet-Perez J, Green SA, Miller WE et al (2004) A primate-dominant third glycosylation site of the *beta2*-adrenergic receptor routes receptors to degradation during agonist regulation. J Biol Chem **279**:38603–38607
15. Ge X, Loh HH and Law PY (2009) *Mu*-opioid receptor cell surface expression is regulated by its direct interaction with ribophorin I. Mol Pharmacol **75**:1307–1316
16. Bond C, LaForge KS, Tian M et al (1998) Single-nucleotide polymorphism in the human *mu* opioid receptor gene alters beta-endorphin binding and activity: possible implications for opiate addiction. Proc Natl Acad Sci USA **95**:9608–9613
17. Garzon J, Juarros JL, Castro MA et al (1995) Antibodies to the cloned *mu*-opioid receptor detect various molecular weight forms in areas of mouse brain. Mol Pharmacol **47**:738–744
18. Huang P, Chen C, Xu W et al (2008) Brain region-specific N-glycosylation and lipid rafts association of the rat *mu* opioid receptor. Biochem Biophys Res Commun **365**:82–88
19. Nagamatsu K, Suzuki K, Teshima R et al (1989) Morphine enhances the phosphorylation of a 58 kDa protein in mouse brain membranes. Biochem J **257**:165–171

20. Pei G, Kieffer BL, Lefkowitz RJ et al (1995) Agonist-dependent phosphorylation of the mouse *delta*-opioid receptor: involvement of G protein-coupled receptor kinases but not protein kinase C. Mol Pharmacol **48**:173–177
21. Arden JR, Segredo V, Wang Z et al (1995) Phosphorylation and agonist-specific intracellular trafficking of an epitope-tagged *mu*-opioid receptor expressed in HEK 293 cells. J Neurochem **65**:1636–1645
22. Appleyard SM, Patterson TA, Jin W et al (1997) Agonist-induced phosphorylation of the *kappa*-opioid receptor. J Neurochem **69**:2405–2412
23. Zhang J, Ferguson SS, Barak LS et al (1998) Role for G protein-coupled receptor kinase in agonist-specific regulation of *mu*-opioid receptor responsiveness. Proc Natl Acad Sci USA **95**:7157–7162
24. El Kouhen R, Burd AL, Erickson-Herbrandson LJ et al (2001) Phosphorylation of Ser363, Thr370, and Ser375 residues within the carboxyl tail differentially regulates *mu*-opioid receptor internalization. J Biol Chem **276**:12774–12780
25. Deng HB, Yu Y, Pak Y et al (2000) Role for the C-terminus in agonist-induced *mu* opioid receptor phosphorylation and desensitization. Biochemistry **39**:5492–5499
26. Kouhen OM, Wang G, Solberg J et al (2000) Hierarchical phosphorylation of *delta*-opioid receptor regulates agonist-induced receptor desensitization and internalization. J Biol Chem **275**:36659–36664
27. Guo J, Wu Y, Zhang W et al (2000) Identification of G protein-coupled receptor kinase 2 phosphorylation sites responsible for agonist-stimulated *delta*-opioid receptor phosphorylation. Mol Pharmacol **58**:1050–1056
28. McLaughlin JP, Xu M, Mackie K et al (2003) Phosphorylation of a carboxyl-terminal serine within the *kappa*-opioid receptor produces desensitization and internalization. J Biol Chem **278**:34631–34640
29. Schulz S, Mayer D, Pfeiffer M et al (2004) Morphine induces terminal *mu*-opioid receptor desensitization by sustained phosphorylation of serine-375. EMBO J **23**:3282–3289
30. Navratilova E, Waite S, Stropova D et al (2007) Quantitative evaluation of human *delta* opioid receptor desensitization using the operational model of drug action. Mol Pharmacol **71**:1416–1426
31. McLaughlin JP and Chavkin C (2001) Tyrosine phosphorylation of the *mu*-opioid receptor regulates agonist intrinsic efficacy. Mol Pharmacol **59**:1360–1368
32. Appleyard SM, McLaughlin JP and Chavkin C (2000) Tyrosine phosphorylation of the *kappa*-opioid receptor regulates agonist efficacy. J Biol Chem **275**:38281–38285
33. Zhang L, Zhao H, Qiu Y et al (2009) Src phosphorylation of *mu*-receptor is responsible for the receptor switching from an inhibitory to a stimulatory signal. J Biol Chem **284**:1990–2000
34. Clayton CC, Bruchas MR, Lee ML et al (2010) Phosphorylation of the *mu*-opioid receptor at tyrosine 166 (Y3.51) in the DRY motif reduces agonist efficacy. Mol Pharmacol 77:339–347
35. Kramer HK, Andria ML, Esposito DH et al (2000) Tyrosine phosphorylation of the *delta*-opioid receptor. Evidence for its role in mitogen-activated protein kinase activation and receptor internalization. Biochem Pharmacol **60**:781–792
36. Kramer HK, Andria ML, Kushner SA et al (2000) Mutation of tyrosine 318 (Y318F) in the *delta*-opioid receptor attenuates tyrosine phosphorylation, agonist-dependent receptor internalization, and mitogen-activated protein kinase activation. Brain Res Mol Brain Res **79**:55–66
37. Papac DI, Oatis JE, Jr., Crouch RK et al (1993) Mass spectrometric identification of phosphorylation sites in bleached bovine rhodopsin. Biochemistry **32**:5930–5934
38. Trester-Zedlitz M, Burlingame A, Kobilka B et al (2005) Mass spectrometric analysis of agonist effects on posttranslational modifications of the *beta*-2 adrenoceptor in mammalian cells. Biochemistry **44**:6133–6143
39. Tobin AB, Butcher AJ and Kong KC (2008) Location, location, location...site-specific GPCR phosphorylation offers a mechanism for cell-type-specific signaling. Trends Pharmacol Sci **29**:413–420
40. Tobin AB (2008) G-protein-coupled receptor phosphorylation: where, when and by whom. Br J Pharmacol **153 Suppl 1**:S167–176
41. Chakrabarti S, Law PY and Loh HH (1998) Distinct differences between morphine- and [D-Ala2,N-MePhe4,Gly-ol5]-enkephalin-*mu*-opioid receptor complexes demonstrated by cyclic AMP-dependent protein kinase phosphorylation. J Neurochem **71**:231–239
42. El Kouhen R, Kouhen OM, Law PY et al (1999) The absence of a direct correlation between the loss of [D-Ala2, MePhe4,Gly5-ol]Enkephalin inhibition of adenylyl cyclase activity and agonist-induced *mu*-opioid receptor phosphorylation. J Biol Chem **274**:9207–9215
43. Zhang L, Yu Y, Mackin S et al (1996) Differential *mu* opiate receptor phosphorylation and

desensitization induced by agonists and phorbol esters. J Biol Chem **271**:11449–11454

44. Xiang B, Yu GH, Guo J et al (2001) Heterologous activation of protein kinase C stimulates phosphorylation of *delta*-opioid receptor at serine 344, resulting in *beta*-arrestin- and clathrin-mediated receptor internalization. J Biol Chem **276**:4709–4716
45. Koch T, Kroslak T, Mayer P et al (1997) Site mutation in the rat *mu*-opioid receptor demonstrates the involvement of calcium/calmodulin-dependent protein kinase II in agonist-mediated desensitization. J Neurochem **69**:1767–1770
46. Lefkowitz RJ (1998) G protein-coupled receptors. III. New roles for receptor kinases and *beta*-arrestins in receptor signaling and desensitization. J Biol Chem **273**:18677–18680
47. Ferguson SS (2001) Evolving concepts in G protein-coupled receptor endocytosis: the role in receptor desensitization and signaling. Pharmacol Rev **53**:1–24
48. Hasbi A, Polastron J, Allouche S et al (1998) Desensitization of the *delta*-opioid receptor correlates with its phosphorylation in SK-N-BE cells: involvement of a G protein-coupled receptor kinase. J Neurochem **70**:2129–2138
49. Bohn LM, Gainetdinov RR, Lin FT et al (2000) *Mu*-opioid receptor desensitization by *beta*-arrestin-2 determines morphine tolerance but not dependence. Nature **408**:720–723
50. McLaughlin JP, Myers LC, Zarek PE et al (2004) Prolonged *kappa* opioid receptor phosphorylation mediated by G-protein receptor kinase underlies sustained analgesic tolerance. J Biol Chem **279**:1810–1818
51. Qiu Y, Loh HH and Law PY (2007) Phosphorylation of the *delta*-opioid receptor regulates its *beta*-arrestins selectivity and subsequent receptor internalization and adenylyl cyclase desensitization. J Biol Chem **282**:22315–22323
52. Wang F, Chen X, Zhang X et al (2008) Phosphorylation state of *mu*-opioid receptor determines the alternative recycling of receptor via Rab4 or Rab11 pathway. Mol Endocrinol **22**:1881–1892
53. Zhang X, Wang F, Chen X et al (2005) *Beta*-arrestin1 and *beta*-arrestin2 are differentially required for phosphorylation-dependent and -independent internalization of *delta*-opioid receptors. J Neurochem **95**:169–178
54. Wolf R, Koch T, Schulz S et al (1999) Replacement of threonine 394 by alanine facilitates internalization and resensitization of the rat *mu* opioid receptor. Mol Pharmacol **55**:263–268
55. Qiu Y, Law PY and Loh HH (2003) *Mu*-opioid receptor desensitization: role of receptor phosphorylation, internalization, and representation. J Biol Chem **278**:36733–36739
56. Zhang X, Wang F, Chen X et al (2008) Post-endocytic fates of *delta*-opioid receptor are regulated by GRK2-mediated receptor phosphorylation and distinct *beta*-arrestin isoforms. J Neurochem **106**:781–792
57. Wang H, Guang W, Barbier E et al (2007) *Mu* opioid receptor mutant, T394A, abolishes opioid-mediated adenylyl cyclase superactivation. Neuroreport **18**:1969–1973
58. Escriba PV, Wedegaertner PB, Goni FM et al (2007) Lipid-protein interactions in GPCR-associated signaling. Biochim Biophys Acta **1768**:836–852
59. Drisdel RC, Alexander JK, Sayeed A et al (2006) Assays of protein palmitoylation. Methods **40**:127–134
60. O'Brien PJ and Zatz M (1984) Acylation of bovine rhodopsin by [3H]palmitic acid. J Biol Chem **259**:5054–5057
61. Ovchinnikov Yu A, Abdulaev NG and Bogachuk AS (1988) Two adjacent cysteine residues in the C-terminal cytoplasmic fragment of bovine rhodopsin are palmitylated. FEBS Lett **230**:1–5
62. O'Dowd BF, Hnatowich M, Caron MG et al (1989) Palmitoylation of the human *beta* 2-adrenergic receptor. Mutation of Cys341 in the carboxyl tail leads to an uncoupled non-palmitoylated form of the receptor. J Biol Chem **264**:7564–7569
63. Chen C, Shahabi V, Xu W et al (1998) Palmitoylation of the rat *mu* opioid receptor. FEBS Lett **441**:148–152
64. Petaja-Repo UE, Hogue M, Leskela TT et al (2006) Distinct subcellular localization for constitutive and agonist-modulated palmitoylation of the human *delta* opioid receptor. J Biol Chem **281**:15780–15789
65. Probst WC, Snyder LA, Schuster DI et al (1992) Sequence alignment of the G-protein coupled receptor superfamily. DNA Cell Biol **11**:1–20
66. Qanbar R and Bouvier M (2003) Role of palmitoylation/depalmitoylation reactions in G-protein-coupled receptor function. Pharmacol Ther **97**:1–33
67. Mouillac B, Caron M, Bonin H et al (1992) Agonist-modulated palmitoylation of *beta* 2-adrenergic receptor in Sf9 cells. J Biol Chem **267**:21733–21737
68. Cherezov V, Rosenbaum DM, Hanson MA et al (2007) High-resolution crystal structure of an engineered human *beta*2-adrenergic G protein-coupled receptor. Science **318**: 1258–1265

69. Shenoy SK (2007) Seven-transmembrane receptors and ubiquitination. Circ Res **100**:1142–1154
70. Bonifacino JS and Traub LM (2003) Signals for sorting of transmembrane proteins to endosomes and lysosomes. Annu Rev Biochem **72**:395–447
71. Shenoy SK, McDonald PH, Kohout TA et al (2001) Regulation of receptor fate by ubiquitination of activated *beta* 2-adrenergic receptor and *beta*-arrestin. Science **294**: 1307–1313
72. Shenoy SK and Lefkowitz RJ (2003) Trafficking patterns of *beta*-arrestin and G protein-coupled receptors determined by the kinetics of *beta*-arrestin deubiquitination. J Biol Chem **278**:14498–14506
73. Chaturvedi K, Bandari P, Chinen N et al (2001) Proteasome involvement in agonist-induced down-regulation of *mu* and *delta* opioid receptors. J Biol Chem **276**: 12345–12355
74. Petaja-Repo UE, Hogue M, Laperriere A et al (2001) Newly synthesized human *delta* opioid receptors retained in the endoplasmic reticulum are retrotranslocated to the cytosol, deglycosylated, ubiquitinated, and degraded by the proteasome. J Biol Chem **276**:4416–4423
75. Li JG, Haines DS and Liu-Chen LY (2008) Agonist-promoted Lys63-linked polyubiquitination of the human *kappa*-opioid receptor is involved in receptor down-regulation. Mol Pharmacol **73**:1319–1330
76. Miggin SM, Lawler OA and Kinsella BT (2002) Investigation of a functional requirement for isoprenylation by the human prostacyclin receptor. Eur J Biochem **269**:1714–1725
77. Miggin SM, Lawler OA and Kinsella BT (2003) Palmitoylation of the human prostacyclin receptor. Functional implications of palmitoylation and isoprenylation. J Biol Chem **278**:6947–6958
78. Muller S, Hoege C, Pyrowolakis G et al (2001) SUMO, ubiquitin's mysterious cousin. Nat Rev Mol Cell Biol **2**:202–210
79. Scheschonka A, Tang Z and Betz H (2007) Sumoylation in neurons: nuclear and synaptic roles? Trends Neurosci **30**:85–91
80. Hay RT (2005) SUMO: a history of modification. Mol Cell **18**:1–12
81. Tang Z, El Far O, Betz H et al (2005) Pias1 interaction and sumoylation of metabotropic glutamate receptor 8. J Biol Chem **280**: 38153–38159
82. Perroy J, Pontier S, Charest PG et al (2004) Real-time monitoring of ubiquitination in living cells by BRET. Nat Methods **1**:203–208

Chapter 8

Discovering G Protein-Coupled Receptor Motifs Mediating Export from the Endoplasmic Reticulum

Chunmin Dong and Guangyu Wu

Abstract

Similar to many other plasma membrane proteins, G protein-coupled receptors (GPCRs) are synthesized in the endoplasmic reticulum (ER). After proper assembly and folding, the receptors are transported from the ER through the Golgi to the cell surface. As the first step in the anterograde trafficking, exit from the ER of nascent GPCRs plays a crucial role in receptor biosynthesis and controls receptor expression at the cell surface, which in turn dictates the functionality of the receptors. Recent studies have revealed that GPCRs possess linear motifs that are required for their export from the ER. In this chapter, we will discuss experimental approaches to identify the GPCR motifs which regulate anterograde transport, specifically exit from the ER, by focusing on α_{2B}-adrenergic receptor.

Key words: GPCR, GPCR trafficking, Anterograde transport, Export, ER, Motif, Cell surface expression, Ligand binding

1. Introduction

G protein-coupled receptors (GPCRs) constitute the largest superfamily of cell surface receptors that modulate a variety of cell functions mainly by coupling to heterotrimeric G-proteins to activate downstream effectors. Similar to many other plasma membrane proteins, the life of GPCRs begins at the endoplasmic reticulum (ER), where they are synthesized, folded, and assembled. Newly synthesized and properly folded receptors export from the ER, via ER-derived COPII-coated transport vesicles, and pass through the ER-Golgi intermediate complex (ERGIC), the Golgi, and the trans-Golgi network (TGN) en route from the ER to the cell surface (1). GPCRs at the plasma membrane may undergo internalization upon stimulation by their ligands.

Craig W. Stevens (ed.), *Methods for the Discovery and Characterization of G Protein-Coupled Receptors*, Neuromethods, vol. 60, DOI 10.1007/978-1-61779-179-6_8, © Springer Science+Business Media, LLC 2011

Internalized receptors in endosomes are sorted either to the recycling endosome for return to the plasma membrane or to the lysosome for degradation. The balance of these dynamic intracellular trafficking dictates the level of receptor expression at the cell surface, which in turn influences the magnitude of cellular response to a given signal.

Over the past decades, most studies on GPCR trafficking have focused on the events involved in receptor internalization, recycling, and degradation. These studies have demonstrated that receptor trafficking plays a crucial role in regulating receptor function and the development of a number of human diseases. In contrast, the molecular mechanisms underlying the export of nascent GPCRs from the ER to the cell surface are just beginning to be revealed. Recent studies from our and other laboratories have demonstrated that, similar to the endocytic trafficking, the anterograde transport of GPCRs is also a highly regulated, dynamic process, which is coordinated by a number of motifs imbedded within the receptors (2–9) as well as many regulatory proteins such as small GTP-binding proteins (10–16). We have focused on family A GPCRs, particularly α_2-adrenergic receptors (α_2-ARs). There are three α_2-ARs, designated as α_{2A}-AR, α_{2B}-AR, and α_{2C}-AR, which all have important roles in regulating sympathetic nervous system, both peripherally and centrally. Our studies have demonstrated that the $F(x)_6LL$ motif in the C-terminus and a single leucine residue in the first intracellular loop are required for α_{2B}-AR to exit from the ER (3, 4, 9), whereas the YS motif in the N-terminus modulates receptor transport at the level of the Golgi (5). In this chapter, we will discuss experimental approaches to identify specific sequences/motifs involved in the regulation of ER export of newly synthesized GPCRs.

2. Experimental Approaches to Identify the GPCR Motifs that Regulate ER Export

In general consideration, there are three major steps in searching for the motifs required for export from the ER of a given GPCR; see Fig. 1 (1) genetically manipulate the receptor gene to generate the constructs in which the domains of interest are truncated or specific amino acid residues are mutated; (2) measure the effect of the deletion or mutation on receptor transport from the ER to the cell surface; (3) elucidate the possible mechanisms underlying the function of identified motifs modulating receptor export. In the following sections, we will discuss the experimental approaches for each of these steps involved to identify the motifs essential for GPCR export and describe the detailed experimental methods to accurately quantitate cell surface numbers and directly visualize the subcellular localization of the receptors in cells by mainly using α_{2B}-AR as a model.

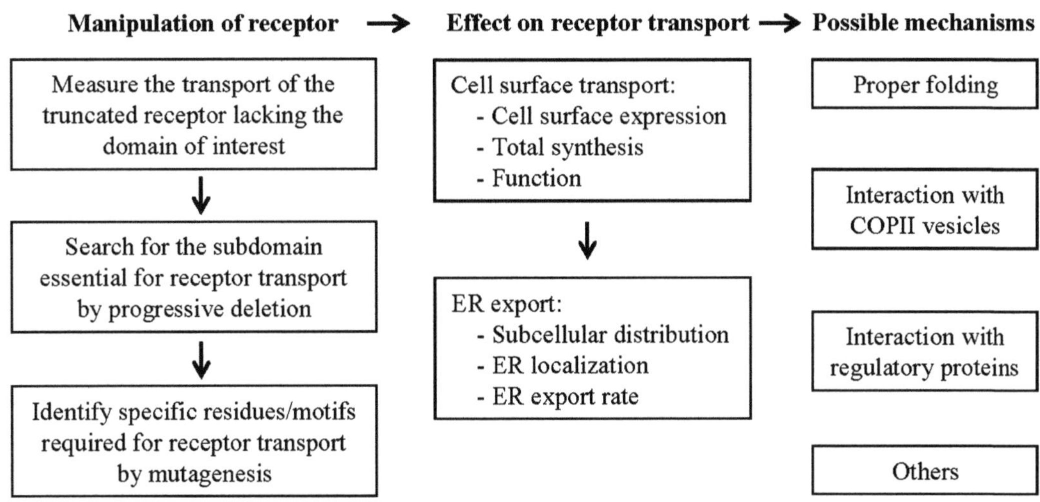

Fig. 1. Experimental approaches to the identification of the motifs involved in the regulation of GPCR export from the ER.

2.1. Manipulation of GPCRs

In order to identify the motifs essential for export from the ER, the first step is to select a domain to be manipulated in a given GPCR. The C-terminus and three intracellular loops are positioned towards the cytoplasm which may provide docking sites for components of the ER-derived COPII transport vesicles, whereas the N-terminus and three extracellular loops are located in the lumen of the ER and are spatially impossible to directly interact with transport machinery in the cytoplasm. Therefore, the C-terminus and the intracellular loops should be primarily considered in searching for the GPCR motifs which are able to physically interact with cytosolic transport machinery.

Once a specific domain is selected to be studied, the next step is to generate the receptor mutant in which the domain of interest is removed and the transport properties of the truncated receptor mutant are compared with those of its wild-type counterpart. Evaluation of the receptor transport is described in the Sect. 8.2.2.

If the deletion mutant indeed has defect in the transport, it may contain signals important for receptor transport. If the domain is not too long, such as the C-terminus of α_{2B}-AR which contains only 20 amino acid residues, site-directed mutagenesis can be directly applied to define specific residues that modulate receptor transport. Each of the residues in the domain of interest is mutated to alanine and then the effect of mutations on receptor transport is determined. If site-directed mutagenesis is too cumbersome for a domain which is relatively large, the progressive deletion strategy can be utilized to search for a subdomain containing the export signal. For example, after demonstrating that deletion of the entire C-terminal 57 amino acid residues abolishes the cell surface transport of angiotensin II type 1 receptor (AT1R), we used the progressive deletion strategy to identify the fragment between residues K307 and I320 essential for AT1R cell surface export (4).

2.2. Analysis of Receptor Export

For each of the truncated or mutated receptors mentioned above, their anterograde transport properties will be analyzed and the results will be compared with the wild-type receptors. Several complementary methods should be utilized to characterize receptor export. The cell surface expression and the export from intracellular compartments, specifically from the ER, have been used in our laboratory as important parameters in defining if a truncated or mutated receptor has defect in the anterograde transport from the ER. Also see Chap. 10 by Sawyer in this book for a novel "catch and release" method for measuring anterograde trafficking of GPCRs.

2.2.1. Measurement of the Cell Surface Expression of GPCRs

If the cell surface expression of a receptor mutant, in which the domain of interest is deleted or individual amino acid residue mutated, is significantly lower than that of its wild-type counterpart, the deleted domain or the mutated residue is important for receptor export to the cell surface. Therefore, to accurately measure the cell surface expression of a given GPCR is a critical step in searching for ER export motifs. The cell surface expression of GPCRs can be quantitated by a number of well-established methods, such as ligand binding, flow cytometry, and biotinylation. Our laboratory has successfully established a ligand binding assay in intact, living cells for quantifying the cell surface numbers of a group of GPCRs, including α_1-AR, α_2-AR, β-AR, and AT1R (3, 9, 12, 14, 15). This method has been used to study the role of Ras-like small GTPases and specific motifs in modulating the intracellular trafficking of GPCRs. For example, we have demonstrated that mutation to alanines of the residues Y12/S13 in the N-terminus, L48 in the first intracellular loop, and I443/F444 and F436 in the C-terminus markedly reduces the cell surface expression of α_{2B}-AR measured by intact cell ligand binding (3, 5, 9). The measurement of the cell surface expression of α_1-AR, β-AR, and AT1R by intact cell ligand binding is described elsewhere (11, 12, 14). The following steps describe the intact cell ligand binding assay to measure the cell surface expression of α_{2B}-AR using the radioligand [^{3}H]-RX821002.

1. Plate HEK293 cells on six-well dishes at a density of 80×10^5 and culture the cells in Dulbecco's modified Eagle's medium (DMEM) with 10% fetal bovine serum, 100 U/mL penicillin, and 100 U/mL streptomycin overnight.
2. Transfect cells with 1.0 μg of wild-type and mutated α_{2B}-AR plasmids using lipofactamine 2000 reagent (Invitrogen) by the standard procedure.

3. After 6 h, split the cells into 12-well dishes precoated with poly-L-lysine at a density of 5×10^5 cells/well.
4. 24–36 h posttransfection, incubate cells with DMEM plus [^{3}H]-RX821002 (specific activity=41 Ci/ mmol from PerkinElmer) at a concentration of 20 nM in a total of 400 μL for 90 min at room temperature.
5. For the nonspecific binding, incubate the cells with [^{3}H]-RX821002 plus rauwolscine (Sigma-Aldrich) at a concentration of 10 μM.
6. Wash the cells twice with 1 mL ice-cold DMEM to remove the excess radioligand.
7. Extract the remained ligand by digesting the cells with 1 M NaOH for 2 h at room temperature.
8. Collect the liquid phase and mix with 5 mL of Ecoscint A scintillation fluid (National Diagnostics Inc., Atlanta, GA). The amount of radioactivity retained is measured by liquid scintillation spectrometry.

Ligand binding of membrane preparations has been long used to quantify GPCR expression. The major advantage of intact cell ligand binding is that it has the ability to accurately measure the numbers of receptors at the plasma membrane, compared with ligand binding of total membrane preparations which contain the receptors expressed in the intracellular compartments, such as the ER and the Golgi. This is particularly important when the mutation or deletion disrupts receptor export, resulting in intracellular retention of the receptor, but does not alter receptor ligand binding properties. However, one issue associated with intact cell radioligand binding assay is that radiolabeled ligands may be able to induce receptor internalization (such as [^{3}H]-Ang II for AT1R). One strategy to limit receptor internalization upon stimulation with the radiolabeled agonists is to carry out the ligand binding assay at low temperature as described (11).

As mutation of the receptors may alter receptor ligand binding affinity, other alternative methods such as flow cytometry should be used to confirm the intact cell ligand binding results. For this purpose, the receptors are tagged with an epitope such as FLAG or HA at their N-termini, which does not alter receptor trafficking and signaling. The cell surface expression of the receptors is measured by flow cytometry following staining with antiepitope antibodies in nonpermeabilized cells. We have used flow cytomtery to demonstrate that mutation of the residue L48 in the first intracellular loop dramatically inhibits the cell surface expression of α_{2B}-AR, consistent with the intact cell ligand binding results (Fig. 2).

In addition to the cell surface expression, it is very important to define if the mutation or deletion influences the overall synthesis of the receptor, which can be accomplished by a number of methods.

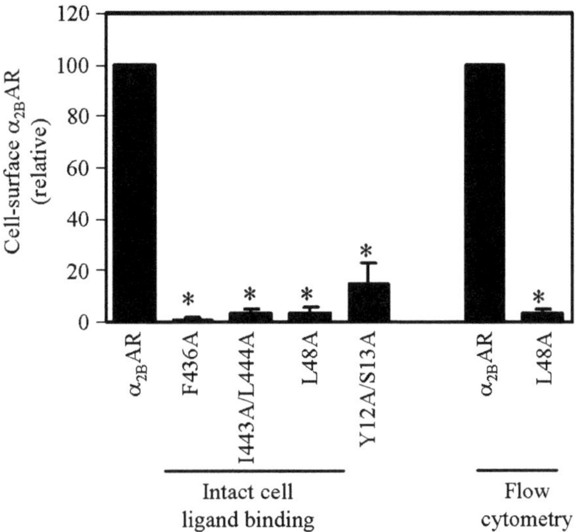

Fig. 2. Inhibition of the cell surface transport of α_{2B}-AR by mutating residues F436, I443/L444, L48, and Y12/S13 as measured by intact cell ligand binding and flow cytometry. For intact cell ligand binding, HEK293 cells transiently transfected with wild-type and mutated α_{2B}-AR were incubated with DMEM plus [^{3}H]-RX821002 at a concentration of 20 nM in a total of 400 μL for 90 min at room temperature. For measurement of α_{2B}-AR cell surface expression by flow cytometry, HEK293 cells were transfected with HA-tagged α_{2B}-AR for 36–40 h. The cells were suspended in PBS containing 1% fetal calf serum at a density of 4×10^6 cells/mL and incubated with high affinity anti-HA-fluorescein (3F10, Roche) at a final concentration of 2 μg/mL for 30 min at 4°C. *$p<0.05$ vs. wild-type α_{2B}-AR. The data are adapted from (3, 5, 9).

If the receptor is tagged with fluorescence proteins such as GFP, the simplest way to measure the overall receptor expression is to directly measure the total fluorescent signal of GFP by flow cytometry or the total GFP expression by immunoblotting using anti-GFP antibodies (3).

It is also necessary to exclude the possibility that the reduction in the cell surface expression of a mutated receptor is caused by constitutive receptor internalization from the plasma membrane to endosomes. For this goal, receptor internalization blockers can be used to inhibit endocytic transport. If the cell surface expression of a mutated receptor is restored upon coexpression with dominant-negative internalization mutants or treatments with pharmacological blockers, defective export of the mutated receptor is likely caused by facilitated endocytic transport. For example, we have demonstrated that transient expression of dominant-negative mutants of arrestin-3, Rab5 and dynamin does not significantly alter the inhibitory effect of mutation of the Y12/S13 motif on the cell surface expression of α_{2B}-AR (5), suggesting that the reduction in the α_{2B}-AR cell surface expression upon mutation of the motif is unlikely caused by promoting constitutive internalization of the receptor.

In addition to measuring the cell surface expression of the receptor, it is also necessary to define if the mutation of the traffic motif could attenuate signaling elicited by that particular receptor. Although it is expected that the effects of the mutation on the cell surface expression and signaling will be parallel, dysfunction of the mutated receptor will not only confirm the defect in the anterograde transport but also indicate the importance of export process in regulating receptor signal propagation in cells. The selection of functional readouts depends on the type of receptor studied. In our studies, the agonist-mediated activation of the mitogen-activated protein kinases ERK1/2 has been used as a functional readout for several family A GPCRs. We have demonstrated that attenuation of receptor expression at the cell surface by mutating the export signals reduces ERK1/2 activation in response to agonist stimulation (3, 5, 9).

2.2.2. Analysis of Receptor Export from the ER

Once a domain or a specific motif is identified to be important for receptor export, the next step is to directly visualize the subcellular distribution of the mutated receptor by microscopy. At least two important pieces of information regarding receptor anterograde transport could be obtained by microscopy. The first is that direct visualization of intracellular distribution of a mutated receptor in cells will confirm the results obtained by measurement of the cell surface expression using intact cell ligand binding or flow cytometry. If a mutated receptor is extensively localized inside the cell but not on the cell surface, this, together with the reduced cell surface expression measured by intact cell ligand binding, will further indicate that the mutated residue is essential for receptor cell surface transport. The second is that microscopic analysis of receptor localization is able to precisely define the intracellular compartment where the mutated residue modulates receptor export. This could be accomplished by colocalization of the mutated receptors with various intracellular organelle markers. For example, by employing the intact cell ligand binding assay, we identified the $F(x)_6LL$ motif in the C-terminus, a single L48 residue in the first intracellular loop, and the YS motif at the N-terminus of α_{2B}-AR playing critical roles in regulating receptor anterograde trafficking (3–5, 9), which are indicated by the decreased cell surface expression of the mutant receptors. Further studies employing microscopy technology to colocalize the mutant receptors with different organelle markers demonstrate that they are trapped in distinct subcellular compartments. In contrast to the ER localization of the $F(x)_6LL$ and L48 mutants, which is reflected by their extensive colocalization with the ER marker DsRed2-ER (3) (Fig. 3a), the YS mutant receptor was colocalized with the cis-Golgi marker GM130 (5) (Fig. 3b). These data suggest that the $F(x)_6LL$ motif and L48 are involved in the ER export of α_{2B}-AR, but the YS motif is involved in

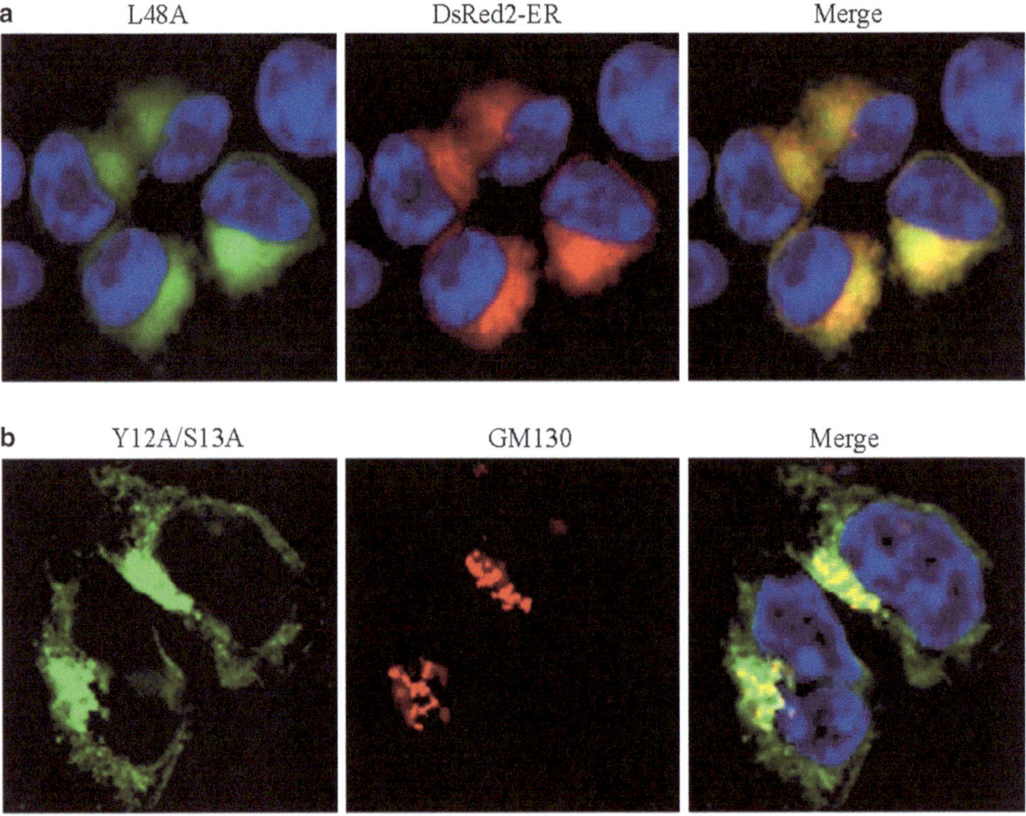

Fig. 3. Colocalization of α_{2B}-AR with ER and Golgi markers. (**a**) Colocalization of the α_{2B}-AR mutant L48A with the ER marker DsRed2-ER. HEK293 cells were transfected with the GFP-tagged L48A mutant together with pDsRed2-ER and the subcellular distribution and colocalization of the receptor with DsRed2-ER were revealed by fluorescence microscopy. (**b**) Colocalization of the α_{2B}-AR Y12A/S13A mutant with the cis-Golgi marker GM130. HEK293 cells were transfected with Y12A/S13A and its colocalization with GM130 was revealed by fluorescence microscopy following staining with antibodies against GM130 at 1:50 dilution. *Top left*, α_{2B}-AR mutants tagged with GFP; *middle*, the ER marker DsRed2-ER (**a**) and the Golgi marker GM130 (**b**); colocalization of the α_{2B}-AR mutants with the ER (**a**) and the Golgi (**b**); DNA staining by 4,6-diamidino-2-phenylindole (nuclei). Scale bars, 10 μm. The data shown in (**a**) and (**b**) are adapted from (3) and (5), respectively.

regulation of the receptor export from the Golgi. Analysis of the subcellular colocalization of GFP-tagged α_{2B}-AR with ER markers in fixed cells is described below.

1. Transfect HEK293 cells on six-well plates with 100 ng of wild-type or mutated α_{2B}-AR tagged with GFP at its C-terminus.
2. 6 h later, split the cells onto cover slips precoated with poly-L-lysine in six-well plates.
3. 24 h following transfection, wash the cells with ice-cold phosphate buffer saline (PBS) twice with gentle shaking (50 rpm for 5 min).

4. Fix the cells with fresh 2 mL of 4% paraformaldehyde-4% sucrose mixture in PBS for 15 min.
5. Wash the cells with 2 mL of ice-cold PBS for 3 × 5 min.
6. Permeablize the cells with 2 mL of PBS containing 0.2% Triton X-100 for 5 min.
7. Wash the cells with 2 mL ice-cold PBS for 3 × 5 min.
8. Incubate the cells with 0.24% normal donkey serum (Jackson Immuno) in PBS for 1 h.
9. Incubate the cells with 50 μL of primary antibodies against the ER marker calregulin or calnexin (Santa Cruz) at 1:50 dilution on parafilm with the cover slips upside down for 2 h. Notes: Antibodies against other intracellular markers should be used as controls, such as ERGIC53 (Alexis Biochemicals) – an ERGIC marker; GM130 (BD Biosciences) – a cis-Golgi marker; p230 (BD Biosciences) – a TGN marker.
10. Gently transfer the cover slips into six-well plates and wash the cells with cold PBS for 3 × 5 min.
11. Incubate the cells with 1 mL of Alexa Fluor 594-labeled secondary antibody (Invitrogen) at a dilution of 1:2,000 for 1 h.
12. Wash the cells with 2 mL ice-cold PBS for 3 × 5 min.
13. Incubate the cells with 4, 6-diamidino-2-phenylindole for 5 min to stain the nuclei.
14. Wash the cells with 2 mL ice-cold PBS for 3 × 5 min.
15. Mount the cover slips with the mounting media.
16. Detect fluorescence with an appropriate microscope. In most of our studies, a Leica DMRA2 epifluorescent microscope was used and images were deconvolved using SlideBook software and the nearest neighbor deconvolution algorithm (Intelligent Imaging Innovations, Denver, CO).

For colocalization of α_{2B}-AR with ER markers in live cells, the cells are cotransfected with α_{2B}-AR-GFP together with pDsRed2-ER (BD Biosciences) in equal amounts (100 ng) for 36–40 h. One hour before imaging, culture medium is replaced with CO_2-independent medium (Invitrogen) and images are obtained in a Zeiss Axiovert microscope (200 M) in our studies.

In addition to colocalization in cells, subcellular fractionation by sucrose gradient can also be used to define the intracellular compartment where the identified motif modulates receptor transport. By using this method, we have shown that the α_{2B}-AR mutant L48A is comigrated with the ER marker calnexin (3), suggesting that the L48 residue modulates α_{2B}-AR export from the ER.

Analysis of the cell surface expression and subcellular distribution of receptors as described earlier will provide important information about receptor anterograde transport from the ER to the cell surface at the steady status. As receptor transport passes through a number of intracellular compartments such as the ERGIC, Golgi, and TGN en route from the ER to the cell surface, the influences of a specific motif on the cell surface expression and export from the ER may not be identical. It is possible that mutation of an ER export motif significantly slows down the rate of receptor export from the ER, without dramatically reducing receptor cell surface expression and disrupting subcellular localization at the steady status. Therefore, after identifying a motif modulating receptor export from the ER as discussed above, one should consider to measuring the effect of mutating the motif on the kinetics of receptor export from the ER.

The kinetics of receptor export from the ER can be analyzed by measuring the formation of complex glycosylated receptors by metabolic pulse-chase labeling, combined with immunoprecipitation and deglycosylation (6, 17). Most GPCRs contain N-linked glycosylation sites at their N-termini and the extracellular loops. For example, AT1R and β_2-AR have putative N-linked glycosylation sites at positions 4, 176, and 188 and positions 6, 15, and 187, respectively. Glycosylation of GPCRs is initiated in the ER and completed during their transport through the Golgi apparatus, resulting in mature receptors, competent for subsequent transport to the cell surface. Therefore, the status of glycosylation can tell us if a receptor is able to exit from the ER and transport to the Golgi.

In a typical experiment, a GPCR is tagged with an epitope such as HA or FLAG at its N-terminus and transiently expressed in cells. The cells are metabolically labeled with [^{35}S]-methionine/cysteine and chased for different periods of time. After the receptors are immunoprecipitated with antiepitope antibodies, the immunoprecipitated receptors are treated with deglycosidases, peptide N-glycosidase F (PNGase), and endoglycosidase H (EndoH), and visualized by autoradiography following SDS-PAGE. Whereas PNGase F removes the glycosylation that occurs in both the ER and the Golgi (complex glycosylation), Endo H selectively removes unprocessed high-mannose oligosaccharides (core glycosylation in the ER). The receptors resistant to Endo H and sensitive to PNGase F treatment are those transported to the Golgi complex. Since pulse chase is carried out for variable periods of time, the rate and extent of receptor transport from the ER to the Golgi will be determined.

2.3. Elucidation of Possible Mechanisms of Motifs in ER Transport

Once a motif is identified to modulate receptor export from the ER, the next step is to delineate the possible molecular mechanisms. This step is much more difficult than the first two steps, because a number of factors, such as proper folding, recruitment onto the transport vesicles, interactions with accessory proteins, and dimerization, potentially modulate the ER export of GPCRs. Disruption of each of these processes could lead to defective export from the ER. For example, constitutive dimerization in the ER that has been well described for many GPCRs may be a prerequisite for receptor export from the ER to the cell surface (17, 18). It is possible that the reduction in the cell surface expression and ER accumulation of a receptor upon mutating a specific motif is induced by disrupting receptor dimerization. In our studies, we have demonstrated that mutation of the $F(x)_6LL$ motif in the C-termini, a single L48 residue in the first intracellular loop and Y12/S13 in the N-terminus does not influence α_{2B}-AR dimerization in the ER (3, 19). In the following sections, we will briefly discuss how proper folding, recruitment onto the COPII-coated transport vesicles and interaction with accessory proteins influence the ER export of GPCRs.

2.3.1. Proper Folding

Misfolding is the most common cause for the loss of ability of GPCRs to transport from the ER to the cell surface. Indeed, a number of loss-of-function GPCRs associated with many hereditary diseases are induced by their misfolding, resulting in intracellular accumulation of the receptors. Circular dichroism, treatments with chemical and pharmacological chaperones, and low temperature culture have been used to determine if a GPCR mutant is correctly folded. For example, the pharmacological chaperones are cell permeable receptor ligands, which stabilize the conformation of misfolded receptors for long enough to evade the scrutiny of the ER resident chaperones thereby promoting their export from the ER. However, these methods may not work for all kinds of misfolding. Therefore, the folding problem has been very difficult to be excluded in studying GPCR export from the ER.

2.3.2. Recruitment onto and Direct Interaction with the COPII Vesicles

Protein export from the ER is exclusively mediated through the COPII-coated transport vesicles. Assembly of the COPII vesicles takes place on the ER membrane at discrete locations called ER exit sites and is initiated by Sar1 activation, which recruits the Sec23/24 complex to the ER membrane and is then clustered by Sec13/31. It has been demonstrated that protein export from the ER is a selective process that may be dictated by short, linear sequences called ER export motifs. The ER export motifs mediate cargo interaction with Sec24 subunits. This interaction facilitates the recruitment of cargo molecules onto the ER exit sites, resulting in the concentration of cargo in the COPII vesicles (20). Of various ER export motifs identified, the diacidic (DxE) and the dihydrophobic (FF) motifs in the cytoplasmic C-termini of several membrane proteins (21–28) have been shown to mediate cargo interaction with Sec24. More importantly, these motifs are able to confer their transport properties to other ER-retained proteins, thus functioning as independent ER export signals (8, 22–27).

Similar to other plasma membrane proteins, GPCR export from the ER is mediated through COPII-coated transport vesicles (15). Although a number of specific sequences have been shown to modulate GPCR export from the ER and transport to the cell surface, none of them has been shown to directly interact with components of the COPII vesicles or to facilitate the recruitment of receptor onto the COPII vesicles. Therefore, whether or not ER export of GPCRs is dictated by independent ER export motifs needs further investigation. Our recent studies have demonstrated that α_{2B}-AR is able to physically associate with Sec24 (Dong C and Wu G. Unpublished data) and the identification of specific amino acid residues mediating receptor binding to Sec24 is underway.

2.3.3. Interaction with Regulatory Proteins

GPCR export from the ER is modulated by direct interactions with multiple regulatory proteins such as ER chaperones, accessory proteins, and receptor activity modifying proteins, which may stabilize receptor conformation, facilitate receptor maturation, and promote receptor delivery to the plasma membrane. The influence of these regulatory proteins on GPCR export trafficking is summarized in our review article (1). It is possible that disruption of the interactions between GPCRs and regulatory proteins leads to the defect in receptor targeting to the cell surface (29). Therefore, if the identified motif modulating receptor ER export does not alter proper folding, interaction with COPII vesicles, and dimerization, one should search for proteins interacting with the motif using any protein–protein interaction assays.

3. Future Directions and Conclusions

The physiological functions of GPCRs are dependent on their precise localization in the cell. Indeed, defective GPCR transport from the ER to the cell surface is associated with the pathogenesis of a variety of human diseases. For example, numerous naturally occurring mutations in GPCRs themselves prevent proper folding and lead to ER retention, which have been implicated in inherited diseases such as nephrogenic diabetes insipidus, retinitis pigmentosa, and male pseudohermaphroditism (30, 31). It is becoming increasingly appreciated that GPCRs may carry specific codes for their exit from intracellular compartments, particularly the ER. However, the molecular mechanisms remain poorly understood. Therefore, an interesting direction in GPCR export trafficking is to search for proteins interacting with these well-defined motifs that modulate receptor export from the ER. Probably, the most exciting experiments are to search for linear, independent ER export motifs which are able to not only mediate GPCR interaction with components of the COPII vesicles but also confer its transport ability to facilitate the ER exit of other ER-retained proteins. Nevertheless, further elucidation of the molecular mechanisms underlying the anterograde transport of nascent GPCRs will provide an important foundation for developing new therapeutic strategies in treating diseases involving abnormal trafficking and functioning of the receptors.

Acknowledgments

This work was supported by National Institutes of Health grant R01GM076167.

References

1. Dong C, Filipeanu CM, Duvernay MT et al (2007) Regulation of G protein-coupled receptor export trafficking. Biochim Biophys Acta **1768**:853–870.
2. Schulein R, Hermosilla R, Oksche A et al (1998) A dileucine sequence and an upstream glutamate residue in the intracellular carboxyl terminus of the vasopressin V2 receptor are essential for cell surface transport in COS.M6 cells. Mol Pharmacol **54**:525–535.
3. Duvernay MT, Dong C, Zhang X et al (2009) A single conserved leucine residue on the first intracellular loop regulates ER export of G protein-coupled receptors. Traffic **10**:552–566.
4. Duvernay MT, Zhou F, Wu G (2004) A conserved motif for the transport of G protein-coupled receptors from the endoplasmic reticulum to the cell surface. J Biol Chem **279**:30741–30750.
5. Dong C, Wu G (2006) Regulation of anterograde transport of *alpha*2-adrenergic receptors by the N termini at multiple intracellular compartments. J Biol Chem **281**:38543–38554.
6. Bermak JC, Li M, Bullock C, Zhou QY (2001) Regulation of transport of the dopamine D1 receptor by a new membrane-associated ER protein. Nat Cell Biol **3**:492–498.
7. Robert J, Clauser E, Petit PX et al (2005) A novel C-terminal motif is necessary for the export of the vasopressin V1b/V3 receptor to the plasma membrane. J Biol Chem **280**:2300–2308.
8. Carrel D, Hamon M, Darmon M (2006) Role of the C-terminal di-leucine motif of 5-HT1A and 5-HT1B serotonin receptors in plasma membrane targeting. *J Cell Sci* **119**: 4276–4284.
9. Duvernay MT, Dong C, Zhang X et al (2009) Anterograde trafficking of G protein-coupled receptors: function of the C-terminal F(X)6LL motif in export from the endoplasmic reticulum. Mol Pharmacol **75**:751–761.
10. Zhang X, Wang G, Dupre DJ et al (2009) Rab1 GTPase and dimerization in the cell surface expression of angiotensin II type 2 receptor. J Pharmacol Exp Ther **330**:109–117.
11. Filipeanu CM, Zhou F, Claycomb WC et al (2004) Regulation of the cell surface expression and function of angiotensin II type 1 receptor by Rab1-mediated endoplasmic reticulum-to-Golgi transport in cardiac myocytes. J Biol Chem **279**:41077–41084.
12. Filipeanu CM, Zhou F, Fugetta EK et al (2006) Differential regulation of the cell-surface targeting and function of *beta*- and *alpha*1-adrenergic receptors by Rab1 GTPase in cardiac myocytes. Mol Pharmacol **69**:1571–1578.
13. Wu G, Zhao G, He Y (2003) Distinct pathways for the trafficking of angiotensin II and adrenergic receptors from the endoplasmic reticulum to the cell surface: Rab1-independent transport of a G protein-coupled receptor. J Biol Chem **278**:47062–47069.
14. Dong C, Wu G (2007) Regulation of anterograde transport of adrenergic and angiotensin II receptors by Rab2 and Rab6 GTPases. Cell Signal **19**:2388–2399.
15. Dong C, Zhou F, Fugetta EK et al (2008) Endoplasmic reticulum export of adrenergic and angiotensin II receptors is differentially regulated by Sar1 GTPase. Cell Signal **20**:1035–1043.
16. Madziva MT, Birnbaumer M (2006) A role for ADP-ribosylation factor 6 in the processing of

G-protein-coupled receptors. J Biol Chem **281**:12178–12186.

17. Petaja-Repo UE, Hogue M, Laperriere A et al (2000) Export from the endoplasmic reticulum represents the limiting step in the maturation and cell surface expression of the human delta opioid receptor. J Biol Chem **275**: 13727–13736.
18. Salahpour A, Angers S, Mercier JF et al (2004) Homodimerization of the *beta*2-adrenergic receptor as a prerequisite for cell surface targeting. J Biol Chem **279**:33390–33397.
19. Zhou F, Filipeanu CM, Duvernay MT et al (2006) Cell-surface targeting of *alpha*2-adrenergic receptors – inhibition by a transport deficient mutant through dimerization. Cell Signal **18**:318–327.
20. Farhan H, Reiterer V, Korkhov VM et al (2007) Concentrative export from the endoplasmic reticulum of the *gamma*-aminobutyric acid transporter 1 requires binding to SEC24D. J Biol Chem **282**:7679–7689.
21. Kappeler F, Klopfenstein DR, Foguet M et al (1997) The recycling of ERGIC-53 in the early secretory pathway. ERGIC-53 carries a cytosolic endoplasmic reticulum-exit determinant interacting with COPII. J Biol Chem **272**:31801–31808.
22. Nishimura N, Balch WE (1997) A di-acidic signal required for selective export from the endoplasmic reticulum. Science **277**:556–558.
23. Nishimura N, Bannykh S, Slabough S et al (1999) A di-acidic (DXE) code directs concentration of cargo during export from the endoplasmic reticulum. J Biol Chem **274**:15937–15946.
24. Nishimura N, Plutner H, Hahn K et al (2002) The *delta* subunit of AP-3 is required for efficient transport of VSV-G from the trans-Golgi network to the cell surface. Proc Natl Acad Sci USA **99**:6755–6760.
25. Nufer O, Guldbrandsen S, Degen M et al (2002) Role of cytoplasmic C-terminal amino acids of membrane proteins in ER export. J Cell Sci **115**:619–628.
26. Nufer O, Kappeler F, Guldbrandsen S et al (2003) ER export of ERGIC-53 is controlled by cooperation of targeting determinants in all three of its domains. J Cell Sci **116**:4429–4440.
27. Votsmeier C, Gallwitz D (2001) An acidic sequence of a putative yeast Golgi membrane protein binds COPII and facilitates ER export. EMBO J **20**:6742–6750.
28. Wendeler MW, Paccaud JP, Hauri HP (2007) Role of Sec24 isoforms in selective export of membrane proteins from the endoplasmic reticulum. EMBO Rep **8**:258–264.
29. Ge X, Loh HH, Law PY (2009) *Mu*-opioid receptor cell surface expression is regulated by its direct interaction with ribophorin I. Mol Pharmacol 75:1307–1316.
30. Aridor M, Hannan LA (2002) Traffic jams II: an update of diseases of intracellular transport. Traffic **3,** 781–790.
31. Morello JP, Bichet DG (2001) Nephrogenic diabetes insipidus. Ann Rev Physiol 63, 607–630.

Chapter 9

Identifying G Protein-Coupled Receptor Escorts, Chaperones, and Intracellular Tethers Regulating Receptor Density at the Cell Surface

Stefano Marullo, Liliana Pardo Lopez, and Lamia Achour

Abstract

G protein-coupled receptor (GPCR) responsiveness is dynamically regulated by various mechanisms, allowing fine-tuning of cell signaling. Modulation of GPCR plasma membrane density, via their release from intracellular compartments, constitutes a recently identified important process in this context. This phenomenon requires a complex network of interactions between GPCRs, "private" chaperones and escort proteins, and gatekeepers, which are directly involved in the retention of GPCRs in the intracellular compartments. The molecular and functional characterization of the players in this game is at its very beginning and requires appropriate quantitative methods of investigation to unravel the mechanisms that are involved.

Key words: GPCR, Chaperone, Escort protein, Secretory pathway, Resensitization

1. Introduction

G protein-coupled receptors (GPCRs) are the largest family of receptors and are involved in nearly all the major physiological functions of the organism (1). Since the vast majority of signaling events elicited by GPCRs take place at the cell surface and since many different experimental approaches have confirmed beyond any doubt that GPCRs are present at the plasma membrane, the common and reasonable assumption is that these receptors must spend most of their time in this compartment. Since GPCRs are synthesized in the endoplasmic reticulum (ER), as all other plasma membrane proteins, and maturate within the secretory pathway to reach the cell surface, it is however expected that a minor

Craig W. Stevens (ed.), *Methods for the Discovery and Characterization of G Protein-Coupled Receptors*, Neuromethods, vol. 60, DOI 10.1007/978-1-61779-179-6_9, © Springer Science+Business Media, LLC 2011

proportion of recently synthesized receptors are present within intracellular compartments at any given time.

An enormous amount of studies have addressed the issue of the quantity and duration of the signal mediated by GPCRs, not only because of pathophysiological considerations but also because these receptors represent by far the largest target for pharmacological intervention (2). GPCR receptivity in tissues depends on various regulatory mechanisms, such as phosphorylation or interaction with arrestins (3, 4) and on the actual density of receptors present at the cell surface when the agonist is delivered. In this context, since GPCRs are supposed to principally reside at the cell surface, the vast majority of studies on GPCR trafficking have been focused on receptor endocytosis and on subsequent cellular events such as recycling or sorting to the degradative compartments (5). Only a few studies have proposed that the number of cell surface receptors might possibly also depend on regulatory phenomena at the level of the secretory pathway. After having represented an almost inaudible voice in the consensual chorus, these atypical observations have been supported by a growing number of studies indicating that both in reconstituted systems and native cells a significant proportion of functional GPCRs may be retained within intracellular stores prior its delivery to the cell surface in a regulated manner. Not only are these new data challenging the current dogma, they could also profoundly change the way of understanding receptor regulation and the strategies of pharmacological intervention.

In this chapter we will first review the experimental evidence for the regulated intracellular retention and delivery of GPCR, as well as current knowledge on molecular mechanisms and players involved in this phenomenon. In a second part we will describe current methods to investigate these issues, as well as predictable methodological evolutions.

2. Intracellular GPCRs: Receptors, Proportion, and Cells Types

The central nervous system accounts for many of the existing examples of GPCRs, for which intracellular retention and/or regulated release from intracellular stores has been experimentally documented. For instance, although the precise cartography of δ-opioid receptors (DOR) in the brain and the mechanisms controlling their regulated release to the cell surface are still matter of debate (6–9), it appears that intracellular pools of functional receptors are present in native neuronal cells. The cell surface translocation of DOR from these intracellular compartments might explain mechanistically the enhanced effect of DOR-targeting drugs during chronic pain (10, 11). Studies of the κ-opioid receptor (KOR) trafficking from the ER/Golgi apparatus to plasma membranes in reconstituted cell-systems, are similarly consistent with the existence of regulated cytoplasmic stores of this receptor (12). Activation of NMDA receptors in primary cultures of rat neostriatal neurons, recruits dopamine D1 receptors (D_1R) from the interior of the cell to the plasma membrane (13), providing a cell biological basis for the effect of NMDA on dopamine signaling. These data confirmed previous studies in tubular renal cells, where recruitment of cytoplasmic D_1R to the plasma membrane was documented after either agonist activation of cell surface receptors (14) or via atrial natriuretic peptide-dependent heterologous activation (15). Since it was also found that the α_{1A}-adrenergic receptors ($\alpha_{1A}R$), which are localized to a large extent in the cytoplasm of tubular renal cells, were rapidly translocated to the plasma membrane after a 10-s stimulation of surface receptors with an α_1-adrenergic receptor agonist, the authors proposed that the recruitment of GPCRs from internal stores may be a ubiquitous mechanism for receptor sensitization. The origin of this concept, however, should be attributed to a pioneering study conducted a few years before on protease-activated receptors PAR1 and PAR2, which are irreversibly activated by thrombin cleavage and then internalized and degraded in lysosomes. Replenishment of plasma membrane thrombin receptors from a large pool of intracellular receptor protected from activation by thrombin and translocated to the plasma membrane upon activation of cell surface receptors, was proposed as a mechanism for the recovery of thrombin responsiveness (16). In humans, the pituitary gland GnRH receptor (GnRHR) is highly susceptible to single-point mutations that result in hypogonadotropic hypogonadism. In most cases, these mutations cause receptor retention within the ER and, consequently, diminished receptor availability at the plasma membrane (17). Small nonpeptide molecules known as "pharmacological chaperones" can promote the functional rescue of these supposedly misfolded mutant receptors (18). Interestingly, pharmacological chaperones also increase the plasma membrane expression of the wild type GnRHR indicating that this receptor is partially retained within intracellular stores (17).

As mentioned earlier, although in all the studies the proportion of intracellular receptors was sufficiently important to play functional roles, the precise quantification of this phenomenon was not performed. Only recently a study on chemokine CCR5 receptor was specifically designed to precisely quantify the proportion of intracellular receptors in natural cells. Surprisingly, about 90% of the receptor was found in the cytoplasm in both naïve T cells from healthy donors and in a human monocytic cell line. Using specific markers for intracellular organelles, it could be established by confocal fluorescence microscopy studies that two major pools of CCR5 are localized in the ER and the Golgi apparatus (19).

Abundant intracellular stores of chemokine receptors, which can rapidly be mobilized to the cell surface, have also been documented for CCR7 in mouse (our unpublished results) and CXCR4 in human lymphocytes (20). It has been proposed that maintaining an intracellular pool of functional and rapidly available chemokine receptors may represent a general mechanism for the sustained sensitivity of leukocytes to chemokine guidance within tissues (19).

In addition to the GPCRs, for which the presence within intracellular compartments has been experimentally documented in natural tissues, a growing list of receptors does not arrive at the cell surface when expressed in heterologous systems. For many of these receptors, which often require interaction with specific chaperones and escorts to reach the plasma membrane (see below), it is therefore likely that pools of intracellular endogenous receptors also exist in natural cells. The adrenomedullin and the calcitonin-like receptors (21), vomeronasal pheromone receptors (22), odorant receptors (23), the α_{1D}-adrenoceptor (24), the LSH receptor (25), the adrenocorticotropin MC2 receptor (26), and the bitter taste (TAS2) receptor (27) fall into this category.

3. Proteins Facilitating GPCR Folding or Progression in the Biosynthetic Pathway

As for all other plasma membrane polytopic proteins (spanning membranes several times), folding and maturation of GPCRs in the secretory pathway require the action of numerous nonspecific chaperones, folding factors, and enzymes, which contribute to the general control-quality system. If nascent GPGRs pass each control checkpoint, they can move to a subsequent step; if not, misfolded receptors are directed to the ER-associated degradation pathway, the ERAD. Recent articles have reviewed these issues in detail (28, 29). Interestingly, in this context, an increasing number of studies have also reported the role of so-called "private" chaperones and escort proteins, which specifically help GPCR folding and maturation in the secretory pathway.

3.1. GPCR Private Chaperones

GPCR private chaperones are proteins, with or without enzymatic activity, which participate in receptor folding or in the removal of unfolded molecules. Chaperones permanently reside in intracellular compartments and do not remain associated with mature receptors at the cell surface. For example, NinaA and RANBP2, are two cyclophilin type II chaperones possessing a peptidyl-prolyl *cis-trans* isomerase activity, which contribute to the maturation of *Drosophila* and vertebrate rhodopsin, respectively (30, 31). HSJ1 proteins, also interact with rhodopsin but participates in routing improperly folded rhodopsin to the ERAD, thus protecting neurons against cytotoxic protein aggregation (32).

DRiP78 is a putative two-transmembrane domain protein with both N and C-terminal extremities localized in the cytosol. Overexpression or down-modulation of DRiP78 causes ER retention of D_1R and impaired kinetics of receptor glycosylation (33). DRiP78 binds to a conserved motif found in the C-terminus of several GPCRs, suggesting that DRiP78 may function as a chaperone for various receptors. Indeed, a subsequent study indicated a role of DRiP78 in the maturation of the AT_1 angiotensin II receptor (AT_1R) (34). DRiP78 may also interact with Gγ subunits of heterotrimeric G proteins, protecting them from degradation, until they form a stable complex with the Gβ subunit (35).

The glandular epithelial cell 1 protein (GEC1) is expressed ubiquitously and is particularly abundant in the central nervous system. GEC1 binds to microtubules and interacts with the *N*-ethylmaleimide-sensitive factor (NSF), an ATPase involved in membrane fusion. Ultrastructural studies revealed that GEC1 is localized in intracellular compartments such as the ER and the Golgi apparatus. In reconstituted systems, GEC1 was reported to enhance cell surface expression of the human KOR, the prostaglandin EP3.f receptor, the GluR1 subunit of the AMPA receptor, and the $GABA_A$ receptor. Chaperone-like effects of GEC1 might be explained, at least in part, by its cytoskeleton-linker functions, which could facilitate anterograde trafficking of receptors (36).

ATBP50 (for AT_2R binding protein of 50 kDa) is a membrane-associated Golgi protein, which binds to the cytoplasmic carboxyterminal tail of the AT_2R and promotes its cell surface expression. Inhibition of ATBP50 expression is associated with receptor retention within intracellular compartments (37), although it is not clear how ATBP50 plays his role of AT_2R chaperone. ATBP50 is one of the 3 splice variants of a same gene, which share two myosin-like coiled-coil regions and form both homo- and heterodimers *in vitro*. These proteins display a much broader distribution than the AT_2R, consistent with additional functions, such as the control of cell proliferation. The splice variant ATIP3, for example, is under-expressed in high-grade breast carcinomas (38).

RACK1 (from Receptor for Activated C-Kinase 1) is another example of an intracellular protein, interacting with and regulating the cell surface expression of GPCR, namely the thromboxane A2 receptor and the chemokine receptor CXCR4. In cells with low RACK1 after specific siRNA treatment, cognate GPCRs are retained in the ER (39). Interestingly, although the targeting function of RACK1 does not operate for all GPCRs, (the ß2-adrenoceptor and the prostanoid DP receptor, for example, are not affected by RACK1 depletion (39)), this chaperone seems to regulate the cell surface export of other transmembrane proteins such as the multidrug resistant protein 3 in hepatocytes (40).

3.2. GPCR Private Escorts

Private escort proteins bind nascent GPCRs in the ER and escort them to the Golgi complex and to the plasma membrane, where they remain associated with the receptor. They were identified more than 10 years ago, first in *C. elegans*, in which odorant receptor localization to olfactory cilia was shown to require interaction with the ODR-4 transmembrane protein (41). The first example of escort proteins for mammalian GPCRs is represented by Receptor-Activity-Modifying-Proteins (RAMPs). RAMPs are a family of type-I single-transmembrane domain proteins with a large N-terminal extracellular domain and a short C-terminus. They were initially described as obligatory interacting partners for the cell surface expression of a Class-B GPCR, the calcitonin-like receptor (CLR). Interestingly, depending on the RAMP isoform the receptor is associated with, the ligand binding properties of the calcitonin-like receptor vary. RAMP1 determines affinity for the calcitonin gene-related peptide (CGRP), whereas association with RAMP2 induces an adrenomedullin receptor phenotype (21). After these pioneering studies, several other GPCR escorts have been described (see Table 1). Since they have been the object of recent reviews (29, 42, 43), only new issues or some common important features of these escorts will be highlighted here.

From a structural point of view, for example, nearly all of GPCR escorts are single-transmembrane domain proteins and some may form homo- or heterodimers. In this context, a unique aspect of the Melanocortin 2 Receptor Accessory Protein (MRAP) is that this escort is found at the plasma membrane as an antiparallel dimer (44). In addition, while helping the MC2R (the melanocortin 2, ACTH receptor) to reach the cell surface, MRAP also seems to have the opposite effect on the MC5R, since it inhibits dimerization and surface localization of this receptor (45).

Dysfunction of several GPCR escort proteins has been associated with disease. Mutations of the MRAP gene are involved the familial glucocorticoid deficiency, characterized by the resistance of the adrenal cortex to adrenocorticotropin (46). Mutations of Receptor Expression Enhancing Protein 1 (REEP1), an escort for odorant receptors, have recently been correlated with autosomal dominant hereditary spastic paraplegia (HSP) type SPG31 (47, 48), although the underlying pathophysiological mechanisms remain to be elucidated. Abnormal serotonin signaling has been implicated in the pathophysiology of depression. A depression-like phenotype is observed in knockout mice for p11 (49) a member of the S100 EF-hand calcium-dependent signaling modulators (50), which specifically interacts with serotonin 5-HT_{1B} receptors (5-HT_{1B}R) at the cell surface. Cell surface density and function of 5-HT_{1B}Rs are decreased in these mice, indicating that p11 is an escort for 5-HT_{1B}R . Interestingly, p11 was increased in

Table1
Private GPCR escort proteins

Escort	Cognate GPCRs	Additional functions	References
RAMP family			
RAMP-1	CLR, which binds CGRP; CaSR; CTR which binds amylin	Binds to VPAC1R	(21, 73–75)
RAMP-2	CLR, which binds adrenomedullin	VPAC1R enhanced signaling; binds to GR and PTH1R	(21, 75)
RAMP-3	CLR, which binds adrenomedullin; CaSR; CTR, which binds amylin	Binds to PTH2R and VPAC1R	(21, 73–75)
REEP family			
REEP-1	OR, TAS2		(23, 27)
REEP-2	OR		(23)
REEP-3	OR		(23)
RTP family			
RTP-1	OR		(23)
RTP-2	OR		(23)
RTP-3	TAS2		(27)
RTP-4	TAS2	Targeting MOR-DOR heterodimer to the plasma membrane	(27, 76)
M10 – M1	V2R	MHC class Ib molecules neuronal plasticity	(22, 77)
MRAP	MC2R (facilitated export) MC5R (inhibited export)		(44, 45) (45)
P11	5-HT_{1B}R	Member of the S100 EF-hand calcium-dependent signaling modulators	(49, 50)
CD4	CCR5	Interaction with class II major histocompatibility complex	(19)

the brain of mice treated with antidepressants and was also found reduced in depressed patients.

The tissue distribution of nearly all private escorts is much broader than that of their cognate GPCR(s). This fact can be interpreted as an indication that each escort protein may facilitate the surface expression of additional GPCRs. This assumption has proven to be true, for instance, in the case of RTP (Receptor Transporting Proteins) and REEP proteins, which were at first described as escorts for odorant receptors (23). Subsequently, RTP and REEP mRNAs were detected in human circumvallate papillae and testis, which are the sites of bitter taste receptor

(TAS2R) expression, and experiments in heterologous cells confirmed the enhancement of TAS2R cell surface targeting upon interaction with RTP3-4 and REEPs (27). Alternatively, targeting GPCRs to the cell surface might be viewed as a "moonlighting job" for some escort proteins. Illustration of this hypothesis, come from a recent study on the control of the CCR5 chemokine receptor expression at the plasma membrane by the CD4 glycoprotein (19). CD4 is a single-membrane spanning domain receptor interacting with class II major histocompatibility complex molecules (51) and with its cognate ligand IL-16. Only a minor fraction (1:100 to 1:10) of CD4 is constitutively associated with CCR5 at the cell surface of lymphocytes and monocytic cells, although this association has a major pathophysiological relevance, since it forms the principal receptor for the Human Immunodeficiency Virus (HIV) (52). Achour et al. demonstrated that the interaction between CCR5 and CD4 occurs in the ER and enhances the export of CCR5 to the cell surface in both reconstituted models and natural blood cells (19). Thus, in addition to its immunological functions, CD4 works like a private escort protein for CCR5.

4. Retention and Release Mechanisms

From what has been summarized earlier, the concepts that functional GPCRs may be stored in various intracellular pools prior to their final targeting to the cell surface and that they may be helped on this journey by multiple specific receptor-interacting proteins are now supported by a large body of experimental evidence. At present, it is difficult to estimate what proportions of GPCRs are concerned by these issues. We anticipate a rapid and collective answer to this question; once more investigators involved in GPCR research decide to study these aspects with the appropriate methods. A number of important questions, however, remain largely unexplored and will require further investigation.

Although the available sampling is relatively limited, it is already clear that a given chaperone or escort protein does not function with any tested GPCR (39, 53), and that a given GPCR is not retained with the same efficacy in all cell types and all subcellular compartments (see Sect. 5). These observations indicate the existence of a complex "retention system" involving several different mechanisms, which are present at variable concentration in each cell type. They also imply the existence of a molecular code, which governs the interaction between each GPCR on one hand, chaperone, escorts, and retention proteins on the other hand. Finally, it will be also critical to understand how these interactions are regulated *via* the stoichiometry of the various components or by activation signals.

4.1. Motifs Regulating Retention or Forward Export

Recent studies have shown that traffic of cell surface membrane receptors can be controlled by specific short and linear sequence signal motifs, which are usually located within the cytoplasmic domains of the receptors. For example, an important role of the C-terminus for ER export has been demonstrated for different GPCRs, including the α_{2B}-adrenergic (α_{2B}AR), the AT_1R, the α_{2B}-adrenergic (α_{2B}AR), the β_2-adrenergic (β_2AR), the M_1-muscarinic receptor (M_1ACHR), the vasopressin V2 receptor (V2R), the D_1R, as well as rhodopsin (54–59). Also, mutagenesis studies of GPCR C- and N-termini have led to the identification of several motifs that play a critical role in GPCR export from the ER. In particular, different classes of di-leucine-based signals have been distinguished. The $F(X)_6LL$ motif in the membrane proximal C-terminus of AT_1R, α_{2B}AR, and M_1ACHR, for example, is required for their release from ER (54, 56) and cannot be fully substituted by other hydrophobic residues. The function of the Phe residue in the $F(X)_6LL$ motif is likely mediated by the interaction of this residue with other hydrophobic residues of the first transmembrane domain (TM1), playing an important role in receptor folding (55). Similarly, the $FN(X)_2LL(X)_3L$ sequence is considered an important motif for the V3R conformation (60). The export of the V3R $TT(X)_2TT(X)_3T$ mutant is perturbed but can be restored by the addition of the pharmacological chaperone SSR149415, suggesting that the $FN(X)_2LL(X)_3L$ motif is more likely involved in a transport-competent folding of the receptor rather than in its export (61). In some cases, the carboxyl tail of the receptor is not sufficient to direct the traffic of the receptor to the cell surface. A di-leucine motif (Leu245-Leu246) within the third intracellular loop of DOR has been identified as a regulator of the intracellular trafficking of the receptors in conjunction with the carboxyl tail sequence (62). Recently, a motif regulating GPCR exit from the ER, based around a conserved Leu residue (L48) in the first intracellular loop, has been identified for the α_{2B}AR, the β_2AR, the AT_1R, and the α_{1B}AR (63).

Motifs regulating receptor export seem also to exist in the luminal (extracellular) side of some GPCRs. Indeed, a Tyr-Ser motif was identified as linear independent export signal directing the α_2AR exit from the Golgi. This motif is highly conserved in the N-terminus of α_{2A}AR, α_{2B}AR, and α_{2C}AR subtypes, it could provide a common signal for the Golgi export of this subgroup of adrenergic receptors (64). Although the molecular significance of these motifs has not been elucidated in most cases, it is clear that some of them are directly involved in the interaction with chaperone proteins. The $F(X)_3F(X)_3F$ sequence, which is highly conserved among almost all rhodopsin-like GPCRs, falls in this category. It is present in the proximal C-terminus of the D_1R, and the substitution of phenylalanine residues, results in trapping of the receptor in the ER and improper maturation. This motif directly binds to the ER chaperone protein DRiP78 (33).

4.2. Signals Promoting the Release from Intracellular Stores

The few experimental data available on this issue already indicate that the release of GPCRs from intracellular stores is a complex phenomenon. The simple stoichiometry of GPCRs, escorts, and retention molecules can regulate the amount of receptor exported to the cell surface. As mentioned in Sect. 3.1, both overexpression or down-modulation of DRIP78 lead to ER retention of D_1R (33). Also, in a reconstituted model, the effect of CD4 on cell surface targeting of CCR5 was concentration-dependent and displayed a bell-shaped pattern (19). Thus, respective concentration of the various proteins involved in the control of the export as well as the concentration of a GPCR in the intracellular compartments likely affects the basal level of receptor surface expression in a given cell. This basal level of surface receptor can be modulated by external signals. As described in Sects. 2 and 4.3, the activation of cell surface receptors with agonist ligands can accelerate the release of several GPCRs from internal stores. In the case of chemokine receptors, a robust increase of the plasma membrane fraction of natural blood cells can also be observed after a few minutes of adhesion (personal unpublished results). Finally, conformational changes of the intracellular receptor itself, promoted for example by the binding of membrane permeable pharmacological chaperones, can promote its release from ER retention and progression to the cell surface (17).

4.3. Gatekeepers Involved in the Intracellular Retention of GPCRs

Only a few proteins, which are directly and specifically involved in the control of GPCR cell surface export via intracellular retention, have been identified so far. The protease-activated PAR2 receptor was shown to interact, by means of its second extracellular loop, with the N-terminal domain of the Golgi-resident type-I transmembrane protein p24A. This interaction determines the retention of the PAR2 in this compartment. Interestingly, upon activation of cell surface PAR2, some signal induces the recruitment of the small G protein ARF1, in its GDP-bound form, to Golgi membranes, where it is activated by a specific exchange factor. This process results in the dissociation of PAR2 from p24A and subsequent receptor trafficking to the plasma membrane (65). In a different context, during development, a GPCR-interacting, ER-resident protein was reported to control the surface density of Frizzled, a GPCR, which promotes caudalizing signals. This protein, known as Shisa, is specifically expressed in head ectoderm, where it inhibits the cell surface trafficking of Frizzled, which is retained in the ER. Shisa-mediated receptor retention thus constitutes a mechanism to control head–tail polarity (66).

Other recent examples of proteins, which regulate plasma membrane targeting via intracellular retention have also been reported for growth-factor receptors (67), the gamma secretase complex (68) or the glutamate transporter (69), suggesting that this mechanism constitutes a more general strategy to control the density of signaling molecules at the cell surface.

5. Methods to Investigate the Regulated Export of GPCRs

Several useful methods are described in this section to quantitatively evaluate the proportion of intracellular receptor, the subcellular compartments in which the receptor is stored, and to compare different cell types for their capacity of exporting a given GPCR to the cell surface. These methods are easily accessible to any laboratory interested in GPCRs and are mostly adapted from basic techniques.

5.1. The GPCR Export Assay

Part of the discrepancy between studies on the cell surface targeting of a given GPCR, likely depends on the cell type where the studies have been conducted. Indeed it is conceivable that, depending on the precise stoichiometry of escort proteins, chaperones and retention molecules interacting with the investigated receptor, the proportion of the receptor reaching the cell surface may vary. To compare the actual capacity of different cell lines to target a given receptor to the cell surface, we have designed the FACS-based receptor export assay, which is described in Fig. 1. This assay can also be used to compare the trafficking of receptor variants in the same cell type.

Various types of either adherent (e.g., HEK-293T, CHO, COS, HeLa cells) or suspension (e.g., THP-1, Jurkat) intact mammalian cell lines can be used for this assay. Cells are seeded in 6-well plates (500,000 cells per well in 2 ml of the appropriated culture medium) and transfected with increasing amounts (typically 50–1,000 ng) of a double-tagged GPCR construct. The receptor displays an extracellular epitope, which is recognized by a sensitive antibody such as Myc, FLAG, HA, or His6 (if available an antibody against the native N-terminus can be used as well). It is also fused upstream of the yellow variant of the green fluorescent protein. To be representative of the wild type receptor, the construct should maintain its principal biological functions, such as ligand binding, signaling, or endocytosis. "Empty" vector (such as pCDNA3.1 or any other cloning vector) is used to maintain identical total amount of transfected DNA. Twenty-four to forty-eight hours after transfection, the cells are detached and incubated for 1 h at 4°C with the appropriate primary antibody directed against the extracellular epitope (e.g., Myc epitope). After washings and staining with the appropriate

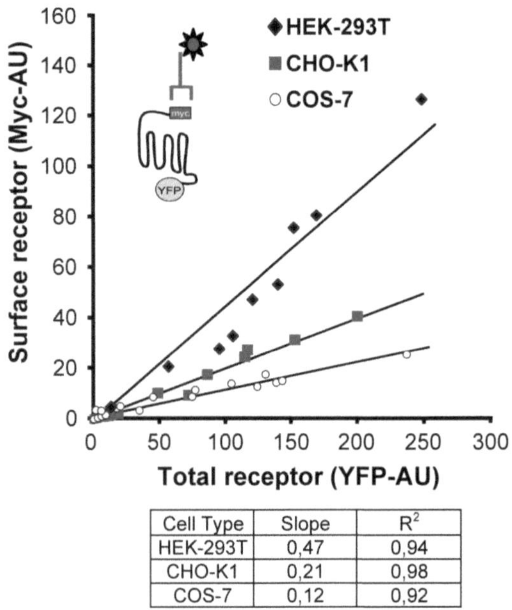

Cell Type	Slope	R^2
HEK-293T	0,47	0,94
CHO-K1	0,21	0,98
COS-7	0,12	0,92

Fig. 1. GPCR cell surface export assay using the β2-adrenoceptor as model. HEK-293T, CHO-K1, or COS-7 cells were transfected with increasing amounts of plasmid DNA coding for Myc-β2AR-YFP. Forty-eight hours after transfection, FACS experiments were performed on each cell type using a mouse monoclonal anti-myc tag antibody (9E10, Roche), as primary antibody, followed by a Cy5-conjugated antimouse secondary antibody (Rockland, Gilbertsville, PA). The values of the Cy5-related fluorescence (corresponding to surface β2AR) were plotted as a function of the YFP signal (corresponding to the total β2AR present in the cells). The calculated slope values and corresponding correlation coefficients (R^2) are shown under the plots.

secondary antibody (e.g., Cy5 or Alexa 647 conjugated antibodies) for 1 h at 4°, cells are then washed, fixed in 2% paraformaldheyde and analyzed by FACS (e.g., Cytomycs FC500 FACS analyzer, Beckman Coulter, Fullerton, CA). For each point of transfection, 10,000 cells are sorted and data analyzed using the available software. The amount of total receptor expressed in cells is estimated by measuring YFP-associated fluorescence. The amount of surface receptor is determined by the signal given by the extracellular epitope immunoreactivity. The values obtained for the surface receptor are plotted as a function of the YFP signal, corresponding to the total expressed receptor. Data are fitted using Excel or GraphPad Prism software. In general these plots are linear: the better a GPCR is expressed at the cell surface the higher the gradient. A typical GPCR export experiment is shown in Fig. 1 using as model the β_2AR, which is expressed in three different cell lines (HEK-293T, COS-7 and CHO-K1). The surface expression of the β_2AR increases linearly with the rising amount of total expressed receptor in all cell lines, although the slope is higher in HEK293 cells. This experiment demonstrates that, for any given total concentration, the β_2AR is better exported to the cell surface in HEK-293 T than in CHO-K1 or COS-7 cells.

5.2. Qualitative and Quantitative Fluorescence Approaches

To determine the subcellular distribution of a GPCR, immunofluorescence studies are carried out using either a confocal microscope or a wide-field microscope allowing the capture of Z images for subsequent deconvolution analysis. When possible, experiments are performed at the same time on intact and permeabilized cells expressing endogenous receptors using native receptor-specific antibodies. If available, it is recommended to perform the staining with two or more specific antibodies, which recognize distinct epitopes to avoid any possible artifact. The same experiment should also be conducted in parallel in control cell line, which does not express the GPCR under investigation. The following protocol has been designed to study the distribution of chemokine receptors in T lymphocytes from healthy donors. After separation and purification with appropriate procedures, T lymphocytes are fixed in 4% paraformaldheyde for 20 min at room temperature followed by a 15-min quenching with 50 mM NH_4Cl in PBS for 15 min. Cells are then incubated with blocking buffer (e.g., PBS containing 1% Bovine Serum Albumine) with or without detergent (e.g., 0.05% tween 20), for permeabilization. After incubation with the specific primary antibodies, cells are washed and incubated with appropriate secondary antibodies. After final washings, cells are mounted on slides and analyzed by fluorescence microscopy (confocal Leica TCS SP2 AOBS confocal microscope, or Zeiss Axiovert 100M microscope equipped with a cooled CCD camera (Orca ER, Hamamatsu) for deconvolution studies). An example of this type of experiment is shown in Fig. 2, which shows the subcellular distribution of endogenous CCR5 in purified human T lymphocytes.

The obtained images are analyzed using appropriate software: Image J (National Institute of Health, Bethesda, MD) or Metamorph 7, Meinel Algorithm with iteration 5 (Molecular Devices, Sunnyvale, CA) for deconvolution. These softwares are particularly useful to determine in which intracellular compart ment(s) the studied GPCR is accumulated, using quantitative colocalization algorithms. For these experiments, antibodies directed against proteins characteristic of a specific organelle (calnexin, calreticulin, Bip for the ER; GM130, giantin for the Golgi apparatus), and recognized by different secondary antibodies are co-incubated with antireceptor antibodies. Also, a rough estimate of the proportion of intracellular versus cell surface receptor can be achieved by comparing the fluorescence signal in specific areas of the cells.

A more precise quantitation of the actual proportion of surface receptor versus total can be obtained by FACS analysis, comparing the signal in unpermeabilized and permeabilized cells. For estimating the surface receptors, the same protocol described in the Sect. 5.1 is used. To quantify total receptor, cells are first fixed in 1% paraformaldheyde for 20 min, washed with PBS, and

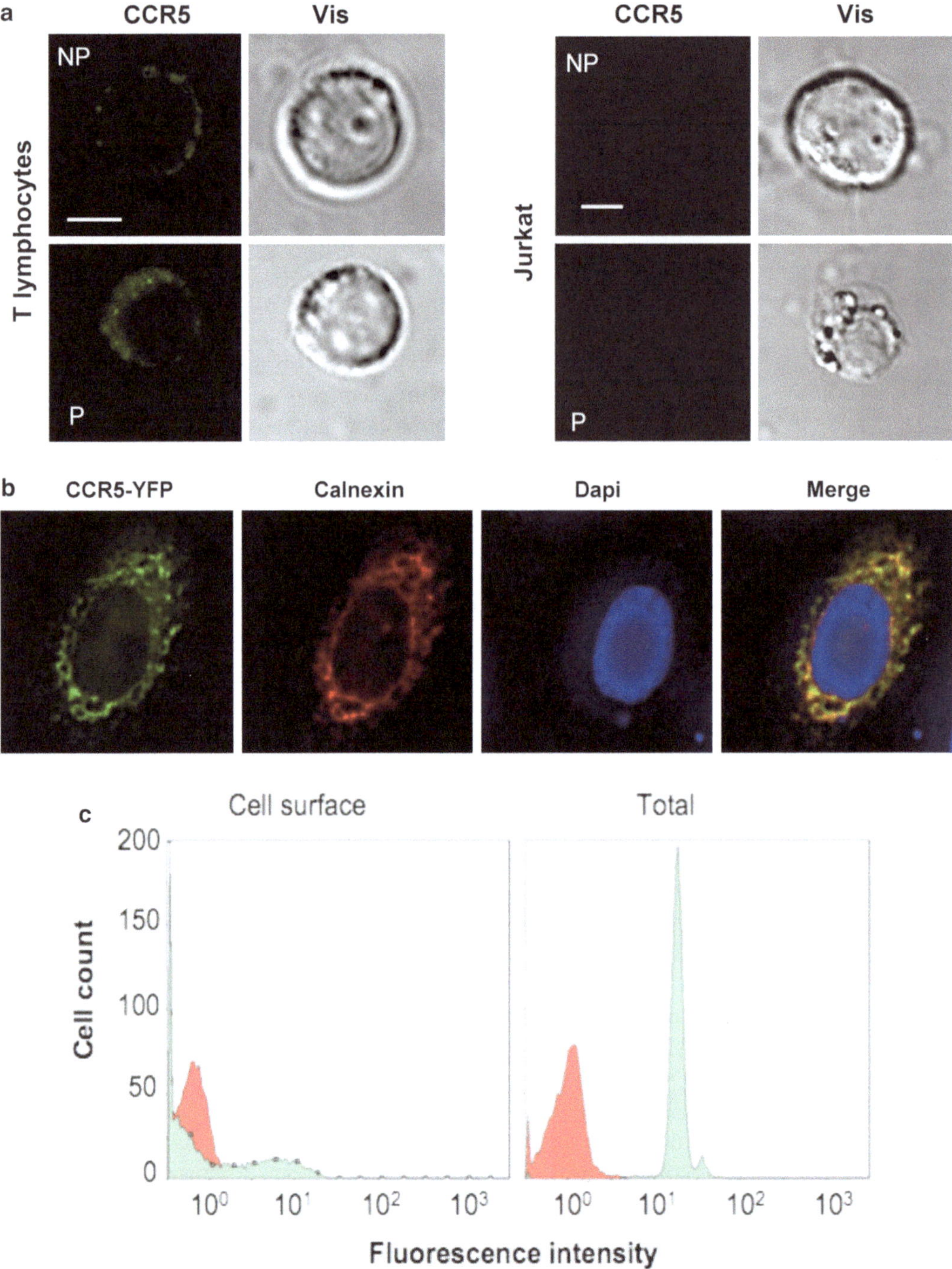

Fig. 2. Subcellular distribution of the human CCR5 receptor. (**a**) Freshly prepared resting T lymphocytes from healthy donors, and control Jurkat cells, which do not express CCR5, were stained for surface (nonpermeabilized, NP) and total (permeabilized, P) CCR5. After incubation with the 2D7 mouse monoclonal antihuman CCR5 antibody (BD Pharmingen, San Diego, CA), cells were incubated with Alexa-Fluor[R]488-conjugated antimouse antibody (Molecular Probes, Invitrogen, Eugen, OR). Images were obtained with a confocal microscope. Scale bar, 10 μM. (**b**) CHO-K1 cells transfected with a CCR5-YFP-coding plasmid were stained using the ER marker calnexin. Stacks of Z images recorded with a CCD camera mounted on an inverted wide-field epifluorescence microscope (immersion objective 100×) were deconvoluted. Colocalization of CCR5-YFP with calnexin is shown in (merge). Nucleus is stained with DAPI (*3rd panel*). (**c**) FACS analysis of T lymphocytes, stained for surface (*left panel*) and total (after cell permeabilization, *right panel*) CCR5, using the rat 1/85a Alexa-fluor[R]647-conjugated antihuman CCR5 (Biolegend, San Diego, CA) or Alexa-Fluor 647[R]-conjugated rat IgG2a-κ, the isotypic control, and analyzed by flow cytometry. Light grey histograms correspond to the specific labeling for CCR5 and dark grey histograms to the isotype control, in intact and permeabilized cells.

then permeabilized for 1 h at 4°C in PBS, containing 2% FBS, and 0.2% triton X-100. Cells are then incubated for 1 h at 4°C with PBS-FBS 2% containing the specific primary antibodies or control isotypes to determine the nonspecific staining. After washing and incubation with the appropriate secondary antibodies, cells are washed again with PBS-FBS 2% and stained for 1 h at 4°C with the appropriate secondary antibodies. After a final washing with PBS and fixation in 2% paraformaldheyde, cells are processed by FACS (Fig. 2). The proportion of receptor at the cell surface is calculated as: ratio of surface signal measured in nonpermeabilized cells over the total signal measured in permeabilized cells.

5.3. Evaluating the Retention Properties or Escort Functions of GPCR-Interacting Proteins

The methods above can be coupled with siRNAs-mediated inhibition of specific cellular proteins or the expression of exogenous proteins to investigate the possible roles that these molecules exert on GPCR export to the cell surface. This approach is particularly useful, when GPCR-interacting proteins, identified by proteomic or two-hybrid studies, appear to be possible regulators of GPCR forward trafficking because of their particular subcellular distribution and/or because they display a structural organization similar to that of known chaperones, escorts, or gatekeepers.

6. Future Directions

The identification of the various components of the complex molecular network, which controls receptor retention within intracellular compartments and forward trafficking, is clearly an important task for the years to come. We anticipate that two main complementary strategies will mostly contribute to this characterization. The first is based on the identification of GPCR-interacting proteins using intact receptors or portions of receptors as baits. Yeast two-hybrid studies, which have already contributed a significant proportion of current knowledge on this topic, will probably identify new partners. More recently, some studies have successfully used proteomic approaches for the same purpose (70) and will become more and more popular due to the fast technical improvements in the field (71). The availability of large shRNA libraries covering the entire genome will soon permit high throughput functional studies (72) specifically addressing the cell surface export of a given GPCR, provided that sufficiently robust and reproducible assays as well as appropriate methods for data analysis are set up.

Quantitative cell biology-based approaches, particularly those conducted in native cell-systems, when appropriate tools are available, will assess the precise role of these novel GPCR interactors.

7. Conclusions

The regulated export of GPCR from intracellular stores to the plasma membrane is a relatively recent area of investigation with promising research avenues. So far, we only know a very limited piece of the story, and most of the work is ahead of us. Moreover, if this concept could be extended from the currently limited number of examples to a substantial proportion of GPCRs, we anticipate important pathophysiological implications as well as novel opportunities for the pharmacological manipulation of GPCRs.

Acknowledgements

The authors thank Drs. Mark GH Scott and Hervé Enslen for the critical reading of the manuscript. This work is supported by a grant (#R09158KK) of the Agence Nationale de la Recherche sur le SIDA (ANRS) to SM. The sabbatical leave of LPL was supported by a grant from CONACyT.

References

1. Bockaert J, Pin JP (1999) Molecular tinkering of G protein-coupled receptors: an evolutionary success. Embo J **18**:1723–1729
2. Pierce KL, Premont RT, Lefkowitz RJ (2002) Seven-transmembrane receptors. Nat Rev Mol Cell Biol **3**:639–650
3. Ferguson SS, Barak LS, Zhang J et al (1996) G-protein-coupled receptor regulation: role of G-protein-coupled receptor kinases and arrestins. Can J Physiol Pharmacol **74**: 1095–1110
4. Premont RT, Gainetdinov RR (2007) Physiological roles of G protein-coupled receptor kinases and arrestins. Ann Rev Physiol **69**:511–534.
5. Hanyaloglu AC, von Zastrow M (2008) Regulation of GPCRs by membrane trafficking and its potential implications. Ann Rev Pharmacol Toxicol **48**:537–568
6. Cahill CM, Morinville A, Lee MCet al (2001) Prolonged morphine treatment targets *delta* opioid receptors to neuronal plasma membranes and enhances *delta*-mediated antinociception. J Neurosci **21**:7598–7607
7. Kim KA, von Zastrow M (2003) Neurotrophin-regulated sorting of opioid receptors in the biosynthetic pathway of neurosecretory cells. J Neurosci **23**:2075–2085
8. Patwardhan AM, Berg KA, Akopain AN et al. (2005) Bradykinin-induced functional competence and trafficking of the *delta*-opioid receptor in trigeminal nociceptors J Neurosci **25**:8825–8832
9. Scherrer G, Imamachi N, Cao YQ et al (2009) Dissociation of the opioid receptor mechanisms that control mechanical and heat pain. Cell **137**:1148–1159
10. Cahill CM, Holdridge SV, Morinville A (2007) Trafficking of *delta*-opioid receptors and other G-protein-coupled receptors: implications for pain and analgesia. Trends Pharmacol Sci **28**:23–31
11. Bie B, Pan ZZ (2007) Trafficking of central opioid receptors and descending pain inhibition. Mol Pain **3**:37
12. Chen C, Li JG, Chen Y et al (2006) GEC1 interacts with the *kappa* opioid receptor and enhances expression of the receptor. J Biol Chem **281**:7983–7993
13. Scott L, Kruse MS, Forssberg H et al (2002) Selective up-regulation of dopamine D1 receptors in dendritic spines by NMDA receptor activation. Proc Natl Acad Sci USA **99**:1661–1664
14. Brismar H, Asghar M, Carey RM et al (1998) Dopamine-induced recruitment of dopamine

D1 receptors to the plasma membrane. Proc Natl Acad Sci USA **95**:5573–5578

15. Holtback U, Brismar H, DiBona GF et al (1999) Receptor recruitment: a mechanism for interactions between G protein-coupled receptors. Proc Natl Acad Sci USA **96**:7271–7275
16. Hein L, Ishii K, Coughlin SR et al (1994) Intracellular targeting and trafficking of thrombin receptors. A novel mechanism for resensitization of a G protein-coupled receptor. J Biol Chem **269**:27719–27726
17. Conn PM, Ulloa-Aguirre A, Ito J et al (2007) G protein-coupled receptor trafficking in health and disease: lessons learned to prepare for therapeutic mutant rescue *in vivo*. Pharmacol Rev **59**:225–250
18. Janovick JA, Knollman PE, Brothers SP et al (2006) Regulation of G protein-coupled receptor trafficking by inefficient plasma membrane expression: molecular basis of an evolved strategy. J Biol Chem **281**:8417–8425
19. Achour L, Scott MG, Shirvani H et al (2009) CD4 – CCR5 interaction in intracellular compartments contributes to receptor expression at the cell surface. Blood **113**:1938–1947
20. Ding Z, Issekutz TB, Downey GP et al (2003) L-selectin stimulation enhances functional expression of surface CXCR4 in lymphocytes: implications for cellular activation during adhesion and migration. Blood **101**:4245–4252
21. McLatchie LM, Fraser NJ, Main MJ et al (1998) RAMPs regulate the transport and ligand specificity of the calcitonin- receptor-like receptor. Nature **393**:333–339
22. Loconto J, Papes F, Chang E et al (2003) Functional expression of murine V2R pheromone receptors involves selective association with the M10 and M1 families of MHC class Ib molecules. Cell **112**:607–618
23. Saito H, Kubota M, Roberts RW et al (2004) RTP family members induce functional expression of mammalian odorant receptors. Cell **119**:679–691
24. Uberti MA, Hague C, Oller H et al (2005) Heterodimerization with *beta*2-adrenergic receptors promotes surface expression and functional activity of *alpha*1D-adrenergic receptors. J Pharmacol Exp Ther **313**:16–23
25. Pietila EM, Tuusa JT, Apaja PM et al (2005) Inefficient maturation of the rat luteinizing hormone receptor: A putative way to regulate receptor numbers at the cell surface. J Biol Chem **280**:26622–26629
26. Noon LA, Franklin JM, King PJ et al (2002) Failed export of the adrenocorticotrophin receptor from the endoplasmic reticulum in non-adrenal cells: evidence in support of a requirement for a specific adrenal accessory factor. J Endocrinol **174**:17–25
27. Behrens M, Bartelt J, Reichling C et al (2006) Members of RTP and REEP gene families influence functional bitter taste receptor expression. J Biol Chem **281**:20650–20659
28. Hebert DN, Molinari M (2007) In and out of the ER: protein folding, quality control, degradation, and related human diseases. Physiol Rev **87**:1377–1408
29. Achour L, Labbe-Juillie C, Scott MGH et al (2008) An escort for G Protein Coupled Receptors to find their path: implication for regulation of receptor density at the cell surface. Trends Pharmacol Sci **29**:528–535
30. Baker EK, Colley NJ, Zuker CS (1994) The cyclophilin homolog NinaA functions as a chaperone, forming a stable complex *in vivo* with its protein target rhodopsin. Embo J **13**:4886–4895
31. Ferreira PA, Nakayama TA, Pak WL et al (1996) Cyclophilin-related protein RanBP2 acts as chaperone for red/green opsin. Nature **383**:637–640
32. Westhoff B, Chapple JP, van der Spuy J et al (2005) HSJ1 is a neuronal shuttling factor for the sorting of chaperone clients to the proteasome. Curr Biol **15**:1058–1064
33. Bermak JC, Li M, Bullock C et al (2001) Regulation of transport of the dopamine D1 receptor by a new membrane- associated ER protein. Nat Cell Biol **3**:492–498.
34. Leclerc PC, Auger-Messier M, Lanctot PM et al (2002) A polyaromatic caveolin-binding-like motif in the cytoplasmic tail of the type 1 receptor for angiotensin II plays an important role in receptor trafficking and signaling. Endocrinology **143**:4702–4710
35. Dupre DJ, Robitaille M, Richer M et al (2007) Dopamine receptor-interacting protein 78 acts as a molecular chaperone for G*gamma* subunits before assembly with G*beta*. J Biol Chem **282**:13703–13715
36. Chen Y, Chen C, Kotsikorou E et al (2009) GEC1-*kappa* opioid receptor binding involves hydrophobic interactions: GEC1 has chaperone-like effect. J Biol Chem. **284**:1673–1685
37. Wruck CJ, Funke-Kaiser H, Pufe T et al (2005) Regulation of transport of the angiotensin AT2 receptor by a novel membrane-associated Golgi protein. Arterioscler Thromb Vasc Biol **25**:57–64
38. Rodrigues-Ferreira S, Di Tommaso A, Dimitrov A et al (2009) 8p22 MTUS1 gene product ATIP3 is a novel anti-mitotic protein underexpressed in invasive breast carcinoma of poor prognosis. PLoS One **4**:e7239

39. Parent A, Laroche G, Hamelin E et al (2008) RACK1 regulates the cell surface expression of the G protein-coupled receptor for thromboxane A(2). Traffic **9**:394–407
40. Ikebuchi Y, Takada T, Ito Ket al (2009) Receptor for activated C-kinase 1 regulates the cellular localization and function of ABCB4. Hepatol Res. **39**:1091–1107
41. Dwyer ND, Troemel ER, Sengupta P et al (1998) Odorant receptor localization to olfactory cilia is mediated by ODR-4, a novel membrane-associated protein. Cell **93**:455–466
42. Cooray SN, Chan L, Webb TR et al (2009) Accessory proteins are vital for the functional expression of certain G protein-coupled receptors. Mol Cell Endocrinol **300**:17–24
43. Ritter SL, Hall RA (2009) Fine-tuning of GPCR activity by receptor-interacting proteins. Nat Rev Mol Cell Biol **10**:819–830
44. Sebag JA, Hinkle PM (2007) Melanocortin-2 receptor accessory protein MRAP forms antiparallel homodimers. Proc Natl Acad Sci USA **104**:20244–20249
45. Sebag JA, Hinkle PM (2009) Opposite effects of the melanocortin-2 (MC2) receptor accessory protein MRAP on MC2 and MC5 receptor dimerization and trafficking. J Biol Chem **284**:22641–22648
46. Metherell LA, Chapple JP, Cooray S et al (2005) Mutations in MRAP, encoding a new interacting partner of the ACTH receptor, cause familial glucocorticoid deficiency type 2. Nat Genet **37**:166–170
47. Zuchner S, Wang G, Tran-Viet KN et al (2006) Mutations in the novel mitochondrial protein REEP1 cause hereditary spastic paraplegia type 31. Am J Hum Genet **79**:365–369
48. Beetz C, Schule R, Deconinck T et al (2008) REEP1 mutation spectrum and genotype/phenotype correlation in hereditary spastic paraplegia type 31. Brain **131**:1078–1086
49. Svenningsson P, Chergui K, Rachleff I et al (2006) Alterations in 5-HT1B receptor function by p11 in depression-like states. Science **311**:77–80
50. Donato R (1999) Functional roles of S100 proteins, calcium-binding proteins of the EF-hand type. Biochim Biophys Acta **1450**:191–231
51. Doyle C, Strominger JL (1987) Interaction between CD4 and class II MHC molecules mediates cell adhesion. Nature **330**:256–259
52. Alkhatib G, Combadiere C, Broder CC et al (1996) CC CKR5: a RANTES, MIP-1*alpha*, MIP-1*beta* receptor as a fusion cofactor for macrophage-tropic HIV-1. Science **272**: 1955–1958.
53. Parameswaran N, Spielman WS (2006) RAMPs: the past, present and future. Trends Biochem Sci **31**:631–638
54. Duvernay MT, Zhou F, Wu G (2004) A conserved motif for the transport of G protein-coupled receptors from the endoplasmic reticulum to the cell surface. J Biol Chem **279**:30741–30750
55. Duvernay MT, Dong C, Zhang X et al (2009) Anterograde trafficking of G protein-coupled receptors: function of the C-terminal F(X)6LL motif in export from the endoplasmic reticulum. Mol Pharmacol **75**:751–761
56. Sawyer GW, Ehlert FJ, Shults CA (2010) A conserved motif in the membrane proximal C-terminal tail of human muscarinic M1 acetylcholine receptors affects plasma membrane expression. J Pharmacol Exp Ther **332**:76–86
57. Preisser L, Ancellin N, Michaelis L et al (1999) Role of the carboxyl-terminal region, di-leucine motif and cysteine residues in signalling and internalization of vasopressin V1a receptor. FEBS Lett. **460**:303–308
58. Heymann JA, Subramaniam S (1997) Expression, stability, and membrane integration of truncation mutants of bovine rhodopsin. Proc Natl Acad Sci USA **94**:4966–4971
59. Pankevych H, Korkhov V, Freissmuth M et al (2003) Truncation of the A1 adenosine receptor reveals distinct roles of the membrane-proximal carboxyl terminus in receptor folding and G protein coupling. J Biol Chem **278**: 30283–30293
60. Robert J, Clauser E, Petit PX et al (2005) A novel C-terminal motif is necessary for the export of the vasopressin V1b/V3 receptor to the plasma membrane. J Biol Chem **280**: 2300–2308
61. Robert J, Auzan C, Ventura MA et al (2005) Mechanisms of cell-surface rerouting of an endoplasmic reticulum-retained mutant of the vasopressin V1b/V3 receptor by a pharmacological chaperone. J Biol Chem **280**:42198–42206.
62. Wang W, Loh HH, Law PY (2003) The intracellular trafficking of opioid receptors directed by carboxyl tail and a di-leucine motif in neuro2A cells. J Biol Chem **278**:36848–36858
63. Duvernay MT, Dong C, Zhang X et al (2009) A single conserved leucine residue on the first intracellular loop regulates ER export of G protein-coupled receptors. Traffic **10**:552–566

64. Dong C, Wu G (2006) Regulation of anterograde transport of *alpha*2-adrenergic receptors by the N termini at multiple intracellular compartments. J Biol Chem **281**:38543–38554
65. Luo W, Wang Y, Reiser G (2007) p24A, a type I transmembrane protein, controls ARF1-dependent resensitization of protease-activated receptor-2 by influence on receptor trafficking. J Biol Chem **282**:30246–30255
66. Yamamoto A, Nagano T, Takehara S et al (2005) Shisa promotes head formation through the inhibition of receptor protein maturation for the caudalizing factors, Wnt and FGF. Cell **120**:223–235
67. Couturier C, Sarkis C, Seron K et al (2007) Silencing of OB-RGRP in mouse hypothalamic arcuate nucleus increases leptin receptor signaling and prevents diet-induced obesity. Proc Natl Acad Sci USA **104**:19476–19481
68. Spasic D, Raemaekers T, Dillen K et al (2007) Rer1p competes with APH-1 for binding to nicastrin and regulates *gamma*-secretase complex assembly in the early secretory pathway. J Cell Biol **176**:629–640
69. Ruggiero AM, Liu Y, Vidensky S et al (2008) The endoplasmic reticulum exit of glutamate transporter is regulated by the inducible mammalian Yip6b/GTRAP3-18 protein. J Biol Chem **283**:6175–6183
70. Daulat AM, Maurice P, Froment C et al (2007) Purification and identification of G protein-coupled receptor protein complexes under native conditions. Mol Cell Proteomics **6**:835–844
71. Daulat AM, Maurice P, Jockers R (2009) Recent methodological advances in the discovery of GPCR-associated protein complexes. Trends Pharmacol Sci **30**:72–78
72. Boutros M, Ahringer J (2008) The art and design of genetic screens: RNA interference. Nat Rev Genet **9**:554–566
73. Muff R, Buhlmann N, Fischer JA et al (1999) An amylin receptor is revealed following co-transfection of a calcitonin receptor with receptor activity modifying proteins-1 or -3. Endocrinology **140**:2924–2927
74. Bouschet T, Martin S, Henley JM (2005) Receptor-activity-modifying proteins are required for forward trafficking of the calcium-sensing receptor to the plasma membrane. J Cell Sci **118**:4709–4720
75. Christopoulos A, Christopoulos G, Morfis M et al (2003) Novel receptor partners and function of receptor activity-modifying proteins. J Biol Chem **278**:3293–3297
76. Decaillot FM, Rozenfeld R, Gupta A et al (2008) Cell surface targeting of *mu-delta* opioid receptor heterodimers by RTP4. Proc Natl Acad Sci USA **105**:16045–16050
77. Olson R, Dulac C, Bjorkman PJ (2006) MHC homologs in the nervous system — they haven't lost their groove. Curr Opin Neurobiol **16**:351–357

Chapter 10

A Novel Method for Determining the Kinetics of G Protein-Coupled Receptor Plasma Membrane Expression

Gregory W. Sawyer

Abstract

A new approach for characterizing the plasma membrane delivery of G protein-coupled receptors is described in this chapter. This approach uses an existing technology, the regulated secretion/aggregation technology (RPD™), to cause the accumulation of G protein-coupled receptors in the ER of cells. The trafficking of accumulated receptor from the ER to the plasma membrane of cells can be initiated by a small, membrane permeable ligand. The plasma membrane delivery of receptors can be monitored using radioligand binding or microscopy or both techniques. Using this new approach it should be possible to determine the rate of plasma membrane delivery of a broad range of G protein-coupled receptors and thus facilitate learning more about their anterograde trafficking mechanisms.

Key words: Anterograde trafficking, Plasma membrane expression, Conditional aggregation domain, Regulated secretion/aggregation technology, Kinetics

1. Introduction

There is considerable interest in learning more about the molecular mechanisms responsible for the sorting and anterograde trafficking of G protein-coupled receptors (GPCRs) in polarized cells (e.g., neurons of the CNS). An important first step in learning more about the anterograde trafficking of GPCRs is to characterize and compare the rates of plasma membrane delivery of different GPCRs. Presumably, the rate of plasma membrane delivery for different receptors would be dissimilar for receptors using disparate anterograde trafficking mechanisms and similar for those using the same trafficking mechanism. This type of comparison would be facilitated by the development and use of a trafficking assay

Craig W. Stevens (ed.), *Methods for the Discovery and Characterization of G Protein-Coupled Receptors*, Neuromethods, vol. 60, DOI 10.1007/978-1-61779-179-6_10,

that can be applied to a broad range of GPCRs and allows for the specific determination of the rate of anterograde trafficking.

To date, several different approaches have been used to characterize the rate of delivery of GPCRs to the plasma membrane. For instance, the plasma membrane delivery of muscarinic acetylcholine receptors has been characterized by measuring the recovery of specific [^{3}H]*N*-methylscopolamine ([^{3}H]NMS; a membrane impermeate radioligand) binding after irreversibly alkylating cell surface expressed receptor with propylbenzilylcholine mustard (PrBCM) (1–4). Using this approach, subtype-specific differences in the rate of plasma membrane delivery of muscarinic M_3 and M_4 receptors were observed in CHO cells (2). Specifically, the rate of M_3 receptor plasma membrane delivery was approximately one-third of that of M_4 receptors (2). One interpretation of these results is that the M_3 receptor is using a different anterograde trafficking mechanism than the M_4 receptor. An alternative interpretation is that both M_3 and M_4 receptors use the same trafficking mechanism, but the difference in the observed rates of plasma membrane delivery result from differences in the kinetics of M_3 and M_4 receptor synthesis or in the kinetics of M_3 and M_4 receptor trafficking between an endosomal pool and the plasma membrane.

The contribution of the rate of receptor synthesis to the rate of receptor plasma membrane delivery might be considered to be negligible. However, when using [^{3}H]NMS and PrBCM to determine the rate of muscarinic receptor plasma membrane delivery, treatment of cells with the protein synthesis inhibitor cycloheximide significantly reduced the rate of [^{3}H]NMS recovery (1–4). This observation demonstrates that newly synthesized muscarinic receptor was responsible for the majority of receptor that trafficked to the plasma membrane after incubation with PrBCM and the kinetics of receptor synthesis contributes in a meaningful way to the rate of receptor plasma membrane delivery. Thus, determining the rate of receptor plasma membrane delivery using methods like PrBCM and [^{3}H]NMS is hampered by this contribution, making it difficult to ascertain the reason (i.e., use of different trafficking mechanisms) for observed differences in the rate of receptor plasma membrane delivery.

Other methods have been developed and used that characterize receptor plasma membrane without a significant contribution of receptor synthesis to the rate of delivery. For instance, Rosenberg and coworkers (5), characterized the plasma membrane delivery of glycine receptors, a ligand-gated ion channel, by using a temperature-induced blockage of receptor trafficking from the *trans* Golgi network to the plasma membrane. Incubation of cells at 20°C causes retention of proteins in the *trans* Golgi network by blocking trafficking to the plasma membrane, whereas trafficking between the ER and the *trans* Golgi network is unaffected (6). Raising the temperature to 37°C releases this blockage and basically synchronizes receptor trafficking to the plasma membrane (6, 7). To distinguish between newly delivered receptor and receptor already present on the plasma membrane, they made a $\alpha 1$ subunit of glycine receptors with an N-terminal c-myc tag that could be selectively cleaved from the subunit using thrombin (7). The c-myc tag was removed from receptors expressed on the plasma membrane of cells by using thrombin prior to transitioning cells from 20 to 37°C. Thus, only newly delivered receptor would have a c-myc tag after shifting cells to 37°C, allowing one to observe their delivery using an anti-c-myc antibody and immunocytochemistry (7).

We wanted to develop an alternative approach that could be applied to any GPCR, which would synchronize receptor trafficking to the plasma membrane, and did not require the removal of receptor already expressed on the plasma membrane. The approach we used relies on the regulated secretion/aggregation technology (RPD™). The RPD uses a conditional aggregation domain (CAD), a domain that forms dimers/oligomers in a ligand-reversible manner, to retain a protein of interest in a particular cellular compartment as an aggregate (8). The CAD we used consists of four tandem mutant FKBP12 (Phe^{36} to Met) sequences fused to the amino terminus of human M_1 (hM_1) and M_2 (hM_2) receptors (8–11). Mutant FKBP12 (F_m) proteins form dimers that can be completely and rapidly dissociated by synthetic F_m-selective ligands like AP21998 (8, 10). When multiple tandem F_m domains are fused to a protein of interest, the resulting fusion protein forms large aggregates in the nucleus or the ER of a cell, depending upon whether a nuclear localization or secretion signal sequence, respectively, was used in the construct (8, 10).

2. Overview of Using RPD to Characterize the Plasma Membrane Delivery of Muscarinic Receptors

We fused a CAD consisting of four tandem F_m domains to the N-terminus of hM_1 and hM_2 receptors (Fig. 1). Based on previous investigations using the RPD, we anticipated that nascent CAD-fused muscarinic receptors would be translated in the ER and would be retained in the ER as large aggregates (Fig. 2a). The size and number of the aggregates presumably increase as function of time and create a pool of receptor trapped in the ER. Receptor trafficking is initiated by incubating cells expressing CAD-fused receptors with the F_m domain-selective ligand AP21998, which causes the dissolution of CAD-fused receptor aggregates, allowing receptors to traffic from the ER to the Golgi (Fig. 2b). A furin cleavage site (FCS) positioned between the CAD and the receptor sequences (see Fig. 1) is cleaved by furin in the *cis*-cisternea of the Golgi. Receptors then traffic to the plasma membrane where they are expressed lacking the CAD (Fig. 2c).

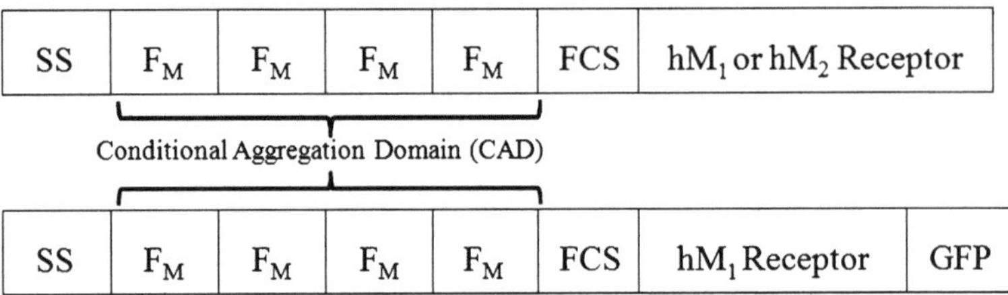

Fig. 1. Illustrations of the CAD fusion proteins expressed in CHO cells. hM_1 and hM_2 receptors were expressed as C-terminal fusion proteins to CAD. The human growth hormone secretion signal sequence (SS) was used to direct the ER insertion and translation of nascent receptor fusion proteins and the furin cleavage site (FCS) was used to remove the CAD from hM_1 and hM_2 receptors as they traffic through the Golgi.

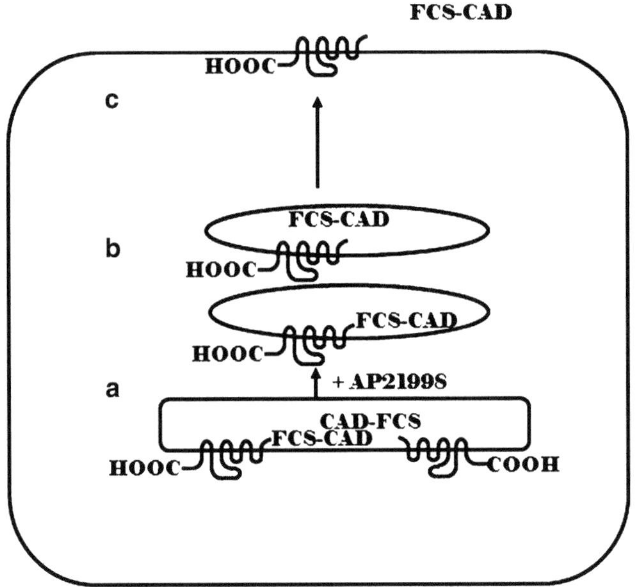

Fig. 2. An illustration of using the RPD to characterize the kinetics of GPCR plasma membrane expression. (**a**) The human growth hormone secretion SS (see Fig. 1) directs the translation of nascent CAD-fused GPCRs in the ER where multivalent interactions between the CADs cause the formation of CAD-fused GPCR aggregates. As a consequence of aggregation, few CAD-fused GPCRs traffic from the ER to the plasma membrane. (**b**) Incubation of cells expressing CAD-fused GPCRs with AP21998 causes the dissolution of aggregates, allowing CAD-fused receptor to traffic to the Golgi. Furin-mediated cleavage of the FCS between the CAD and the receptor (see Fig. 1) occurs in the *cis*-cisternea of the Golgi. (**c**) The GPCR is expressed on the plasma membrane of the cell lacking the CAD. Due to the luminal orientation of CAD, fusion of anterograde vesicles with the plasma membrane releases CAD into the medium.

2.1. CAD-Fused hM_1 Receptors are Retained in an Intracellular Compartment as Aggregates

CAD-fused hM_1 receptors were GFP-tagged (see Fig. 1) to visualize their cellular localization and aggregation state using epi-fluorescence microscopy. In transiently transfected CHO cells, GFP-tagged CAD-hM_1 receptors appeared as large aggregates and did not appear to be expressed on the plasma membrane (Fig. 3a). Based on observations of Rivera and coworkers (8), we surmise that these aggregates were formed in the ER. In their study, a

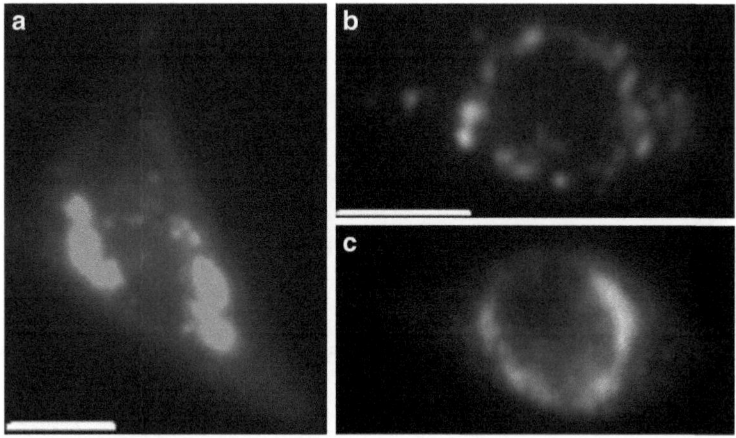

Fig. 3. The localization and aggregation state of GFP-tagged CAD-hM_1 receptors. GFP-tagged CAD-hM_1 receptors were transiently expressed in CHO cells and visualized using epi-fluorescence microscopy as described below in "*Methods for using the RPD to characterize GPCR plasma membrane expression.*" GFP-tagged CAD-hM_1 receptors appear as large aggregates 24 h after transfection (**a**). The localization and aggregation state of GFP-tagged CAD-hM_1 receptors in a CHO cell before (**b**) and after (**c**) a 6-h incubation with the CAD ligand AP21998 (5 μM). Original magnification = 60×; bar = 25 μM. Adapted from (11).

human growth hormone secretion signal sequence was used to direct nascent CAD-fused polypeptides to the ER, where the CAD caused the aggregation of CAD-fused protein via dimerization of the F_m domains making up the CAD (8). As shown in Fig. 1, we expressed hM_1 receptors in CHO cells with a CAD consisting of four F_m domains and a human growth hormone secretion signal sequence.

The cellular localization and aggregation state of GFP-tagged CAD-hM_1 receptors changed when cells were incubated with the F_m domain-selective ligand AP21998. AP21998 has a nanomolar affinity for the F_m domains in a CAD (10). Binding of AP21998 to F_m domains precludes F_m domain dimerization and dissolution of CAD-fused polypeptide aggregates occurs (8, 10). As seen in Fig. 3b and c, a 6-h incubation with AP21998 (5 μM) caused a decrease in the number of discrete GFP aggregates and the receptor obtained a perinuclear localization.

2.2. The CAD is Released into the Medium After AP21998 Treatment

Our cloning strategy allowed for the retention of the FCS in pC4S1-F_M4-hGH (CAD containing plasmid vector obtained from Ariad Pharmaceuticals), which is positioned between the CAD and the hM_1 and hM_2 receptor sequences (see Fig. 1). We accessed the cleavage of the FCS by determining whether the CAD was present in medium harvested from untreated and AP21998 treated CHO cells transiently expressing CAD-hM_1 receptors (see Fig. 2). Anti-FKBP12 did not recognize any polypeptides in the medium harvested from untransfected CHO cells, regardless of whether or not they had been treated with AP21998 (5 μM) for

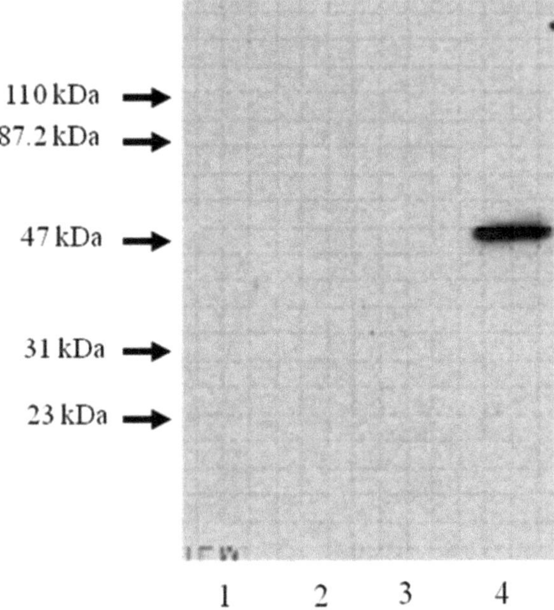

Fig. 4. The CAD is released into the medium of CAD-hM$_1$ transfected CHO cells after treatment with AP21998. Medium was harvested from untransfected (lanes 1 and 2) and CAD-hM$_1$ transfected (lanes 3 and 4) after an 8-h incubation in the absence (lanes 1 and 3) and presence of AP21998 (lanes 2 and 4). The presence of the CAD in harvested medium was visualized after separation using PAGE and blotting onto PVDF in preparation by probing with an anti-FKBP12 antibody. Adapted from (11).

8 h (Fig. 4, lanes 1 and 2). Anti-FKBP12 also did not recognize any polypeptides in the medium harvested from untreated CAD-hM$_1$ transfected CHO cells (Fig. 4, lane 3). In contrast, anti-FKBP12 recognized a single 51 kDa polypeptide in the medium harvested from AP21998 treated (5 μM) CHO cells transiently expressing CAD-hM$_1$ receptors (Fig. 4, lane 4). The mass of this polypeptide was consistent with that predicted for a CAD consisting of four tandem F$_m$ domains (8). Thus, the presence of an anti-FKBP12 reactive polypeptide in the media of CHO cells expressing CAD-fused hM$_1$ receptors after treatment with AP21998 was consistent with the postulated function of the FCS. The furin-mediated cleavage of this site occurs in the *cis*-cisternae of the Golgi, allowing hM$_1$ and hM$_2$ receptors to be expressed on the plasma membrane lacking the CAD (8).

2.3. AP21998 Causes an Increase in the Plasma Membrane Expression of hM$_1$ and hM$_2$ Receptors and the Potency AP21998-Mediated Release is Similar

Prior to incubation with AP21998, little specific [^{3}H]NMS binding was detected in CHO cells transiently expressing CAD-fused hM$_1$ and hM$_2$ receptors (Fig. 5a). [^{3}H]NMS is a membrane-impermeable muscarinic receptor-selective radioligand that can be used to determine the amount of receptor expressed on the plasma membrane of intact, whole CHO cells. Thus, little specific [^{3}H]NMS binding would be indicative of a low amount of plasma membrane expressed receptor. A low level of specific [^{3}H]NMS binding prior

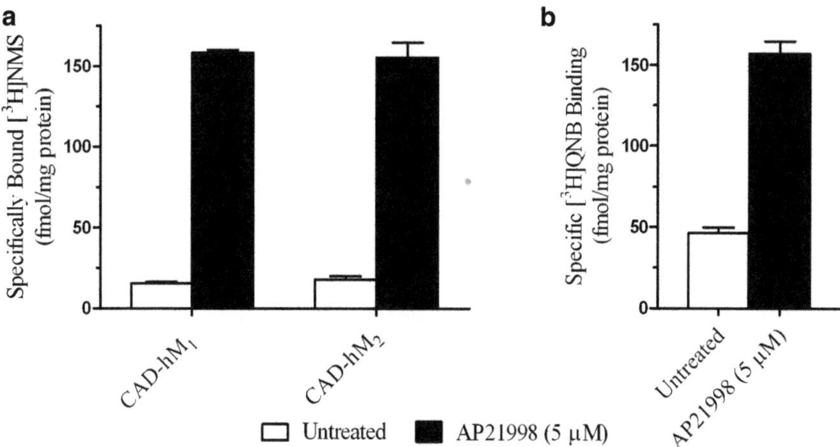

Fig. 5. The effect of AP21998 treatment on [^{3}H]NMS and [^{3}H]QNB binding in CHO cells transiently expressing CAD-fused receptor constructs. (**a**) CHO cells were transiently transfected with CAD-hM$_1$ or CAD-hM$_2$ receptor constructs and then incubated for 8 h in the absence and presence of AP21998 (5 μM). Cells were used in intact, whole cell [^{3}H]NMS binding assays as described in "*Methods for using the RPD to characterize GPCR plasma membrane expression.*" Specific [^{3}H] NMS binding in untreated cells expressing CAD-hM$_1$ and CAD-hM$_2$ receptors was 15.3 ± 1.6 and 17.7 ± 2.3 fmol/mg protein, respectively. Specific binding was 158.4 ± 5.4 and 155.4 ± 9.4 fmol/mg protein in AP21998 treated cells expressing CAD-hM$_1$ and CAD-hM$_2$ receptors, respectively. Each bar represents the mean ± S.E.M. of three experiments conducted in triplicate. Data for CAD-hM$_1$ from (11). (**b**) CHO cells transiently expressing CAD-hM$_1$ receptors were incubated in the absence and presence of AP21998 (5 μM) and then used in intact, whole cell [^{3}H]QNB binding assays as described in "*methods for using the RPD to characterize GPCR plasma membrane expression.*" Specific [^{3}H]QNB binding was 46.2 ± 3.6 fmol/mg protein in untreated cells and 156.9 ± 7.6 fmol/mg protein in AP21998 treated cells. Each bar represents the mean ± S.E.M. of four experiments conducted in triplicate.

to AP21998 treatment is consistent with the postulate that CAD-fused hM$_1$ and hM$_2$ receptors are retained in an intracellular compartment, most likely the ER. Presumably, formation of large CAD aggregates in the ER prevents the sorting of receptors into vesicles that traffic to the Golgi, and thus, impairs receptor plasma membrane expression (8).

Treatment of CHO cells expressing CAD-fused hM$_1$ and hM$_2$ receptors with the F$_m$ domain-selective ligand AP21998 resulted in hM$_1$ and hM$_2$ receptor plasma membrane expression, as determined in intact, whole cell [^{3}H]NMS binding assays. An 8-h incubation with AP21998 (5 μM) caused a ten and ninefold increase in [^{3}H]NMS binding in CHO cells transiently expressing CAD-hM$_1$ and CAD-hM$_2$ receptors, respectively (Fig. 5a). Therefore, the RPD can be used to cause the aggregation and accumulation of numerous nascent CAD-fused hM$_1$ and hM$_2$ receptors in the ER and then synchronize the trafficking of receptors to the plasma membrane using AP21998.

The effect of AP21998 treatment on specific [^{3}H]NMS binding in intact whole CHO cells expressing CAD-fused hM$_1$ and hM$_2$ receptors was consistent with the observed affect of AP21998 on the aggregation state and cellular localization of GFP-tagged CAD-hM$_1$ receptor (see Fig. 3b and c). Prior to incubation with AP21998, GFP-tagged CAD-hM$_1$ receptors were observed as numerous discrete puncta that did not appear to be associated with the plasma membrane (see Fig. 3b). However, there were few discrete puncta in the cell after a 6-h incubation with AP21998 (5 μM) and the GFP-tagged CAD-fused hM_1 receptors appeared to obtain a perinuclear localization (see Fig. 3c). Collectively, these data suggest that AP21998-mediated dissolution of CAD-fused hM_1 and hM_2 receptor aggregates allows the receptors to traffic from the ER and ultimately be expressed on the plasma membrane.

An 8-h treatment with AP21998 (5 μM) also produced a 3.4-fold increase in specific [^{3}H]3-quinuclidinyl benzilate ([^{3}H]QNB) binding in cells expressing CAD-hM_1 receptors (Fig. 5b). [^{3}H]QNB is a membrane permeable muscarinic receptor selective radioligand that should penetrate to intracellular compartments including the ER. The affect of AP21998 on [^{3}H]QNB binding suggests that CAD-fused receptors within receptor aggregates may not be readily accessible to ligand, that the receptor is immature and thus has a low affinity for [^{3}H]QNB, or that the presence of the CAD on the receptor affects the affinity of the receptor for [^{3}H]QNB, or a combination of any one of the three. Consequently, it was not possible to use [^{3}H]QNB to determine the size of the ER retained pool of CAD-fused hM_1 receptor at the beginning of an experiment.

The potency of AP21988 for causing an increase in the plasma membrane expression of hM_1 and hM_2 receptors in CHO cells was determined. The concentration of AP21998 eliciting half-maximal hM_1 and hM_2 receptor plasma membrane expression (EC_{50}) was comparable and was 1.6 ± 0.9 and 0.8 ± 0.9 μM, respectively (Fig. 6). Over the range of concentrations tested, the maximal expression (E_{max}) of hM_1 and hM_2 receptors to the plasma membrane was eight and tenfold above basal, respectively (Fig. 6). We anticipate that the potency of AP21998 for causing the plasma membrane expression of CAD-fused integral membrane proteins, like different subtypes of muscarinic receptors, would be similar for each CAD fusion protein made unless a particular protein influences the tertiary structure of the CAD, and thus alters the affinity for AP21998. In our study, the potency of AP21998 mediated hM_1 and hM_2 receptor plasma membrane expression was similar, suggesting that AP21998 had a similar affinity for the repeated F_M domains of the CAD regardless of whether it was fused to hM_1 or hM_2 receptors.

2.4. The Receptor Plasma Membrane Delivery of hM_1 and hM_2 Receptors During a Short Incubation with AP21998

The expression of hM_1 and hM_2 receptors on the plasma membrane of CHO cells was measured at various times during a 2-h incubation with AP21998 (5 μM). During this incubation, specific [^{3}H]NMS binding increased 3.8-fold over basal (15.1 fmol/mg protein) for hM_1 receptors and 11.2-fold over basal (10.3 fmol/mg protein) for hM_2 receptors (Fig. 7). The slope of the plot of specific [^{3}H]NMS binding against time was used to calculate the initial rate of hM_1 and hM_2 receptor plasma membrane expression. The initial

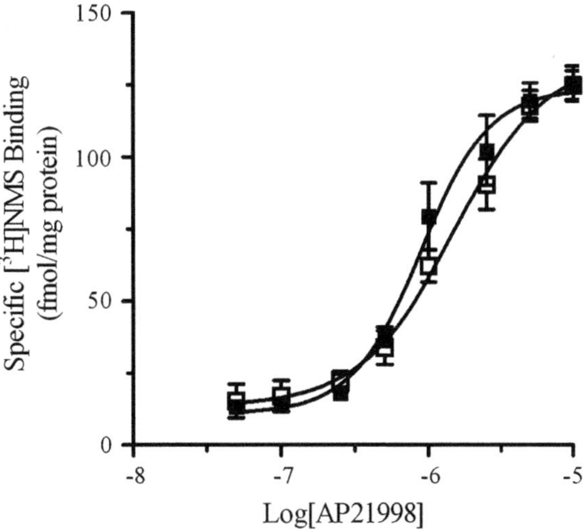

Fig. 6. The potency of AP21998 for causing the plasma membrane expression of hM_1 and hM_2 receptors. CHO cells transiently expressing CAD-hM_1 (*filled square*) and CAD-hM_2 (*open square*) receptors were incubated with equally spaced concentrations of AP21998 (0.33-log unit) for 8 h and then used in intact, whole cell [^{3}H]NMS binding assays as described in "*Methods for using the RPD to characterize GPCR plasma membrane expression.*" Over the range of AP21998 tested, specific [^{3}H]NMS binding increased from 15.3 fmol/mg protein to 124.7 fmol/mg protein for CAD-hM_1 receptor expressing cells and from 13.2 fmol/mg protein to 125.6 fmol/mg protein for CAD-hM_2 receptor expressing cells. Data for CAD-hM_1 receptors is from (11). Each data point represents the mean ± S.E.M. of three experiments conducted in triplicate.

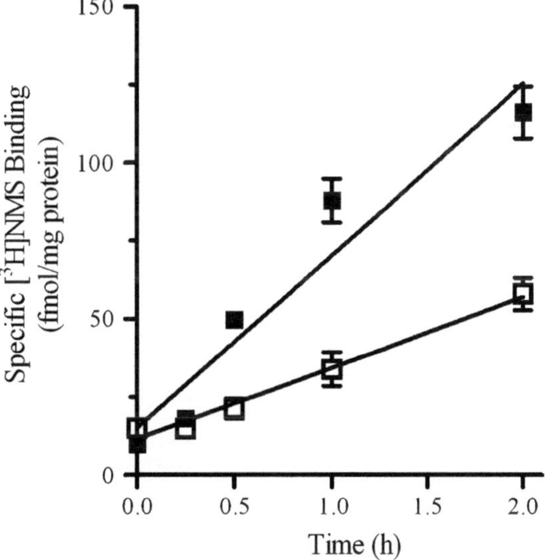

Fig. 7. The plasma membrane delivery of hM_1 and hM_2 receptors in CHO cells during a short incubation with AP21998. CHO cells transiently expressing CAD-fused hM_1 (*open square*) or CAD-fused hM_2 (*filled square*) receptors were incubated with AP21998 (5 μM) for up to 2 h at 37°C in an atmosphere of 5% CO_2/95% air before conducting intact whole cell binding assays with [^{3}H]NMS as described in "*Methods for using the RPD to characterize GPCR plasma membrane expression.*" Each data point represents the mean ± S.E.M. of three experiments performed in triplicate.

rate for hM_2 receptor plasma membrane expression was 2.4-fold greater (55.2 fmol/mg protein h^{-1}) than that for hM_1 receptors (22.8 fmol/mg protein h^{-1}).

The plasma membrane expression of both hM_1 and hM_2 receptors began within 30 min of AP21998 treatment and this time is consistent with that observed for the trafficking of both secreted and membrane proteins across the Golgi stack (see Fig. 7) (11). However, the initial rates for hM_1 and hM_2 receptor plasma membrane expression during this 2 h treatment with a maximal concentration of AP21998 (5 μM) were significantly different from one another; with the initial rate of hM_2 receptor delivery being ~2-fold greater than that for hM_1 receptors. Since the trafficking of CAD-fused hM_1 and hM_2 receptors from the ER to the plasma membrane is synchronized by exposure to AP21998, it is unlikely that the difference in initial rate of hM_1 and hM_2 receptor plasma membrane expression stems from differences in the rate of receptor synthesis. Overall, the difference in the initial rate of hM_1 and hM_2 receptor plasma membrane expression suggests that these receptors use different anterograde trafficking mechanisms.

2.5. The Receptor Plasma Membrane Delivery of hM_1 and hM_2 Receptors During a Long Incubation with AP21998

The plasma membrane expression of hM_1 and hM_2 receptors was measured in CHO cells transiently expressing CAD-fused hM_1 and CAD-fused hM_2 receptors at various times during a 72-h incubation with AP21998. In CHO cells expressing CAD-hM_1 receptors, specific [^{3}H]NMS binding to intact whole cells peaked at approximately 17 h and was 13.8-fold greater than basal (11.5 fmol/mg protein) (Fig. 8). After peaking, specific [^{3}H]NMS binding then decreased to only 2.9-fold over basal at 72 h (Fig. 8). In contrast to cells expressing CAD-hM_1 receptors, specific [^{3}H]NMS binding to intact whole cells expressing CAD-hM_2 receptors peaked at approximately 6 h and was 9.7-fold greater than basal (16.5 fmol/mg protein) (Fig. 8). Specific binding to CAD-hM_2 receptor expressing cells then steadily declined to 1.4-fold above basal at 72 h (Fig. 8).

The observed kinetics for hM_1 and hM_2 receptor plasma membrane expression during the continuous presence of AP21998 were consistent with the trafficking of an accumulated pool of CAD-fused receptor that slowly decays to a steady-state pool (11). Using a mathematical model previously described by us (see (11)), we predict that the rate of CAD-fused hM_1 and hM_2 receptor synthesis is negligible and that the steady-state level of hM_1 and hM_2 receptors is small. Consistent with this prediction, the plasma membrane expression of both hM_1 and hM_2 receptors decayed from peak levels to levels just above basal levels (see Fig. 8). This interesting result may indicate that the accumulation of numerous CAD-fused hM_1 or hM_2 receptors in the ER of CHO cells inhibits either receptor transcription or translation. Regardless of the mechanism, a slow rate of CAD-fused hM_1 and hM_2 receptor synthesis

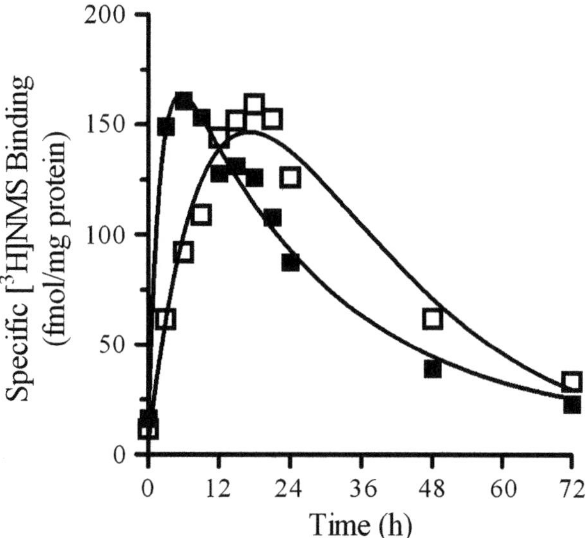

Fig. 8. hM_1 and hM_2 receptor plasma membrane delivery during a long incubation with AP21998 (5 μM). CHO cells transiently expressing CAD-fused hM_1 (*open square*) or CAD-fused hM_2 (*filled square*) receptors were incubated with AP21998 (5 μM) for up to 72 h at 37°C in an atmosphere of 5% CO_2/95% air before conducting intact whole cell binding assays with [^{3}H]NMS as described in "*Methods for using the RPD to characterize GPCR plasma membrane expression.*" Each data point represents the mean of one experiment performed in triplicate.

simplifies kinetic analysis and facilitates the estimation of rate constants and pool sizes (11). In a previous investigation, we used two different mathematical models to fit specific [^{3}H]NMS binding data observed during a 72-h incubation with AP21998 to obtain estimates of rate constants for receptor plasma membrane expression and removal, and for pool size. These models are described in detail in Sawyer et al. (11) and are illustrated in Fig. 9.

Just like the plasma membrane expression of hM_1 and hM_2 receptors during a 2-h treatment with AP21998 (5 μM), the kinetics of hM_1 and hM_2 receptor plasma membrane expression during a 72-h incubation with AP21998 (5 μM) were significantly different from one another (see Figs. 7 and 8). Peak hM_2 receptor plasma membrane delivery occurred 2.8-fold faster than peak hM_1 plasma membrane delivery. This difference supports our conclusion that hM_1 and hM_2 receptors are using disparate trafficking mechanisms in CHO cells.

2.6. The Plasma Membrane Delivery of hM_1 Receptors Elicited to Different Concentrations of AP21998

AP21998 caused both a time- and concentration-dependent increase in specific [^{3}H]NMS binding in CHO cells transiently expressing CAD-hM_1 receptors. CHO cells expressing CAD-hM_1 receptors were incubated for up to 8 h with 0.22, 0.675, 1.25, 5, and 10 μM AP21998. An 8-h incubation with 0.22, 0.675, and 1.25 μM AP21998 produced 1.8±0.03-, 4.5±0.09-, and 6.9±0.4-fold increases in specific [^{3}H]NMS binding (Fig. 10 and Table 1).

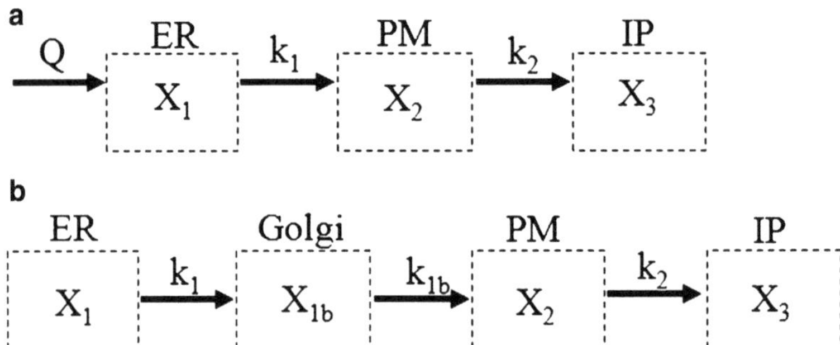

Fig. 9. Illustrations of the models used to estimate the kinetics of receptor plasma membrane delivery following treatment of cells expressing CAD-fused receptor with AP21998. (**a**) In the first model, the rate of CAD-fused receptor synthesis in transiently transfected cells is denoted by *Q*. Newly synthesized CAD fusion protein presumably accumulate in the ER and the amount accumulated there is denoted by X_1. Incubation with the F_m-selective ligand AP21998 disrupts these interactions, allowing CAD-fused receptor to exit the ER and traffic to the plasma membrane according to the first-order rate constant k_1. The amount of receptor accumulating at the plasma membrane is denoted by X_2. The loss of receptors from the plasma membrane is described by the first-order rate constant k_2 and the amount of receptor accumulating in an intracellular pool after loss from the plasma membrane is denoted by X_3. (**b**) In the second model, the ER and the Golgi, which are involved in the synthesis and maturation of integral membrane proteins, were considered. The amount of CAD-fused receptor accumulating in the ER is denoted by X_1. X_{1b} denotes the amount of CAD-fused receptor in the Golgi. The first-order rate constants k_1 and k_{1b} describe the trafficking of receptors between the ER and the Golgi, and the Golgi and the plasma membrane, respectively. The loss of receptors from the plasma membrane is described by the first-order rate constant k_2. The amount of receptor accumulating in an intracellular pool after loss from the plasma membrane is denoted by X_3 like in the first model (**a**). *ER* endoplasmic reticulum; *PM* plasma membrane; *IP* intracellular pool. Adapted from (11).

The 8 h incubation with 5 and 10 μM AP21998 produced a 10.1 ± 0.7- and 10.1 ± 0.2-fold increase, respectively, in specific [^{3}H]NMS binding (Fig. 10 and Table 1).

The slope of the plot of specific [^{3}H]NMS binding against time for each concentration of AP21998 used was used to calculate the initial rate of hM_1 receptor plasma membrane delivery (see Table 1). The initial rates for hM_1 receptor plasma membrane delivery obtained for 5 and 10 μM AP21998 were similar to one another and were 18 ± 0.42 and 18 ± 0.54 fmol/mg protein h^{-1}, respectively (Table 1). The initial rates obtained for 0.22, 0.675, and 1.25 μM AP21998 were different from one another and less than those obtained for 5 and 10 μM AP21998 (Table 1).

As shown in Fig. 10 and Table 1, the rate of hM_1 receptor plasma membrane expression in CHO cells expressing CAD-fused hM_1 receptors was a saturable process and was a function of AP21998 concentration. Half-maximal increase in the rate of hM_1 receptor plasma membrane expression occurred at an AP21998 concentration of 1.34 μM and the maximal rate of delivery was obtained at 5 μM AP21998. These observations suggest that the rate of plasma membrane delivery of hM_1 receptors is dependent on the available free CAD-hM_1 receptor fusion proteins that have dissociated from aggregates and that a consistent rate of delivery can

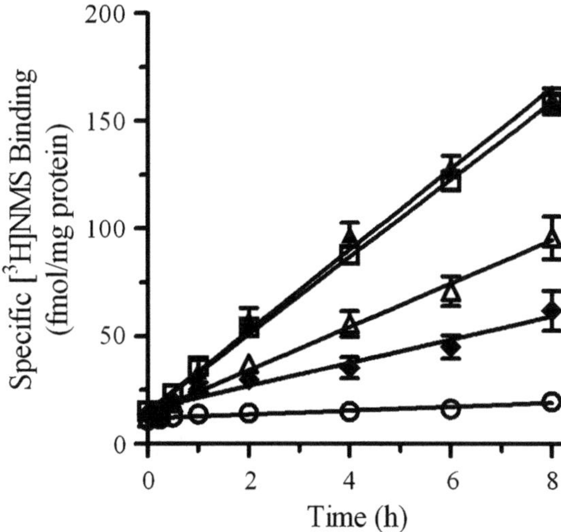

Fig. 10. The plasma membrane delivery of hM_1 receptors elicited to various concentrations of AP21998. CHO cells transiently expressing CAD-fused hM_1 receptors were incubated with 0.22 μM (*open circle*), 0.675 μM (*filled diamond*), 1.25 μM (*open triangle*), 5 μM (*filled triangle*), or 10 μM (*open square*) AP21998 for various periods of time up to 8 h in a humidified incubator set at 37°C in an atmosphere of 5% CO_2/95% air. Cells were then used in intact whole cell [^{3}H]NMS binding assays as described in "*Methods for using the RPD to characterize GPCR plasma membrane expression.*" Each data point represents the mean ± S.E.M. of four to five experiments performed in triplicate. Adapted from (11).

Table 1
The rate of and maximal hM_1 receptor plasma membrane expression elicited to various concentrations of AP21998[a]

[AP21998] (μM)	Maximal hM_1 receptor plasma membrane expression (fmol/mg protein)[b]	Maximal increase (fold over basal)[c]	Rate (fmol/mg protein h^{-1})[d]
0.22	16.3 ± 1.1	1.8 ± 0.03	1.1 ± 0.05
0.675	62.0 ± 9.1	4.5 ± 0.09	6.0 ± 1.2
1.25	95.8 ± 9.2	6.9 ± 0.4	10.2 ± 1.2
5	160.5 ± 4.5	10.1 ± 0.7	18.0 ± 0.42
10	158.3 ± 4.9	10.1 ± 0.2	18.0 ± 0.54

[a] Data is presented as the mean ± S.E.M. and are derived from data shown in Fig. 7. Data from [11]

[b] Maximal hM_1 receptor plasma membrane expression was determined after an 8-h incubation with AP21998 using intact, whole cell [^{3}H]NMS binding assays

[c] Fold-increase above basal was calculated by dividing maximal hM_1 receptor plasma membrane expression by the amount of receptor expressed on the plasma membrane of CHO cells incubated in the absence of AP21998 for 8 h

[d] The rate was determined using linear regression analysis.

be obtained using concentrations of AP21998 that are equal to or greater than the maximally effective concentration (i.e., the initial rate of delivery was the same for 5 and 10 μM AP21998). Due to the limited solubility of AP21998, the highest concentration tested was 10 μM.

2.7. CAD-hM$_1$ Receptors are Functional

Using CAD-fused receptor, it is possible to determine the functional response elicited to different numbers of receptors expression on the plasma membrane of cells. We incubated CHO cells transiently expressing CAD-hM$_1$ receptors with either 1 or 5 μM AP21998 for 8 h and then conducted phosphoinositide hydrolysis assays. For cells treated with 1 μM AP21998, the concentration of carbachol eliciting half-maximal response (EC_{50}) and the Hill coefficient were 40 ± 23 μM and 0.7 ± 0.02, respectively (Fig. 11). For cells treated with 5 μM AP21998, carbachol was 2.5-fold more potent (EC_{50} = 16 ± 6.6 μM) and the maximal response (E_{max}) was 1.2-fold greater. These differences in phosphoinositide hydrolysis elicited to carbachol can be attributed to the different levels of hM$_1$ receptor plasma membrane expression elicited by the different concentration of AP21998. The amount of hM$_1$ receptor expressed on the plasma membrane of CHO cells in 1 and 5 μM AP21998 treated cells was 87.5 and 166.1 fmol/mg protein, respectively (see Fig. 10).

We calculated the theoretical ratio of functional binding capacities from these phosphoinositide hydrolysis concentration-response curves using a modification of Furchgott analysis as described below in Sect. 3. A value of 0.37 was obtained for the

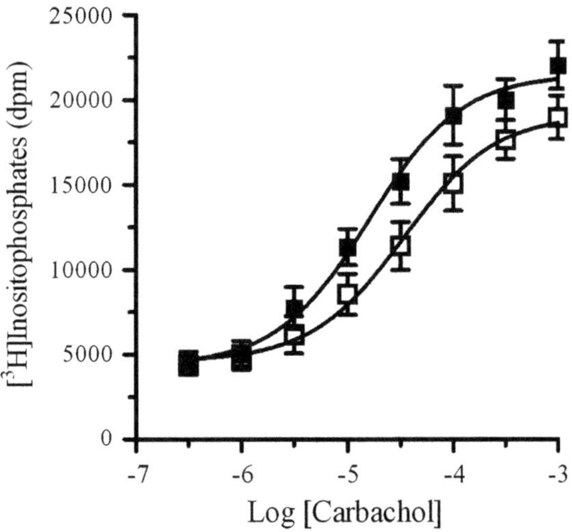

Fig. 11. Once expressed on the plasma membrane, hM$_1$ receptor elicit phosphoinositide hydrolysis. CHO cells transiently expressing CAD-hM$_1$ receptor fusion proteins were incubated with either 1 μM (*open square*) or 5 μM (*filled square*) AP21998 for 8 h, then washed and incubated with [^{3}H]*myo*-inositol in preparation for phosphoinositide assays. Each data point represents the mean ± S.E.M. of three experiments performed in triplicate. Adapted from (11).

theoretical ratio of functional binding capacities. This theoretical ratio was comparable to the ratio of binding capacities calculated from the data shown in Fig. 10 (0.53). The equilibrium dissociation constant (K_D) of carbachol was also estimated from these phosphoinositide data and was 0.19 mM. This estimate is in close agreement with the binding affinity of carbachol at the rat M_1 receptor expressed in CHO-K_1 cells (K_D = 0.17 mM) measured in a buffer containing GTP (0.1 mM) (12).

Using this approach, we found that the theoretical ratio of functional binding capacities calculated from the phosphoinositide response in CHO cells expressing CAD-fused hM_1 receptors following treatment with 1 and 5 μM AP21998 (0.37) was similar to that calculated from the binding capacity of [^{3}H]NMS in cells expressing CAD-fused hM_1 receptors treated with 1 and 5 μM AP21998 (0.54). The small discrepancy between the theoretical and actual binding capacities may be due to experimental variation. Alternatively, this discrepancy may indicate that some of the receptors expressed on the plasma membrane by 1 μM AP21998 are not functional or were not expressed in an appropriate cellular domain with the correct signal transduction mechanisms.

3. Methods for Using RPD to Characterize GPCR Plasma Membrane Expression

3.1. Receptor Constructs

PCR and primers were used to copy and amplify hM_1 and hM_2 (hM_2) receptor sequences lacking a start codon and with 5 *Afl* II and 3 *Bam*H I restriction sites (13). The sequences were ligated in-frame with the F_m sequences (CAD) and the FCS of pC4S1-F_M4-hGH (pCAD) (generously provided by Ariad Pharmaceuticals at www.ariad.com/regulationkits) using *Afl* II and *Bam*H I restriction sites and standard molecular techniques. The resulting constructs, pCAD-hM_1 and pCAD-hM_2 were used to express hM_1 and hM_2 receptors as C-terminal fusion proteins to CAD (see Fig. 1). If one wanted to express a GPCR as a fusion protein to CAD lacking the FCS, *Spe* I and *Bam*H I restriction sites present in pCAD could be used.

To visualize the cellular localization and trafficking of CAD-fused receptors using epi-fluorescence or confocal microscopy, a fluorescence protein tag can be incorporated into the CAD-fused receptor construct. To express CAD-fused hM_1 receptors with a C-terminal GFP tag, we used PCR and primers to copy and amplify the CAD-fused hM_1 receptor sequence from pCAD-hM_1 lacking a stop codon and with 5 *Eco*R I and 3 *Bam*H I restriction sites. This sequence was ligated into pEGFP-N3 (Clontech) using *Eco*R I and *Bam*H I restriction sites and standard molecular techniques. The resulting construct pCAD-hM_1-GFP was used to express hM_1 receptors as C-terminal and N-terminal fusions to CAD and GFP, respectively (see Fig. 1).

3.2. Cell Culture and Transfections

Presumably, any transfectable cell type can be used to investigate GPCR trafficking using the RPD. To date we have characterized the kinetics of hM_1 and hM_2 receptor trafficking in Chinese Hamster Ovary (CHO) cells and observed hM_1 receptor trafficking in primary cortical neurons (data not shown). For experiments described in this chapter, we used CHO cells which were maintained in growth medium (F-12K supplemented with 10% FBS, 100 U/mL penicillin, and 100 mg/mL streptomycin) in a humidified incubator set at 37°C and 5% CO_2/95% air. For experiments, cells were plated in 24-well plates at 2×10^5 cells/well in 0.5 mL transfection medium (F-12K supplemented with 10% FBS) and were transiently transfected with receptor constructs using Lipofectamine 2000 (Invitrogen) and the product protocol. Cells were incubated for 6 h in a humidified incubator at 37°C and the medium was exchanged with fresh transfection medium (0.5 mL). Cells were incubated an additional 18 h (24 h total) in a humidified incubator set at 37°C and 5% CO_2/95% air before conducting experiments.

To date, we have not varied the incubation time posttransfection to determine whether the pool size of ER retained aggregates of CAD-fused receptor influences either receptor trafficking or cell viability. Presumably, the pool size of CAD-fused receptor should not influence the kinetics of AP21998-mediated receptor plasma membrane expression. Fitting the [^{3}H]NMS binding data shown in Fig. 8 to the mathematical model illustrated in Fig. 9, yielded an estimate of 381 and 190 fmol/mg protein for the pool sizes in CAD-fused hM_1 and hM_2 receptor expressing CHO cells, respectively. In light of the earlier mentioned comments in Sect. 2.6, regarding the steady-state level of CAD-fused receptor expression, it would be interesting to incubate transiently transfected cells for 48 or 72 h (instead of 24 h) prior to treatment with AP21998. If our predications are correct, the pool size should neither change appreciably from that estimated at 24 h nor should the kinetics of receptor plasma membrane expression. Additionally, based on our observations with [^{3}H]QNB (see Fig. 5b), a technique other than radioligand binding, like quantification of GFP in the ER using microscopy, will have to be used to directly measure the pool size.

With regards to the effect of ER retained aggregates of CAD-fused receptor on cell viability, Rivera and coworkers (8) state that cells containing aggregates of CAD-fused protein grow indefinitely and at the same rate as control cells. They also state they observed no reduction in the amount of protein stored over a 6-month period [8). In their experiments, they fused the CAD to human growth hormone and insulin, both of which are secreted proteins. To date, the effect of storing CAD-fused integral membrane proteins as aggregates in the ER of cells for a long period of time on cell viability has not been determined. Therefore, it is possible that ER retained aggregates of CAD-fused GPCR may affect cell viability, making it difficult to make a stable cell line.

3.3. Intact, Whole Cell Binding Assays

We chose to monitor the plasma membrane expression of muscarinic receptors using intact whole cell binding assays using [^{3}H]NMS, a membrane impermeable radioligand. We also tried to determine the pool size of CAD-fused hM$_1$ receptors using [^{3}H]QNB, a membrane permeable radioligand. Briefly, transiently transfected CHO cells were washed 2 times with ice-cold PBS (0.5 mL) to remove medium. Intact cells were incubated with either 1.6 nM [^{3}H]NMS or 1.2 nM [^{3}H]QNB in the absence (total binding) and presence (nonspecific binding) of atropine (10 μM) overnight at 4°C in binding buffer (25 mM HEPES, 113 mM NaCl, 6 mM dextrose, 3 mM $CaCl_2$, 3 mM KCl, 2 mM $MgSO_4$, 1 mM NaH_2PO_4, pH 7.4). Unbound radioligand was aspirated from each well and the wells were rapidly washed 2 times with 1 mL ice-cold PBS. Bound radioligand was recovered as described previously (14). Specifically bound [^{3}H]NMS and [^{3}H]QNB was expressed as fmol/mg protein where appropriate. The amount of protein was estimated for each experiment using six wells of a 24-well plate, plated and transfected at the same time as experimental plates, using the BCA reagent.

Using the RPD and microscopic techniques like total internal reflection fluorescence (TIRF) and confocal microscopy, it should be possible to visualize and to characterize GFP-tagged receptor trafficking from the ER, through the Golgi, to the plasma membrane in real-time.

3.4. Phosphoinositide Hydrolysis Assays

To access the function of hM$_1$ receptors expressed in CHO cells expressing CAD-fused hM$_1$ receptors after AP21998 treatment, we used phosphoinositide hydrolysis assays using the method described by Griffin and coworkers (14). Other methods (e.g., agonist-stimulated [^{35}S]GTPγS binding) could be used to access GPCR function.

3.5. Western Blot

The furin-mediated cleavage of the FCS of CAD-fused hM$_1$ receptors was accessed using Western blots and an anti-FKBP12 antibody. Briefly, medium from untransfected and pCAD-hM$_1$ transfected CHO cells was harvested after an 8-h incubation in the absence or presence of AP21998 (5 μM) in F-12K medium (0.5 mL) in a humidified incubator at 37°C in an atmosphere of 5% CO_2/95% air. Polypeptides in the medium were separated on a 10% polyacrylamide gel and then transferred to PVDF. The blot was probed using standard techniques and an anti-FKBP12 primary antibody. Immunoreactive polypeptides were visualized using ECL plus Western Blotting Detection System and BioMax film. Using immunocytochemistry, it should be possible to determine if any receptor is expressed on the plasma membrane of cells with the CAD after treatment with AP21998.

3.6. Epi-Fluorescence Images

We used epi-fluorescence microscopy to determine the cellular localization and aggregation state of GFP-tagged CAD-fused hM_1 receptors. Briefly, CHO cells (0.25×10^5/well) were placed on poly-D-lysine treated plates in transfection medium (0.5 mL) and maintained overnight in a humidified incubator set at 37°C in an atmosphere of 5% CO_2/95% air. The following day, cells were transfected with pCAD-hM_1-GFP using Lipofectamine 2000. Images of CHO cells expressing fluorescence proteins were captured 24-h after transfection using an Olympus IX 71 epi-fluorescence microscope fitted with FITC filter and a Coolsnap Monochrome digital camera. The images were colored green using Metamorph 5.0r7 imaging software. Alternatively, confocal microscopy could be used to visualize the cellular localization and aggregation state of GFP-tagged CAD-fused GPCRs before and after AP21998 treatment. If the microscope is fitted with an incubation chamber, these observations could be made in real-time.

3.7. Data Analysis

The Hill slope, maximal response (E_{max}), and the concentration of either AP21998 or carbachol eliciting half-maximal response (EC_{50}) were estimated using nonlinear regression analysis according to a logistic equation.

Using a modification of Furchgott's (15) method described by Ehlert (16), the equilibrium K_D and the theoretical ratio of functional binding capacities were calculated based on the ability of carbachol to elicit phosphoinositide hydrolysis in cells expressing different levels of hM_1 receptors on the plasma membrane. The concentration-response curve to carbachol of cells expressing the lower level of hM_1 receptors on the plasma membrane was considered to be analogous to the condition of partial receptor inactivation in Furchgott analysis. In brief, equieffective concentrations of carbachol were obtained from the curves from high and low hM_1 receptor expressing CHO cells. These values were fitted to the following equation by nonlinear regression analysis:

$$\log A = \log([A' K_A q] / [K_A + (1 - q)A']).$$

In the equation above, A and A' are the concentrations of carbachol eliciting equivalent phosphoinositide hydrolysis in CHO cells expressing lower and higher levels of hM_1 receptor on the plasma membrane, q corresponds to the theoretical ratio of functional binding capacities, and K_A is the affinity of agonist for the receptor.

3.8. An Alternative Method for Characterizing the Plasma Membrane Delivery of GPCRs

Many GPCRs have a conserved $F(x)_6LL$ motif (x indicates any amino acid residue and L indicates either Leu or Iso) in their membrane proximal C-terminal tail (17). In hM_1 receptors, this motif ^{423}FRDTFRLLL^{431} is structurally analogous to the hydrophobic folding motif $h(x)_3h(x)_2hh$ (^{423}FRDTFRLLL^{431}, where h is any hydrophobic amino acid) described by Krause and coworkers, (18, 19). Mutation of either the $F(x)_6LL$ or $h(x)_3h(x)_2hh$ motif in a particular GPCR may result in its ER retention. To date, mutation of the $F(x)_6LL$ motifs in 5-HT_{1A}, 5-HT_{1B}, α_{1B}- and α_{2B}-adrenergic, angiotensin II type IA, β_2-adrenergic, and hM_1 receptors has been shown to cause ER retention (17, 19–21). Similarly, mutation of the $h(x)_3h(x)_2hh$ motif in the vasopressin V_2 receptor causes ER retention (18).

Mutation of $^{430}LL^{431}$ to $^{430}AA^{431}$ in the $F(x)_6LL$ motif of hM_1 receptors caused ER retention in CHO cells (19). When incubated with the muscarinic receptor selective antagonist atropine (0.1 μM), mutant hM_1 receptors trafficked from the ER to the plasma membrane (19). Using this approach, we found that the mutant hM_1 receptor appeared on the plasma membrane of CHO cells at an initial rate of 115 ± 6 fmol/mg h^{-1} (19). A rate constant for the plasma membrane delivery of mutant hM_1 receptors (0.12 h^{-1}) was obtained by dividing the initial rate by the maximal receptor expressed after an 18-h treatment with atropine (971 fmol/mg protein) (19). Using the RPD regulated secretion/aggregation technology at a lower level of hM_1 receptor expression, we estimated an initial rate of delivery of the wild-type hM_1 receptor to the plasma membrane of 18 fmol/mg protein h^{-1} (see Table 1). When normalized relative to receptor expression at 18 h, the rate constant for the plasma membrane delivery of wild-type hM_1 receptors (0.10 h^{-1}) was approximately the same as that of the mutant receptor (19).

Based on the trafficking of hM_1 receptors possessing a $^{430}LL^{431}$ to $^{430}AA^{431}$ mutation in the $F(x)_6LL$ motif, we think that $F(x)_6LL$ mutations might be useful for investigating the trafficking of other GPCRs possessing it. In our case, a pulse of mutant hM_1 receptor could be synchronously released from the ER with a period of atropine treatment. The trafficking of these receptors was then observed using intact whole cell [^{3}H]NMS binding assays (plasma membrane expression) or fluorescence microscopy (with a GFP-tagged receptor). A similar approach has been described using a GFP-tagged thermoreversible VSVG folding mutant (22).

4. Future Directions

We are interested in using the RPD secretion/aggregation technology and TIRF microscopy to characterize the trafficking of GFP-tagged muscarinic receptors to the plasma membrane of cells in real-time. We are also interested in using lentivirus packaged with either CAD-fused or GFP-tagged CAD-fused muscarinic receptor constructs to transduce primary cortical neurons and then characterize CAD-fused receptor trafficking using either intact whole cell [^{3}H]NMS binding or TIRF microscopy. Lastly, we are planning to make CAD-fused hM_1/hM_2 receptor chimeras to determine whether there is a particular receptor domain that is responsible for the observed differences in the kinetics of hM_1 and hM_2 receptor plasma membrane delivery.

5. Conclusions

Based on our observations, we think that the RPD secretion/aggregation technology can be used to characterize the trafficking of a broad range of GPCRs from the ER to the plasma membrane of cells. Additionally, we anticipate that characterizing GPCR trafficking using this technology is amiable to a wide range of methods used to determine cellular localization (i.e., microscopic techniques) or receptor expression (i.e., binding or functional assays). To date, we have used epi-fluorescence microscopy, phosphoinositide hydrolysis assays, and intact, whole cell [^{3}H] NMS binding assays to characterize the cellular localization, function, and kinetics of receptor plasma membrane expression, respectively, in CHO cells expressing CAD-fused hM_1 receptors. We found that each of these methods complemented one another (e.g., GFP-tagged CAD-fused hM_1 receptors did not appear to be expressed on the plasma membrane in untreated cells and, consistently, there was little specific [^{3}H]NMS binding observed in CHO cells expressing CAD-fused hM_1 receptor prior to AP21998 treatment) and, thus, a more complete picture of hM_1 receptor trafficking could be ascertained.

Overall, using the RPD secretion/aggregation technology, it is possible to characterize the kinetics of receptor trafficking without interference from constitutively expressed receptor, enabling one to compare the kinetics of different GPCR plasma membrane expression without the contribution of receptor synthesis.

Acknowledgments

The activities described were supported by an OHRS award for project number HR03-107S, from the Oklahoma Center for the Advancement of Science and Technology.

References

1. Haddad EB, Rousell J (1995) Muscarinic M2 receptor synthesis: study of receptor turnover with propylbenzilylcholine mustard. Eur J Pharm **290**:201–205.
2. Koenig JA, Edwardson JM (1996) Intracellular trafficking of the muscarinic acetylcholine receptor: importance of subtype and cell type. Mol Pharm **49**:351–359.
3. Koenig JA, Edwardson JM (1994) Routes of delivery of muscarinic acetylcholine receptors to the plasma membrane in NG108-15 cells. Br J Pharm **111**:1023–1028.

4. Koenig JA, Edwardson JM (1994) Kinetic analysis of the trafficking of muscarinic acetylcholine receptors between the plasma membrane and intracellular compartments. J Biol Chem **269**:17174–17182.

5. Rosenberg M, Meier J, Triller A et al (2001) Dynamics of glycine receptor insertion in the neuronal plasma membrane. J Neurosci **21**:5036–5044.

6. Griffiths G, Pfeiffer S, Simons K et al (1985) Exit of newly synthesized membrane proteins from the trans cisterna of the Golgi complex to the plasma membrane. J Cell Biol **101**:949–964.

7. Rosenberg M, Meier J, Triller A et al (2001) Dynamics of glycine receptor insertion in the neuronal plasma membrane. J Neurosci **21**:5036–5044.

8. Rivera VM, Wang X, Wardwell S et al (2000) Regulation of protein secretion through controlled aggregation in the endoplasmic reticulum. Science **287**:826–830.

9. Volchuk A, Amherdt M, Ravazzola M et al (2000) Megavesicles implicated in the rapid transport of intracisternal aggregates across the Golgi stack. Cell **102**:335–348.

10. Rollins CT, Rivera VM, Woolfson DN et al (2000) A ligand-reversible dimerization system for controlling protein-protein interactions. Proc Natl Acad Sci USA **97**:7096–7101.

11. Sawyer GW, Ehlert FJ, Hart JP (2006) Determination of the rate of muscarinic M1 receptor plasma membrane delivery using a regulated secretion/aggregation system. J Pharmacol Toxicol Methods **53**:219–233.

12. Savarese TM, Wang CD, Fraser CM (1992) Site-directed mutagenesis of the rat M1 muscarinic acetylcholine receptor. Role of conserved cysteines in receptor function. J Biol Chem **267**:11439–11448.

13. Bonner TI, Buckley NJ, Young AC et al (1987) Identification of a family of muscarinic acetylcholine receptor genes. Science **237**:527–532.

14. Griffin MT, Hsu JC, Shehnaz D et al (2003) Comparison of the pharmacological antagonism of M2 and M3 muscarinic receptors expressed in isolation and in combination. Biochem Pharmcol **65**:1227–1241.

15. Furchgott RF (1966) The use of *beta*-haloalkylamines in the differentiation of receptors and in the determination of dissociation constants of receptor-agonist complexes. Adv Drug Res **3**:21–55.

16. Ehlert FJ (1987) Coupling of muscarinic receptors to adenylate cyclase in the rabbit myocardium: effects of receptor inactivation. J Pharmcol Exp Ther **240**:23–30.

17. Duvernay MT, Zhou F, Wu G (2004) A conserved motif for the transport of G protein-coupled receptors from the endoplasmic reticulum to the cell surface. J Biol Chem **279**:30741–30750.

18. Krause G, Hermosilla R, Oksche A et al (2000) Molecular and conformational features of a transport-relevant domain in the C-terminal tail of the vasopressin V(2) receptor. Mol Pharm **57**:232–242.

19. Sawyer GW, Ehlert FJ, Shults CA (2010) A conserved motif in the membrane proximal C-terminal tail of human muscarinic M1 acetylcholine receptors affects plasma membrane expression. J Pharmcol Exp Ther **332**:76–86.

20. Duvernay MT, Dong C, Zhang X et al (2009) Anterograde trafficking of G protein-coupled receptors: function of the C-terminal F(X)6LL motif in export from the endoplasmic reticulum. Mol Pharm **75**:751–761.

21. Carrel D, Hamon M, Darmon M (2006) Role of the C-terminal di-leucine motif of 5-HT1A and 5-HT1B serotonin receptors in plasma membrane targeting. J Cell Sci **119**: 4276–4284.

22. Hirschberg K, Miller CM, Ellenberg J et al (1998) Kinetic analysis of secretory protein traffic and characterization of Golgi to plasma membrane transport intermediates in living cells. J Cell Biol **143**:1485–1503.

Part III

The G Protein-Coupled Receptor on the Membrane

Chapter 11

Characterizing the Pharmacology of G Protein-Coupled Receptors in Transfected Cell Lines

Kathryn A. Seely and Paul L. Prather

Abstract

A remarkable potential exists for current and future development of therapeutic drugs acting at GPCRs. As one of the initial steps in GPCR drug development, *in vitro* assays are required to characterize the pharmacology of new ligands acting at distinct GPCRs. This is routinely accomplished by first employing cellular models to establish high affinity, selectivity, and efficacy of a test compound for a specific GPCR involved in a disease process of interest. However, several limitations are encountered when native cell lines or isolated tissues expressing low levels of endogenous GCPRs are employed for receptor characterization. To overcome many of these issues, cells are routinely transfected with cDNA of a desired GPCR to create cell lines stably expressing a sufficient receptor density to allow for adequate pharmacological studies. Although several commercial suppliers offer stably transfected cell lines expressing various GPCRs, as well as kits to examine several signal transduction pathways regulated by GPCRs, the cost and specialized equipment required to conduct these essential studies are often out of reach for many laboratories. Therefore, the purpose of this chapter is to provide a brief, simple, economical, and straightforward guide for the production of a cell line stably expressing a GPCR of interest and the methods required to characterize the basic pharmacology of ligands acting at that receptor. Specifically, methods will describe how to stably transfect cells and to conduct receptor binding studies for determination of receptor density and ligand affinity. Finally, methods will be presented to subsequently characterize the functional signaling of the expressed GPCR, from G-protein activation to regulation of two distinct intracellular effectors.

Key words: Transfection, Saturation binding, Competition binding, GTPγS binding, cAMP, ERK-MAPK, Affinity, Efficacy, Signal transduction

1. Introduction

It has been estimated that approximately 40–50% of currently available drugs used clinically produce their effects by acting at GPCRs (1). However, of the 323 GPCRs identified in the human

Craig W. Stevens (ed.), *Methods for the Discovery and Characterization of G Protein-Coupled Receptors*, Neuromethods, vol. 60, DOI 10.1007/978-1-61779-179-6_11, © Springer Science+Business Media, LLC 2011

genome as potential drug targets, only 46 are the known therapeutic target of drugs on the market today (2). Therefore, a tremendous potential exists for current and future development of drugs acting at GPCRs. As one of the initial steps in GPCR drug development, *in vitro* assays are required to characterize the pharmacology of new ligands acting at distinct GPCRs. This is routinely and most efficiently accomplished by employing cellular models to first establish high affinity, selectivity, and efficacy of a test compound for a specific GPCR involved in a disease process of interest.

1.1. Problematic Issues Associated with Quantification of GPCR Signaling in Isolated Tissues and Native Cell Lines

Although the pharmacology of novel ligands for specific GPCRs can be investigated by employing various isolated tissue preparations or native cell lines expressing endogenous receptors, in many instances the density of the GPCR of interest is often low and highly regulated. Furthermore, in addition to the receptor of interest, many tissues often also express several other closely related GPCRs. Since even very selective ligands for a specific receptor will also bind to other structurally similar GPCRs at higher concentrations, data interpretation for evaluation of novel compounds with unknown selectivity is often confounded under these conditions. Lastly, the ability to obtain sufficient quantities of cells or tissue required for extensive *in vitro* investigation of a given GPCR is often limited by slow growth of native cell lines or small tissue size and the availability of animals required for tissue preparation.

1.2. Advantages for Use of Transfected Cell Lines to Examine GPCR Signaling

To overcome many of these limitations, a very useful technique is often employed whereby cells are transfected (either transiently or stably) with cDNA encoding for a specific GPCR. The transfected cells subsequently express a sufficient density of the desired GPCR and the pharmacology of ligands binding to the GPCR can easily be characterized. Study of cells transfected with GPCRs offer several distinct advantages to use of isolated tissues and/or native cells. First, depending on the type of cDNA constructs employed, the density of GPCR expression can be tightly controlled from low to very high levels. Second, selection of a proper background cell line for transfection (e.g., devoid of closely related GPCRs) will assure sole expression of the receptor for investigation and avoid potential confounding of data interpretation. Third, homologous cDNA recombination techniques can be employed to engineer and express virtually any form or mutated version of the GPCR in question for examination. Lastly, stable cell lines (produced by transfection of cDNA constructs that also provide specific antibiotic resistance) can be developed that allow for production of unlimited quantities of cells for *in vitro* characterization.

1.3. CB_2 Cannabinoid Receptors as an Example

Δ^9-Tetrahydrocannabinol (Δ^9-THC), the main psychoactive ingredient in marijuana, mediates its effects primarily through activation of two GPCRs, CB_1 and CB_2 (3). CB_1 receptors are primarily localized in central and peripheral nervous tissue (4) while CB_2 receptors are predominantly found in the spleen and leukocytes (5). Selective CB_2 receptor ligands are being developed as therapeutics for use as immune modulators of inflammation (6). Although a potentially very important drug target, CB_2 receptors are expressed in relatively low levels throughout the body and are almost always coexpressed with CB_1 receptors. To allow for efficient investigation of CB_2 receptor signaling, our laboratory has developed a Chinese Hamster Ovary (CHO) cell line stably expressing human CB_2 receptors (e.g., CHO-hCB2). This cell line has been extremely useful for pharmacological characterization of novel CB_2 ligands (7) and examination of unique signaling properties of CB_2 receptors (8).

1.4. Summary and Outline for this Chapter

Although several commercial suppliers offer stably transfected cell lines expressing various GPCRs, as well as kits to examine several signal transduction pathways regulated by GPCRs, the cost and specialized equipment required to conduct these essential studies are often out of reach for many laboratories. Therefore, based on our past experience in this area, the purpose of this chapter will be to provide a brief, simple, economical, and straightforward guide to the production of a cell line stably expressing a GPCR of interest and the methods required to characterize the basic pharmacology of ligands acting at that receptor. This chapter is aimed at a target audience that has little, or no, prior experience with these type of techniques. A basic transfection protocol will first be described, followed by sections outlining receptor binding protocols detailing how to determine the density of the expressed GPCRs and the affinity of test ligands for those receptors. Finally, methods will be described to characterize the functional signaling of the expressed GPCR, from G-protein activation to regulation of two distinct intracellular effectors.

2. Methods to Determine Pharmacological Characteristics of GPCRs in Transfected Cells

Methods used to characterize receptors in transfected cells are summarized in Table 1 and described below.

2.1. Transfection

2.1.1. Overview and Objective

Cells expressing a specific GPCR can be created by transiently or stably tranfecting the cDNA encoding for the receptor. The oldest and least expensive transfection technique utilizes coincubation of cDNA with calcium phosphate, resulting in the formation of complexes that are subsequently endocytosed by cells (9).

Table 1
Review of protocols for characterizing GPCRs in transfected cells

Steps	Notes
Transfection	
Seed cells – 80% confluence, Form lipid/cDNA complexes	Wild-type cell selection important
Add to cells, culture 48–74 h, subculture, add selection antibiotic	
Pick colonies, expand colonies, harvest, verify GPCR expression	
Membrane preparation	
Homogenize cells in 50 mM HEPES, pH 7.4 with 0.32 M sucrose	Cell/buffer ratio important
Centrifuge at 1,000 × *g*, repeat 2 times; pool supernatants	
Centrifuge at 1,00,000 × *g*; homogenize pellet; aliquot; store −80°C	
Saturation binding	
Assay buffer 50 mM Tris, pH 7.4, 0.1% BSA, 5 mM $MgCl_2$	K_D and B_{max} determination, expensive
Many radioligand concentrations + membranes	
90 min RT incubation, vacuum filtration, scintillation counting	
Competition binding	
Assay buffer 50 mM Tris, pH 7.4, 0.1% BSA, 5 mM $MgCl_2$	IC_{50} and K_i determination, economical
One radioligand + many cold ligand concentrations + membranes	
90 min RT incubation, vacuum filtration, scintillation counting	
GTPγS binding	
Assay buffer 20 mM HEPES, pH 7.4, 0.1% BSA, 100 mM NaCl, 10 μM GDP, 10 mM $MgCl_2$	Mainly limited for use with Gi/Go-class of G-proteins
Many test ligand concentrations + [^{35}S]GTPγS + membranes	
1 h 30°C incubation, vacuum filtration, scintillation counting	
Adenylyl cyclase assay	
Seed cells – 80% confluence; [^{3}H]adenine 1–2 h 37°C incubation	Seed cells evenly
Many test ligand concentrations + 10 μM forskolin, 15 min 37°C	
[^{3}H]cAMP by column chromatography, scintillation counting	

(continued)

Table 1 (continued)

Steps	Notes
pERK-MAPK assay	
Seed cells – 90% confluence; many test ligand concentrations	pERK-Ab quality most important
1–15 min 37°C incubation, fix cells, pERK-Ab overnight at 4°C	
2° HRP Ab 1 h at RT, quantify chemiluminescence	

This method results in relatively low transfection efficiency. A second method, electroporation, achieves higher levels of transfection by applying an electrical pulse which opens transient pores in the plasma membrane through which cDNA enters cells (10). A third, and perhaps most common method, is chemical transfection involving the use of liposomal agents. This section will describe this method to create cells that transiently or stably express any GPCR of interest. We have used this technique to create CHO cells stably expressing CB2 receptors (e.g., CHO-hCB2) (7).

2.1.2. Materials Summary

Wild-type cells (e.g., CHO cells, ATCC, Manassas, VA), normal growth medium (e.g., Dulbecco's Modified Eagle's Medium, DMEM, Mediatech Cellgro, Herndon, VA), serum-free medium (e.g., Optimem, Invitrogen, Carlsbad, CA), a cDNA construct subcloned into an appropriate plasmid (e.g., CNR2-pcDNA3.1 obtained from the Missouri S&T cDNA Resource Center), a cationic-lipid (e.g., Lipofectamine-2000, Invitrogen, Carlsbad, CA), a suitable selection antibiotic (e.g., Geneticin, G418, Mediatech Cellgro, Herndon, VA), a humidified and heated CO_2 incubator, and various tissue culture supplies.

2.1.3. General Method

The day prior to transfection, cells are seeded into dishes and cultured in normal growth medium in a humidified atmosphere of 5% CO_2/95% O_2 at 37°C without antibiotics at a density that will result in 50–80% confluence the following day (e.g., 3×10^6 CHO cells in 100 mm dishes). On the day of transfection, the liposomal reagent (e.g., Lipofectamine-2000) is diluted in a 1:1 v/v ratio in serum-free medium (e.g., Optimem) and allowed to stand at room temperature (RT) for 5 min. Drug the 5 min incubation, the cDNA/plasmid construct is also diluted similarly in optimem. The diluted liposomal agent and plasmid are then gently mixed together and allowed to stand at RT for an additional 15 min to form stable complexes. A proper ratio of lipid agent to plasmid is one the most critical variables required for success of this technique. Thus, the optimal ratio should be determined empirically for each cell line and plasmid examined. Manufacturers suggest a range of liposomal agent to be employed that should be used as a starting point for optimization. To begin transfection, the normal growth medium is replaced with optimem and the lipid/plasmid complexes are slowly added in a dropwise fashion to the medium, gently swirling the dish after addition of each drop to insure even distribution. Cells are then cultured for 48–72 h in optimem. For assays involving transient transfection, at this point cells may be employed for various assays of choice. However, to create a cell line stably expressing a GPCR, the following series of steps are required. Transfected cells are first subcultured in normal growth medium by seeding additional dishes at a relatively low density (e.g., a 1:10 dilution of transfected cells). Importantly, at this time an appropriate selection antibiotic is then added to all dishes (e.g., geneticin, 1 mg/mL) to eliminate all cells not expressing the transfected plasmid providing antibiotic resistance. A low density of cells is employed to assure that surviving colonies will theoretically arise from growth of a single cell. Dishes are left undisturbed for 2 weeks, or until appearance of visible colonies. Surviving colonies are then transferred to individual wells of a 24-well culture plate utilizing large orifice pipet tips and subsequently cultured in the continued presence of the selection antibiotic. Once confluent, the entire contents of the individual wells are transferred to a single well of a six-well culture plate for growth expansion. When each well of the six-well plate reaches confluence, half of the cells are harvested and used to screen for the presence of the expressed GPCR by receptor binding techniques (see following sections). The remaining half of cells are maintained in culture until the transfection efficiency for that colony has been determined (e.g., to decide if the colony will be retained or discarded).

2.1.4. Advantages, Disadvantages, and Special Considerations

A major advantage of this technique is that no special equipment is required and thus it can be performed by virtually any laboratory that performs cell culture experiments. The only real disadvantage is that unless specific advanced transfection techniques are employed to allow regulation of receptor expression levels (see following sections), the density of GPCR obtained in a given colony is random and dependent on the amount and location of cDNA integration into the cellular genome. Therefore, many colonies must be screened to obtain suitable cell lines. Prior to transfection, it is important to carefully consider the selection of the wild-type cell line to be used. Ideally, the cell line selected should not contain the endogenously expressed receptor or closely related receptors. Further, it should endogenously express G-proteins and intracellular effectors that are normally regulated by the receptor under physiological conditions. Finally, the growth characteristics and the transfection efficiency of the wild-type cells should also be considered.

3. Receptor Binding

Receptor binding experiments can be used to determine the presence, density, and pharmacological characteristics of GPCRs expressed in transfected cells. This section will present general protocols, describing how to conduct two important and commonly used types of receptor binding experiments; saturation (Fig. 1a) and competition (Fig. 1b) binding. Prior to these descriptions, however, a general method to produce partially purified plasma membranes for use in receptor binding assays will be described.

3.1. Membrane Preparation

3.1.1. Overview and Objective

GPCRs are integral membrane proteins (spanning the membrane 7 times) and thus are localized primarily to the plasma membrane of cells. As such, purified plasma membrane preparations are most often used in binding experiments as a concentrated source of receptors which reduces nonspecific binding of radioligands and improves the signal-to-noise ratio of the assay. The following is a brief, relatively simple protocol to produce partially purified plasma membranes for receptor binding experiments (11).

3.1.2. Materials Summary

Pelleted transfected cells (e.g., CHO-hCB2), homogenization buffer (0.32 M sucrose, 3 mM $MgCl_2$, 1 mM EDTA, 50 mM HEPES pH 7.4), Dounce glass homogenizer (Wheaton, Philadelphia PA) – a high speed and an ultracentrifuge.

3.1.3. General Method

Harvested pellets of frozen/thawed cells are resuspended in the ice-cold homogenization buffer in a ratio of approximately 1×10^5 cells per 1 mL of buffer. The cell suspension is then subjected to ten complete strokes in a glass Dounce homogenizer maintained on ice. The cell homogenates are centrifuged at $1{,}000 \times g$ for 10 min at 4°C and the supernatant collected in a glass beaker immersed in ice. The remaining pellet is resuspended in one-half the original volume of ice-cold homogenization buffer and the homogenization procedure repeated twice. Supernatants from all centrifugations are pooled in the original chilled beaker maintained on ice. The pooled supernatants are subjected to a final 90 min centrifugation at $100{,}000 \times g$ at 4°C. The resulting pellet of partially purified membranes should be resuspended in ice-cold homogenization buffer (minus sucrose) to an approximate concentration of 1 mg/mL. Finally, the preparation should be divided into aliquots of appropriate sizes for use in subsequent assays and stored at –80°C. One aliquot should be used to determine the actual protein concentration of the preparation.

3.1.4. Advantages, Disadvantages, and Special Considerations

This simple protocol produces a consistent yield of partially purified membranes. However, if yields are low, optimization of the ratio of cells to homogenization buffer and/or the addition of a

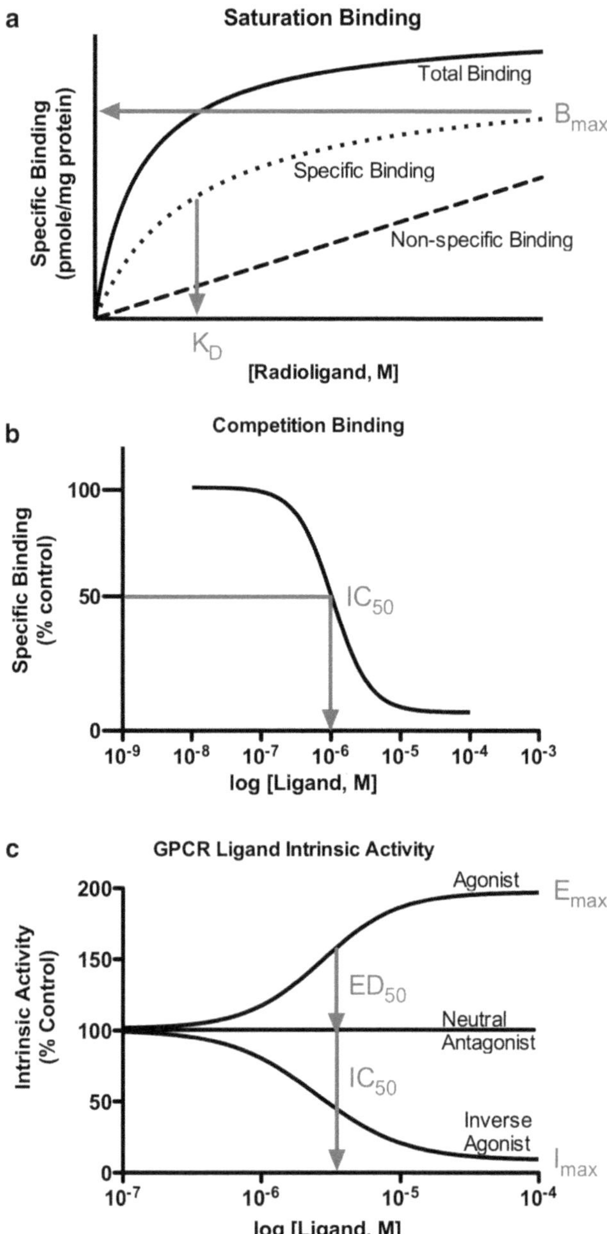

Fig. 1. Examples of the graphic presentation of (**a**) saturation and (**b**) competition receptor binding and (**c**) intrinsic activity produced by ligands acting at GPCRs.

protease inhibitor cocktail to the buffer should be considered. Lastly, the number of freeze/thaw cycles for membrane preparations should be limited.

3.2. Saturation Binding

3.2.1. Overview and Objective

Saturation binding is performed by determining the amount of specific binding occurring in response to incubation of a GPCR source (e.g., membranes) with increasing concentrations of a radioligand until a "saturation" (or plateau) in the amount of binding is obtained (see Fig. 1a). Nonlinear regression analysis of data obtained from these types of experiments allows for an accurate determination of the B_{max} (the density of receptor sites) of a GPCR expressed in a transfected cell line and the K_D (a measure of affinity) of the radioligand for the expressed receptor. This section will describe a standard, detailed, and straightforward protocol to conduct saturation binding experiments (12).

3.2.2. Materials Summary

Purified membranes prepared from a cell line transfected with a GPCR of interest (e.g., CHO-hCB2), binding buffer (5 mM $MgCl_2$, 0.1% bovine serum albumin, 50 mM Tris pH 7.4), a high-affinity tritiated radioligand selective for the GPCR of interest (e.g., [^{3}H]CP-55,940 for CB1/CB2, PerkinElmer, Boston, MA), a nonradioactive ligand also selective for the GPCR of interest (e.g., WIN-55,212-2 for CB1/CB2, Tocris Bioscience, Ellisville, MO), glass test tubes, glass filter papers, a cell harvester apparatus for vacuum filtration (e.g., Brandel, Gaithersburg, MD), and a graphics software program capable of conducting nonlinear regression analysis (e.g., GraphPad Prism, San Diego, CA).

3.2.3. General Method

The first step to conduct a successful saturation binding assay is determination of the proper concentrations of the radioligand to be examined. The concentrations selected should span a range of approximately tenfold less, to tenfold higher, than the anticipated K_D for the radioligand. Relatively few receptors will be occupied at concentrations tenfold less, while near full receptor occupancy occurs at concentrations tenfold higher, than the K_D of the radioligand. Examining a large number of radioligand concentrations will increase the accuracy of the binding curve. However, due to high cost of radioligand, we have found that employing six concentrations results in consistently accurate results.

For each radioligand concentration tested, four assay tubes are required. Two duplicate assay tubes to determine the amount of "total binding" and two tubes to define the amount of "nonspecific binding" for each concentration are prepared. In addition to the standard assay buffer, the two total binding tubes contain membranes and a specific radioligand concentration to be examined. Radioactivity present following filtration of these tubes reflects the amount of radioligand binding to all targets, both specific (e.g., the target GPCR) and nonspecific (e.g., other cell constituents, filter paper, etc.). The two nonspecific binding tubes not only include membranes and the radioligand, but also contain a receptor saturating concentration of a competing nonradioactive ligand. Inclusion of a high concentration of the nonradioactive ligand (~100–1,000-fold greater that its K_D) should totally displace the radioligand from all specific binding sites (e.g., the GPCR of interest). Therefore, the radioactivity remaining after filtration of these tubes reflects the amount of radioligand that binds to only nonreceptor targets. Finally, the amount of "specific binding" at each radioligand concentration is calculated by subtracting the nonspecific binding from the total binding values.

In summary, each radioligand concentration examined will require four assay tubes, two to determine total binding and two to define nonspecific binding. Since a minimum of six concentrations should be tested, a typical saturation binding curve will consist of 24 assays tubes (e.g., 4 tubes × 6 concentrations). This number is very convenient because Brandel cell harvesters, used to separate bound from free radioligand (see following), simultaneously filter samples in multiples of 24 (e.g., 24, 48, or 72).

3.2.4. Experimental Protocol

To conduct the binding experiment it is convenient to prepare the binding reaction in glass test tubes in a final volume of 1 mL, both for addition of 10× drug stocks and to minimize ligand depletion. Each of the 24 tubes should contain the following ingredients all prepared in a common binding buffer (50 mM Tris, 0.1% bovine serum albumin, 5 mM of $MgCl_2$, pH 7.4): (1) membrane protein (25–100 μg, see consideration following), (2) a specific concentration of radioligand to be examined, and (3) either a receptor saturating concentration of a competing nonradioactive ligand selective for the GPCR in question (for nonspecific binding) or vehicle (for total binding). We find it useful to prepare and add all assay constituents as 10× stock solutions so that when 100 μL of each is added and diluted to the final 1 mL volume, a 1× concentration is achieved. Once all ingredients are added, tubes are mixed well by vortexing and then incubated for 90 min at RT with gentle agitation to reach equilibrium binding (see consideration following).

Following incubation, binding reactions are terminated by rapid vacuum filtration through presoaked Whatman GF/B glass fiber filters employing a cell harvester. Importantly, the filtration step also separates the bound radioligand (to the GPCR) from the free radioligand (in the buffer). To ensure complete removal of excess free radioligand, the initial filtration step is routinely followed by three of more washes with ice-cold binding buffer. The filters are then placed into 7 mL vials and an appropriate volume of scintillation fluid (e.g., 4 mL) is added. After overnight incubation at RT, the samples are shaken and counted for a minimum of 5 min per sample in a liquid scintillation counter to quantify the amount of radioactivity present.

3.2.5. Data Graphing and Analysis

Binding data are analyzed by utilizing a graphics software package capable of conducting nonlinear regression analysis such as GraphPad Prism® v4.0b (GraphPad Software, Inc., San Diego, CA). First, the amount of specific binding (e.g., total minus nonspecific binding) is plotted on the *Y*-axis as a function of the specific radioligand concentration plotted on the *X*-axis (see Fig. 1a). Both measures are presented on a linear scale. The units for specific binding are most often presented as moles per mg of protein (e.g., pmol/mg) and the units for the radioligand concentration are presented as nanomolar (e.g., nM). The extrapolated amount of radioligand bound by the receptor when the plateau (or saturation) of binding is reached is defined as the number or density of receptor sites (e.g., B_{max}). In addition, the concentration of radioligand required to attain half-maximal levels of saturation binding is defined as the affinity (e.g., K_D) of that radioligand for the receptor. Both of these values are provided in the "best fit" analysis calculated by nonlinear regression.

3.2.6. Advantages, Disadvantages, and Special Considerations

A major advantage of saturation binding is that it provides the most accurate measure of receptor density (B_{max}) and affinity (K_D) for a radioligand, relative to other available methods. However, since relatively high concentrations of expensive radioligands are required to attain saturable levels of binding, a significant disadvantage associated with this technique is cost. Several additional specific considerations should also be noted. First, the optimal amount of membranes employed should be determined empirically for each cell line. Cells expressing higher receptor densities will require less membrane. To reduce nonspecific binding, the least amount of membranes that produce a consistent, quantifiable level of binding should be used. Second, all calculations for saturation binding are based on the assumption that measurements are made at conditions of equilibrium binding (e.g., where radioligand association and dissociation are equal and a steady state level of binding is observed). Therefore, initial experiments for a given radioligand and cell line should be performed to demonstrate the time required for a plateau in the specific binding of a single radioligand concentration to occur. Third, to prevent radioligand dissociation during the filtration process, ice-cold buffer should be employed for all washes. Finally, due to the critical function of the cell harvester for these type of experiments, it is important that the manufacturer's suggested maintenance for the apparatus is strictly followed.

3.3. Competition Receptor Binding

3.3.1. Overview and Objective

Competition binding is performed by determining the ability of increasing concentrations of nonradioactive (cold) ligands to compete with (and thus decrease) the amount of specific binding of a single, fixed concentration of a radioligand to a GPCR (see Fig. 1b). The half-maximal inhibitory concentration (e.g., IC_{50}) derived from these type of experiments is converted to a measure of affinity (K_i) for the cold ligand. This section will describe a standard protocol to conduct competition binding experiments (7).

3.3.2. Materials Summary

The materials required for competition binding are identical to those listed previously for saturation binding (see Sect. 11.3.2.2).

3.3.3. General Method

The first step for conducting a successful competition binding experiment is to select the single, fixed concentration of the radioligand, as well as the range of concentrations of the cold ligand to be examined. The radioligand is usually employed at a concentration near its K_D value. This concentration results in approximately 50% receptor occupancy by the radioligand, and for most transfected cell lines, this amount of radioactivity is sufficient to demonstrate consistent, quantifiable levels of binding. The concentrations of the cold ligand to be examined should be selected such that the concentration predicted to produce a half-maximal reduction in specific binding (e.g., IC_{50}) will be positioned close to the middle of the concentration-effect curve (see Fig. 1b). Furthermore, to cover an adequate span, it is useful to select concentrations one log unit apart. For example, a typical range of seven concentrations that might be tested are 0.01, 0.1, 1, 10, 100, 1,000, and 10,000 nM.

Competition binding experiments can routinely be conducted utilizing 24 assay tubes. Each concentration of cold ligand to be tested should be performed in triplicate. All 24 tubes will contain the previously determined, single concentration of the radioligand. The first triplicate of tubes will have no cold ligand. The second triplicate will contain the lowest concentration of cold ligand, and so on, until the seventh triplicate is prepared to contain the highest concentration of cold ligand to be examined. Importantly, the eighth and final triplicate will be used to define nonspecific binding (see previous saturation binding section for preparation). Since only a single radioligand concentration is being utilized, only one triplicate of tubes to measure nonspecific binding is required. In summary, each competition binding curve will consist of 24 assay tubes, with each condition performed in triplicate. This experimental design will allow for examination of seven cold ligand concentrations (the initial one being a "0" concentration) and the final triplicate reserved to define nonspecific binding.

3.3.4. Experimental Protocol

The actual preparation of the assay tubes and filtration process are identical to that described previously for saturation binding (see Sect. 11.3.2.4).

3.3.5. Data Graphing and Analysis

Competition receptor binding is analyzed by first subtracting the nonspecific binding from the total binding for all of triplicates and graphing the percent of control specific binding as a function of the log concentrations of the cold ligand being examined (see Fig. 1b). Although the affinity for the receptor cannot be determined directly utilizing competition receptor binding, the following Cheng-Prusoff equation can be employed to convert the IC_{50} (calculated by nonlinear regression) to a measure of affinity, K_i (equation 1): $K_i = IC_{50}/(1 + [L]/K_D)$. The IC_{50} defined by these experiments is the half-maximal inhibitory concentration of the competitor, [L] is the concentration of the radioligand, and K_D is the affinity of the radioligand for the receptor (13).

3.3.6. Advantages, Disadvantages, and Special Considerations

There are two major advantages of competition binding. First, much less radioligand is required for these type of studies relative to saturation binding experiments. Therefore, competition binding experiments are much more economical. Second, with use of only a single radioligand for a GPCR of interest, competition binding can be employed to determine the affinity of a virtually unlimited number of nonradioactive ligands. All "Special Considerations" cited for saturation binding are applicable for competition binding as well.

4. Functional Characterization of GPCRs

Receptor binding assays importantly identify the affinity of ligands that bind to GPCRs, but regrettably provide no information about the intrinsic activity of ligands acting at these receptors. The next sections will describe three functional assays used to predict whether novel ligands act as agonists (stimulating receptor activity), antagonists (having no effect on receptor activity), or inverse agonists (inhibiting constitutive receptor activity) at GPCRs (Fig. 1c).

4.1. G Protein Activation ([^{35}S]GTPγS Binding Assay)

4.1.1. Overview and Objective

Agonists bind to GPCRs, producing conformational changes in the receptors, resulting in an association with, and activation of, G-proteins. Therefore, G-protein activation is the first step in the signaling cascade produced by agonist stimulation of GPCRs. Activation of G-proteins results in an exchange of GDP for GTP on the of G-protein α-subunit. The activated Gα-subunit bound by GTP is quickly hydrolyzed to the inactive GDP, "turning-off" signaling. A useful assay to measure G-protein activation monitors the increase in the amount of a radioactively labeled GTP analog ([^{35}S]GTPγS) that occurs when membranes are coincubated with agonists selective for GPCRs. Since [^{35}S]GTPγS is not hydrolyzed, once it binds to the activated G-protein present in membranes, it does not dissociate and can be quantified by rapid vacuum filtration to separate bound from free [^{35}S]GTPγS. Constitutively active GPCRs also stimulate GDP/GTP exchange, thus this assay can likewise be used to identify the activity of inverse agonists by their ability to reduce the constitutive activity of GPCRs. This section will describe a simple assay to detect the functional activation (by agonists) or inhibition (by inverse agonists) of G-proteins by GPCRs in transfected cell lines (7).

4.1.2. Materials Summary

Purified membranes prepared from a cell line transfected with a GPCR of interest (e.g., CHO-hCB2), assay buffer (100 mM NaCl, 10 mM $MgCl_2$, 10 μM GDP, 0.1% BSA, 20 mM HEPES pH 7.4), wash buffer (0.1% BSA, 20 mM HEPES pH 7.4), $[^{35}S]$GTPγS, cold GTPγS, GPCR ligands to be evaluated, glass test tubes, glass filter papers, a cell harvester apparatus for vacuum filtration (e.g., Brandel, Gaithersburg, MD), and graphics software to conduct nonlinear regression analysis (e.g., GraphPad Prism, San Diego, CA).

4.1.3. General Method

The experimental design for conducting $[^{35}S]$GTPγS binding assays is very similar to that previously described for competition receptor binding assays (see Sect. 11.3.3.3). First, the range of concentrations of the test ligand are selected such that the concentration predicted to produce a half-maximal stimulation (e.g., ED_{50}) or inhibition (e.g., IC_{50}) of $[^{35}S]$GTPγS binding should be positioned close to the middle of the concentration-effect curve (see Fig. 1c). Furthermore, to cover an adequate span, it is useful to select concentrations one log unit apart. A total of 24 tubes can be used to generate a single concentration-effect curve for $[^{35}S]$GTPγS binding. Each concentration of test ligand should be examined in triplicate. All tubes will contain 10–50 μg of membrane protein (to be determined empirically for each cell line) and 0.1 nM $[^{35}S]$GTPγS, a concentration producing a level of binding that is easily quantifiable for the majority of transfected cell lines. The first triplicate of tubes will have no test ligand. The second triplicate will contain the lowest concentration of the test ligand, and so on, until the seventh triplicate is prepared to contain the highest concentration of test ligand to be examined. Importantly, the eighth and final triplicate will be used to define nonspecific binding, by including a high concentration (10 μM) of cold GTPγS. After addition of all ingredients and vigorous mixing, reactions are incubated for 1–2 h at 30°C, followed by rapid filtration to terminate the reaction and separate bound from free $[^{35}S]$GTPγS. Nonspecific binding is subtracted from all samples and G-protein activity is presented graphically as percent $[^{35}S]$GTPγS binding relative to control binding (e.g., basal activity) as a function of ligand concentration, presented on a log scale (Fig. 1c). Measures of both potency (e.g., ED_{50} or IC_{50} values) and efficacy (e.g., E_{max} or I_{max}) of ligands acting at GPCRs are obtained from nonlinear regression analysis of the curves presented.

4.1.4. Advantages, Disadvantages, and Special Considerations

The [^{35}S]GTPγS binding assay is versatile in that experiments can examine potential effects produced by agonists, antagonists, or inverse agonists. However, a limitation of this assay is that it is most useful only for GPCRs coupled to the Gi/Go-class of G-proteins because this class of Gα-subunits possesses a high guanine nuleotide exchange rate relative to other classes of G-proteins (e.g., Gs, Gq or G12/13). Importantly, proper controls must be performed to prove that the modulation of [^{35}S]GTPγS binding produced by a test ligand occurs in response to action via the predicted GPCR. For example, the effects on [^{35}S]GTPγS binding produced by a test ligand should be reversed by selective, neutral antagonists for the target GPCR. Furthermore, the test ligand should not alter G-protein activity in membranes prepared from nontransfected cells.

4.2. Adenylyl Cyclase Activity (Intracellular cAMP Assay)

4.2.1. Overview and Objective

Activation of many GPCRs produces stimulation of Gi/Go- or Gs-classes of G-proteins which then proceed to inhibit or simulate the activity, respectively, of the intracellular effector adenylyl cyclase. Modulation of this second step in the signal transduction pathway is reflected by alterations in the levels of intracellular cAMP. Therefore, the ability of test ligands to increase or decrease levels of intracellular cAMP in cultured whole cells via interaction with transfected GPCRs is a functional measure that is easily monitored. The adenylyl cyclase assay also allows for the characterization of agonists, neutral antagonists, and even inverse agonists at GPCRs expressed in transfected cells. This section will describe a relatively straightforward assay designed to detect inhibition of adenylyl cyclase activity produced by activation of GPCRs coupled to the Gi/Go-class of G-proteins (8). Slight modifications of this assay (see discussion following) also allow for monitoring of activation of GPCRs coupled to the Gs-class of G-proteins, which produce increases in cAMP levels.

4.2.2. Materials Summary

GPCR-transfected whole cells (e.g., CHO-hCB2 cells), incubation buffer (DMEM containing 0.9% NaCl, 500 μM IBMX, 2 μCi/well [^{3}H]adenine), KRHB Assay Mix (110 mM NaCl, 5 mM KCl, 1 mM $MgCl_2$, 1.8 mM $CaCl_2$, 25 mM glucose, 55 mM sucrose, 500 μM IBMX, 10 μM forskolin, 10 mM HEPES, pH 7.4), GPCR ligands to be examined, acidic alumina (Sigma-Aldrich, St. Louis, MO), and Poly-Prep Chromatography Columns (Bio-Rad, Hercules, CA).

4.2.3. General Method

The first step to conduct a successful adenlylyl cyclase assay is to determine the range of concentrations of the test ligand to be examined. The concentrations are selected such that the concentration predicted to produce a half-maximal inhibition (e.g., IC_{50}) of adenylyl cyclase activity will be positioned close to the middle of the concentration-effect curve (see Fig. 1c). Furthermore, to cover an adequate span, it is useful to select concentrations one log unit apart. A total of 24 assay tubes, corresponding to treatment of cells seeded in a 24-well plate, can be used to generate a single concentration-effect curve for modulation of adenylyl cyclase activity by test ligands. To increase the levels of intracellular cAMP, so that the inhibition of adenylyl cyclase activity can easily and consistently be observed, 10 μM forskolin (a direct activator of adenylyl cyclase activity) is also included in all tubes except those designated to define basal activity. Each concentration of test ligand examined should be performed in triplicate. The first triplicate of tubes should have no test ligand. The second triplicate will contain the lowest concentration of the test ligand, and so on, until the seventh triplicate is prepared to contain the highest concentration test ligand to be examined. Importantly, the eighth and final triplicate will be used to define basal adenylyl cyclase activity by omission of forskolin. All drug dilutions are prepared in the Krebs-Ringer-HEPES buffer (KRHB) assay mix containing 500 μM 3-isobutyl-1-methylxathine (IBMX) to prevent breakdown of cAMP. In summary, each adenylyl cyclase inhibition curve will consist of 24 assay tubes, with each condition performed in triplicate. This design allows for testing of seven ligand concentrations (the initial one being a "0" concentration) and the final triplicate reserved to define basal adenylyl cyclase activity.

The assay utilizes plated cells, which take up [^{3}H]adenine (upon incubation) to form intracellular-labeled pools of [^{3}H]ATP that are converted [^{3}H]cAMP upon activation of adenylyl cylcase. To begin the assay, cells are seeded (as consistently as possible) in a 24-well plate such that approximately 80% confluence will be attained on the day of the assay. To begin the assay, normal growth media is replaced with an 500 μL (per well) of an incubation mix containing 500 μM IBMX and [^{3}H]adenine (2 μCi/well). Cells are then cultured with the incubation mix for 1–2 h at 37°C and 5% CO_2/95% O_2 to allow for sufficient [^{3}H]adenine uptake. Following the incubation period, the radioactive medium is replaced with 500 μL (per well) of a KRHB-based assay mix containing various concentrations of the test ligand. The 24-well plate is then floated on a 37°C water bath for 15 min and the reaction is terminated and cells lysed by the addition of 50 μL of 2.2 N HCl. The entire 500 μL assay volume of each well is next loaded onto an individual Poly-Prep chromatography column containing 0.9 g of acidic alumina. Each column is washed with 8 mL of 0.001 N HCl, followed by a 2-mL wash with 0.1 M ammonium acetate. Columns are then placed over 20 mL scintillation vials filled with 10 mL of scintillation fluid and [^{3}H]cAMP is eluted by addition of a final 4 mL of 0.1 M ammonium acetate. Following complete elution, vials are vigorously shaken and radioactivity quantified utilizing a liquid scintillation counter. Basal

[^{3}H]cAMP levels are subtracted from all other (forskolin-stimulated) samples and the data are presented graphically as percent of control intracellular cAMP levels as a function of test ligand concentration, presented on a log scale (Fig. 1c). Measures of both potency (e.g., IC_{50} values) and efficacy (e.g., I_{max}) of test ligands acting at GPCRs are obtained from nonlinear regression analysis of the concentration-effect curves presented.

4.2.4. Advantages, Disadvantages, and Special Considerations

The adenylyl cyclase assay is versatile in that experiments can examine potential effects produced by agonists, antagonists, or inverse agonists for GPCRs coupled to either the Gi/Go- or Gs-class of G-proteins. To examine Gs-coupled receptors, similar experiments are conducted with the omission of forksolin. Importantly, proper controls must be performed to prove that the modulation of adenylyl cyclase activity produced by a test ligand occurs in response to action via the predicted GPCR. For example, the observed effects produced by a test ligand should be reversed by selective, neutral antagonists for the target GPCR. Furthermore, no effect of the test ligand on adenylyl cyclase activity should be observed in nontransfected wild-type cells.

4.3. pERK-MAPK Assay (Phosphorylated-Extracellular Signal-Regulated Kinase Subgroup of the Mitogen-Activated Protein Kinases)

4.3.1. Overview and Objective

A third and final assay to characterize the functional activation of GPCRs in tranfected cells is regulation of the ERK-MAPK system. Almost all GPCRs, including those coupled to the Gi/Go-, Gs-, Gq- and G12/13-classes of G-proteins, have been shown to converge to regulate the activity of the ERK-MAPK system. For example, any GPCR altering the function of protein kinase A (PKA), protein kinase C (PKC), or β-arrestin will also modulate the activity of ERK-MAPK. Therefore, monitoring ERK-MAPK activity might serve as a universal assay to monitor GPCR function (1). Highly selective antibodies recognizing the active phosphorylated form of ERK1/2 have been developed (e.g., pERK1/2-MAPK) and are thus often used to monitor ERK-MAPK activity. This section will describe, in a step-by-step manner, an assay to monitor GPCR regulation of ERK-MAPK activity in whole cells (8).

4.3.2. Materials Summary

GPCR-transfected whole cells (e.g., CHO-hCB2 cells), serum-free DMEM, phosphate buffered saline (PBS), 4% formaldehyde solution, wash buffer (0.1% Triton X-100 in PBS), quench buffer (1% hydrogen peroxide and 1% sodium azide solution in wash buffer), blocking buffer (5% milk in wash buffer), poly-l-lysine (Sigma, St. Louis, MO), *active* p44/42 ERK-MAPK (Cell Signaling Technology, Beverly, MA), donkey anti-rabbit immunoglobin HRP (Pierce, Rockford, IL), SuperSignal West Femto reagent (Pierce, Rockford, IL), crystal violet (Fisher Diagnostics, Middletown, VA), and a SPECTRAFluor Plus luminometer (Tecan U.S., Durham, NC).

4.3.3. General Method

The method described is a modified version (8) of the fast activated cell-based ELISA (FACE™) assay, developed by Active Motif (Carlsbad, CA). The first step for this assay involves selection of concentrations of the test ligands to be evaluated. The concentrations should be selected such that the concentration predicted to result in half-maximal stimulation (e.g., ED_{50}) of ERK-MAPK activity is near the middle of the concentration-effect curve (see Fig. 1c). To cover an adequate span, concentrations should be selected one log unit apart. Treatment of 24 wells of cells seeded in a 96-well plate (see following), can be used to generate a single concentration-effect curve for modulation of pERK-MAPK activity by test ligands. For each concentration tested, three wells should be treated. This design allows for testing of one vehicle and seven drug concentrations.

Cells are seeded in serum-free DMEM into a 96-well plate (precoated with 10 μg/mL poly-l-lysine) 24 h prior to the assay, at a density that will produce 90% confluence on the day of the experiment. Immediately prior to the assay, cells are rinsed once with warmed (37°C) serum-free DMEM. Drugs are diluted in serum-free DMEM and exposed to cells floating on a 37°C water bath for times ranging from 1 to 15 min (see Special Considerations). The plates are placed on ice after the incubation, rinsed 3 times with PBS and fixed with 4% formaldehyde in PBS. Fixed cells are then rinsed 3 times for 5 min with wash buffer and quenched for 20 min in quench buffer. After 3 rinses of 5 min with wash buffer, nonspecific binding is blocked by a 1-h incubation with 5% milk in wash buffer at RT while shaking. Cells are then rinsed twice for 5 min with wash buffer and incubated overnight at 4°C while shaking with a primary antibody (1:500) recognizing the phosphorylated form of ERK-MAPK (e.g., *active* p44/42 ERK-MAPK). Cells are rinsed 3 times for 5 min with wash buffer and incubated with the secondary antibody (donkey anti-rabbit immunoglobin horseradish peroxidase, 1:2,000) for 1 h at RT in wash buffer while shaking. Cells are rinsed 3 times for 5 min with wash buffer, followed by 2 rinses of 5 min with PBS. SuperSignal West Femto reagent (50 μL) is added to each well while plates are floating on an ice bath. Plates are warmed to RT and immediately scanned utilizing a SPECTRAFluor Plus luminometer. Plates are rinsed twice for 5 min with wash buffer and twice for 5 min with PBS. Protein concentration is determined by adding 100 μL of crystal violet to each well and incubating plates at RT for 30 min. Cells are then rinsed 3 times for 5 min with PBS and incubated for 1 h with 100 μL of 1% SDS in PBS. Plates are scanned utilizing a microplate reader set to detect fluorescent emission at 595 nm. Data are presented graphically as percent of control pERK-MAPK levels as a function of test ligand concentration, presented on a log scale (Fig. 1c). Measures of both potency (e.g., ED_{50} values) and efficacy (e.g., E_{max}) of ligands acting at GPCRs are obtained from nonlinear regression analysis of the concentration-effect curves presented.

4.3.4. Advantages, Disadvantages, and Special Considerations

The greatest advantage of this assay is that the functional activation of virtually any GPCR desired can be measured. Initially, a time-course for ERK-MAPK phosphorylation by test ligands should be determined for each cell line and GPCR examined to select the optimal time for peak ERK-MAPK activation. Furthermore, proper controls should be performed to show that the test ligand does not alter total ERK1/2 levels. Lastly, success of the assay depends heavily on the quality of the pERK-MAPK antibody employed. We found that the primary antibody marketed by Cell Signaling Technology produces very reliable and consistent results.

5. Future Directions

Several novel and exciting "state of the art" techniques, beyond the scope of this introductory chapter, are also currently available to investigate GPCR signaling. Many of these are presented in detail in other chapters of this book, however, a few will briefly be mentioned here. First, several commercially available transfection systems are available that allow for careful regulation of the levels of GPCRs expressed in transfected cells (e.g., Tet-On and Tet-Off Gene Expression Systems, Clontech, Mountain View, CA). Second, although receptor binding studies most commonly employ the use of radioligands, nonradioactive assays using fluorescent ligands have been developed that offer many distinct advantages (e.g., Cisbio Bioassays, Bedford, MA). Third, to similarly avoid the use of radioactivity, fluorescently labeled, nonhydrolysable GTP analogs such as Eu-GTPγS are also available to measure GPCR regulation of G-protein activity (PerlinElmer, Waltham, MA). Fourth, cell lines expressing novel GPCR-Gα-subunit fusion constructs have been developed that provide models with a unique 1:1 receptor to G-protein stoichiometry, suggested to more closely resemble signaling occurring in physiological settings (e.g., Jena Bioscience, Jena, Germany). Lastly, since almost all GPCRs interact with β-arrestin, a recent technique monitoring the translocation of fluorescently tagged β-arrestin in cells following receptor activation has been proposed as a universal assay to asses GPCR function (Transfluor Technology, Molecular Devices, Sunnyvale, CA).

6. Conclusions

Based on the tremendous therapeutic potential and need for future development of drugs acting at GPCRs, it is truly an exciting time for those interested in this quickly advancing field. It is hoped that the methods described in this short chapter will provide some basic information for investigators, relatively unfamiliar with *in vitro* techniques, to begin characterization of GPCRs in their own laboratories.

References

1. Eglen RM, Reisine T (2009) New insights into GPCR function: implications for HTS. Meth Mol Biol **552**:1–13.
2. Lagerstrom MC, Schioth HB (2008) Structural diversity of G protein-coupled receptors and significance for drug discovery. Nat Rev Drug Discov 7:339–357.
3. Howlett AC (1995) Pharmacology of cannabinoid receptors. Ann Rev Pharmacol Toxicol **35**:607–634.
4. Herkenham M, Lynn AB, Little MD et al (1990) Cannabinoid receptor localization in brain. Proc Nat Acad Sci USA **87**: 1932–1936.
5. Galiegue S, Mary S, Marchand J et al (1995) Expression of central and peripheral cannabinoid receptors in human immune tissues and leukocyte subpopulations. Eur J Biochem **232**:54–61.
6. Conti S, Costa B, Colleoni M et al (2002) Anti-inflammatory action of endocannabinoid palmitoylethanolamide and the synthetic cannabinoid nabilone in a model of acute inflammation in the rat. Br J Pharmacol **135**: 181–187.
7. Shoemaker JL, Joseph BK, Ruckle MB et al (2005) The endocannabinoid noladin ether acts as a full agonist at human CB2 cannabinoid receptors. J Pharmacol Exp Ther **314**: 868–875.
8. Shoemaker JL, Ruckle MB, Mayeux PR et al (2005) Agonist-directed trafficking of response by endocannabinoids acting at CB2 receptors. J Pharmacol Exp Ther **315**: 828–838.
9. Graham FL, van der Eb AJ (1973) A new technique for the assay of infectivity of human adenovirus 5 DNA. Virology **52**:456–467.
10. Taghian DG, Nickoloff JA (1995) Electrotransformation of Chinese hamster ovary cells. Methods Mol Biol **48**:115–121.
11. Martin NA, Prather PL (2001) Interaction of co-expressed *mu*- and *delta*-opioid receptors in transfected rat pituitary GH_3 cells. Mol Pharmacol **59**:774–783.
12. Martin NA, Terruso MT, Prather PL (2001) Agonist Activity of the *delta*-antagonists TIPP and TIPP-*psi* in cellular models expressing endogenous or transfected *delta*-opioid receptors. J Pharmacol Exp Ther **298**:240–248.
13. Cheng Y, Prusoff W (1973) Relationship between the inhibition constant (Ki) and the concentration of inhibitor which causes 50 per cent inhibition (IC50) of an enzymatic reaction. Biochem Pharmacol **22**:3099–3108.

Chapter 12

Novel Assay Technologies for the Discovery of G Protein-Coupled Receptor Drugs

Elisa Alvarez-Curto, Richard J. Ward, and Graeme Milligan

Abstract

The development of new therapeutic drugs acting at G protein-coupled receptors (GPCRs) whose ligand specificity is known is of great importance to the pharmaceutical industry and to the population at large. It is also vital that surrogate ligands can be identified for GPCRs at which the endogenous ligand(s) remain unknown so that their potential as drug targets can be assessed. It is against this background that we consider a selection of technologies that are emerging to meet these challenges.

Key words: Orphan GPCR, Protein complementation, Cell-based assay, Homogeneous, Label-free assay, Fluorescent ligand, High-throughput screening

1. Introduction

The requirement for novel assay technologies to drive the discovery of new drugs, whose mechanism of action is via G protein-coupled receptors (GPCRs), is highlighted by the recent decline in the rate at which such drugs have been approved for use. This is mirrored by a decline in the rate at which GPCRs whose endogenous ligands are unknown (orphan GPCRs) are having their natural ligands identified (deorphanisation). The deorphanisation of GPCRs remains an important element in the drug discovery process because of the 356 human genes encoding non-chemosensory GPCRs, 131 remain unliganded (http://www.iuphar-db.org/index.jsp). The identification of endogenous or surrogate ligands, and an understanding of their signalling pathways, may allow the validation of these GPCRs as potential therapeutic targets.

Craig W. Stevens (ed.), *Methods for the Discovery and Characterization of G Protein-Coupled Receptors*, Neuromethods, vol. 60, DOI 10.1007/978-1-61779-179-6_12, © Springer Science+Business Media, LLC 2011

This slowing of the pace of discovery has a number of possible causes, of which perhaps the most obvious is that those GPCRs which are most amenable to the application of small molecule drugs have already been subject to the development of an extensive pharmacology; as it were, "the low hanging fruit has already been picked". In addition to this, the (comparatively) recent revelation that GRCRs exist not solely as monomeric entities, but often as units of dimers or larger multimers of receptors (both homo and heteromers), which are themselves part of large signalling complexes also adds to the complexity of drug design. However, the concept of receptor dimers may be a potentially useful one as well, since it could be that combinations of drugs able to act at both partners of a dimer or a larger "bivalent" ligand able to do the same thing may be therapeutically advantageous.

Novel assay technologies which will assist in the search for such drugs must be robust, that is, work reproducibly under a range of conditions and be tolerant of additives or conditions. They will preferably be homogeneous, of a format which requires only a few additions to a plate and does not require extensive washing. Finally, they must be amenable to high-throughput screening (HTS) and miniaturisation, though it has been suggested (1) that the rate at which the latter proceeds will slow and be overtaken by an increased emphasis upon more rigorous hit validation, probably by some of the techniques described below.

2. Methods to Identify GPCR Drugs

In this chapter, we discuss cutting edge technologies such as label-free assays, followed by receptor structure-based analysis and finally fluorescence-based techniques.

2.1. Label-Free Assays

The CellKey™ (MDS Analytical Technologies, http://www.cellkey.com) and Epic® (Corning Incorporated, http://www.corning.com) systems are label-free, non-invasive, real-time direct measurement systems for detecting overall, integrated changes in cells expressing GPCRs (or other biomolecules of interest) when treated with ligands. In both cases cells are plated onto specially designed microplates (in a variety of formats from 96 well) and allowed to form a monolayer, after which the cells are washed, drug treatments applied and measurements made.

2.1.1. CellKey™

The CellKey™ system is based upon cell impedance technology or cellular dielectric spectroscopy (CDS). This is the measurement of complex impedance changes which occur as the result of receptor activation by ligand binding. Experiments are carried out in microplates with electrodes set into the base of each well, cells are allowed to plate down and impedance changes in response to ligand addition are measured by the CellKey™ instrument. The CellKey™ system applies a constant voltage across the cell monolayer which causes current to flow, at low frequency, around and between the cells (extracellular current) and at higher frequencies, through the cells (transcellular current). The current (transcellular and extracellular components) changes when the cells modify their shape, volume, (internal structure and electrolyte levels), cell–cell contacts and cell-substrate adherence. The current measured by the CellKey™ instrument is the result of one, some of or all of these factors which determine the characteristics and scale of the signal detected (Fig. 1a). Characteristic profiles of the response to receptors coupling via different G_{alpha} subunits can be generated and used to identify the coupling of "new" receptors. The CellKey™ has important applications in the screening of endogenous GPCRs in native tissues, primary cells and stem cells.

The utility of the CellKey™ system for use in structure–activity relationship (SAR) studies was evaluated with three G_i coupled GPCRs: the dopamine D_{2S}, dopamine D_{2L} and the muscarinic M_4 (2). It was found that the CellKey™ assay results correlated with classical methods such as [^{35}S]GTPγS binding and cAMP production for agonist and antagonist responses over a 1,000-fold range of compound potencies. It was further observed that CellKey™ is insensitive to colour quenching and can be easily integrated into a compound screening scheme based upon microtitre plates. Peters and Scott (3) demonstrated the use of the CellKey™ with regard to screening on the basis of "functional selectivity", that is, a receptor may activate different signalling pathways depending upon which receptor conformation(s) are stabilised by the ligand and also cell specific signalling, where a ligand/receptor combination can activate different signal transduction pathways in different cell types. As described above, cellular impedance technology is able to distinguish between signalling via G_s, $G_{i/o}$ and G_q subunits on the basis of differing impedance profiles. It was determined that the melanocortin MC_4 receptor couples through the G_s and G_q subunits, while cannabinoid CB_1 couples through G_i and G_s as would be expected (http://www.iuphar-db.org). Peters and Scott (3) further showed that endogenous dopamine D_1 receptor in two different cell types (U-2 and SK-N-MC) is able to produce cell-specific signalling, with regard to impedance profiles, while in the same cell lines, full and partial agonists gave potency and efficacy values similar to those found elsewhere in the literature.

2.1.2. Epic®

The Epic® system measures changes in the local index of refraction, which result from ligand-induced dynamic mass redistribution (DMR) within the bottom 150 nm region of the cell monolayer. These measurements are suitable for both high-throughput ligand

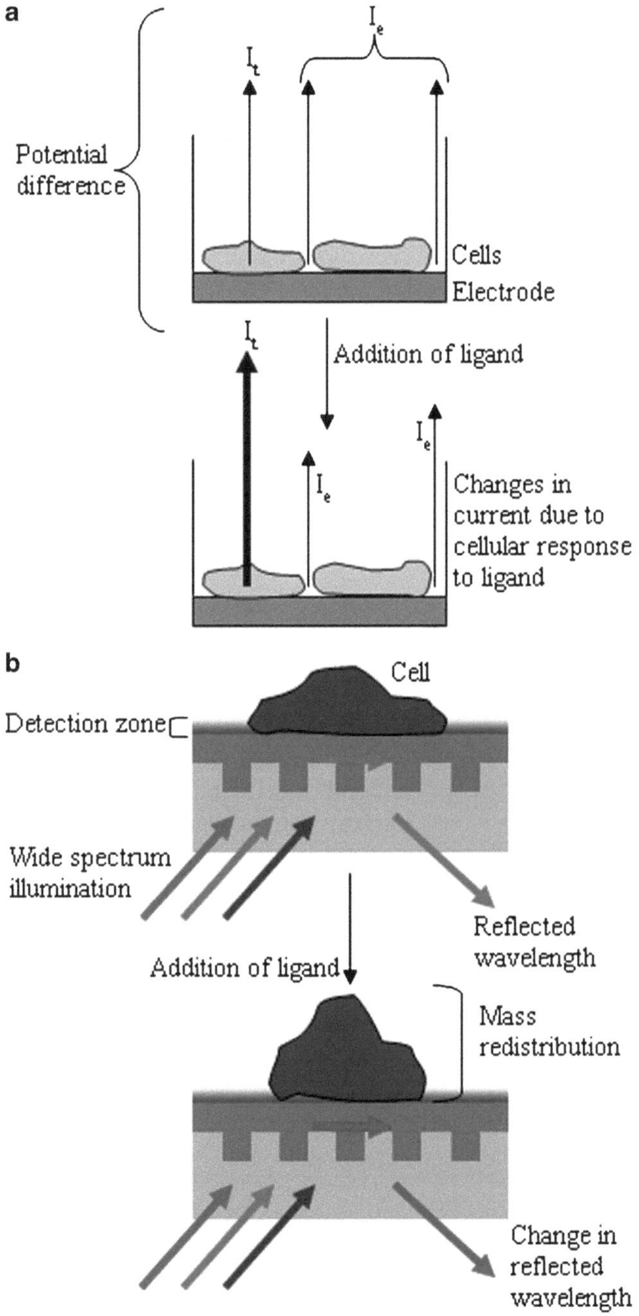

Fig. 1. Cell-based direct measurement assays. (**a**) The CellKey™ system measures changes in current, both extracellular (I_e) and transcellular (I_t) current when the cellular impedance changes due to mass redistribution upon ligand stimulation. (**b**) Epic® measures changes in the wavelength of reflected light caused by the cellular changes in response to ligand which occur in a detection zone above the surface of the sensor.

screening and the observation of functional responses to GPCRs activated by ligands. The basis of the Epic® system is the microplates whose well bases contain refractive waveguide grating optical biosensors able to make highly sensitive refractive index measurements in a detection zone which extends to 150 nm above the surface of the sensor. Each sensor is considered to have three layers, a glass substrate with grating, a thin waveguide film (dielectric layer) and the cell or biomolecule layer. The sensor is illuminated with broadband light consisting of many wavelengths, only a small part of which, that is resonant with the waveguide grating structure, is strongly reflected and detected by the Epic® system. If the cells, which form the third layer of the sensor, respond to an applied ligand, then intracellular changes caused by binding events (DMR) alter the index of refraction and thus the wavelength of the reflected light as detected by the Epic® system (Fig. 1b).

To perform an Epic® assay, cells plated down overnight onto an Epic® assay plate have their media removed and replaced with assay buffer. These plates are equilibrated for 1 h and an initial baseline measurement taken, following which the diluted ligands are injected and further measurements made. Wavelength shifts due to added ligand may then be analysed and used to plot, for example, dose response curves or changes over a time-course.

The application of the Epic® system has been shown by Antony et al. (4) and Kebig et al. (5) who used DMR to demonstrate that "dualsteric" agonists induce pathway-specific signalling. This work was based upon the muscarinic receptor subtypes which have a conserved orthosteric binding site and a less well-conserved allosteric one. In order to achieve good binding and signalling specificity, which are otherwise lacking in muscarinic receptors, hybrid ligands were synthesised between a non-selective orthosteric agonist and an M_2 selective allosteric fragment. This was found to be functional and M_2 selective. DMR experiments were able to show that the "dualsteric" ligand is pathway selective in that cells expressing the M_2 receptor were able to signal via $G_{i/o}$ and G_s when activated with a conventional orthosteric ligand and only $G_{i/o}$ (as shown by pertussis toxin treatment) when activated with the "dualsteric" ligand.

2.2. Second Messenger-Based Assays

The HitHunter™ system (DiscoveRx Corporation, http://www.discoveRx.com) is a collection of second messenger assays designed to detect the activation of receptors which couple through $G_{\alpha i}/G_{\alpha s}$ (HitHunter™ cAMP) or G_q/G_{11} (HitHunter™ Inositol(1,4,5) trisphosphate). Both assays are homogeneous and suitable for HTS applications.

The HitHunter™ cAMP assay is based upon enzyme complementation and involves competition between cAMP labelled with a small enzyme fragment "ED" (cAMP-ED) and cAMP generated as a second messenger for an anti-cAMP antibody. That is, when no additional cAMP is present in the cell lysate, the cAMP-ED is bound to anti-cAMP antibody, preventing it from complementing the larger "EA" enzyme fragment, which is present in solution. However, when additional cAMP is present it displaces some of the "ED" labelled cAMP and this then becomes available to complement the "EA" enzyme fragment, allowing hydrolysis of substrate and signal generation. Applications of this assay have been described by Lee et al. (6) in which it was used to measure the effect of allosteric agonists of FFA2 (Free fatty Acid receptor 2) upon cAMP generation and by Shemesh et al. (7) who used it to identify and validate new peptide agonists for GPCRs.

The HitHunter™ Inositol(1,4,5) trisphosphate (IP_3) assay is based upon fluorescence polarisation. After cell lysis, IP_3 with a fluorescent tracer (IP_3-tracer) and an IP_3 binding protein are added. The IP_3-tracer binds the IP_3 binding protein which causes it to "tumble" more slowly in solution, which results in a polarised fluorescent signal. When the IP_3-tracer is displaced from the binding protein by free IP_3 from the lysed cells, the degree of polarisation is reduced in a manner proportional to the amount of additional IP_3. The use of this assay in high-throughput screens has been reviewed by McLoughlin et al. (8). Gossman and Zhao (9) measured IP_3 release in the cochlea using this assay also.

A popular alternative to the HitHunter™ IP_3 assay is the Cisbio (http://www.htrf.com) IP-One HTRF® (Homogeneous Time Resolved FRET) assay which measures IP_1, a downstream metabolite of IP_3. The advantage of this is that IP_1 is stable in the presence of LiCl, unlike IP_3 which has a very short half-life and so is less ideal for quantification (10). IP_1 is measured by a competitive immunoassay based upon HTRF (see final section), where IP_1 labelled with the HTRF acceptor d2 competes with endogenous IP_1 for a cryptate (Eu^{+3} or Tb^{+2} HTRF donor) labelled IP_1 antibody. Hence an increase in endogenous IP_1 displaces the d2 labelled IP_1 from the antibody and so reduces the HTRF signal generated between the cryptate labelled antibody and d2 labelled IP_1. This assay is simple, rapid and highly accurate while being amenable to HTS in a variety of plate formats. Ayoub et al. (11) describe a use of this assay to compare the functionality of PAR1 (Protease Activated Receptor 1) with a modification of the receptor in which it is fused to enhanced yellow fluorescent protein (eYFP).

These assays share with the PathHunter™ β-arrestin assay, (next section), the advantage of needing no specialised instrumentation as they can be read on standard plate readers.

2.3. β-Arrestin Recruitment-Based Assays

Unexpected receptor pharmacology and the effects of "biased ligands" highlight the fact that classical views of receptor activation that result in G-protein activation and second messenger signalling might not provide the whole picture (12). Furthermore, there is mounting evidence of receptor signalling that is independent from G-protein complexes (13). β-Arrestins (β-arrestin 1 and 2) are cytosolic proteins classically associated with desensitisation. They bind to agonist-activated receptor making β-arrestin recruitment one of the earlier events that can be monitored in the signalling cascade. A less amplified signal is an advantage in itself during screening as the occurrence of negative or false positive readouts is greatly reduced. This has been seen as an opportunity to develop novel assays that exploit β-arrestin recruitment as an indirect measure of receptor activation. It is also important to highlight the fact that β-arrestins have been shown to bind and regulate unconventional GPCRs, such as the Frizzled receptor as well as non-7TM (7 transmembrane) receptors including IGF1R, offering many new drug screening possibilities (14–16).

The PathHunter™ β-arrestin assay (DiscoveRx Corporation, http://www.discoveRx.com) is an enzyme complementation assay which detects the recruitment of β-arrestin to the receptor upon activation and so is unaffected by the G_{alpha} coupling specificity. In order to detect this interaction a 42 amino acid, weakly complementing, fragment of β-galactosidase (called Prolink) is fused to the carboxy-terminal of the GPCR, while a 42 amino acid amino-terminal deletion mutant of β-galactosidase (EA or enzyme acceptor) is fused to the β-arrestin. When these two β-galactosidase fragments are brought together by a receptor-β-arrestin interaction, they complement to produce a functioning enzyme which is able to hydrolyse a substrate to produce a chemiluminescent signal. If no interaction occurs between the receptor and β-arrestin, there is little background signal due to the low affinity of the fragments (Fig. 2).

The PathHunter™ β-arrestin assay is a homogeneous, gain of function assay which can detect agonists, antagonists, partial agonists, inverse agonists, allosteric modulators and G_{alpha} independent signalling events. However, β-arrestin-independent signalling events are not detected. Assays can be carried out in HTS or high content screening (HCS) formats using standard plate readers. This assay is stoichiometric, produces no amplification effect of the signal and generates high quality data without recourse to non-standard, high investment instruments.

Similarly to the PathHunter™ assay, Invitrogen (http://www.invitrogen.com) developed a new technique to study β-arrestin-receptor interactions: the Tango™ GPCR Assay System. This is based on reporter gene expression induced by the recruitment of a protease-linked β-arrestin to an activated receptor that releases a transcription factor (GAL4-VP16 chimeric protein) fused to the

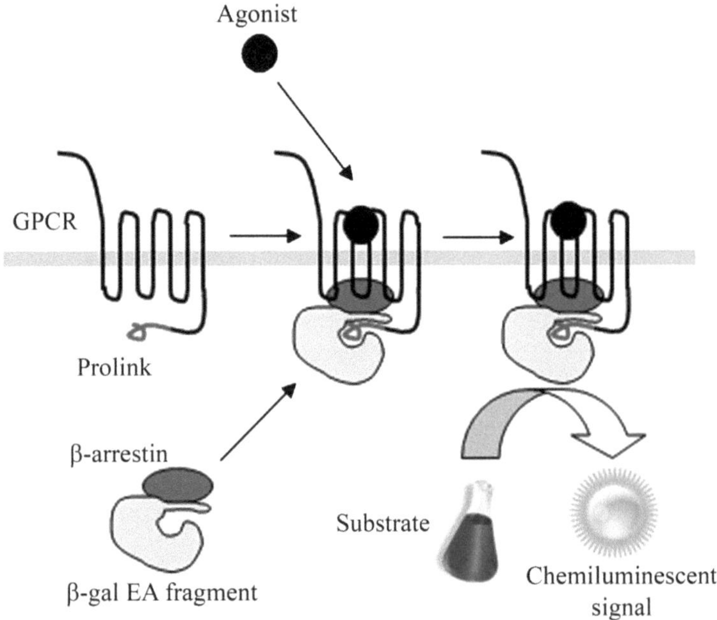

Fig. 2. The PathHunter™ β-arrestin assay involves complementation between the Prolink fragment of β-galactosidease fused to the GPCR and the EA fragment fused to β-arrestin upon recruitment of β-arrestin to the ligand-activated GPCR. The complementing fragments modify a substrate to generate a chemiluminescent signal.

C-terminal of the receptor upon β-arrestin interaction. The system has been used to identify a series of compounds in the Tango cell line expressing the *kappa* opioid receptor (17).

There are a number of recent examples in which the PathHunter™ method has been used to identify ligands ranging from glucocorticoid receptor agonists (18) to lipids which act as ligands of GPCRs (19). Evaluation of receptor and ligand pairings was particularly successful in the case of a collection of nine newly de-orphaned, but poorly characterised receptors (19). Using this assay, the authors were able to confirm the authenticity of several pairings such as the receptor G2A and 9S-hydroxyoctadecadienoic acid (9-HODE) or GPR92 and lysophosphatidic acid, LPA. Cannabinoid receptor pharmacology has been extensively scrutinised using this method. The unusual biology of this group of receptors and new additions to the family, such as the incompletely characterised GPR55, believed to be a third cannabinoid receptor, make them ideal for this type of assay format (19–21).

An alternative to the use of the PathHunter™ β-arrestin recruitment assay is one which involves the generation of a BRET (Bioluminescence Resonance Energy transfer) signal in response to the recruitment of β-arrestin (linked to a luminescent donor, *Renilla* luciferase or Rluc) to a receptor fused to enhanced yellow fluorescent protein (eYFP) (22). This assay requires the addition of the substrate colentrazine, which is oxidised by the luciferase, causing the emission of light energy which is transferred to the eYFP resulting in a characteristic yellow light signal from the eYFP. The assay can be carried out using standard plate readers and adapted to extended kinetic assays by the use of alternate enzyme substrates such as Enduren (Promega Corporation) (23). The efficacy and potency of anti-psychotic and anti-parkinsonian drugs at the D_2 receptor were investigated with respect to β-arrestin recruitment using this approach by Masri et al. (24) and Klewe et al. (25).

Translocation of β-arrestin to the plasma membrane can also be monitored in a high content screen format using a fluorescent-labelled form of β-arrestin (β-arrestin-GFP) and the Transfluor system (MDS Analytical Technologies, http://www.moleculardevices.com). This can be analysed in a variety of confocal readers such as the Evotec Technology Opera (Evotec, http://www.evotec.com) (26) or the ArrayScan reader from Cellomics (http://www.cellomics.com). The assay is compatible with HTS platforms such as 384 well plates and uses β-arrestin-GFP redistribution as a measurement of receptor activation (27). Other processes such as receptor internalisation and trafficking may be studied using this approach. An elegant study on the responsiveness of GPR55 to cannabinoid ligands has recently been reported using this technology (28). Cannabinoid and non-cannabinoid ligands were tested for their agonist activity at GPR55 by measuring β-arrestin-2-GFP trafficking, helping define further aspects of the very interesting yet atypical pharmacology of this receptor.

2.4. Phage Display

Phage display is a well-established, rapidly evolving technique which was developed by Smith (29) for the study of protein–protein interactions. It has applications in, among other areas, the identification of peptide ligands, in affinity chromatography, the mapping of antibody epitopes and defining protein–protein interactions.

The basic principle of this technique is the modification of a (usually filamentous, non-lytic, M13 or fd) bacteriophage to express the protein, peptide (linear or cyclic) or antibody libraries as an element of the bacteriophage protein coat. The phage are then used to "probe" a protein of interest and those which interact are eluted and amplified by using them to infect bacteria from which the phage are then recovered. This generates a population of phage that are enriched with those which interact with the target protein and which can then be used for subsequent rounds of "panning" to further amplify the proportion of interacting bacteriophage. Finally, the eluted phage particles are used to infect bacteria to allow the isolation of individual plaques of bacteriophage which may then be analysed with respect to their DNA sequence, revealing the sequence of the interacting protein or peptide. This physical link between the isolated interacting protein and the DNA which encodes it is one of the most important advantages of the technique.

An application of this technique has been described (Xavier Leroy, Actelion Pharmaceuticals, conference communication) to isolate GPCR peptide ligands by a process of "reverse pharmacology". That is, a phage library displaying random 12 amino acid peptides was incubated with HEK293 cells expressing the GPCR of interest and any bound phage eluted. This eluted phage was then used to infect bacteria, titrated and identified by a cell-based ELISA. Interacting phage were then used for subsequent rounds of panning to enrich the library further. Finally, DNA was extracted from interacting phage and sequenced to identify the interacting peptide. The peptides can be synthesised and evaluated in cell lines expressing the GPCR of interest using binding, GTPγS, calcium mobilisation and cell migration assays.

Examples of the application of this technique include the works of Szardenings et al. (30) and also Houimel et al. (31) who made use of phage display to identify peptides interacting with the human melanocortin 1 and CCR3 receptors respectively. Phage display has been used to identify peptide agonists acting at the level of the G-protein, rather than the receptor, as described in (32). Here, the functional relevance of the peptide-G protein interaction was demonstrated by the greater sensitivity to guanine nucleotides of high affinity agonist binding to the Adenosine A1 receptor in the presence of the peptide. More recently the work of Bikkavilli et al. (33) described the use of a random phage displayed library to identify peptide ligands for the orphan GPCR MAS1. This work resulted in the identification of a peptide agonist for MAS1 and two peptide motifs, expected to be found in proteins interacting with MAS1.

This appears to be a useful technique, particularly for the identification of surrogate ligands for orphan receptors, in that it is inexpensive and as such, widely accessible. It is potent, allowing the discovery of specific peptides of high affinity to GPCRs, which may be orthosteric or allosteric ligands.

2.5. Structure-Based Drug Design and Stabilised Receptors (StaRs™)

The development of new drugs is increasingly based on structural information about the way ligands bind to their target. For this reason, knowledge about the 3D structure of proteins and protein–ligand complexes is paramount (34). The crystal structure of a number of GPCRs has been resolved in the last few years (35–38) but despite these structures being useful to build models for other GPCRs, they are far from ideal or practical to use as a basis for HTS. Many soluble targets, such as kinases or proteases, have benefited greatly from drug discovery based on crystal structures and biophysical interactions of ligands and fragments (39), whereas this has proven hard to apply to GPCR drug discovery. As membrane proteins, the inherent detergent instability of

GPCRs when purified and the flexibility and heterogeneity of receptor conformations found in purified samples are some of the problems that have to be solved before good quality crystals are obtained. Studies carried out by several groups demonstrate that it is possible to artificially "evolve" membrane proteins by introducing mutations so that they become stable in a detergent solution compatible with crystallisation (40–42). Commensurate with this, a rather elegant way to overcome these issues with GPCRs is to "lock" the receptor in a particular conformation such that the receptor remains functional and it is thermally stable. Heptares Therapeutics (http://www.heptares.com) has developed a cutting edge technology that follows this approach by generating stabilised receptors (StaR™). StaRs™ are generated by introducing a small number of point mutations and deletions that will reduce receptor flexibility but preserve receptor functionality in detergent-solubilised form. The "locked" conformation is likely to reduce heterogeneity within the purified sample by fixing the receptor in, for example, a homogenous antagonist bias conformation, therefore encouraging the formation of crystals. The process of generating a StaR™ from the unstable native receptor to crystallisation involves iterations of mutagenesis of the receptor, thermostability and functionality testing and recombination of mutants. Once the receptor is stable, it can be produced in large quantities by recombinant methods and can be purified for crystallography and screening applications. The works of Serrano-Vega, Warne and colleagues are good examples of the use of thermally stable receptors to obtain crystals, in this case of the turkey β_1-adrenergic receptor (43, 44). In these studies, the resulting receptor was stabilised preferentially in an antagonist-bound conformation, suitable for crystallographic studies.

StaR™ receptors can be immobilised using a variety of platforms and coupling strategies for high-throughput applications. Since they retain GPCR function and expected pharmacology when immobilised, the kinetics, affinity and selectivity of compound libraries can be tested. These can help defining new models for compound binding as well as leads on intractable targets.

An interesting alternative application for StaR™ receptors is to use them as antigens to raise antibodies which can be applied to a number of GPCRs involved in cancer, inflammation and immune cell responses, as they might be susceptible to being treated by or allosterically modulated with antibodies as opposed to small molecules (45–47). StaR™ receptors immobilised on plates may also be used for screening antibody libraries *in vitro*.

2.6. Fluorescent and Chemiluminescent Protein Complementation

Fluorescent protein complementation is a means of detecting protein–protein interactions within a living cell by utilising the reassembly (complementation) of two parts of a fluorescent protein, which have been fused individually to the proteins of interest

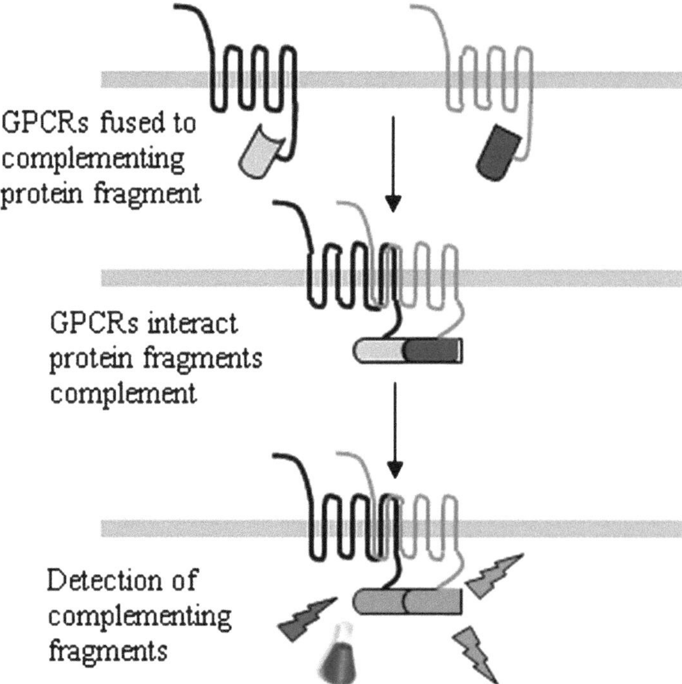

Fig. 3. The formation of GPCR dimers is used to illustrate the principle of BiFC or BiLC, thus GPCRs are fused to the fluorescent protein or luminescent enzyme fragments. When the GPCRs dimerise, the fragments complement one another and are able to generate a fluorescent or chemiluminescent signal.

and are to be found in a largely unfolded state. When these proteins of interest interact, the parts of the fluorescent protein also interact and fold sufficiently to complement one another, generating a specific fluorescent signal. The technique is also known as Bimolecular Fluorescence Complementation (BiFC) (Fig. 3). The related technique of Bimolecular Luminescence Complementation (BiLC) is similar but utilises the reformation of *Renilla reniformis* or *Gaussia princeps* luciferases from split, non-luminescent fragments upon interaction of their fusion partners. It is also possible to extend this approach by using the complemented fluorescent protein as the donor or acceptor of a resonance energy transfer experiment (48). As with all recombinant protein tagging approaches, it is important that appropriate controls should be carried out to account for the possibility that the fluorescent protein fragments may affect the function, localisation or trafficking of the protein to which they are fused. In addition the possibility that the fluorescent protein fragments may associate without the proteins of interest interacting, or that the fluorescent protein fragments may, by their own association, encourage an interaction between the proteins of interest, (which may not otherwise happen), should also be controlled for. In addition, when designing an assay or screen it should be born in mind that a reconstituted fluorescent protein is likely to have less intense fluorescence than that of a fluorescent protein consisting of a single polypeptide (48). The complementation of fragments of a fluorescent protein appears to be irreversible (unlike that of the "luminescent" proteins) (49) and so this precludes the use of BiFC for applications which involve dynamic changes to protein complexes, though for other applications this could conceivably be an advantage.

A use of BiFC for the identification of previously unknown drug effects at targets other than those intended and also the discovery of additional efficacies has been described in (50). This work involved the screening of 107 drugs from 6 different therapeutic groups against a panel of 49 fluorescent protein complementation reporters for 10 cellular processes. It was shown that this strategy could form the basis for identification of new, unpredicted and potentially therapeutically useful, modes of action of drugs and also provide a warning of possible undesirable effects.

It seems likely that BiFC/BiLC and their incorporation into resonance energy transfer strategies (BRET or FRET) may provide a basis for future drug screens, particularly multicolour BiFC, incorporating different split fluorescent proteins as described in (48).

2.7. Fluorescent Screening Assays

Fluorescence-based techniques have been at the front line of GPCR pharmacology research for many years and they probably are the ones that have seen major advances in the last 10 years. Many pharmaceutical companies and academic laboratories are investing much time and effort in elucidating all the intricacies of receptor–ligand binding complexes and of receptor and regulating protein interactions. Some of the fluorescence-based techniques used in drug-discovery efforts use real-time analysis of ligand–receptor interactions (Fluorescence Correlation Spectroscopy or FCS), some concentrate on receptor localisation and distribution in either single cells or tissues using confocal microscopy, whereas others measure directly ligand–receptor interaction and ligand binding kinetics by using fluorescent ligands. Some of the advantages of the aforementioned techniques are that they may solve issues relating to the biology of GPCRs, such as specific microdomain localisation in rafts or caveolae, and represent a safer alternative to traditional radioactive methods such as radioligand binding. Furthermore, the techniques included and discussed in the following sections can be used in single cell analysis where more information can be obtained, as well as being suitable for studies in either primary cells or differentiated stem cells, which is an obvious advantage when screening novel drug targets.

2.8. Fluorescent Ligand Binding

Fluorescent ligands have been used since the 1970s as labelling of a specific ligand with a fluorophore represents a very good alternative to radioactive assays. The improvements in scanning confocal microscopy and FCS detection techniques have given a great boost to the use of these types of ligands, as they enable resolution of receptor–ligand complexes to the level of single cell or even a single molecule. In theory, there are no limitations as to which GPCR ligands can be labelled with a fluorescent molecule; however, in practice there are a number of considerations that must be taken into account. Peptide ligands seem to be more suitable for this type of approach than small monoamine molecules, something which is directly related to the size of the molecule and the receptor–ligand mode of interaction. During the design of a fluorescent ligand one must consider the properties of both fluorophore and pharmacophore. Larger peptide ligands tend to bind to the most amino terminal region of the receptor or to the extracellular loops whereas small molecules bind deep into the pocket created by the transmembrane domains. Due to these characteristics of the ligands, the introduction of a sizable fluorophore in the case of a small amine has greater chances of disrupting the binding of the pharmacophore than in the case of a peptide (Fig. 4a). A way of circumventing this is to add a linker between the two; however, this might also affect the properties of the ligand, so the physical and chemical properties of the fluorophore have to be carefully considered as this will determine the behaviour of the labelled drug during the assay from both binding and chemical points of view. Ultimately, the fluorescent-labelled compound must retain the properties of the native ligand. Obviously, the more information there is about the SAR and the selectivity of the ligand of interest the easier it will be to design the new molecule. Substitutions in the molecule that are well tolerated when generating a radiolabelled ligand (if available) can be used as a guide to decide what part of the molecule is more resilient to changes. Regarding the fluorophore, there is a variety of dyes to choose from ranging from different Alexa Fluor versions to BODIPY-based ones (51). Companies such as CellAura Technologies Ltd (http://www.cellaura.com) have specialised in such services, marketing and generating custom fluorescent ligands.

There is a wide range of commercially available multimode readers that can be adapted to read not only fluorescence but also luminescence and absorbance. Both confocal and non-confocal-based readers such as Opera (Perkin Elmer) and FLIPRtetra (MDS Analytical Technologies) among others are suitable for processing assays using fluorophore-tagged ligands. This represents an obvious advantage for HTS where many assays can be run in parallel: fluorescent–ligand saturation binding, displacement with unlabelled compounds and, more interestingly, it offers

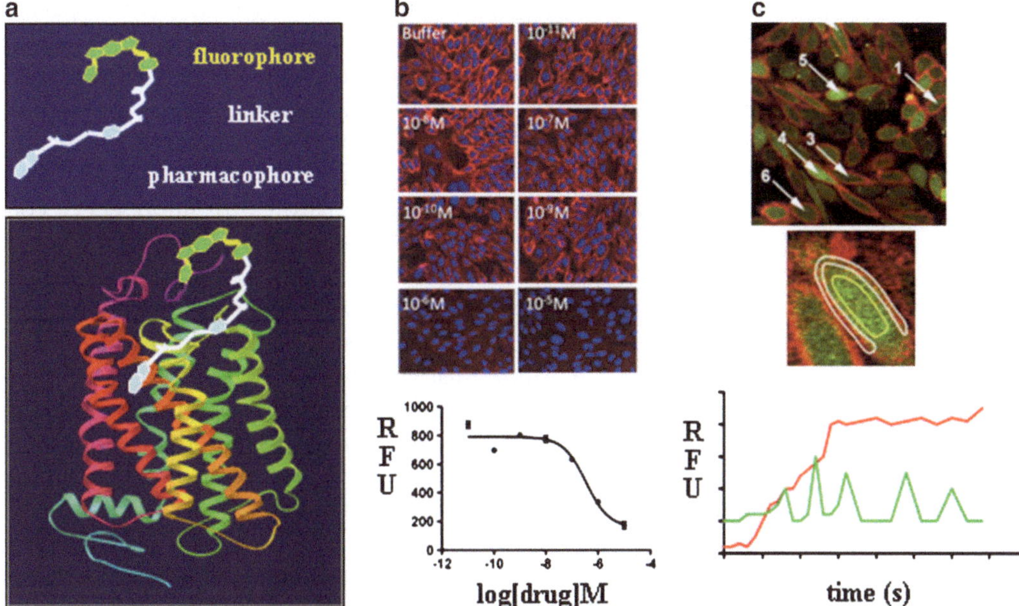

Fig. 4. (**a**) Schematic representation of a fluorescent ligand (*top panel*) showing the fluorophore, linker and pharmacophore components and how it might fit in the receptor-binding pocket (*bottom panel*). (**b**) Displacement by unlabelled ranitidine of 30 nM H_2-633-AN fluorescent ligand bound to membranes of CHO cells expressing the H_2 receptor (*top panel*) as seen using the Opera[R] plate reader (Perkin Elmer) and quantification of the loss of fluorescent signal (*bottom panel*) (adapted from CellAura Technologies). (**c**) Example of simultaneous confocal imaging of the binding of fluorescent ABEA-X-BY630 to CHO cells expressing the adenosine A_3 receptor and changes in $[Ca^{+2}]_i$ upon ligand stimulation (*top panel*); selection of regions of interest (ROI) used to record changes of intensity (*middle*), where the cytoplasmatic region indicates changes in $[Ca^{+2}]_i$ and the membrane region records variation in ligand binding; the corresponding fluctuations as plotted through time (top trace, membrane; bottom trace, cytoplasm; *bottom panel*). Modified from (52).

the possibility of high-throughput double-read out assays such as simultaneous ligand binding and measurement of, for example, mobilisation of intracellular calcium in response to the ligand (Fig. 4b and c). Ligand binding assays are undoubtedly very specific but they are less information-rich than any functional assay. Therefore, double-readout assays are useful not just for offering multiple readouts from a single plate, a fact that is rather valuable from the HTS perspective, but because they will define differences between agonist, antagonist or inverse agonist effects with no second follow-on assay required. Cordeaux and co-workers show a good example of elegant simultaneous imaging of ligand binding using the receptor agonist ABEA-X-BY630, labelled with BODIPY630, and calcium signalling in cells expressing the A_3 adenosine receptor (A_3-AR). Using this approach, combined with FCS (discussed below) they show that there is quite a degree of heterogeneity in both the distribution of the fluorescent ligand in the membrane of cells (a fact reflected on the calcium responses on the same cells) and furthermore, they showed that the agonist-bound human A_3 adenosine receptor exists in at least two different complexes with different motilities (52).

The fact that the labelled ligand may have different properties to the native one has been a source of worry and it might be that due to this concern that this type of assay is less suitable to perform screening with large numbers of compounds as the optimisation for each of them will probably be too time consuming. However, this particular type of approach is well suited to screen primary cells, particularly in combination with fluorescence correlation spectrometry techniques (see next section).

2.9. Fluorescence Correlation Spectroscopy

It is becoming more apparent that signalling from membrane receptors is highly compartmentalised, not only at the subcellular level but also at a molecular level within the membrane environment where receptors seem to be in specific microdomains in which they interact with different regulatory and signalling protein partners (53, 54). This undoubtedly introduces a new dimension to the pharmacology of the receptor and, therefore, it is vital to develop techniques that allow study of these events at the molecular level. Generally, in all macroscopic fluorescent methods performed in 96, or even 384 wells, the output data is an average of the intensity of the emission signal from all excited molecules present in each well; as the assay format is miniaturised, the fluorescent signal decreases due to a decreased sensitivity and background effects which in turn increases assay variability. The only way to solve this problem is to focus on the single molecule level in a small enough volume. Fluorescence correlation spectrometry (FCS) is the only fluorescent technique able to do this as it is practically immune to the effects of miniaturisation.

FCS was developed in the 1950s and 1960s as a way to measure the kinetics of chemical reactions under zero perturbation conditions and one of its first applications was to measure the kinetics of DNA untwisting using absorbance changes (55). FCS has come a long way since then and now it is widely used in the field of GPCR pharmacology and HTS drug screening mainly due to the advances in fluorescent technologies, optics and detection methods. FCS exploits the specificity of the fluorescent signal and is based, similarly to other diffusion techniques such as fluorescence recovery after photobleaching (FRAP), on recording fluctuations in photon emissions as a fluorescent molecule diffuses through a limited excitation volume (~0.2–0.5 fL) created by a highly focused laser beam in a confocal microscope (Fig. 5). However, a fundamental difference between those two techniques is that FCS only detects fluctuations on moving molecules, whereas FRAP also takes into account those which are immobile. The scale of the FCS excitation volume in the range of femtolitres is, as discussed above, one of the parameters that make this technology amenable to HTS, as samples can be miniaturised to only a few microlitres. Fluorescent molecules, either a fluorescent-tagged receptor or more commonly fluorescent ligands at

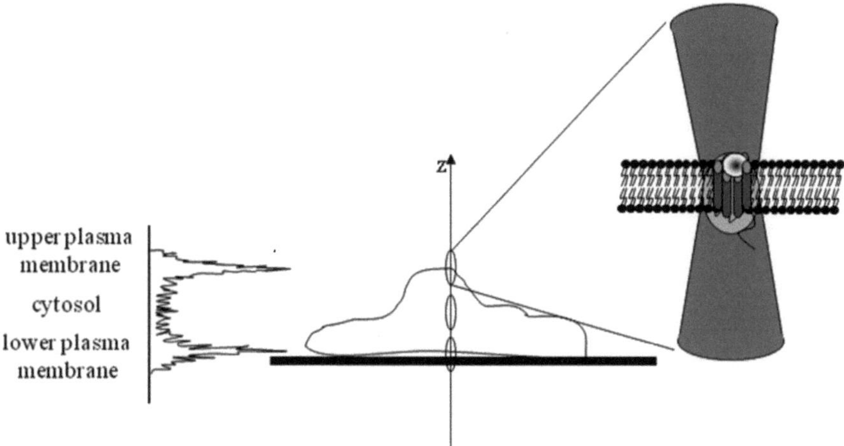

Fig. 5. Schematic drawing showing a fluorescent correlation spectrometry experiment. Cells are scanned in the z-dimension after being labelled with the fluorescent ligand of interest. The fluorescence intensity fluctuations of the receptor–ligand complexes are recorded at different points on the cell and at different time points. The excitation volume, in the range of femtolitres, (represented here by an ellipse), is generated by a laser beam, while the intensity fluctuations are recoded by highly sensitive avalanche photodiodes.

tracer concentrations (nM), diffusing through this defined field are excited and their photon emissions and subsequent fluctuations are recorded by highly sensitive photon detectors in a time-resolved manner. Using autocorrelation statistical methods, the data analysis yields information about the diffusion coefficient (diffusion speed) of the fluorescent particle from which its molecular state can be deduced. The amplitude of the correlation function (amplitude of the photon burst or brightness) provides a measurement for the number of molecules (concentration) in the sample at every given time. Consequently, through this technology one can establish differences between the diffusion rates of large vs. small molecules, as the length of each photon burst corresponds to the time that each molecule spends in the excitation confocal volume.

This is a particularly useful feature when performing ligand-binding assays as it is possible not only to differentiate but also to quantify the amount of faster moving free ligand vs. that of the slower receptor-bound complex. When the fluorescent agonist or antagonist is bound to the receptor, it causes a change in the diffusion coefficient. This can be measured by fixing the measurements to specific areas of the upper membrane of cells growing in monolayer (receptor-bound ligand) and comparing with the area just above it (free-ligand) (56). The coefficient known as diffusion time or dwell time (τ_D) is used to compare diffusion rates between receptors and receptor complexes. This parameter is calculated from the autocorrelation curve and is a specific for each agonist-bound receptor. Most ligand-binding studies performed on GPCRs using FCS show two diffusion components, τ_{D2} and $\tau_{D3,}$ which most likely reflect differently diffusing species of receptor that might be interpreted as ligand-bound receptors in different signalling microdomains, or associated to different scaffolding or regulatory proteins. There is a third component τ_{D1} that is also always recorded in this assays and it represents fast-moving free-ligand. FCS-based studies by Briddon and collaborators on the A_1-adenosine receptor (A_1-AR) have shown using quantitative τ_D data that the number of receptor–ligand complexes increased with increasing concentrations of free ligand in the medium, an effect that could be reversed when cells were treated with a selective A_1-AR antagonist (57). Although the true nature of the complexes associated with any of the components is still not known, they appear to reveal significant differences between agonist and antagonist-bound receptors, with the agonist-occupied receptors being faster than those bound to antagonist (58). Finally, as FCS is a highly sensitive, non-invasive and cell-based technique it has also been used to study the dynamics of receptor and receptor complexes in live primary cells. An example of this can be found in the work of Hegener et al. (59) where they investigate β_2-adrenergic receptor–ligand interactions in hippocampal neurons using a fluorescent-labelled arterenol derivative.

Quite an exciting application of FCS in conjunction with the fluorescent technique of bi-molecular fluorescent complementation is to look at the formation of homodimers or heterodimers in live cells and assess the differences in diffusion constants therefore shedding some new light on the molecular plasticity of GPCR complexes (56). However, as both binding partners in this case are diffusing over the lipid bilayer, only a significant increase in mass will yield a measurable effect on the diffusion complexes.

Somewhat similar to FCS, fluorescence intensity distribution analysis (FIDA) and fluorescence anisotropy (FA) are methods that can be adapted to HTS platforms and provide useful insights on ligand screening using receptor-binding formats.

2.10. Other Confocal Imaging Applications

As discussed in the previous section, high-content cell-based confocal fluorescence imaging has become a commonly used tool in GPCR drug discovery. There is a myriad of readouts for confocal-based HTS or HCS: GPCR ligand-induced internalisation, receptor–arrestin interaction, and ligand-induced conformational changes. Receptor interaction with arrestins, as discussed before, has become a very popular assay particularly as there is growing evidence of more receptors activating G protein-independent signalling cascades. Proximity and translocation assays using fluorescent receptors and arrestins are successfully used in HCS formats (60). Receptor internalisation and cellular redistribution can be monitored after labelling the receptor with some of the specific fluorescent ligands discussed in previous sections (61). Furthermore, large collections of medicinal chemistry compounds can be screened to discover novel agonist or antagonists targeting modified GPCRs. Using confocal imaging, O'Dowd et al. (62) described a method that exploits the nuclear trafficking pathway and protein translocation to the nucleus. Importin proteins, when binding to a nuclear localisation sequence (NLS) in the receptor, promote its translocation to the nucleus. Introduction of an NLS in a GPCR lacking this endogenous signal mediates ligand-independent receptor translocation to the nucleus. By introducing this NLS in a GPCR that does not normally have it in a conformation-dependent position of the GPCR (helix8), ligand-independent receptor translocation to the nucleus from the surface is mediated. This GPCR-NLS is then used to identify compounds that when bound to the receptors have the capability of preventing translocation of the receptor from the cell surface to the nucleus. In this work, the dopamine D_1 receptor (D_1) is used and stable cells expressing D_1-NLS-GFP show receptor expression in the cytoplasm and nucleus until treatment with increasing concentrations of antagonist re-localise it to the cell surface.

2.11. Homogeneous Time-Resolved Förster Resonance Energy Transfer (HTRF)

Lanthanides are the key component in time-resolved FRET. They are particularly useful due to their spectral and photophysical properties, specifically that they have large Stoke's shifts, of as much as 300 nm (difference between the excitation maxima and emission wavelengths) and extremely long emission half-lives (from μs to ms) when compared to more traditional fluorophores. The most commonly used lanthanides in HTRF are europium (Eu^{+3}) and terbium (Tb^{+2}) and in order to be able to improve their fluorescence, they tend to be complexed with organic chelates or cryptates. Complexes such as europium or terbium-cryptate are regularly used now as donors in HTRF assays. The long life of these types of fluorophores means that one can carry out detection in a time-resolved manner, which greatly reduces any background noise produced by auto-fluorescence from the sample. As with any other resonance energy transfer technique, there is a second fluorophore that can interact with the lanthanide and accept energy from it. Clearly, the excitation spectrum of the acceptor must overlap with that of the donor. The most commonly used acceptor, XL665, which was first purified from red algae, has an overlapping spectrum with that of the Eu^{+3} with a maximum emission peak at 665 nm. There is a new second generation acceptor molecule, d2, that also shows ideal spectral properties, but in addition it has the advantage of a much smaller mass, thus avoiding any steric hindrance problems.

HTRF technology has multiple applications, for example, it can be adapted to second messenger assays such as cAMP or inositol phosphate (IP) production (see earlier section). These immuno-competitive assays are cell-based, accurate, sensitive and non-radioactive, with the added advantage that they can be miniaturised and run in high-throughput formats.

HTRF in combination with SNAP-CLIP technology is a rather more interesting application of this technology to study receptor–ligand binding and receptor–receptor interaction. The SNAP tag (63), initially commercialised by Covalys (http://www.covalys.com), is a 20-kDa enzyme based on the O^6-alkylguanine-DNA-alkyltransferase that can be fused to the N-terminal end of a GPCR of interest. SNAP reacts specifically with benzylguanine substrates which can be in turn labelled with a FRET partner, terbium cryptate donor or d2 acceptor. The CLIP tag is a modified version of SNAP and it reacts specifically with benzylcytosine substrates. SNAP-CLIP tags and HTRF combined technologies have been recently described as a way to detect and quantify receptor homodimerisation and heterodimerisation at the cell surface (64). This is a rather exciting and novel way of studying the pharmacology of receptor complexes, particularly since the existence of such complexes has opened the gates to new drug development opportunities. Data obtained using this technology might shed some new light on the consequences of receptor oligomerisation such as cooperative binding with different ligands for the different receptors in the complex, or signalling cross-talk. CisBio Bioassays (http://www.htrf.com) have extensively developed this technology with their Taglite™ cell surface receptor platform and ligand binding assays.

3. Future Directions and Conclusions

The main two directions for novel drug discovery can be summarised as protein-based and cell-based assays. Protein-based biochemical and biophysical studies such as structure-based drug design have the advantage of studying GPCRs in a controlled environment with limited variables to control. There are advantages and disadvantages to all of the techniques and applications described above and these are summarised in Table 1. Advances in GPCR structures obtained using nuclear magnetic resonance (NMR) combined with the use of surface-plasmon resonance will provide new information about the different receptor conformations at different states (65). However, although this is far from being high-throughput it might set the path for novel drug design, potentially in combination with *in silico* approaches. The second group of assays, cell-based assays, seems to be gaining momentum, particularly label-free set ups, such as cell impedance assays. The main advantage of these is the possibility of screening primary cells or even tissues. Cell-based assays are evolving towards being able to screen large compound libraries using simple functional readouts and taking down new routes such as G protein-independent pathways, such as the β-arrestin recruitment assays.

Table 1
Summary of assay technologies for the discovery of GPCR drugs

	HTS/HCS compatible	Cell based	Label free	Compatible with miniaturisation
Phage display	No	n/a	n/a	No
β-arrestin recruitment	Yes	Yes	No	Yes
CellKey/Epic	Yes	Yes	Yes	Yes[a]
Structure-based screening	No	No	n/a	No
Fluorescent/chemiluminescent complementation	Yes	Yes	No	Yes
Fluorescent correlation spectrometry	Yes[b]	Yes	No	Yes
Fluorescent-labelled ligand binding	Yes	Yes	No	Yes
Homogeneous time-resolved FRET	Yes	Yes	No	Yes

[a] Compatible with miniaturisation within the limits of the plate design

[b] Compatible with 96-well format using a Evotec Zeiss Confocor system or similar platform; n/a: not applicable

References

1. Mayr LM, Bojanic D (2009) Novel trends in high-throughput screening. Curr Opin Pharmacol **9**:580–588.
2. Peters MF, Knappenberger KS, Wilkins D et al (2007) Evaluation of cellular dielectric spectroscopy, a whole-cell, label-free technology for drug discovery on Gi-coupled GPCRs. J Biomol Screen **12**:312–319.
3. Peters MF, Scott CW (2009) Evaluating cellular impedance assays for detection of GPCR pleiotropic signaling and functional selectivity. J Biomol Screen **14**:246–255.
4. Antony J, Kellershohn K, Mohr-Andra M et al (2009) Dualsteric GPCR targeting: a novel route to binding and signaling pathway selectivity. FASEB J **23**:442–450.
5. Kebig A, Kostenis E, Mohr K et al (2009) An optical dynamic mass redistribution assay reveals biased signaling of dualsteric GPCR activators. J Recept Signal Transduct Res **29**:140–145.
6. Lee T, Schwandner R, Swaminath G et al (2008) Identification and functional characterization of allosteric agonists for the G protein-coupled receptor FFA2. Mol Pharmacol **74**:1599–1609.
7. Shemesh R, Toporik A, Levine Z et al (2008) Discovery and validation of novel peptide agonists for G-protein-coupled receptors. J Biol Chem **283**:34643–34649.
8. McLoughlin D, Bertelli F, Williams C (2007) The A, B, Cs of G-protein-coupled receptor pharmacology in assay development for HTS. Expert Opin Drug Discov **2**:603–619.
9. Gossman DG, Zhao HB (2008) Hemichannel-mediated inositol 1,4,5-trisphosphate (IP3) release in the cochlea: a novel mechanism of IP3 intercellular signalling. Cell Commun Adhes **15**:305–315.
10. Trinquet E, Fink M, Bazin H et al (2006) D-myo-inositol 1-phosphate as a surrogate of D-myo-inositol 1,4,5-tris phosphate to monitor G protein-coupled receptor activation. Anal Biochem **358**:126–135.
11. Ayoub MA, Maurel D, Binet V et al (2007) Real-time analysis of agonist-induced activation of protease-activated receptor 1/Galphai1 protein complex measured by bioluminescence resonance energy transfer in living cells. Mol Pharmacol **71**:1329–1340.
12. Violin JD, Lefkowitz RJ (2007) *Beta*-arrestin-biased ligands at seven-transmembrane receptors. Trends Pharmacol Sci **28**:416–422.
13. DeWire SM, Ahn S, Lefkowitz RJ et al (2007) *Beta*-arrestins and cell signalling. Ann Rev Physiol **69**:483–510.

14. Chen W, Hu LA, Semenov MV et al (2001) *Beta*-arrestin 1 modulates lymphoid enhancer factor transcriptional activity through interaction with phosphorylated dishevelled proteins. Proc Natl Acad Sci USA **98**:14889–14894.
15. Schulte G, Schambony A, Bryja V (2010) *Beta*-arrestins – scaffolds and signalling elements essential for WNT/Frizzled signalling pathways? Br J Pharmacol **159**:1051–1058.
16. Lin F T, Daaka Y, Lefkowitz RJ (1998) *Beta*-arrestins regulate mitogenic signaling and clathrin-mediated endocytosis of the insulin-like growth factor I receptor. J Biol Chem **273**:31640–31643.
17. Doucette C, Vedvik K, Koepnick E et al (2009) *Kappa* opioid receptor screen with the Tango *beta*-arrestin recruitment technology and characterization of hits with second-messenger assays. J Biomol Screen **14**:381–394.
18. Patel A, Murray J, McElwee-Whitmer S et al (2009) A combination of ultrahigh throughput PathHunter and cytokine secretion assays to identify glucocorticoid receptor agonists. Anal Biochem **385**:286–292.
19. Yin H, Chu A, Li W et al (2009) Lipid G protein-coupled receptor ligand identification using beta-arrestin PathHunter assay. J Biol Chem **284**:12328–12338.
20. van der Lee MM, Blomenrohr M, van der Doelen AA et al (2009) Pharmacological characterization of receptor redistribution and *beta*-arrestin recruitment assays for the cannabinoid receptor 1. J Biomol Screen **14**:811–823.
21. McGuinness D, Malikzay A, Visconti R et al (2009) Characterizing cannabinoid CB2 receptor ligands using DiscoveRx PathHunter *beta*-arrestin assay. J Biomol Screen **14**:49–58.
22. Kocan M, Pfleger KD (2009) Detection of GPCR/*beta*-arrestin interactions in live cells using bioluminescence resonance energy transfer technology. Methods Mol Biol **552**: 305–317.
23. Pfleger KD, Seeber RM, Eidne KA (2006) Bioluminescence resonance energy transfer (BRET) for the real-time detection of protein-protein interactions. Nat Protoc **1**:337–345.
24. Masri B, Salahpour A, Didriksen M et al (2008) Antagonism of dopamine D2 receptor/*beta*-arrestin 2 interaction is a common property of clinically effective antipsychotics. Proc Natl Acad Sci USA **105**: 13656–13661.
25. Klewe IV, Nielsen SM, Tarpo L et al (2008) Recruitment of *beta*-arrestin 2 to the dopamine D2 receptor: insights into anti-psychotic and anti-parkinsonian drug receptor signalling. Neuropharmacology **54**:1215–1222.
26. Garippa RJ, Hoffman AF, Gradl G et al (2006) High-throughput confocal microscopy for *beta*-arrestin-green fluorescent protein translocation G protein-coupled receptor assays using the Evotec Opera. Methods Enzymol **414**:99–120.
27. Barak LS, Ferguson SS, Zhang J et al (1997) A *beta*-arrestin/green fluorescent protein biosensor for detecting G protein-coupled receptor activation. J Biol Chem **272**:27497–27500.
28. Kapur A, Zhao P, Sharir H et al (2009) Atypical responsiveness of the orphan receptor GPR55 to cannabinoid ligands. J Biol Chem **284**: 29817–29827.
29. Smith GP. (1985) Filamentous fusion phage: novel expression vectors that display cloned antigens on the virion surface. Science **228**:1315–1317.
30. Szardenings M, Tornroth S, Mutulis F et al (1997) Phage display selection on whole cells yields a peptide specific for melanocortin receptor 1. J Biol Chem **272**:27943–27948.
31. Houimel M, Loetscher P, Baggiolini M et al (2001) Functional inhibition of CCR3-dependent responses by peptides derived from phage libraries. Eur J Immunol **31**:3535–3545.
32. Hessling J, Lohse MJ, and Klotz KN (2003) Peptide G protein agonists from a phage display library. Biochem Pharmacol **65**:961–967.
33. Bikkavilli RK, Tsang SY, Tang WM et al (2006) Identification and characterization of surrogate peptide ligand for orphan G protein-coupled receptor mas using phage-displayed peptide library. Biochem Pharmacol **71**:319–337.
34. Congreve M, Murray CW, Blundell TL (2005) Structural biology and drug discovery. Drug Discov Today **10**:895–907.
35. Palczewski K, Kumasaka T, Hori T et al (2000) Crystal structure of rhodopsin: A G protein-coupled receptor. Science **289**:739–745.
36. Cherezov V, Rosenbaum DM, Hanson MA et al (2007) High-resolution crystal structure of an engineered human *beta* 2-adrenergic G protein-coupled receptor. Science **318**: 1258–1265.
37. Rasmussen SG, Choi HJ, Rosenbaum DM et al (2007) Crystal structure of the human *beta*2 adrenergic G-protein-coupled receptor. Nature **450**:383–387.
38. Jaakola VP, Griffith MT, Hanson MA et al (2008) The 2.6 angstrom crystal structure of a human A2A adenosine receptor bound to an antagonist. Science **322**:1211–1217.
39. Congreve M, Marshall F (2010) The impact of GPCR structures on pharmacology and structure-based drug design. Br J Pharmacol **159**:986–996.

40. Bowie JU (2001) Stabilizing membrane proteins. Curr Opin Struct Biol **11**:397–402.
41. Zhou Y, Bowie JU (2000) Building a thermostable membrane protein. J Biol Chem **275**:6975–6979.
42. Lau FW, Nauli S, Zhou Y et al (1999) Changing single side-chains can greatly enhance the resistance of a membrane protein to irreversible inactivation. J Mol Biol **290**:559–564.
43. Serrano-Vega MJ, Magnani F, Shibata Y et al (2008) Conformational thermostabilization of the *beta* 1-adrenergic receptor in a detergent-resistant form. Proc Natl Acad Sci USA **105**:877–882.
44. Warne T, Serrano-Vega MJ, Baker JG et al (2008) Structure of a *beta* 1-adrenergic G-protein-coupled receptor. Nature **454**: 486–491.
45. Bond RA, Ijzerman AP (2006) Recent developments in constitutive receptor activity and inverse agonism, and their potential for GPCR drug discovery. Trends Pharmacol Sci **27**:92–96.
46. Gupta A, Heimann AS, Gomes I et al (2008) Antibodies against G-protein coupled receptors: novel uses in screening and drug development. Comb Chem High Throughput Screen **11**:463–467.
47. Peter JC, Wallukat G, Tugler J et al (2004) Modulation of the M2 muscarinic acetylcholine receptor activity with monoclonal anti-M2 receptor antibody fragments. J Biol Chem **279**:55697–55706.
48. Vidi PA, Watts VJ. (2009) Fluorescent and bioluminescent protein-fragment complementation assays in the study of G protein-coupled receptor oligomerization and signalling. Mol Pharmacol **75**:733–739.
49. Hu CD, Chinenov Y, Kerppola TK (2002) Visualization of interactions among bZIP and Rel family proteins in living cells using bimolecular fluorescence complementation. Mol Cell **9**:789–798.
50. MacDonald ML, Lamerdin J, Owens S et al (2006) Identifying off-target effects and hidden phenotypes of drugs in human cells. Nat Chem Biol **2**:329–337.
51. Middleton RJ, Kellam B (2005) Fluorophore-tagged GPCR ligands. Curr Opin Chem Biol **9**:517–525.
52. Cordeaux Y, Briddon SJ, Alexander SP et al (2008) Agonist-occupied A3 adenosine receptors exist within heterogeneous complexes in membrane microdomains of individual living cells. FASEB J **22**:850–860.
53. Chini B, Parenti M (2004) G-protein coupled receptors in lipid rafts and caveolae: how, when and why do they go there? J Mol Endocrinol **32**:325–338.
54. Ostrom RS, Insel PA (2004) The evolving role of lipid rafts and caveolae in G protein-coupled receptor signaling: implications for molecular pharmacology. Br J Pharmacol **143**:235–245.
55. Elson EL (2004) Quick tour of fluorescence correlation spectroscopy from its inception. J Biomed Opt **9**:857–864.
56. Briddon SJ, Gandia J, Amaral OB et al (2008) Plasma membrane diffusion of G protein-coupled receptor oligomers. Biochim Biophys Acta **1783**:2262–2268.
57. Briddon SJ, Middleton RJ, Cordeaux Y et al (2004) Quantitative analysis of the formation and diffusion of A1-adenosine receptor-antagonist complexes in single living cells. Proc Natl Acad Sci USA **101**:4673–4678.
58. Briddon SJ, Hill SJ (2007) Pharmacology under the microscope: the use of fluorescence correlation spectroscopy to determine the properties of ligand-receptor complexes. Trends Pharmacol Sci **28**:637–645.
59. Hegener O, Prenner L, Runkel F et al (2004) Dynamics of *beta* 2-adrenergic receptor-ligand complexes on living cells. Biochemistry **43**:6190–6199.
60. Haasen D, Schnapp A, Valler MJ et al (2006) G protein-coupled receptor internalization assays in the high-content screening format. Methods Enzymol **414**:121–139.
61. Beaudet A, Nouel D, Stroh T et al (1998) Fluorescent ligands for studying neuropeptide receptors by confocal microscopy. Braz J Med Biol Res **31**:1479–1489.
62. O'Dowd BF, Alijaniaram M, Ji X et al (2007) Using ligand-induced conformational change to screen for compounds targeting G-protein-coupled receptors. J Biomol Screen **12**: 175–185.
63. Gautier A, Juillerat A, Heinis C et al (2008) An engineered protein tag for multiprotein labeling in living cells. Chem Biol **15**:128–136.
64. Maurel D, Comps-Agrar L, Brock C et al (2008) Cell-surface protein-protein interaction analysis with time-resolved FRET and SNAP-tag technologies: application to GPCR oligomerization. Nat Methods **5**:561–567.
65. Kofuku Y, Yoshiura C, Ueda T et al (2009) Structural basis of the interaction between chemokine stromal cell-derived factor-1/ CXCL12 and its G-protein-coupled receptor CXCR4. J Biol Chem **284**:35240–35250.

Chapter 13

Discovering Cell Type-Specific Patterns of G Protein-Coupled Receptor Phosphorylation

Kok Choi Kong, Andrew B. Tobin, and Adrian J. Butcher

Abstract

Phosphorylation of G protein-coupled receptors (GPCRs) occurs within seconds of agonist stimulation and is one of the most prevalent mechanisms through which signalling of this super receptor family is regulated. Although traditionally associated with receptor desensitisation and internalisation, there is an increasing body of evidence that suggests that GPCRs employ phosphorylation as a mechanism of coupling the receptor to non-G protein signalling pathways. Recently, it has become clear that GPCRs can be differentially phosphorylated in different cell types or tissues, possibly by tissue- or cell type-specific employment of a variety of receptor kinases to generate specific "phosphorylation patterns" that might encode signalling properties on the receptor. Although hampered by low levels of expression and high hydrophobicity, GPCR phosphorylation can be studied using various methods including two-dimensional (2D) phosphopeptide mapping, mass spectrometry, production of phospho-specific antibodies and site-directed mutagenesis. In this chapter, we discuss the first three methods which are employed in our laboratory for the studies of M_3-muscarinic receptor phosphorylation.

Key words: Antibodies, G-protein-coupled receptor, Immunoprecipitation, Phosphorylation, Proteomics, 2D phosphopeptide maps, SDS-PAGE, Mass spectrometry

1. Introduction

The importance of G protein-coupled receptors (GPCRs) in signal transduction and regulation of biological functions is beyond doubt. They represent the largest family of mammalian cell-surface receptors containing about 800 members and more than 1% of the human genome (1). They mediate virtually all aspects of physiological functions ranging from sensory recognition (vision, odours, taste, and pain) to complex behavioural events (e.g. memory and learning) with an exceptional diversity of endogenous ligands, which include biogenic amines, peptides, glycoproteins, bioactive

Craig W. Stevens (ed.), *Methods for the Discovery and Characterization of G Protein-Coupled Receptors*, Neuromethods, vol. 60, DOI 10.1007/978-1-61779-179-6_13, © Springer Science+Business Media, LLC 2011

lipids, nucleotides, ions and proteases (2, 3). Moreover, more than 50% of the current therapeutic agents target GPCRs (4, 5).

Hence, the diversity of GPCRs has allowed this family of cell-surface receptors to be involved in the regulation of nearly every mammalian biological process. This complexity is further highlighted by the fact that a single receptor subtype can show wide spread tissue expression and be involved in very different physiological processes depending on the tissue in which the receptor is expressed. This raises the intriguing question of the nature of the mechanism employed by a single receptor subtype in tissue-specific signalling. This question is exemplified in the case of the Gq/11-coupled M_3-muscarinic receptor (M3R) which is expressed not only in neurons throughout the brain but also in smooth muscle cells of the trachea and bladder as well as insulin secreting β-cells of the islet of Langerhans. In these tissues, the M3R is known to regulate defined physiological responses such as memory and learning, smooth muscle contraction and insulin secretion (6–8).

Recent studies from our laboratory as well as others have provided evidence that suggests that GPCRs employ phosphorylation as a mechanism to generate different biological responses (9–11). In essence, a receptor can be phosphorylated by a variety of different kinases in different cell types generating specific “phosphorylation patterns” that might encode signalling properties on the receptor (12). The tissue-specific phosphorylation code might be the result of the tissue-specific employment of receptor kinases to provide a unique phosphorylation pattern on the receptor (11, 12).

The challenges in studying GPCR phosphorylation, as with the studies of GPCRs themselves, lie in their low levels of expression (even in transfected systems) as well as their highly hydrophobic nature which renders them problematic for biochemical extraction and analysis. In spite of this, a number of techniques have been successfully applied in the determination of the phosphorylation sites of GPCRs. These techniques include phosphopeptide mapping (11, 13), mass spectrometry (9, 14), production of phospho-specific antibodies (9, 10, 15, 16) and site-directed mutagenesis (17–19).

Our laboratory has employed strategic combinations of three of the methods in the study of site-specific phosphorylation of the M3R. First, we determine the phosphorylation patterns (we referred to as the “phosphorylation signature”) of the receptor in different cell types using two-dimensional (2D) phosphopeptide maps. This is achieved by (^{32}P)-labelling of the cells, solubilisation and immunoprecipitation of the receptor, resolving the radiolabelled-receptor by SDS-PAGE and transfer to nitrocellulose membrane followed by tryptic digest of the receptor into phosphopeptides and finally resolving the phosphopeptides by 2D phosphopeptide mapping. The major advantage of this method is its ability to analyse the full extent of phosphorylation of the receptor. This is because every phosphopeptide derived from the radiolabelled-receptor after protease digestion is able to be visualised (as spots) on the autoradiograph of the phosphopeptide map. Thus, any changes in the phosphorylation status of the receptor are readily detected by a change in the phosphopeptide "spots" in the map. In addition, since this method does not rely on the expression levels of the receptor but rather on the incorporation of the (^{32}P)-orthophosphate into the cells, it is useful in native primary cells with low receptor expression levels. However, the disadvantage is that the precise sites of phosphorylation are not determined.

The second approach is to determine the precise sites of phosphorylation of the receptor by using mass spectrometry. However, this is only practically feasible with high expression systems where a large amount of receptors can be purified. The major advantage to this approach is that it allows for the design of phospho-specific antibodies. These antibodies can be used in the third strategy adopted by our laboratory which involves use of these phospho-specific antibodies in Western blots and immunocytochemistry studies that establish site-specific receptor phosphorylation in native tissues.

In this chapter the application of these three approaches (i.e. 2D tryptic phosphopeptide mapping, mass spectrometry and generation of phosphorylation site-specific antibodies) in our studies of the M3R are described. These methods can be readily adapted for similar studies on other GPCRs.

2. Evidence for Differential Phosphorylation of GPCRs

It is widely accepted that most GPCRs are rapidly phosphorylated after agonist stimulation and that this phosphorylation occurs mostly at serine and threonine residues within the third intracellular loop and/or C-terminal tail. In addition, there are reports of phosphorylation in the first and second intracellular loops (20, 21) and in some cases phosphorylation by tyrosine kinases has been observed (22). Underlying this complex pattern of phosphorylation is an expanding list of protein kinases responsible for GPCR phosphorylation. Prominent among these kinases are members of the G protein-coupled receptor kinase (GRK) family which, together with the second messenger activated kinases protein kinase A (PKA) and protein kinase C (PKC), were among the first protein kinases identified to phosphorylate GPCRs (23–25). However, with an appreciation of the complexity of GPCR phosphorylation has come an increasing understanding of the true extent of the GPCR kinases. Recent studies have now demonstrated that casein kinase 1α (10, 26), protein kinase CK2 (11), protein kinase B (PKB/akt) (27) and receptor-tyrosine kinases can also mediate phosphorylation of defined GPCR subtypes (22). The complexity of GPCR phosphorylation, mediated by the combination of multiple protein kinases, has suggested a sophisticated mechanism of receptor regulation where specific patterns of phosphorylation encode defined signalling outcomes (12). For example, the kinetics of arrestin recruitment to the β_2 adrenergic receptor differ depending on which GRK phosphorylates the receptor (28). Furthermore, in the case of V_2 vasopressin and AT_{1A} angiotensin receptors, GRK2 and GRK3 appear to mediate phosphorylation that results in arrestin recruitment and receptor internalisation, whereas it is GRK5 and GRK6 activity that results in the receptor coupling to the MAP-kinase pathway (29, 30). These studies suggest, but do not actually demonstrate, that the GRKs phosphorylate distinct sites on the receptor and that this drives differential signalling outcomes.

In support of the notion that differential phosphorylation of GPCRs results in differential signalling have been our studies on the M3R. Here, we have demonstrated that protein kinase CK2-mediated phosphorylation regulates coupling of the receptor to the JUN kinase pathway. In contrast, phosphorylation by CK1α regulates M3R signalling to ERK1/2 whereas GRK-mediated phosphorylation is associated with receptor internalisation and desensitisation (26, 31, 32).

3. Methods to Study Cell Type-Specific Phosphorylation of GPCRs

3.1. Phosphorylation Site Identification by 2-Dimensional (2D) Phosphopeptide Mapping

Though originally developed in the laboratory of Tony Hunter in the late 1980s (33), 2D phosphopeptide mapping was not applied to the studies of GPCRs until the mid-1990s, until Ozcelebi and Miller conducted a comprehensive study of the phosphorylation patterns of cholecystokinin receptors (34). Since then, it has been used in conjunction with Edman sequencing to try to identify specific phosphorylation sites of a number of GPCRs (11, 13).

The first part of this technique involves the radiolabelling of cells expressing the GPCR of interest with (^{32}P)-orthophosphate. The receptors can either be expressed as recombinant proteins in model cell lines such as Chinese hamster ovary (CHO) and human embryonic kidney (HEK) cells or it can be endogenously expressed native proteins in primary cell cultures such as cerebellar granule neurones or airway smooth muscle cells. Cells are usually grown to 80–95% confluency in multi-6-well plates and incubated at 37°C with 100 μCi/ml (see Note 1) of (^{32}P)-orthophosphate (PerkinElmer) for 60–90 min (see Note 2) in phosphate-free Krebs/HEPES buffer (see Table 1). The cells are then stimulated with the relevant ligand for 5–15 min and lysed with cold RIPA buffer (see Table 1).

Table 1
Recipes for buffers used in methods to determine phosphorylation of GPCRs

Buffer	Recipe
Phosphate-free Krebs/HEPES	(in mM) HEPES, 10 at pH 7.4; NaCl, 118; KCl, 4.3; $MgSO_4$, 1.17; $CaCl_2$, 1.3; $NaHCO_3$, 25; glucose, 11.7
RIPA buffer	(in mM) Tris–HCl, 10 at pH 7.4; EDTA, 2; β-glycerophosphate, 20, NaCl, 160; 1% Nonidet P-40 (NP-40) 0.5% deoxycholate
TEG buffer	(in mM) Tris–HCl, 10 at pH 7.4; EDTA, 2; β-glycerophosphate, 20
Protein A-sepharose	1.5 g from GE-Healthcare (cat. no. 17-0780-01)re-suspended in 50 ml of TEG buffer
Laemmli buffer	(in mM) Tris–HCl, 125 at pH 6.8; β-mercaptoethanol, 10; 4% sodium dodecyl sulphate (SDS), 20% glycerol; 0.05% bromophenol blue
Trypsin solution	1 μg of sequencing grade modified trypsin from Promega (cat. no. V5111) dissolved in 50–150 μl freshly prepared 50 mM NH_4HCO_3 solution
pH 1.9 buffer	88% formic acid:acetic acid:water 25:78:897 (v/v/v)
Isobutyric acid chromatography buffer	Isobutyric acid:n-butanol:pyridine:acetic acid:water 1250:38:96:58:558 (v/v/v/v/v)

The second part of the technique is the solubilisation and immunoprecipitation of the receptor. The cells are allowed to be solubilised for at least 10 min on ice in RIPA buffer. All subsequent steps in this part should be carried out with ice-cold buffer and centrifuges set at 4°C. Ensure all cellular material is in suspension by pipetting up and down and transfer all material to a 1.5-ml microcentrifuge tube. Sealed screw-top tubes should be used to prevent radioactivity contamination. Pellet down particulate material by centrifuging at maximum speed (~16,000 × *g*) for 10 min at 4°C. Transfer 900 μl of supernatant into a fresh tube and add 100 μl of diluted anti-receptor antibody (1–5 μg per sample) in TEG buffer (see Table 1). Mix by vortexing and leave on ice for 60–90 min. Add 180 μl of protein A-Sepharose (GE-Healthcare, see Table 1) slurry to each sample. Rock gently (~20 rpm) in 4°C room for 15 min to allow binding of the antibodies to the protein A beads. The microcentrifuge tubes can be placed in a large 50ml Falcon tube while rocking to prevent accidental radioactivity contamination. The beads are then washed 3 times with cold TEG buffer. During each wash, centrifuge at 500 × *g* for 30 s at 4°C to pellet down the beads and aspirate off supernatant carefully without loosing any beads. On the final wash, remove all the supernatant by aspiration with a fine-tipped gel loading pipette. Add 20 μl of Laemmli sample buffer (see Table 1) to the Protein A pellet and mix the sample by flicking the tube and then heating at 50–60°C for 2–3 min. Do not boil the sample as the receptor may aggregate. Centrifuge the sample briefly at high speed to pellet down the beads and load onto a sodium dodecyl sulphate-polyacrylamide gel electrophoresis (SDS-PAGE) gel.

The third part of the technique involves resolving the receptor by SDS-PAGE and transfer onto nitrocellulose membrane. The sample is applied onto a SDS-PAGE gel (usually 8% or 10% for GPCRs) and run until the bromophenol blue front (diaphragm) reaches about 1 cm from the bottom of the gel (this usually take about 40–45 min for a 8 × 10 cm minigel). Transfer the proteins on the gel to nitrocellulose membrane by using wet transfer apparatus. Though semi-dry transfer systems can also be used, wet transfer is recommended as the transfer efficiency is higher. Polyvinylidene fluoride (PVDF) membrane binds more strongly to the proteins which may affect the yield of digested protein in subsequent steps and therefore should be avoided. Ponceau S staining should also be avoided as it may affect the digestion of the proteins in subsequent steps. After transfer, wrap the membrane in cling film and expose to Hyperfilm (GE-Healthcare). The membrane must not be dried as this will fix the proteins onto the membrane and render digestion impossible. An overnight X-ray film exposure should be sufficient to visualise the bands. The cassette can be stored in −80°C with an intensifying screen to improve the signal. A fluorescent ruler marker can be used as a guide to the position and orientation of the membrane.

The last part of this technique is the tryptic digest of the resolved receptor protein and subsequent 2D separation of the digested tryptic peptides. The receptor protein is cut out from the nitrocellulose membrane by superimposing the autoradiograph obtained from above with the membrane. Contours of the band may be marked using a needle or a pen. Precision of the cuts should be verified by re-exposing the membrane to X-ray film. The individual band is further cut into small fine pieces and block in microcentrifuge tube with 200 μl of 0.5% polyvinylpyrrolidon K30 (PVP) in 0.6% acetic acid at 37°C for 30 min. This step is to block protein binding sites on the nitrocellulose that may otherwise capture trypsin. The membrane pieces are then washed 3 times with distilled water and the proteins on the membrane pieces are digested by incubating in 50–150 μl of trypsin solution (see Table 1) overnight at 37°C. Proteins attached to the nitrocellulose will be cleaved and peptides will be released from the membrane. The next day, the supernatant is transferred to a new microcentrifuge tube and the membranes are washed once or twice with distilled water for 15–30 min with shaking at 1,500 rpm at room temperature. The supernatants are pooled and dried down completely with a SpeedVac at room temperature (2–6 h). If the membrane pieces still contain a high amount of radioactivity, they can be digested again and the samples pooled. The dried pellet is re-dissolved with 25–50 μl of pH 1.9 buffer (see Table 1) and dried again with a SpeedVac. This is to get rid of the ammonium salt in the trypsin solution which can affect electrophoretic separation of the phosphopeptides. The pellet is then re-dissolved in 5–10 μl of pH 1.9 buffer, vortexed vigorously and centrifuged at maximum speed for 1 min before being spotted, in small portions (0.5 μl), onto a 20 cm × 20 cm cellulose thin layer chromatography (TLC) plate. A hair-drying fan with no heating may be used to dry every portion after being spotted onto the TLC plate to accelerate the process. Hot air will "bake" the peptides to the plate and it is essential to get the smallest spot possible as this will ensure generation of high quality maps. The spot should be 2 cm from one end of an edge of the plate and 4 cm from the adjacent other. After all the sample is being spotted, the plate is wetted with pH 1.9 buffer using a same size (20 cm × 20 cm) Whatman paper with a circular hole of 1 cm in diameter at the position of the phosphopeptide spot. The areas around the sample should be wetted first so that the buffer converges to the centre of the circle and acts to concentrate the spotted sample. The rest of the plate is wetted by pressing the paper over the entire surface of the plate. The plate should look dull-grey and not shiny because of too much buffer on the plate. Areas where the buffer has puddled should be blotted carefully with a KimWipe. The plate is then electrophoresed for 30–40 min at 2,000 V in Hunter thin layer peptide mapping electrophoresis unit (CBS Scientific Inc, HTLE-7002) and air-dried extensively in a fume hood. This first dimension separates the positively charged or neutral phosphopeptides, which migrate to the cathode on the right side, from the free phosphate and highly negatively charged peptides, which migrate to the anode at the left side. Before performing the second dimension, a lane of cellulose 3 cm from top of the plate is scraped off to stop solvent migration before it reaches the top of the plate. The ascending chromatography is run overnight in isobutyric acid chromatography buffer (see Table 1). The plate is dried extensively in a fume hood the next day, and wrapped in cling film and exposed to X-ray film or a phosphorimager for 2–5 days, depending on the intensity of the radioactivity on the plate. The phosphopeptide map can then be obtained by developing the X-ray film or scanning the phosphorimager (Fig. 1).

As mentioned previously, these maps allow the visualisation of the full extent of phosphorylation of a GPCR, which is showed as a distinct pattern on the TLC plate as different phosphopeptides generated from the tryptic (or other proteases) digest of the phosphorylated receptor migrate differently on the plate. Furthermore, the changes in phosphorylation of a GPCR upon

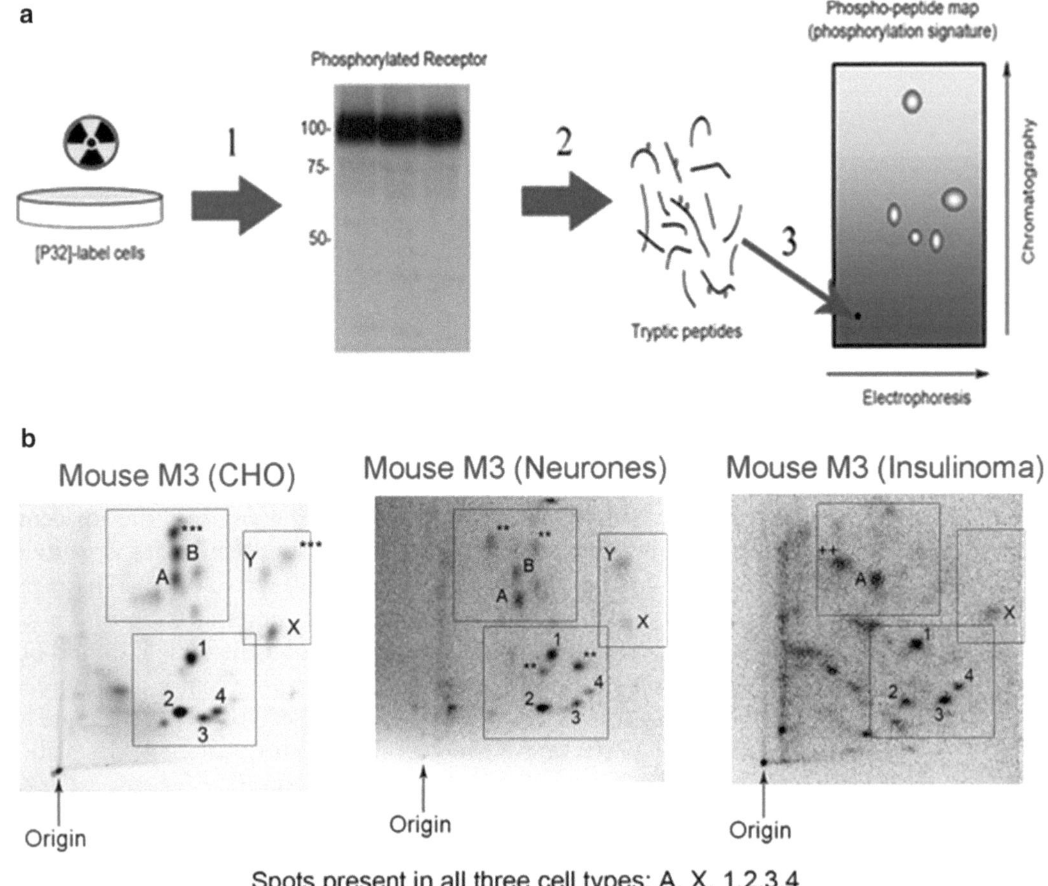

Fig. 1. Work flow for generating a tryptic phosphopeptide map – An example showing differential phosphorylation of the M_3-muscarinic receptor in three different cell types. (**a**) Cells expressing the receptor of interest are labelled with (^{32}P)-orthophosphate and either stimulated with agonist or vehicle. (1) The receptors are then solubilised and immunoprecipitated with receptor specific antibodies. The receptors are resolved by SDS-PAGE and the position of the phosphorylated receptor revealed in an autoradiograph. (2) The receptor band is cut from the gel and digested with trypsin. (3) The tryptic peptides are spotted onto a cellulose plate and subjected to 2D chromatography to resolve the peptides. The positions of the phosphopeptides are determined from an autoradiograph of the plate. This generates a phosphopeptide map that reveals the phosphorylation profile or *phosphorylation signature* of the receptor. (**b**) An example of three phosphorylation profiles of the M_3-muscarinic receptor. The receptor is expressed as a recombinant protein in CHO cells, or endogenously expressed in cerebellar granule neurones or the insulinoma cell line MIN6. It can be seen that some phosphopeptides appear in all three profiles (for example those labelled 1,2,3,4, A and X). In contrast, some phosphopeptides are only seen in one cell type and still others are restricted to just two of the three cell types. These studies demonstrate that the overall phosphorylation profile of the M_3-muscarinic receptor is dependent on the cell type in which the receptor is expressed.

treatment with specific agonists can also be detected as particular phosphopeptide spots present on the map change in intensity. However, in order to be able to determine the identity of a particular phosphopeptide on the map, a large amount (~1–10 pmol) of that phosphopeptide needs to be generated, isolated from the TLC plate, purified by HPLC and identified by Edman sequencing or mass spectrometry. This is usually difficult if not impossible to achieve. Thus, the following two approaches are employed to solve this problem.

Notes

1. There is a natural tendency to increase the amount of protein used for the 2D mapping procedure while there is often an intuitive inhibition to use sufficiently high ^{32}P-orthophosphate concentrations. However, ^{32}P-labelling of the cellular phosphate pool, its incorporation in the protein of interest, and quality of the antibody used for immunoprecipitation are the most important parameters for obtaining good quality maps. Therefore, in optimising experimental conditions for a specific GPCR, it is highly advisable to increase the concentration of ^{32}P-orthophosphate used, decrease the number of cells used and optimise the immunoprecipitation protocol.
2. Although isotopic equilibrium point is not reached within the short-term labelling described here, longer labelling times was found to have no beneficial effects for the studies of the phosphorylation of the receptor but rather produced detrimental effects to the cells. Thus, we find 1–2 h labelling time is sufficient for the purpose of the studies of receptor phosphorylation.

3.2. Phosphorylation Site Identification by Mass Spectrometry

For over 2 decades, the discovery of "soft ionisation" techniques of matrix assisted laser desorption ionisation (MALDI) and electrospray ionisation have provided valuable tools for the study of individual protein and entire proteomes (35). Spectra resulting from analysis of proteolytic digests can now be quickly searched against databases of proteins and in silico proteolytic digests to identify unknown protein and posttranslational modifications (36). Recently, with the introduction of extremely fast and accurate mass spectrometers such as the 4000 Q trap (Applied Biosystems) or the LTQ orbitrap (Thermo) which can offer up to 1 part per million mass accuracy, mass spectrometry has become an increasingly viable alternative to traditional methods of analysis of protein phosphorylation such as those mentioned earlier in this chapter. First, no radioactive label is required for identification of phosphorylation and second, the identity of the protein and the exact location of the phosphorylated residue can be obtained from one experiment. The technique can also be used in combination with isotopic labelling of peptides using iTRAQ (isobaric tag for relative and absolute quantitation) or SILAC (stable isotope labelling with amino acids in cell culture) such that relative changes in phosphorylation at specific sites can be measured in response to stimulation with agonist (37). In the field of GPCRs, mass spectrometry has been used to identify sites of phosphorylation in the β_2 adrenergic receptor (14), the V_2 vasopressin receptor (38), the M3R (39) and the CXCR4 chemokine receptor (9). The following describes the purification and preparation of a GPCR for analysis by mass spectrometry.

3.2.1. Receptor Purification

The first part of the procedure involves purification of approximately ~1μg of the receptor of interest in its phosphorylated form. Cells should be stably transfected with the GPCR of interest and should express at levels greater than 0.5–1 pmol/mg total protein. An epitope tag, usually HA or FLAG appended to the N-terminus or C-terminal tail of the receptor will greatly aid purification which can usually be achieved in one step. Cells should be grown to 80–90% confluency in T175 flasks or expanded surface roller bottles in sufficient quantities that yield about 1 μg of purified receptor. In our case, we have purified sufficient quantities of M3R from 20 T175 flasks containing CHO cells expressing the M3R at 2 pmols/mg protein. If the phosphorylation status of the stimulated receptor is required, then cells should be washed and incubated in Krebs/HEPES buffer for 1 h and incubated with the appropriate ligand for 5–10 min. Cells should be immediately harvested and lysed by addition of TE buffer containing protease and phosphatase inhibitors in a Polytron homogeniser. Large cellular debris and un-lysed cells can be removed by centrifugation at 1,000 × *g* for 5 min before membranes are collected from the resulting supernatant by centrifugation at 40,000 × *g* for 1 h at 4°C.

Receptors are readily solubilised from the membranes by addition of phosphate buffered saline (PBS) containing 1% NP-40 with phosphatase and protease inhibitors and incubation on ice for 1 h. Insoluble material can be removed by centrifugation at 20,000 × *g* for 20 min. The resulting supernatant is diluted 1:1 with PBS and receptors are immunoprecipitated by addition of 200 μl of packed agarose beads coupled to antibody (either anti-HA (Roche) or anti-FLAG (Sigma)). To obtain a clean preparation, beads should be extensively washed (e.g. 4 × 15 min) with PBS containing 0.5% NP-40 prior to elution of the beads with a volume of Laemmli sample buffer that is equivalent to the volume of the packed antibody-agarose beads.

3.2.2. Preparation of Tryptic Peptides for Mass Spectrometry

The purified receptors should be separated by SDS-PAGE on 1.5 mm thick 8% or 10% gels and stained with a colloidal coomassie blue or other mass spectrometry friendly protein stain. After destaining, the receptor band can be extracted from the polyacrylamide with a clean scalpel. The band should be cut into 1–2 mm squares and washed 3 × 15 min in 100 mM ammonium bicarbonate. Prior to digestion the protein is reduced and alkylated to covalently modify any cysteines and prevent them from forming disulphide bonds. Discard the wash and re-suspend the gel pieces in 10 mM DTT in 50 mM ammonium bicarbonate (enough to cover the gel) and incubate for 30 min at 65°C. Discard the DTT solution and replace with 100 mM iodoacetamide dissolved in 50 mM ammonium bicarbonate and incubate for 30 min in the dark. Shrinking of the gel pieces prior to re-swelling with the protease solution is achieved by washing 3 × 15 min with 50% acetonitrile in 50 mM ammonium bicarbonate and finally with 100% acetonitrile, residual solvent should be removed by spinning the sample for 5 min in a rotary evaporator. To digest the protein, prepare 100 μl of 50 mM ammonium bicarbonate containing 1 μg of protease (usually trypsin) and re-hydrate the gel pieces, if additional buffer is required to completely cover the gel pieces then additional ammonium bicarbonate can be added. The digest should be incubated for 12–18 h at 37°C.

When the digest is complete, vortex thoroughly and remove and keep the supernatant. The gel pieces can be washed with 80% acetonitrile containing 0.1% TFA to recover additional peptides. The supernatants can be pooled and concentrated in a rotary evaporator and re-suspended in 0.1% TFA. For analysis by MALDI, samples prepared in this way may need to be desalted using a reverse phase cartridge or tip containing C18 resin. Eluates are then mixed 1:1 with matrix solution such as 10 mg/ml 2,5-dihydroxybenzoic acid (DHB) in 50% acetonitrile/1% phosphoric acid or 10 mg/ml *alpha*-cyano-4-hydroxycinnamic acid in 50% actonitrile/0.1% TFA and 0.5 μl is spotted and dried onto a stainless steel target plate with the appropriate peptide standards.

The resulting MALDI spectrum are compared to *in silico* proteolytic digests of the protein of interest either manually or using software such as MASCOT (Matrix Science). Phosphorylated peptides are observed as having an increased mass of 79.9 Da ($+PO_4^{3-}$) and may also show an additional peak corresponding to approximately a 98-Da decrease in mass caused by elimination of phosphoric acid (H_3PO_4) although the 98 Da value can vary from instrument to instrument.

More detailed analysis can be carried out using for example an ion trap mass spectrometer where ms/ms spectra can be obtained. Here, peptides are separated by reverse phase chromatography before individual peptides are fragmented and the masses of the resulting fragments are accurately measured. The MS/MS spectra can be interrogated manually or using MASCOT software and from this, the precise sites of phosphorylation can be determined.

3.3. Generation and Characterisation of Phosphorylation Site-Specific Antibodies

3.3.1. Antibody Production

Based upon data obtained from mass spectrometric analysis of receptors overexpressed in cell lines or *in vitro* assays, antibodies can be raised to study phosphorylation of receptors expressed at very low levels in native tissue by techniques such as western blotting or immunohistochemistry. Unlike regular polyclonal antibody production, the choice of immunising peptide is limited as it should be centred around the phosphorylation site of interest. Generally, a peptide of 15–16 amino acid residues is sufficient with the phospho-amino acid of interest around the centre. It is better to keep the sequence as short as possible to restrict the production of non-phospho-specific antibodies and a corresponding non-phosphorylated version of the peptide is always made to aid purification of the resulting anti-serum. As with regular anti-peptide polyclonal antibodies, a cysteine residue can be added to the N- or C-terminus to aid conjugation to the carrier protein such as keyhole limpet haemocyanin (KLH). Design and synthesis of the immunising peptide is often offered as a service and can be offered as a package which includes immunisation and bleeding of the host animals and can in some cases include purification of the resulting anti-serum. These programmes are an attractive and usually cheaper alternative to raising these antibodies in house and although they may differ slightly from programme to programme, the basic principles are similar. For each antibody, 20 mg of each of 2 peptides are synthesised, one phosphopeptide and one non-phosphopeptide. The phosphopeptide is coupled to a carrier protein, mixed with an adjuvant and 2 rabbits are immunised over a 3 month programme with 4 immunisations and 4 bleeds.

Once the programme is finished, it is necessary to capture the phosphorylation-specific antibodies and remove antibodies which are specific for the non-phosphorylated peptide. This can be achieved by passing the serum through a column containing the immobilised phosphorylated peptide which will trap all antibodies specific to both the phosphorylated and non-phosphorylated sequences. The eluate from this column is then loaded onto a column containing the immobilised non-phosphorylated peptide. This removes antibodies which recognise only the non-phosphorylated form and phosphorylation-specific antibodies are collected in the flowthrough.

3.3.2. Characterisation of Phosphorylation-Specific Antibodies

Once the phosphorylation-specific antibodies have been isolated, it is worthwhile performing a basic characterisation before the antibodies are used experimentally. This will confirm that phosphorylation-specific antibodies are not contaminated with non-phosphorylation-specific antibodies and vice versa. A simple way to do this is to isolate the target receptor by immunoprecipitation and treat with calf intestinal alkaline phosphatase (CIAP) to remove phosphorylation. It is then possible to compare the phospho-specific antibody's ability to detect phosphorylation before and after CIAP treatment. Cells expressing the receptor of interest should be grown to 80–90% confluence in 6-well plates, washed and stimulated with the appropriate agonist. Cells should be lysed by addition of 300 μl RIPA buffer containing protease and phosphatase inhibitors and insoluble material removed by centrifugation at 20,000 × *g* for 15 min. The receptor of interest by can be immunoprecipitated from the resulting supernatant by addition of 40 μl of packed agarose bead coupled to anti-HA (Roche) or anti-FLAG (Sigma) antibodies followed by incubation at 4°C on a rolling platform for 1–2 h. After extensive washing in RIPA buffer without phosphatase inhibitors, wash the beads twice with CIAP buffer supplemented with 0.25% n-octylglucoside. Re-suspend the beads in 50 μl CIAP buffer containing 0.25% n-octylglucoside and protease inhibitors, add 40 units of CIAP and incubate at 37°C overnight. After phosphatase treatment, wash the beads 3 times with CIAP buffer supplemented with 0.25% n-octylglucoside and re-suspend in 50 μl Laemmli sample buffer and analyse the CIAP treated and un-treated samples by SDS-PAGE and western blotting with the appropriate antibodies.

A second method of characterisation can be carried out by constructing fusions of the receptor region with proteins such as glutathione-s-transferase (GST) or maltose binding protein (MBP) and expressing these in a bacterial system. Although bacterial kinase may phosphorylate the fusion protein, the fact that the bacteria contain a very limited kinome that bears little relationship to the eukaryotic kinome means that the likelihood that the fusion protein is phosphorylated at the site of interest is very low. Thus, the resulting fusion proteins will very likely not be phosphorylated and so should not react with the phosphorylation-specific antibodies when tested in western blotting (Fig. 2). Either of these steps will confirm that the phosphorylation-specific antibodies are not contaminated with non-phosphorylation-specific antibodies.

a

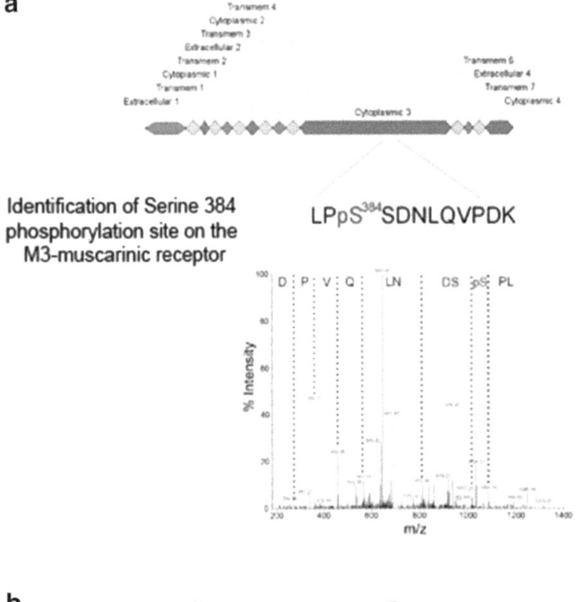

Purification of the receptor is followed by mass spectometry to determine the sites of phosphorylation. Phospho-peptides are then synthesied from which phospho-specific antibodies can be raised.

b

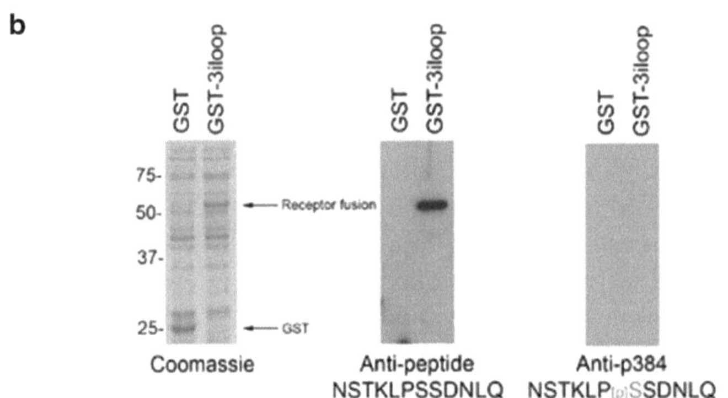

Antibodies that recognise the phosphorylated receptor are purified from those that recognise the non-phosphorylated receptor. These purified antibodies can initially be characterised using bacterial fusion proteins of the receptor as these fusion proteins are unlikely to be phosphorylated at the site of interest.

c

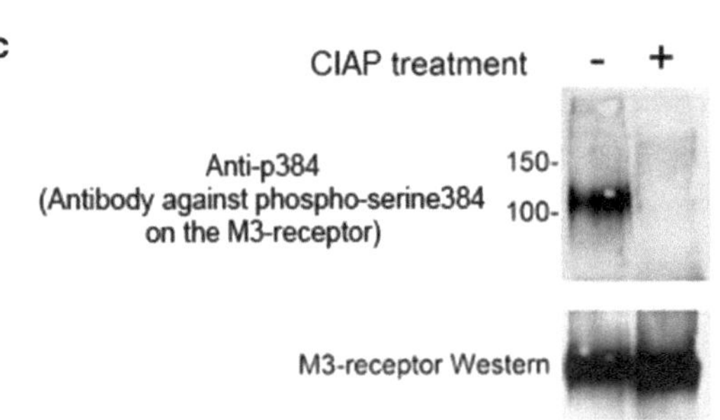

Antibody charactisation should also include testing the ability of the phospho-specific antibody to recognise the immunoprecipiated receptor and monitoring the loss of immunoreactivity following treatment with phosphatse (CIAP).

Fig. 2. Work flow for phospho-acceptor site identification – An example using the phosphorylation status of the M_3-muscarinic receptor on serine^{384}. (**a**) The recombinant M_3-muscarinic receptor containing a C-terminal HA tag was expressed in CHO cells. Cells were stimulated with methacholine (100 μM) for 5 min harvested and membranes prepared. The receptor was then solubilised from the membranes and purified on an HA column. The purified receptor was subjected to tryptic digestion and the peptides analysed by mass spectrometry. Shown is the LC MS/MS trace defining

4. Future Directions

Nuclear magnetic resonance (NMR) spectroscopy has been used for the study of protein phosphorylation for quite some time. A non-radioactive ^{31}P NMR spectroscopy protocol has been developed (40) and used to investigate autophosphorylation of the insulin receptors *in vitro* (41). Recently, high-resolution NMR spectroscopy has been used to identify phosphorylated protein residues in a fragment of the human protein Tau (42). This method relies on using a uniform ^{15}N-labelled protein sample to record a 2D Heteronuclear Single Quantum Coherence (HSQC) spectrum, where each signal represents the amide moiety of a single amino acid of a given protein. As the position of each peak in this HSQC spectrum (i.e. its chemical shift) is extremely sensitive to H^N proton environment, a serine or threonine side chain phosphorylation leads to chemical shift change. Therefore, with this method, NMR spectroscopy can be used to monitor the incorporation of a phosphate, both qualitatively, by the assigning peaks shifted, and quantitatively, by integrating peak intensity.

Selenko et al. applied a similar approach to study the phosphorylation of a small peptide derived from the viral SV40 T antigen by protein kinase CK2 (43). They even went a step further by exploring the function of CK2 in a more cell-like environment, by injecting ^{15}N-labelled protein into *Xenopus* oocytes and monitoring the phosphorylation of the peptide in situ. However, with all the NMR methods mentioned above, a very high protein concentration (in micromolar range) of the protein of interest was used.

Though it is theoretically possible to apply high-resolution NMR techniques to the characterisation of GPCR phosphorylation, the low sensitivity of these methods coupled with the need for high concentrations of both the receptor and kinase are serious obstacles.

A more immediate challenge in the GPCR field is the understanding of the physiological role of receptor phosphorylation. To this end, the generation of transgenic animals expressing a phosphorylation-deficient or phosphorylation-defective form of

Fig. 2. (continued) serine^{384} in the third intracellular loop as a phosphoserine. (**b**) Antibodies to a peptide containing pSerine^{384} were generated. Crude bacterial lysates containing either GST or GST fused to the third intracellular loop of the mouse M_3-muscarinic receptor (GST-3iloop) were probed with either the antibody that recognised the non-phosphorylated peptide or the phospho-specific antibody. (**c**) CHO cells expressing the recombinant mouse M_3-muscarinic receptor were stimulated with methacholine (100 μM) for 5 min. The receptor was then solubilised and immunoprecipitated with an M_3-muscarinic receptor-specific monoclonal antibody. The nitrocellulose was then probed with the anti-phosphoserine 384 antibody. Where indicated the immunoprecipitated receptor was treated with calf intestinal phosphatase (CIAP) to remove phosphates from the receptor. The nitrocellulose was then stripped and probed for the M_3-muscarinic receptor as a loading control.

the GPCR of interest can reveal the true physiological role(s) of a particular phosphorylation site or phosphorylation pattern of the receptor. Transgenic animals selectively expressing a phosphorylation-defective receptor at a specific tissue or cell type can be generated for the study of tissue- or cell type-specific phosphorylation. Although this is a long process (approximately 18 months to generate and initially characterise a transgenic mouse) the outcomes can be profound. In the case of the M3R, for example, we have determined a role for the phosphorylation of this receptor subtype in fear conditioning memory and learning and in the regulation of insulin secretion.

5. Conclusions

It is clear that GPCRs employ phosphorylation in the regulation of receptor coupling to downstream signalling pathways. If we are truly to understand the nature of GPCR signalling we will, therefore, need to fully understand the precise nature of receptor phosphorylation. Up to this point, most of the research in this area has focused on the nature of the protein kinases involved in receptor phosphorylation. We now have the instruments and the techniques that will allow for the detailed examination of the sites on GPCRs that are phosphorylated. Our current working model is that the sites that are phosphorylated on a GPCR determine the signalling properties of the receptor. Since there are a large number of phospho-acceptor sites on GPCRs, we hypothesis that in any given cell type only those sites that impart signalling properties to the receptor that allow for a specific cellular role will be employed. Hence, understanding the nature of the phosphorylation sites on the receptor in a cell type-specific manner will be essential in understanding the signalling role of any receptor subtype in a specific physiological setting. The techniques and approaches described in this chapter are aimed at allowing us to carry out this task.

References

1. Pierce KL, Premont RT, Lefkowitz RJ (2002) Seven-transmembrane receptors. Nat Rev Mol Cell Biol **3**:639–650.
2. Gainetdinov RR, Premont RT, Bohn LM et al (2004) Desensitization of G protein-coupled receptors and neuronal functions. Ann Rev Neurosci **27**:107–144.
3. Gether U (2000) Uncovering molecular mechanisms involved in activation of G protein-coupled receptors. Endocr Rev **21**:90–113.
4. Drews J (2000) Drug discovery: a historical perspective. Science **287**:1960–1964.
5. Hopkins AL, Groom CR (2002) The druggable genome. Nat Rev Drug Discov **1**:727–730.
6. Eglen RM (2006) Muscarinic receptor subtypes in neuronal and non-neuronal cholinergic function. Auton Autacoid Pharmacol **26**:219–233.
7. Eglen RM, Hegde SS, Watson N (1996) Muscarinic receptor subtypes and smooth muscle function. Pharmacol Rev **48**:531–565.
8. Wess J, Eglen RM, Gautam D (2007) Muscarinic acetylcholine receptors: mutant mice provide new insights for drug development. Nat Rev Drug Discov **6**:721–33.

9. Busillo JM, Armando S, Sengupta R et al (2010) Site-specific phosphorylation of CXCR4 is dynamically regulated by multiple kinases and results in differential modulation of CXCR4 signaling. J Biol Chem **285**:7805–7817.

10. Luo J, Busillo JM, Benovic JL (2008) M3 muscarinic acetylcholine receptor-mediated signaling is regulated by distinct mechanisms. Mol Pharmacol **74**:338–347.

11. Torrecilla I, Spragg EJ, Poulin B et al (2007) Phosphorylation and regulation of a G protein-coupled receptor by protein kinase CK2. J Cell Biol **177**:127–137.

12. Tobin AB, Butcher AJ, Kong KC (2008) Location, location, location...site-specific GPCR phosphorylation offers a mechanism for cell-type-specific signalling. Trends Pharmacol Sci **29**:413–420.

13. Blaukat A, Pizard A, Breit A et al (2001) Determination of bradykinin B2 receptor *in vivo* phosphorylation sites and their role in receptor function. J Biol Chem **276**:40431–40440.

14. Trester-Zedlitz M, Burlingame A, Kobilka B et al (2005) Mass spectrometric analysis of agonist effects on posttranslational modifications of the *beta*-2 adrenoceptor in mammalian cells. Biochemistry **44**:6133–6143.

15. Jones BW, Song GJ, Greuber EK et al (2007) Phosphorylation of the endogenous thyrotropin-releasing hormone receptor in pituitary GH3 cells and pituitary tissue revealed by phosphosite-specific antibodies. J Biol Chem **282**:12893–12906.

16. Tran TM, Friedman J, Qunaibi E et al (2004) Characterization of agonist stimulation of cAMP-dependent protein kinase and G protein-coupled receptor kinase phosphorylation of the *beta*2-adrenergic receptor using phosphoserine-specific antibodies. Mol Pharmacol **65**:196–206.

17. Kara E, Crepieux P, Gauthier C et al (2006) A phosphorylation cluster of five serine and threonine residues in the C-terminus of the follicle-stimulating hormone receptor is important for desensitization but not for *beta*-arrestin-mediated ERK activation. Mol Endocrinol **20**:3014–3026.

18. Mendez A, Burns ME, Roca A et al (2000) Rapid and reproducible deactivation of rhodopsin requires multiple phosphorylation sites. Neuron **28**:153–164.

19. Seibold A, Williams B, Huang ZF et al (2000) Localization of the sites mediating desensitization of the *beta*(2)-adrenergic receptor by the GRK pathway. Mol Pharmacol **58**:1162–1173.

20. Kim KM, Valenzano KJ, Robinson SR et al (2001) Differential regulation of the dopamine D2 and D3 receptors by G protein-coupled receptor kinases and *beta*-arrestins. J Biol Chem **276**:37409–37414.

21. Nakamura K, Hipkin RW, Ascoli M (1998) The agonist-induced phosphorylation of the rat follitropin receptor maps to the first and third intracellular loops. Mol Endocrinol **12**:580–591.

22. Paxton WG, Marrero MB, Klein JD et al (1994) The angiotensin II AT1 receptor is tyrosine and serine phosphorylated and can serve as a substrate for the src family of tyrosine kinases. Biochem Biophys Res Commun **200**:260–267.

23. Benovic JL, Pike LJ, Cerione RA et al (1985) Phosphorylation of the mammalian *beta*-adrenergic receptor by cyclic AMP-dependent protein kinase. Regulation of the rate of receptor phosphorylation and dephosphorylation by agonist occupancy and effects on coupling of the receptor to the stimulatory guanine nucleotide regulatory protein. J Biol Chem **260**:7094–7101.

24. Kristiansen K (2004) Molecular mechanisms of ligand binding, signaling, and regulation within the superfamily of G-protein-coupled receptors: molecular modeling and mutagenesis approaches to receptor structure and function. Pharmacol Ther **103**:21–80.

25. Lefkowitz RJ (2004) Historical review: a brief history and personal retrospective of seven-transmembrane receptors. Trends Pharmacol Sci **25**:413–422.

26. Budd DC, Willars GB, McDonald JE et al (2001) Phosphorylation of the Gq/11-coupled M3-muscarinic receptor is involved in receptor activation of the ERK-1/2 mitogen-activated protein kinase pathway. J Biol Chem **276**:4581–4587.

27. Doronin S, Shumay E, Wang HY et al (2002) Akt mediates sequestration of the *beta*(2)-adrenergic receptor in response to insulin. J Biol Chem **277**:15124–15131.

28. Violin JD, Ren XR, Lefkowitz RJ (2006) G-protein-coupled receptor kinase specificity for beta-arrestin recruitment to the *beta*2-adrenergic receptor revealed by fluorescence resonance energy transfer. J Biol Chem **281**:20577–20588.

29. Kim J, Ahn S, Ren XR et al (2005) Functional antagonism of different G protein-coupled receptor kinases for *beta*-arrestin-mediated angiotensin II receptor signaling. Proc Natl Acad Sci USA **102**:1442–1447.

30. Ren XR, Reiter E, Ahn S et al (2005) Different G protein-coupled receptor kinases govern G protein and *beta*-arrestin-mediated signaling of V2 vasopressin receptor. Proc Natl Acad Sci USA **102**:1448–1453.

31. Willets JM, Challiss RA, Nahorski SR (2002) Endogenous G protein-coupled receptor kinase 6 Regulates M3 muscarinic acetylcholine receptor phosphorylation and desensitization in human SH-SY5Y neuroblastoma cells. J Biol Chem **277**:15523–15529.
32. Willets JM, Mistry R, Nahorski SR et al (2003) Specificity of G protein-coupled receptor kinase 6-mediated phosphorylation and regulation of single-cell M3 muscarinic acetylcholine receptor signaling. Mol Pharmacol **64**: 1059–1068.
33. Boyle WJ, van der Geer P, Hunter T (1991) Phosphopeptide mapping and phosphoamino acid analysis by two-dimensional separation on thin-layer cellulose plates. Methods Enzymol **201**:110–149.
34. Ozcelebi F, Miller LJ (1995) Phosphopeptide mapping of cholecystokinin receptors on agonist-stimulated native pancreatic acinar cells. J Biol Chem **270**:3435–3441.
35. Lane CS (2005) Mass spectrometry-based proteomics in the life sciences. Cell Mol Life Sci **62**:848–869.
36. Perkins DN, Pappin DJ, Creasy DM et al (1999) Probability-based protein identification by searching sequence databases using mass spectrometry data. Electrophoresis **20**:3551–3567.
37. Ong S-E, Mann M (2005) Mass spectrometry-based proteomics turns quantitative. Nat Chem Biol **1**:252–262.
38. Wu S, Birnbaumer M, Guan Z (2008) Phosphorylation analysis of G protein-coupled receptor by mass spectrometry: identification of a phosphorylation site in V2 vasopressin receptor. Anal Chem **80**:6034–6037.
39. Butcher AJ, Torrecilla I, Young KW et al (2009) N-methyl-D-aspartate receptors mediate the phosphorylation and desensitization of muscarinic receptors in cerebellar granule neurons. J Biol Chem **284**:17147–17156.
40. Hirai H, Yoshioka K, Yamada K (2000) A simple method using 31P-NMR spectroscopy for the study of protein phosphorylation. Brain Res Brain Res Protoc **5**:182–189.
41. Ge Y, Peng H, Huang K (2006) 31P NMR study on the autophosphorylation of insulin receptors in the plasma membrane. Anal Bioanal Chem **385**:834–839.
42. Landrieu I, Lacosse L, Leroy A et al (2006) NMR analysis of a *Tau* phosphorylation pattern. J Am Chem Soc **128**:3575–3583.
43. Selenko P, Frueh DP, Elsaesser SJ et al (2008) *In situ* observation of protein phosphorylation by high-resolution NMR spectroscopy. Nat Struct Mol Biol **15**:321–329.

Chapter 14

Quantifying Allosteric Modulation of G Protein-Coupled Receptors

Frederick J. Ehlert and Hinako Suga

Abstract

In this chapter, methods for the analysis of allosterism in equilibrium binding and functional assays are described. The functional response to activation of a G protein-coupled receptor is usually measured at a point downstream from receptor activation in the signaling pathway, and the effects of allosteric modulation on the response can be attributed to scalar changes in the observed affinity (α) and intrinsic efficacy (β) of the agonist-receptor complex. If the concentration-response curve of the agonist is measured in the absence and presence of a range of concentrations of the allosteric modulator, then it is always possible to estimate the affinity constant of the modulator (K_2) and the product (γ) of the modulatory changes in the observed affinity and efficacy of the agonist-receptor complex ($\gamma = \alpha\beta$). In many instances, it is impossible through analysis of receptor function only to determine the α and β components of the modulation. Regardless, the parameter γ is unique in that it represents the ratio of the microscopic affinity constants of the modulator for the active and inactive states of the receptor in situations where there is one orthosteric activation site directly linked to the allosteric site. Radioligand binding assays can be used to estimate affinity modulation between the allosteric modulator and the orthosteric radioligand or another nonlabeled orthosteric ligand. By measuring allosterism in both functional and binding assays, it is possible to measure the affinity and efficacy components of allosteric modulation.

Key words: Allosterism, Affinity, Efficacy, Modulation, Pathway selectivity, Ligand-directed signaling

1. Introduction

When an endogenous ligand interacts with the orthosteric site of its receptor, it generates a vector of information having components of affinity and efficacy. This signal is then translated through a series of steps into the final response. An allosteric ligand can modify this process by changing the affinity and efficacy components of the receptor complex. In this chapter, we describe how to measure this modulation.

Craig W. Stevens (ed.), *Methods for the Discovery and Characterization of G Protein-Coupled Receptors*, Neuromethods, vol. 60, DOI 10.1007/978-1-61779-179-6_14, © Springer Science+Business Media, LLC 2011

Allosteric modulation can be analyzed at different hierarchical levels (1). At the deepest level (level 3), allosteric and orthosteric ligands have effects that are measured in terms of how they change the probability that the receptor will undergo an abrupt transition between resting and active states. The output of the receptor is digital and represents a series of on and off transitions. When viewed from the perspective of a large population of receptors, the summation of these random single receptor events yields a constant level of activation in time known as the ensemble average or stimulus. At this level of analysis (population analysis, level 2), receptor activation is defined by the stimulus function, which represents activation expressed as a function of the ligand concentration (2, 3). The maximum of this function represents observed intrinsic efficacy (ε), and the concentration of agonist eliciting half-maximal activation is defined as the observed dissociation constant (K_{obs}). An allosteric ligand can cause concentration-dependent changes in the observed affinity and efficacy components of the stimulus function (4–6). For many GPCRs it is possible to measure the modulation in affinity using radioligand binding methods, but direct measurement of receptor activation are difficult and rarely measured. Rather, a downstream response is usually measured (level 1 measurement), and the EC_{50} and E_{max} values of the response can be much different from the K_{obs} and ε values of the stimulus function. In this chapter, we describe how to estimate the allosteric modulation in the level 2 parameters, K_{obs} and ε, through analysis of a downstream response (level 1 response).

Investigating the effects of allosteric modulators on the functional response to an agonist is a powerful means of detecting allosterism because it enables one to measure changes in both the affinity and efficacy of the agonist-receptor complex. This modulation can vary for a single allosteric agent depending on the orthosteric ligand with which it is interacting. Thus, functional assays provide a simple means to measure the allosteric modulation in the action of the endogenous ligand, and thus, enable one to predict how the drug might alter physiological mechanisms *in vivo*.

Although functional assays are useful for detecting allosterism in the first place, they are not the best way to *prove* that the mechanism is allosteric. There are many ways that the putative modulator could alter the signaling pathway at a site downstream from receptor activation that could mimic an upstream allosteric effect on the receptor. The demonstration that the drug modulates the binding of a ligand to the receptor itself is the most unequivocal evidence for allosterism. For this reason, we present methods for the analysis of allosterism in both radioligand binding and functional assays.

Most measurements of receptor function are obtained under equilibrium conditions because the interacting ligands often equilibrate with the receptor in seconds, provided that they do not require substantial time to diffuse to the receptor in a thick tissue, for example. Thus, we focus on an equilibrium analysis for both functional and binding assays. Ligands with high affinity (observed dissociation constant $<10^{-8}$ M), however, may require substantial time to equilibrate with the receptor. Radioligands often have high affinity for their target receptors, and consequently, their rate constants for equilibration can be on the order of minutes or hours. Measuring how allosteric drugs modulate the kinetics of radioligand binding is also a powerful approach for detecting and confirming allosterism. The approach is not without its limitations, particularly in situations where the allosteric site is in the extracellular loops of the GPCR and occupancy of the site prevents access and egress of the orthosteric ligand to the central orthosteric binding pocket within the TM domains. An excellent description of the analysis of allosteric modification of orthosteric radioligand binding kinetics has been presented by Lazareno and Birdsall (7). The reader will also find the reviews by Christopoulos (8) and Christopoulos and Kenakin (9) which are helpful in understanding the principles and significance of allosterism.

2. Analysis of Allosterism at GPCRs in Functional and Binding Assays

2.1. The Allosteric Ternary Complex Model

To detect or measure allosterism in functional or radioligand binding studies, it is necessary to have a model for allosteric interactions. From this model, equations are developed expressing response (functional assay) or binding (binding assay) as a function of the concentration of one of the interacting ligands. The equation is then fitted to the data by global nonlinear regression analysis to estimate allosteric parameters or to determine whether the data are significantly different from a simpler model, like competitive inhibition. In this section, we describe the model and list some of the equations for analyzing allosterism, the derivations of which have been presented in other publications.

The basis of our analysis is the allosteric ternary complex model (5, 10) shown in Fig. 1. The affinity constants of the orthosteric (D) and allosteric (A) ligands are denoted by K_1 and K_2, respectively. These have inverse concentration units (e.g., M^{-1}) and represent the reciprocal of the concentration of ligand required for half-maximal receptor occupancy in the absence of the other interacting ligand. The constant α represents the cooperativity of the system, and its value is equivalent to the maximum factor by which one ligand modifies the affinity constant of the other. For example, if the affinity constant of D increases by a factor of 10 when all of the allosteric sites are saturated with A, then

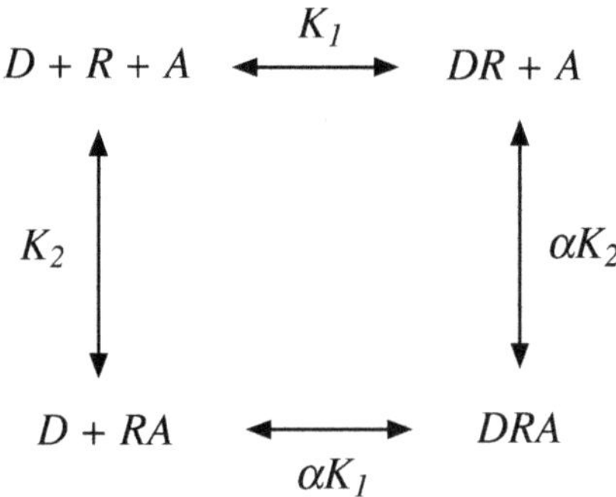

Fig. 1. The allosteric ternary complex model describes the interaction between orthosteric (*D*) and allosteric (*A*) ligands on a single receptor with distinct binding sites for the two ligands. The affinity constants of orthosteric and allosteric ligands for the free receptor are denoted by K_1 and K_2, respectively, and α denotes the cooperativity factor for their interaction. In our analysis, we have considered the condition where the intrinsic activities of the *R* and *RA* complexes are negligible. The activities of the *DR* and *DRA* complexes are denoted by ε and ε^*, respectively, with their ratio ($\varepsilon^*/\varepsilon$) denoted as β. The product of the change in the affinity (α) and efficacy (β) of the *DR* receptor complex caused by the binding of the allosteric ligand is denoted as γ.

the value of α is 10. A value of 0.1 for α refers to the situation when the modulator causes the affinity of *D* to be reduced to one-tenth of its initial value. These two situations refer to positive ($\alpha > 1$) and negative ($0 < \alpha < 1$) cooperativity. One of the consequences of the model in Fig. 1 is that allosteric effects are reciprocal; that is, if *A* modulates the affinity of *D* by the scalar α, then *D* modulates the affinity of *A* by the same amount.

Each of the four types of receptor species (*R*, *DR*, *RA* and *DRA*) can exhibit a different level of activity even though there may only be two quantal states of the receptor (resting and active). This is because the activity of a population of the receptor species depends on the set point of the equilibrium between the states. In the absence of ligands, there may be some basal activity associated with the ligand-free state because spontaneous collisions with the G protein can allosterically activate *R*. The equilibrium between the states shifts in the direction of activation when an agonist is bound to the receptor (*DR*), and it can switch to a new equilibrium when both the agonist and allosteric modulator are bound (*DRA*). The change in the activity of the receptor associated with the binding of the ligand represents the difference in the activity of the *R* and *DR* states and is defined as the observed intrinsic efficacy of the agonist (ε). In our examples, we will assume that the activity of *R* is negligible so that ε is proportional to the activity of *DR*. The activity of the *DRA* complex is denoted as ε^*, and the constant β represents the ratio $\varepsilon^*/\varepsilon$. The binding of the allosteric modulator can also shift the equilibrium between ground and active states. If the allosteric modulator has high selectivity for the active state, it could activate the receptor just as an orthosteric agonist can. In our examples, we will focus on the condition where the modulator has little activity by itself (i.e., the activity of *RA* is negligible).

The solution to the allosteric ternary complex model (4, 11) yields the following equation for receptor activation (S) by the agonist (D) in the presence of an allosteric modulator (A):

$$S = \frac{Dq\varepsilon R_{\mathrm{T}}}{D + \frac{1}{pK_1}} \qquad (1)$$

in which R_{T} represents the total receptor population, and p and q represent the scalar quantities by which the allosteric modulator changes the observed dissociation constant and intrinsic efficacy of the agonist. These are defined as:

$$p = \frac{1 + A\alpha K_2}{1 + AK_2} \qquad (2)$$

$$q = \frac{1 + A\alpha\beta K_2}{1 + A\alpha K_2} \qquad (3)$$

In many instances, it is impossible to discriminate between affinity α and efficacy β modulation in functional assays, so the product of these terms in the numerator of (3) can be replaced with the single parameter γ ($\gamma = \alpha\beta$):

$$q = \frac{1 + A\gamma K_2}{1 + A\alpha K_2} \qquad (4)$$

This change will facilitate global nonlinear regression analysis as described below. As mentioned above, it is usually beyond the current state of the art to measure receptor activation at GPCRs easily so we need to modify the equation to account for the situation where a downstream response is measured instead of receptor activation. The reverse engineering approach developed by Black and Leff (11) solves this problem. These investigators showed that if the input to a system can be described by a one-site model (e.g., (1)) and the output by a logistic equation of the type normally used in analyzing concentration-response curves, then the function that converts the input (S) into the output is (R):

$$R = \frac{S^m M_{\mathrm{sys}}}{S^m + K_E^m} \qquad (5)$$

In this equation, R represents the response, M_{sys} the maximum response of the system, m the transducer-slope factor, and K_E the sensitivity of the transduction mechanism. To obtain the equation for analyzing the functional response, (1) is substituted into (5):

$$R = \frac{D^m M_{sys}}{D^m + \left(\frac{(D + p(1/K_1))^m}{(q\tau)^m} \right)} \tag{6}$$

In this equation, the parameters p and q are defined as in (2) and (3) or (4), respectively, and the parameters ε, R_T and K_E have been replaced with the single parameter τ, which is defined as:

$$\tau = \frac{\varepsilon R_T}{K_E} \tag{7}$$

We describe how to use this equation in the following section on analysis of function. It is also possible to use a null method for analyzing the functional response as described previously (4, 5).

The allosteric ternary complex model can also be solved to yield an equation for the binding of the radioligand (L) in the presence of various concentrations of the allosteric modulator (5, 10). The following equation is in a form that is useful for analyzing a series of saturation curves for the radioligand that were measured in the presence of various fixed concentrations of the allosteric modulator in addition to the control condition where no modulator is added.

$$B = \frac{LB_{max}}{L + \frac{1}{pK_1}} \tag{8}$$

In this equation, B denotes the specific binding of the radioligand, p is defined in (2), B_{max} represents the total receptor density, and K_1 represents the affinity constant of the radioligand. The parameter B_{max} is analogous to R_T in (1). But the two parameters are not identical, because B_{max} represents the density of all of the receptors and R_T only those that can couple with the G proteins that are eliciting the measured response.

Another way to analyze the effect of an allosteric modulator in radioligand binding assays is to measure a series of modulation curves, each measured in the presence of a fixed concentration of radioligand and various concentration of the modulator. Equation (8) can also be used for this analysis, but in this instance, the modulator concentration (A in (2)) is the independent variable.

If each binding value of modulation curve (B) is normalized relative to the binding of the radioligand measured in the absence of the modulator B', then the following equation can be used for the analysis:

$$\frac{B}{B'} = T\frac{L + 1/K_1}{L + 1/pK_1} \quad (9)$$

In this equation, p is defined as in (2), and T denotes the fractional estimate of binding in the absence of the modulator.

To estimate the cooperative interactions between the binding of the allosteric modulator (A) and an orthosteric ligand (X) other than the radioligand (L), it is necessary to measure the binding of the radioligand at a fixed concentration in the presence of increasing concentrations of X and a single concentration of the modulator (A). It is possible to estimate the cooperativity (α_X) between the binding of nonlabeled orthosteric ligand X and the modulator A using the following equation (1):

$$\alpha_X = \frac{1}{R}\left(\frac{1-R}{AK_2} + \frac{1+\alpha_L LK_{1-L}}{1+LK_{1-L}}\right) \quad (10)$$

In this equation, K_{1-L} denotes the affinity constant of the radioligand, K_2 denotes the affinity constant of the allosteric modulator, α_L denotes the cooperativity between the binding of the radioligand and the allosteric modulator, and R denotes the IC_{50} value of X measured in the absence of A divided by that measured in its presence. The IC_{50} value represents the concentration of X required for half-maximal inhibition of the binding of L.

2.2. Analysis of Allosterism in Functional Assays

Practically any response to an agonist can be measured in the absence and presence of an allosteric modulator to determine the parameters of the latter. A useful format is to measure a complete concentration-response curve to the agonist in the presence and absence of various concentrations of the putative allosteric modulator. A hallmark feature of allosterism is that the effect of the modulator reaches a limit at high concentrations (ceiling effect). This maximal effect is determined by γ (product of the allosteric change in affinity and efficacy ($\gamma = \alpha\beta$)). The goal in designing the experiment, therefore, is to collect data over a range of concentrations of the modulator to establish the concentration zone of the ceiling effect. Without sampling data in this region, it would be impossible to estimate log γ accurately. In addition, if the compound in question is a negative allosteric modulator that does not influence the E_{max} of the agonist, then it would be impossible to discriminate its behavior from that of a competitive inhibitor unless the ceiling effect is measured.

2.2.1. Positive Allosteric Modulation of the mGluR4

Metabotropic glutamate receptors exhibit abundant allosteric interactions, and the data in Fig. 2 published by Marino et al. (12) serve as a useful example for the analysis of a positive allosteric modulator at the human mGluR4. The figure illustrates the effects of increasing concentrations of *N*-phenyl-7-(hydroxyimino)cyclopropa-[*b*]chromen-1a-carboxamide (PHCCC) on the Ca^{2+} response to L-(+)-2-amino-4-phosphonobutyric acid (L-AP4) in Chinese hamster ovary (CHO) cells expressing mGluR4 and a chimeric G protein consisting mainly of G_q and the C-terminal tail of G_i. As shown in the figure, increasing concentrations of L-AP4 cause an increase in Ca^{2+} mobilization, and both the potency and E_{max} of this response increase with increasing concentrations of PHCCC.

To estimate the affinity constants of L-AP4 (K_1) and the positive allosteric modulator (K_2), global nonlinear regression analysis is used to fit (6) to the data with (2) and (4) substituted for values of p and q, respectively. Essentially, all of the curves are fitted simultaneously, sharing the estimates M_{sys}, m, K_1, K_2, α, and γ among the curves. This can be done using Microsoft Excel application or GraphPad Prism, for example. Below, we describe the procedure for using Prism. We assume that the reader is already familiar with the use of Prism.

All of the data are entered into a datasheet in Prism. First, the log concentrations of the agonist are entered into the column on the far left of the datasheet as the independent variable, *X*. For the data in Fig. 2a, we simply estimated the mean response values for each curve from a published figure of the data by Marino et al. (12). The mean response values for the control condition are entered into the appropriate cells in column A. The mean data for the curves measured in the presence of 1, 3, 10, and 30 μM PHCCC are entered into the appropriate cells in columns B–E. To simplify the process, we label the columns with the log concentration of the modulator, and use these entries as variable constraints in the regression analysis. This way it is unnecessary to modify the equation for different types of experiments. The column header for the control data is set to an arbitrarily large negative number (e.g., −20) to represent the log of zero, which is actually negative infinity. The column headers for B–E are labeled with −6, −5.523, −5, and −4.523 to represent the log concentrations of 1, 3, 10, and 30 μM, respectively. In situations where there are replicate measurements of a given concentration-response curve, these are entered in sub-columns. For example, replicate experiments of the control concentration-response curve are entered into sub-columns of column A. Replicate measurements of the concentration-response curves measured in the presence of the lowest concentration of modulator are entered into sub-columns of column B, and the process is repeated for the additional concentrations of modulator.

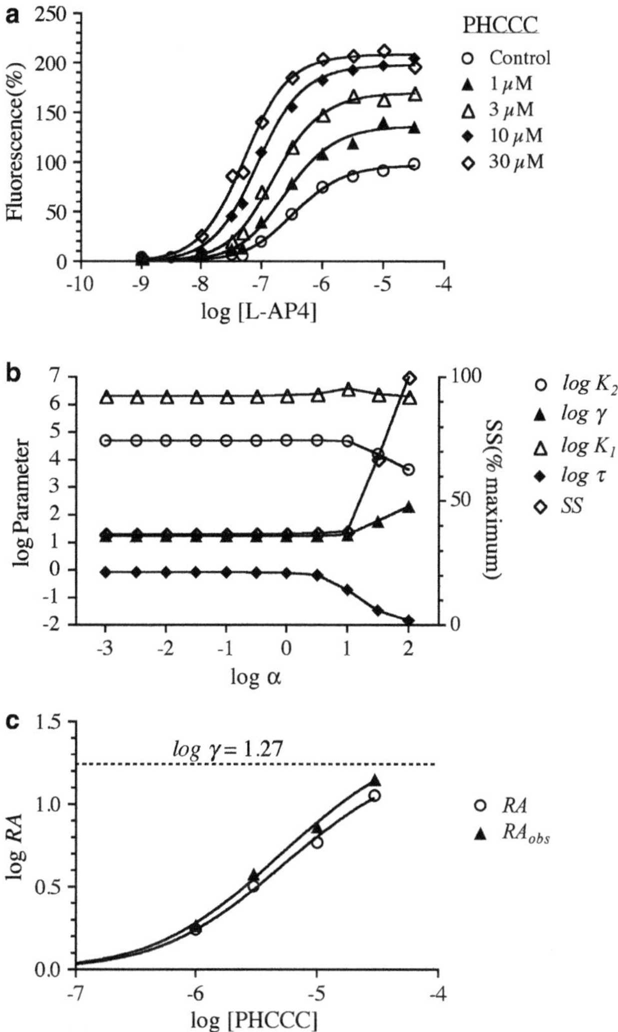

Fig. 2. *Analysis of the allosteric effect of PHCCC on the Ca^{2+} response to L-AP4 at the mGluR4*. (**a**) The effects of increasing concentrations of PHCCC on the Ca^{2+} response to L-AP4 in CHO cells expressing the human mGluR4 and a chimeric G protein consisting of G_q and the C-terminal tail of G_i. The mean response values have been estimated from a published figure of the data by Marino et al. (12). The theoretical curves represent the least squares fit of (6) to the data. (**b**) Summary of global nonlinear regression analysis of the data in (**a**). The logarithm of the parameter α was constrained to various values within the domain −3 – 2, and the estimates of the other parameters that yielded a least squares fit of (6) were determined. When the value of log α is in the range log $\alpha \leq 0.5$, the residual sum of squares (*SS*) is a minimum, and the values of the other parameters are constant. (**c**) The affinity constant of the allosteric modulator K_2 and the product of its affinity and efficacy modulation (γ) can be estimated by analysis of the RA values. The RA values were estimated from the data in (**a**) using (14) (RA_{obs}) or by using a method described previously (RA, (4)). The theoretical curves represent the least squares fit of (22) to the data. The *dashed line* indicates the asymptote for the maximum log RA value, which is equivalent to log γ.

Next, a user-defined equation is entered into Prism. This is (6) above, with p and q defined as in (2) and (4). To simplify the process, we enter (2) and (4) to define the variables p and q, and then (6) is entered using the variables p and q. We modify the equations so that parameters, K_1, K_2, α, γ, and τ, are expressed as logarithms. Thus, the following three equations are entered into Prism for nonlinear regression analysis:

$$P = \frac{1+10\wedge(LOGA+LOGK2)}{1+10\wedge(LOGA+LOGALPHA+LOGK2)} \tag{11}$$

$$Q = \frac{1+10\wedge(LOGA+LOGGAMMA+LOGK2)}{1+10\wedge(LOGA+LOGALPHA+LOGK2)} \tag{12}$$

$$Y = \frac{MSYS*(10\wedge X)^m}{(10\wedge X)^m + \dfrac{(10\wedge X + P*(10\wedge LOGK1))^m}{(Q*(10\wedge LOGTAU))^m}}. \tag{13}$$

The log variables entered into Prism, *LOGA*, *LOGK*1, *LOGK*2, *LOGALPHA*, *LOGGAMMA*, and *LOGTAU*, correspond to the log values of the variables A, K_1, K_2, α, γ, and τ, respectively, in (2), (4), and (6). Also, the variables *MSYS*, *P*, and *Q* in Prism represent M_{sys}, p, and q in (2), (4), and (6).

Before continuing with our description of the regression analysis, it is necessary to have estimates of the EC_{50} and E_{max} values of the concentration-response curves, because these are useful for making the initial parameter estimates for regression analysis. The EC_{50} and E_{max} values can be estimated in Prism using the variable slope dose-response curve equation. The data in Fig. 2a were analyzed in this manner, and the E_{max} and EC_{50} values are listed in Table 1. When the Hill slopes of the concentration-response curves are similar to one, it is possible to estimate the product (pq) of the change in affinity (p) and efficacy (q) caused by each concentration of the modulator from the EC_{50} and E_{max} values using (14) below. This product is known as relative activity (RA):

$$RA = \frac{E_{max} EC_{50\text{ - control}}}{E_{max\text{ - control}} EC_{50}} \tag{14}$$

In this equation, E_{max} and EC_{50} represent the values estimated in the presence of the modulator, and $E_{max\text{-control}}$ and $EC_{50\text{-control}}$ represent those measured in the absence of the modulator. The RA values are also listed in Table 1 for each concentration of the modulator.

Table 1
Effect of PHCCC on the concentration-response curve of L-AP4 for eliciting Ca^{2+} mobilization in CHO cells expressing mGluR4

Condition	E_{max}	pEC_{50}	Hill slope	Log RA_{obs}[a]	Log RA[b]
Control	95.6	6.48	1.13	0	0
PHCCC (1 μM)	136	6.59	1.06	0.27	0.25
PHCCC (3 μM)	168	6.81	1.17	0.58	0.51
PHCCC (10 μM)	199	7.02	1.13	0.86	0.77
PHCCC (30 μM)	208	7.29	1.13	1.15	1.04

The parameters were estimated from the data in Fig. 2
[a]The RA_{obs} values were calculated from the E_{max} and EC_{50} values using (22)
[b]The RA values were calculated as described previously (4)

Now let's return to the estimation of the initial parameter values. These are calculated using (15)–(21) below, in which the initial estimate of a given parameter is indicated with an asterisk. The initial estimate of log K_1 is assigned the $p\mathrm{EC}_{50}$ value of the curve with the lowest E_{max} (control curve, $p\mathrm{EC}_{50} = 6.5$):

$$LOGK1^* = p\mathrm{EC}_{50\,-\,\mathrm{control}} \tag{15}$$

The initial estimate of log τ (0) is calculated as:

$$LOGTAU^* = p\mathrm{EC}_{50\,-\,\mathrm{control}} - LOGK1^* \tag{16}$$

The initial estimate of log γ (1.15) is approximated by the log RA_{obs} value estimated in the presence of the highest concentration of modulator ($\mathrm{RA}_{A\text{-}max}$):

$$LOGGAMMA^* = \mathrm{RA}_{A\,-\,max} \tag{17}$$

The initial estimate of log α (0.4) is calculated as:

$$LOGALPHA^* = \frac{p\mathrm{EC}_{50-A-max} - p\mathrm{EC}_{50-control}}{2} \tag{18}$$

in which, $p\mathrm{EC}_{50\text{-}A\text{-}max}$ denotes the negative logarithm of the EC_{50} value measured in the presence of the highest concentration of the modulator. The initial value of log K_2 is estimated as the negative log concentration of modulator that caused a two-fold change in the RA value (–log A_2), which is approximately equal to –log 10^{-6} M (i.e., 6):

$$LOGK2^* = -\log \mathrm{A}_2 \tag{19}$$

The initial estimate of M_{sys} (208) is assigned a value equal to the largest E_{max} value (E_{max}'):

$$MSYS^* = E_{max}' \quad (20)$$

Finally, the initial estimate of the transducer-slope factor m is assigned a value of 1:

$$M^* = 1 \quad (21)$$

Having assigned the initial parameter estimates, it necessary to define their constraints for global nonlinear regression analysis in Prism. The variable *LOGA* is assigned a value equal to the column heading (see above). All the other parameter estimates are designated as shared parameters. The preparation for global nonlinear regression is complete, and the best fitting parameters of the model can be obtained by initiating the regression.

When this was done, we obtained parameter estimates and a good fit to the data in Fig. 2a. The standard error in the estimate of log α, however, was colossal (log α, -1.20 ± 18). To establish the accuracy and the correlation among parameter estimates, we repeated the analysis several times, each time constraining log α to a constant over the range −3 to 2 in half log increments. Figure 2b summarizes the parameter estimates from this analysis. It can be seen that the residual sum of squares (*SS*) reaches a minimum when the value of log $\alpha \leq 0.5$. Over this range, the estimates of all of the other parameters are constant and the residual *SS* is at a constant minimum. Hence, despite our inability to estimate log α with any accuracy, it is possible to estimate the other parameters. These are log K_1, 6.31 ± 0.094; log K_2, 4.70 ± 0.068; log γ, 1.24 ± 0.076; log τ, -0.068 ± 0.089; m, 1.19 ± 0.092, and M_{sys}, 214 ± 3. Although we are unable to obtain individual estimates of α and β, it is possible to estimate their product γ. The theoretical curves in Fig. 2a represent the least squares fit of (6) to the data.

It is impossible to obtain individual values for α and β because the change in the output of model caused by reducing the value of one parameter (α) can be eliminated by increasing the value of the other (β) so that the model behaves the same as long as the product of the constants (γ) remains at the best fitting value. In this situation, it is not uncommon for a nonlinear regression application to fail to converge on a solution. If this happens, it is necessary to constrain log α to a constant value and determine if the regression routine converges on a solution. If so, one needs to explore parameter space to identify the domain of log α values for which the residual *SS* is a minimum as illustrated in Fig. 2b. Having found this domain, one can simply constrain log α to any value within this domain, and determine the best estimates of the other shared parameters. Obviously, in this situation it is impossible to estimate α. While the process may seem complicated, this type of analysis is frequently done in other branches of mathematics in one step on the computer. As the appreciation of the analysis of allosteric interactions increases, it is likely that computer applications will be developed that do the requisite searching of parameter space automatically, so that the operator does not have to do this manually in repetitive trials of regression analysis.

It is possible to estimate γ and the affinity constant of the allosteric modulator using a much simpler approach based on RA analysis. Recall that when the Hill slopes of the concentration-response curves are similar to one, it is possible to estimate the RA values from the EC_{50} and E_{max} values using (14). These log values (RA_{obs}) are plotted against the modulator concentration in Fig. 2c. A simple nonlinear regression analysis is done to fit the following equation to the data:

$$\log \mathrm{RA} = \log\left(\frac{1 + A\gamma K_2}{1 + AK_2}\right) \tag{22}$$

When the regression analysis is done with Prism, the corresponding log form of (22) is used:

$$Y = \log\left(\frac{1 + 10^{\wedge}(X + LOGGAMMA + LOGK2)}{1 + 10^{\wedge}(X + LOGK2)}\right) \tag{23}$$

In this equation, X represents the log concentration of the modulator (log A), and Y represents the corresponding log RA value. The parameters *LOGGAMMA* and *LOGK2* refer to log γ and log K_2 as described earlier. The initial estimates of these parameters are calculated in the same manner as that described earlier in (17) and (19). Regression analysis yields estimates of log K_2, 4.61 ± 0.086 and log γ, 1.39 ± 0.058. These are similar to those estimated above using global nonlinear regression analysis on the primary data of the concentration-response curves shown in Fig. 2a. It is also possible to estimate the RA values using a procedure described previously, which does not depend on the Hill slopes being equivalent to one (4). Regression analysis of these values yielded the following estimates: log K_2, 4.65 ± 0.085 and log γ, 1.27 ± 0.055. The best fit of (22) to the two estimates, RA and RA_{obs}, is illustrated in Fig. 2c. As the concentration of modulator increases, the function asymptotically approaches the maximum RA value, which is equivalent to γ.

2.2.2. Negative Allosteric Modulation at the mGluR1a

An example of the analysis of negative allosteric modulation is illustrated in Fig. 3. These results show the effect of 4-[1-(2-fluoropyridin-3-yl)-5-methyl-1*H*-1,2,3-triazol-4-yl]-*N*-isopropyl-*N*-methyl-3,6-dihydropyridine-1(2*H*)-carboxamide (FTIDC) on the Ca^{2+} response to L-glutamate in CHO cells expressing the human mGluR1a. The mean data points have been estimated

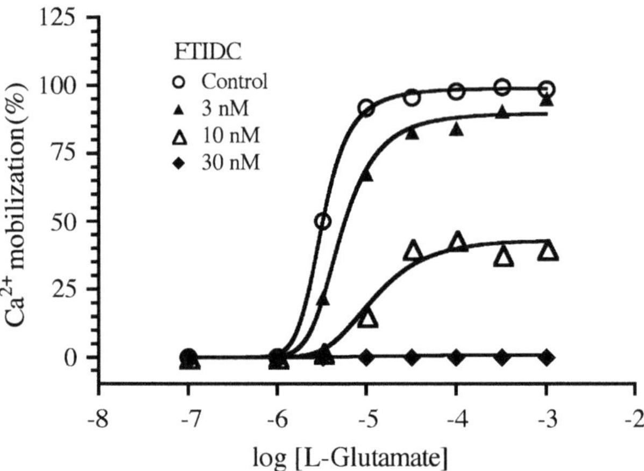

Fig. 3. Analysis of the allosteric effect of FTIDC on the Ca^{2+} response to L-glutamate at the mGluR1a. The effects of increasing concentrations of FTIDC on the Ca^{2+} response to L-glutamate in CHO cells expressing the human mGluR1a. The mean response values have been estimated from a published figure of the data by Suzuki et al. (13). The theoretical curves represent the least squares fit of (6) to the data.

Table 2
Effect of FTIDC on the concentration-response curve of L-glutamate for eliciting Ca^{2+} mobilization in CHO cells expressing mGluR1a

Condition	E_{max}	pEC$_{50}$	Hill slope	Log RA$_{obs}$[a]
Control	97.2	5.49	3.02	0
FTIDC (3 nM)	89.2	5.24	1.97	-0.29
FTIDC (10 nM)	40.5	4.94	3.56	-0.93
FTIDC (30 nM)	na	na	na	na

The parameters were estimated from the data in Fig. 3
[a]The RA$_{obs}$ values were calculated from the E_{max} and EC$_{50}$ values using (22)

from a graph of the data published by Suzuki et al. (13). The analysis of the data is analogous to that described for the positive modulator shown in Fig. 2.

To begin the process, the EC$_{50}$ and E_{max} values of the concentration-response curves are determined using the dose-response equation with a variable slope in the Prism application. Table 2 shows the estimates from this analysis as well as the RA values that were calculated from the parameter estimates using (14). Global nonlinear regression analysis is done using (2), (4), and (6) as described earlier in connection with the analysis of the data in Fig. 2a. Initial parameter values for the regression analysis are estimated using (15)–(21), which yields, *MSYS**, 97; *LOGK1* *, 5.5; *LOGK2* *, 8.5; *LOGALPHA* *, −0.3, and *M**, 1. The value of γ is constrained to zero for the regression analysis, because there is no response in the presence of the highest concentration of the FTIDC. This is done by constraining *LOGGAMMA* to an arbitrarily large negative value (−20), because the actual logarithm of zero is negative infinity. Regression analysis yielded an estimate of log α, 0.17 ± 0.63. Taking the antilogs yields an estimate of 1.5 for α and an SEM corresponding to a 4.3-fold variation. Thus, the data are not inconsistent with the postulate that FTIDC has no affect on the observed affinity of L-glutamate, but causes a concentration-dependent decrease in its observed intrinsic efficacy. The data were reanalyzed, constraining log α to zero (i.e., $\alpha = 1$) to reflect this hypothesis, and the analysis yielded the following parameter estimates log K_1, 5.82 ± 0.17; log τ, 0.16 ± 0.06; log K_2, 7.71 ± 0.18; m, 8.5 ± 2.6, and M_{sys}, 103 ± 2.0. There was no significant increase in residual error when the data were analyzed constraining the value of log α to 0. In other words, under the latter condition, the model provided a good fit to the data. In any case, there were little differences in parameter estimates between the two analyses.

2.2.3. Negative Allosteric Modulation at the M2 Muscarinic Receptor

Another example of the analysis of negative allosteric modulation is shown in Fig. 4a. Here, the effects of gallamine on the inhibition of adenylate cyclase activity elicited by the highly efficacious muscarinic agonist, oxotremorine-M, were measured in homogenates of the rat myocardium (14). This response is known to be mediated by the M_2 muscarinic receptor (15). The data were analyzed by nonlinear regression analysis using the variable slope

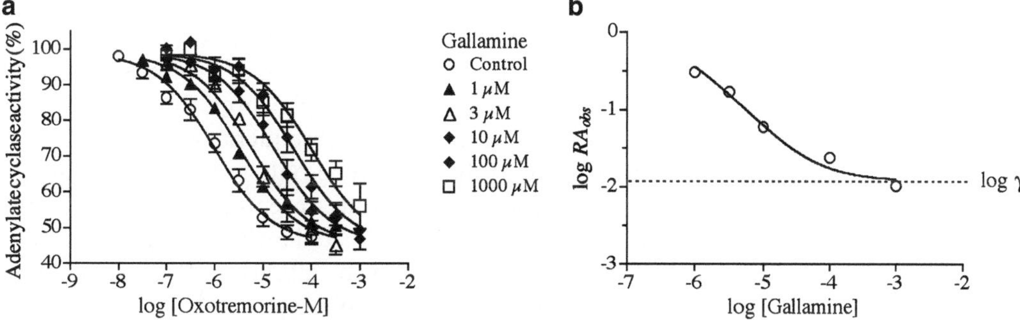

Fig. 4. Analysis of the allosteric effect of gallamine on M_2 muscarinic receptor-mediated inhibition of adenylate cyclase activity in the rat myocardium. (**a**) The effects of increasing concentrations of gallamine on the concentration-response curve to oxotremorine-M was measured in homogenates of the rat myocardium. The data are from Ehlert (14). (**b**) The affinity constant of the allosteric modulator K_2 and the product of its affinity and efficacy modulation (γ) can be estimated by analysis of the RA values. The RA_{obs} values were estimated from the data in (**a**) using (14). The theoretical curves represent the least squares fit of (22) to the data. The *dashed line* indicates the asymptote for the maximum log RA value, which is equivalent to log γ.

Table 3
Effect of gallamine on oxotremorine-M-mediated inhibition of adenylate cyclase activity in the rat myocardium

Condition	pEC_{50}	Log RA[a]
Control	6.00	0
Gallamine 1 μM	5.48	-0.52
Gallamine 3 μM	5.23	-0.77
Gallamine 10 μM	4.78	-1.23
Gallamine 100 μM	4.38	-1.63
Gallamine 1,000 μM	4.02	-1.99

The parameters were estimated from the data in Fig. 4
[a]The RA_{obs} values were calculated from the E_{max} and EC_{50} values using (22)

dose-response equation in Prism to estimate the E_{max} and EC_{50} values of the concentration-response curves, and these estimates are listed in Table 3. The E_{max} values have been normalized relative to adenylate cyclase activity measured in the absence of oxotremorine-M. Gallamine had no effect on adenylate cyclase activity in the absence of oxotremorine-M. It is clear from Fig. 4a and the summary of the parameters in Table 3 that gallamine has little or no effect on the E_{max} value of oxotremorine-M. In this situation, the best fitting values of K_1, τ and α in (6) represent infinite domains that are unbounded on at least one side of the parameter scale. While the data could be analyzed with (6), there are simpler methods.

First, let us first review what we can glean from the data. It is known that oxotremorine-M is highly efficacious, so it is possible that gallamine could cause some inhibition of the intrinsic efficacy of the oxotremorine-M-receptor complex even though it appears to have no effect on the E_{max}. A subsequent binding and functional analysis showed that gallamine only reduces the affinity of oxotremorine-M for M_2 muscarinic receptor (1). But from the data in Fig. 4a, all that can be estimated is the affinity of gallamine and the product of its change in the affinity and efficacy components of the agonist-receptor complex (γ).

The simplest way to analyze the data is to use the approach outlined in connection with the analysis shown in Fig. 2c. First, the RA values are calculated for each concentration of the modulator using (14) as described above. This calculation is accurate if the Hill slopes are similar to one or if there is little change in the Hill slope with addition of modulator. These estimates are listed in the Table 3. In this case, the RA value is simply the reciprocal of the shift in the concentration-response curve caused by gallamine because gallamine has little effect on the E_{max}. A plot of the log RA value against the log modulator concentration is shown in Fig. 4b. Regression analysis of the data according to the (22) yields the best fitting curve shown in Fig. 4b and parameter estimates of log K_2, 6.27 ± 0.08 and log γ, 1.92 ± 0.09.

Another way to analyze the data is to fit all of the concentration-response curves simultaneously to the following equation:

$$R = T - \frac{D^n * E_{max}}{D^n + \left(\frac{EC_{50}}{RA}\right)^n} \quad (24)$$

in which, R represents the response, T represents adenylate cyclase activity in the absence of the agonist (D), n represents the Hill slope, and RA is defined by taking the antilog of both sides of (22). The general strategy for fitting the data involves global nonlinear regression analysis using an approach similar to that described in connection with the analysis of the data in Fig. 2a with (6). First, the data are entered into columns, with replicate measurements entered into sub-columns. The log concentration of allosteric modulator is used as the column title. Thus, for the control curve and the curves measured in the presence of 1, 3, 10, 30, and 100 μM gallamine, the columns are entitled with −20, −6, −5.52, −5, −4.52, and −4. Next, the requisite equations are entered into Prism in logarithmic form. The two equations define log RA (22) and the response (24):

$$LOGRA = LOG\left(\frac{1+10\wedge(LOGA+LOGGAMMA+LOGK2)}{1+10\wedge(LOGA+LOGK2)}\right) \quad (25)$$

$$Y = T - \frac{((10\wedge X\wedge)N) * EMAX}{((10\wedge X)\wedge N) + (10\wedge(LOGEC50 - LOGRA))\wedge N} \quad (26)$$

The log variables, *LOGA*, *LOGK2*, *LOGGAMMA*, *LOGEC50* and *LOGRA*, correspond to the log values of the A, K_2, γ, EC_{50}, and RA, respectively, in (22) and (24). The variables *T*, *N*, and *EMAX* represent T, n, and E_{max} in (24) (Note that (25) is nearly identical to (23) except that the variables *Y* and *X* in (23) have been changed to *LOGRA* and *LOGA* in (25) because (23) is used as a regression equation, whereas (25) defines the variable *LOGRA* in regression (26)). The initial estimate of log γ (*LOGGAMMA**) is assigned a value equivalent to the log RA value measured at the highest concentration of gallamine (−1.8). *LOGK2** is estimated as the -log concentration of *A* that caused a twofold increase in RA (6.0). *EMAX** and *LOGEC50** are estimated as the corresponding

E_{max} and log EC_{50} values of the control concentration-response curve. Finally, T^* is estimated as adenylate cyclase activity in the absence of the agonist, and N^* is assigned a value of 1.

During global nonlinear regression analysis, the parameter *LOGA* is constrained to the values of the column titles, and all of the other parameters are shared. The input into Prism is now complete and global nonlinear regression analysis can be initiated, which yielded the following parameter estimates: log K_2, 6.26 ± 0.09; log γ, −1.88 ± 0.08; T, 99.1 ± 1.1; log EC_{50}, −6.00 ± 0.07; E_{max}, 53 ± 2.2%; and n, 0.73 ± 0.06. The estimates of log K_2 and log γ are nearly identical to those estimated from the simple regression of the log RA values in Fig. 4b.

2.2.4. Simulation of Negative Allosteric Modulation

An example of negative allosteric modulation associated with a reduction in both the observed potency and E_{max} of the agonist is shown in Fig. 5a. These data were generated using (6) followed by the addition of random error having a mean of zero and a maximum of ±10%. The initial analysis involves estimating the E_{max}, EC_{50} and Hill slope values of the data using the sigmoidal dose-response equation with a variable slope. The resulting parameter estimates are listed in Table 4. The control concentration-response curve exhibits a Hill slope of 1.67, whereas that measured in the presence of the highest concentration of the modulator (10 mM) exhibits a lower slope of 1.29. Under these conditions, the simple calculation of RA_{obs} using the EC_{50} and E_{max} values of the agonist has moderate error. Nonetheless, these values were calculated and are listed in Table 4. It is also possible to calculate precise estimates of RA using methods described previously, and these are also listed in Table 4.

Ultimately, the data in Fig. 5a are analyzed by global nonlinear regression analysis according to (6) with the variable p and q defined by (2) and (4). The three equations are entered into Prism in log form as shown in (11)–(13). The data are entered into a data sheet in Prism as described above in connection with the analysis of the data in Figs. 2 and 3. The titles of the data columns are labeled with the log concentration of the modulator. The control column of data is labeled with −20, whereas those for the 0.01, 0.1, 1.0, and 10 mM concentrations of modulator are labeled with −5, −4, −3, and −2, respectively.

The initial estimate of log K_1 and the other parameters can be gleaned by careful analysis of the curves shown in Fig. 5a. At least part of the modulation involves a reduction in intrinsic efficacy because the E_{max} of the agonist is clearly depressed at the two highest concentrations of modulator. At the second lowest concentration of the modulator (0.1 mM), negative efficacy modulation probably also occurs, but it is not associated with a reduction in the E_{max} of the agonist. This implies that the agonist is highly efficacious and elicits a maximal response at a submaximal level of

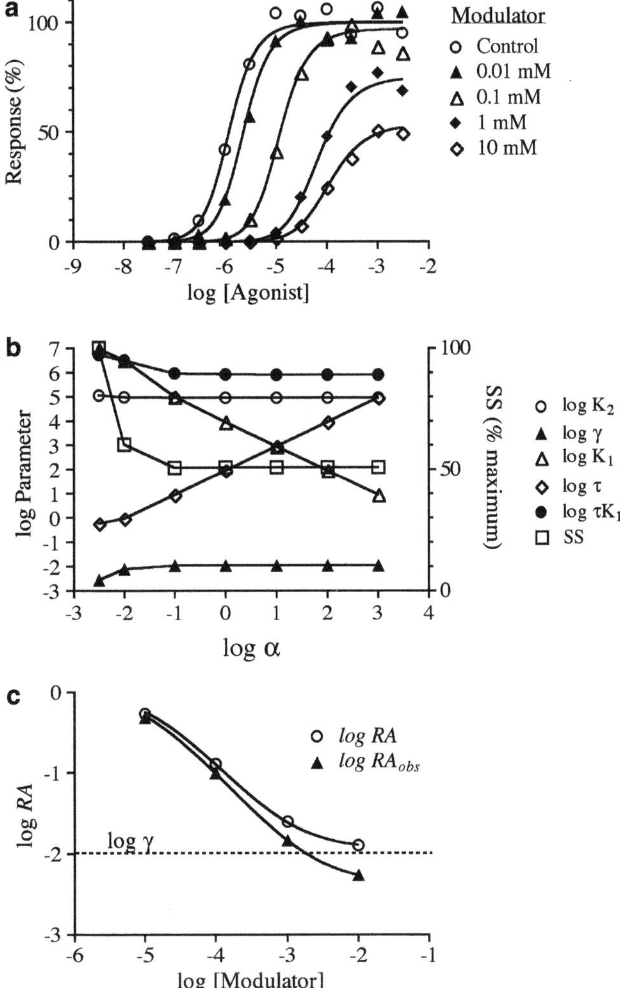

Fig. 5. Simulation of the effects of a negative allosteric modulator on the response to an agonist. (**a**) The theoretical data were generated using (6) with the following parameters: log K_1, 5; log K_2, 5; log τ, 1; log α, −1; log γ, −2; m, 1.8 and M_{sys}, 100. A 10% random error was added to the theoretical values to generate the simulated data. The theoretical curves represent the least squares fit of (6) to the data. (**b**) Summary of global nonlinear regression analysis of the data in (**a**). The logarithm of the parameter α was constrained to various values within the domain −2.5 – 3, and the estimates of the other parameters that yielded a least squares fit of (6) were determined. When the value of log α is in the range log $\alpha \geq -1$, the residual *SS* is a minimum, and the values of the other parameters are constant. (**c**) The affinity constant of the allosteric modulator K_2 and the product of its affinity and efficacy modulation (γ) can be estimated by analysis of the RA values. The RA values were estimated from the data in (**a**) using (14) (RA_{obs}) or by using a method described previously (RA) (4). The theoretical curves represent the least squares fit of (22) to the data. The *dashed line* indicates the asymptote for the maximum log RA value, which is equivalent to log γ.

Table 4
Simulation of the effect of a negative allosteric modulator on the concentration-response curve of an agonist

Condition	E_{max}	pEC_{50}	Hill slope	Log RA_{obs}[a]	Log RA[b]
Control	102	5.91	1.66	0	0
Modulator (0.01 mM)	99.1	5.61	1.65	-0.305	-0.264
Modulator (0.1 mM)	92.1	4.95	1.69	-0.998	-0.890
Modulator (1 mM)	73.9	4.22	1.51	-1.83	-1.60
Modulator (10 mM)	51.3	3.94	1.29	-2.26	-1.89

The parameters were estimated from the data in Fig. 5
[a]The RA_{obs} values were calculated from the E_{max} and EC_{50} values using (22)
[b]The RA values were calculated as described previously (4)

receptor occupancy in the absence of the modulator. If so, then the control $p\text{EC}_{50}$ value of the agonist overestimates its log affinity constant ($\log K_1 < p\text{EC}_{50}$). If the entire effect of the modulator involves efficacy modulation, then the $p\text{EC}_{50}$ measured in the presence of the highest concentration of the modulator would be a good approximation of $\log K_1$. But, it is unknown how much of the effect of the modulator involves efficacy modulation. A suitable compromise for estimating $\log K_1$ is to take the average of control $p\text{EC}_{50}$ ($p\text{EC}_{50\text{-control}}$, 5.9) and that measured in the presence of the highest concentration of the modulator ($p\text{EC}_{50\text{-max}}$, 3.9):

$$LOGK1^* = \frac{p\text{EC}_{50\,-\,\text{control}} + p\text{EC}_{50\,-\,\text{max}}}{2} \quad (27)$$

This calculation yields about –5 for the estimate of $LOGK1^*$. The initial estimate of $\log \gamma$ is equal to the maximum observed RA value (–1.8):

$$LOGGAMMA^* = \log \text{RA}_{\text{obs}\,-\,\text{max}} \quad (28)$$

The initial estimate of Log K_2 can be determined by estimating the negative log concentration of modulator that reduces the RA_{obs} value by one-half (i.e., $LOGK_1^* = 5$). The initial estimates of M^* and M_{sys} are simply equivalent to the Hill slope and E_{max} of the control concentration-response curve (i.e., $M^* = 1.7$; $MSYS^* = 100$). Finally, initial estimate of $\log \tau$ is calculated from (16) (i.e., $LOGTAU^* = 1.0$).

After the initial parameter estimates are entered into Prism, the variable, $LOGA$, is constrained to the values of the data column titles, and all of the other parameters are shared. Initiating regression analysis yielded a good fit of the model to the data, but the estimates of log α (−0.73 ± 1.25), log K_1 (4.72 ± 1.20) and log τ (1.22 ± 1.14) exhibited substantial variation. To explore parameter space and examine the correlation between parameter estimates, we set log α to a constant over the range −2.5 to 3 and estimated the values of the other parameters that minimized the residual *SS*. Figure 5b shows the results of this analysis. When the value of log α was constrained to a constant over the domain log $\alpha \geq -1$, then the residual *SS* was at a minimum. Over the same domain, the values of log K_2, log γ, m, and M_{sys} were also constant. As the value of log α increased, however, the best estimate of log K_1 decreased while that for log τ increased. As a consequence, the log of the product τK_1 remained constant. Thus, it was possible to obtained a least squares fit to the data by constraining log α to a constant ≥ −1 during regression analysis. This strategy yielded the following estimates, log K_2, 4.96 ± 0.049; log γ, −1.96 ± 0.086; m, 1.83 ± 0.17 and M_{sys}, 101.8 ± 1.5. These values are similar to those used in the generation of the theoretical data in the first place, log K_2, 5; log γ, −2; m, 1.8; M_{sys}, 100. Although it was impossible to estimate log τ, log K_1, and log α, the values used in the simulation of the data (1, 5 and −1, respectively) are in the domain of values that generated a minimum residual *SS* during regression analysis.

The simplest way to estimate log K_2 and log γ is to analyze the plot of log RA_{obs} against log A (Fig. 5c) following the procedure described above for the analysis of the data in Figs. 2c and 4b. This method involves nonlinear regression analysis using (22), which yields estimates of log K_2, 4.98 ± 0.01 and log γ, −2.35 ± 0.02. The estimate of log γ, however, differs a little from that estimated above (1.96 ± 0.086) and that used to simulate the data (−2). This is because efficacy modulation of the response to a highly efficacious agonist often involve a change in the Hill slope of the concentration-response curve when the transducer function has a slope factor greater than one ($m > 1$). In this situation, the simple calculation of RA_{obs} using (14) has moderate error, and using the approach described previously provides an accurate estimate of RA (4). When the accurate estimates of log RA were analyzed, regression analysis yields estimates of log K_2, 4.88 ± 0.02 and log γ, −1.94 ± 0.02 that are in close agreement with those used to simulate the data.

2.3. Analysis of Allosterism in Radioligand Binding Experiments

Radioligand binding experiments can be used to determine the affinity of the modulator for its allosteric site on the receptor (K_2) and its cooperative effects on the binding of the orthosteric ligand (α). An advantage is the simplicity of the system because it seems likely that the only way the modulator could influence the binding of the radioligand is by directly binding to the receptor, whereas in functional experiments, the putative modulator could inhibit or amplify the signaling process downstream from the receptor yielding artifactual allosteric effects. An obvious disadvantage is that the approach does not directly provide information about the allosteric modulation in the activity of the receptor population. Our analysis focuses on measuring the affinity constant of the modulator (K_2) and its cooperative effect on the binding of the radioligand (α_L) or another orthosteric ligand (α_X).

2.3.1. The Modulator-Radioligand Interaction

The characterization of the interaction between an orthosteric radioligand and allosteric modulator at equilibrium involves estimating the affinity constants of the radioligand (K_1) and modulator (K_2) and the cooperativity constant (α_L). To estimate these parameters, the binding of the radioligand is measured at various concentrations in the absence and presence of various concentrations of the allosteric modulator. The simulated data in Fig. 6 are an example of the type of measurements required for this analysis. These data were generated by calculating theoretical binding values using (8) with p defined by (2) and then adding a random error having a mean of zero and a range of ±10%. The theoretical data are listed in tabular form in Table 5, and they are plotted in Fig. 6a as a series of saturation curves measured in the presence of increasing concentration of the modulator. Each saturation curve represents a column of data in Table 5. Another way to look at the same data is to plot them as a series of inhibition curves, each at a fixed concentration of the radioligand and various concentrations of the modulator (Fig. 6b). Each inhibition curve represent of row of data in Table 5. In both cases, the data are analyzed by global nonlinear regression analysis using (8) with the variable p defined by (2). The difference in the two analyses depends on whether the log radioligand (log L) or modulator (log A) is treated as the independent variable.

For the analysis of saturation curves (Fig. 6a), the data are entered into Prism in a manner analogous to that described above in connection with Figs. 2 and 3. The log concentration of radioligand is the independent variable, and the binding values for each saturation curve are entered into different columns in the datasheet. The titles of the columns are labeled with the log concentration of the modulator. The control column of data is labeled with –20, whereas those for the 0.01, 0.1, 1.0, and 10 mM concentrations of modulator are labeled with –5, –4, –3, and –2, respectively. The regression equation used to define the variable log p is entered (11) followed by the log form of (8):

$$Y = \frac{10^{\wedge} X + BMAX}{10^{\wedge} X + 10^{\wedge} (LOGP - LOGK1)} \tag{29}$$

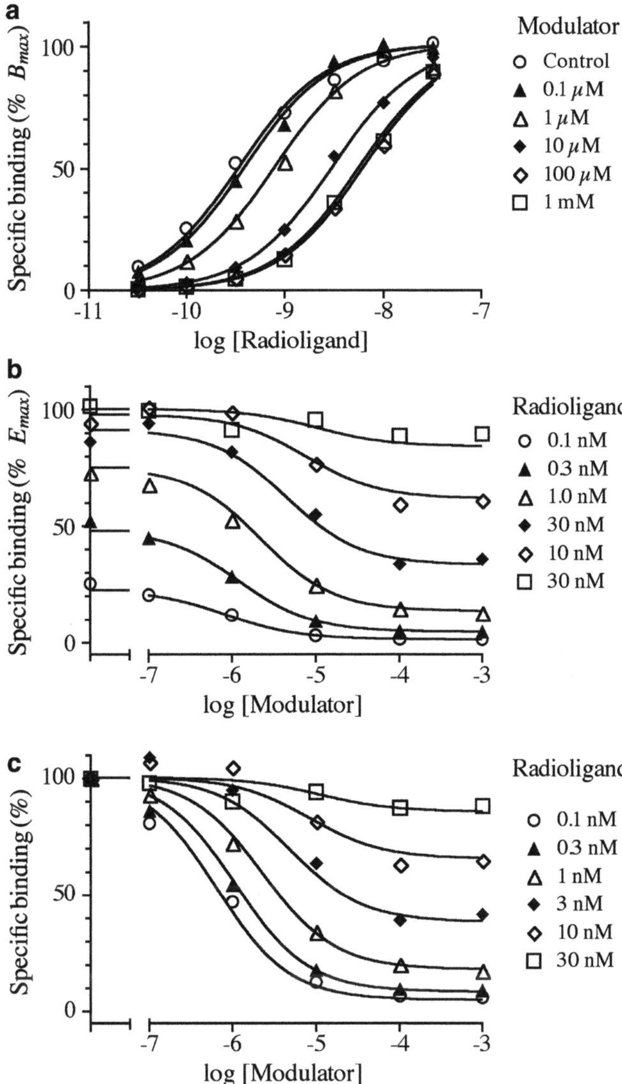

Fig. 6. Simulation of the effect of a negative allosteric modulator on the binding of a radioligand. The binding data were simulated using (8) and the following parameters: B_{max}, 100; log K_1, 9.5; log K_2, 6.2 and log α, −1.3. A 10% random error was added to the theoretical data to generate the simulated data shown in the figure. The simulated data are listed in Table 5. (**a**) The simulated data have been plotted as a series of saturation curves for the radioligand measured in the absence (control) and presence of increasing concentrations of the modulator. The theoretical curves represent the best fit of (8) to the data with *L* as the independent variable. (**b**) The same simulated data have been plotted as a series of inhibition curves for the modulator, each measured at a different fixed concentration of the radioligand. The theoretical curves represent the least squares fit of (8) to the data with *p* defined by (2). *A* in (2) is the independent variable. (**c**) The inhibition curves in (**b**) have been normalized relative to the binding measured in the absence of *A*. The theoretical curves represent the least squares fit of (9) to the data with *p* defined by (2).

Table 5
Simulated radioligand binding data

Radioligand	Log [modulator]					
(nM)	$-\infty$	−7	−6	−5	−4	−3
0.032	9.64	7.73	4.15	1.07	0.54	0.54
0.10	25.4	20.6	12.0	3.24	1.74	1.63
0.32	52.3	45.1	28.5	9.44	5.12	4.83
1.0	73.0	67.9	52.7	24.86	14.74	12.8
3.2	86.3	94.1	82.0	55.04	33.9	36.0
10	94.3	101	98.8	76.92	59.5	61.2
32	102	99.6	91.5	95.65	88.8	89.6

The theoretical data were generated using (8) with p defined by (2). A random 10% error was then added to the data

In the latter equation, X denotes the log concentration of the radioligand (log L), Y the specific binding of the radioligand, and $BMAX$ the maximal binding capacity. The initial value for log K_1 ($LOGK1$ *) can be estimated from the plot in Fig. 6a as the log of the reciprocal of the molar concentration of radioligand required for half-maximal receptor occupancy by the radioligand in the absence of modulator (i.e., log $1/[3.2\times10^{-10}]=9.5$). The initial value of B_{max} ($BMAX$*) can be easily estimated as the maximal plateau in the binding values (i.e., 100). The initial value of log α is estimated as the inverse of the maximal shift (about tenfold) in the saturation curve caused by the highest concentration of the modulator (i.e., log $1/10=-1$). Next, $LOGA$ is constrained to the values of the column titles, and all of the other parameters are shared. Initiation of regression analysis yielded the following estimates for the simulated data in Fig. 6a: B_{max}, 101 ± 1; log K_1, 9.46 ± 0.03; log K_2, 6.18 ± 0.06 and log α, -1.27 ± 0.03. These are similar to those used to simulate the data in the first place (B_{max}, 100; log K_1, 9.5; log K_2, 6.2 and log α, −1.3).

When the data are analyzed as a series of inhibition curves (Fig. 6b), the independent variable is the allosteric modulator concentration (log A). The various log concentrations of the modulator are entered into a datasheet as the X values, beginning with −20 as an approximation for the log of zero followed by −7, −6, −5, −4, and −3. The corresponding binding values of the radioligand are entered into columns with the titles of the columns labeled with the log molar concentration of the radioligand used in the inhibition experiment. These values are −11, −10.5, −10, −9.5, −9, −8.5, −8, −7.5, and −7. Although the equations for regression are the same as those used in the analysis of the plot in Fig. 6a, they must be changed to reflect the difference in the independent variable:

$$P = \frac{1+10\text{^}(X + LOGK2)}{1+10\text{^}(X + LOGALPHA + LOGK2)} \quad (30)$$

$$Y = \frac{10\text{^}\,LOGL + BMAX}{10\text{^}\,LOGL + 10\text{^}(LOGP - LOGK1)} \quad (31)$$

In (30), X denotes the log concentration of the modulator, and in (31), *LOGL* denotes the log concentration of the radioligand. For this regression, the values of *LOGL* are constrained to those of the column titles, and all of the other parameters are shared. Initiation of regression analysis yields the following parameters estimate: B_{max}, 101 ± 1; log K_1, 9.46 ± 0.03; log K_2, 6.18 ± 0.07 and log α, -1.26 ± 0.04. These are about the same as those estimated when the data were analyzed as a series of saturation curves.

Finally, the data can be analyzed as a series of normalized inhibition curves as shown in Fig. 6c by global nonlinear regression analysis using (9) with p defined by (2). The analysis is done in a manner analogous to that described above for Fig. 6a, b and yields estimates of the maximum binding in the absence of allosteric modulator (T, 101 ± 1) as well as the other parameters: log K_1, 9.72 ± 0.14; log K_2, 6.36 ± 0.08 and log α, -1.47 ± 0.13.

The format of the experiment in Fig. 6 clearly shows that the effect of the allosteric modulator reaches a limit at high concentrations. When the data are displayed as a series of saturation curves, it is clear that the modulator only shifts the radioligand occupancy curve to the right by a factor equivalent to $1/\alpha_L$. In contrast, a competitive inhibitor would cause the curve to shift to the right in a manner proportional to the inhibitor concentration. When displayed as a series of inhibition curves, the allosteric modulator is unable to displace the radioligand completely at high concentrations because both ligands can bind to the receptor simultaneously.

If the affinity constant of the radioligand is already known, it is possible to estimate the parameters of the allosteric modulator from a single inhibition curve. This may be premature unless there is confirming evidence of the type shown in Fig. 6 that the modulator has been fully characterized and proven to be allosteric. Figure 7 shows an example of the inhibition of the binding of the muscarinic antagonist $[^3H]$*N*-methylscopolamine ($[^3H]$NMS) by the allosteric modulator, gallamine. The measurements were made on CHO cells stably expressing the human M_2 muscarinic receptor (1). Gallamine has previously been shown to modify the binding

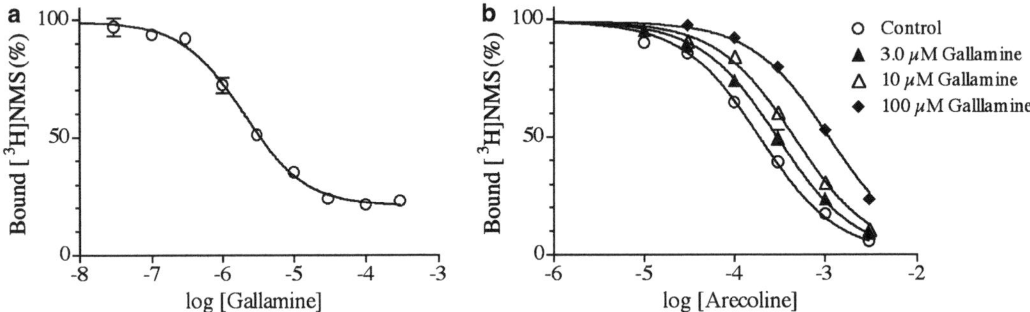

Fig. 7. Analysis of the allosteric interaction between the binding of the arecoline and gallamine to the human M_2 muscarinic receptor expressed in CHO cells. (**a**) The inhibition of the binding [^{3}H]NMS by various concentrations of the allosteric modulator gallamine was measured. The concentration of [^{3}H]NMS was 1 nM. The theoretical curve represents the least squares fit of (32) to the data. (**b**) The competitive inhibition of the binding of [^{3}H]NMS by the orthosteric agonist arecoline was measured in the absence (control) and presence of various concentrations of gallamine. Each competition curve has been normalized relative to the binding measured in the absence of arecoline. The data are from Ehlert and Griffin (1).

of orthosteric ligands to muscarinic receptors allosterically and to cause an allosteric inhibition of the functional responses to muscarinic agonists (10, 16). A simple way to analyze the data in Fig. 7a is with the following one-site inhibition curve:

$$Y = \text{Top} - \left(\frac{\text{Top} - \text{Bottom}}{1 + \frac{\text{IC}_{50}}{A}} \right) \tag{32}$$

In this equation, A denotes the modulator concentration, Top, the binding of radioligand in the absence of A, Bottom, the plateau level of binding that occurs at high concentrations of A, and IC_{50}, the concentration of A causing half-maximal inhibition of binding. Analysis of the data in this manner yields the following parameters: log IC_{50}, -5.71 ± 0.04; Top, $99.3 \pm 1.1\%$ and Bottom, $20.9 \pm 1.2\%$. When the latter is normalized relative to Top (100*Bottom/Top) a value of 21.05% is obtained. When expressed as a fraction (0.21), this parameter (Y') is related to the cooperativity factor (α_L) according to the following equation (5):

$$\alpha_L = \frac{1 + LK_1(1 - Y')}{Y'} \tag{33}$$

Substituting in 6.9×10^9 M^{-1} for the affinity constant of [^{3}H]NMS (K_1), 0.46×10^{-9} M for the concentration of [^{3}H]NMS (L) yields an estimate of 0.060 for α_L (log $\alpha_L = -1.22$). It is also possible to estimate the affinity constant of gallamine (K_2) for the allosteric site using the following equation (5):

$$K_2 = \frac{1}{\mathrm{IC}_{50}\left(\dfrac{1 + X\alpha_L K_1}{1 + XK_1}\right)} \tag{34}$$

Substituting in 0.060 for α_L and the appropriate values for the IC_{50} of gallamine and K_1 yields an estimate of $1.8 \times 10^6\ \mathrm{M}^{-1}$ for K_2 ($\log K_2 = 6.25$).

Another way to analyze the data is to use (9) for the regression analysis with parameter p defined by (2) and the value of log K_1 constrained to a constant (i.e., a previously determined estimate).

2.3.2. The Interaction of a Modulator with a Nonlabeled Orthosteric Radioligand

Binding assays are not limited to the analysis of the modulator-radioligand interaction; one can also investigate the interaction between a nonlabeled orthosteric ligand and an allosteric modulator. This analysis has been described in detail by Lazareno and Birdsall (7). A simple variation of their analysis is shown in Fig. 7b. Here, the competitive inhibition of the binding of [^{3}H]NMS by the agonist arecoline was measured in the absence and presence of various concentrations of gallamine. The measurements were made on CHO cells stably expressing the human M_2 muscarinic receptor (1). The concentration of [^{3}H]NMS was 1 nM for all of the binding measurements. Each arecoline/[^{3}H]NMS competition curve has been normalized relative to these values. These data can be analyzed with Prism using a logistic equation (variable slope):

$$Y = \mathrm{Top}\left(\frac{1}{1 + \left(\dfrac{X}{\mathrm{IC}_{50}}\right)^n}\right) \tag{35}$$

In this equation, n denotes the Hill slope and the other parameters are defined in relation to (34). For this analysis, the Top and n are shared among the curves. Regression analysis yielded global estimates of 99% and 1.01 for Top and n, respectively. The log IC_{50} values for arecoline in the absence and presence of 3, 10, and 100 μM gallamine translate into gallamine-induced log-shifts in the arecoline/[^{3}H]NMS curve of 0.27, 0.55, and 0.97, respectively. The estimate of the cooperativity between the binding of gallamine and arecoline (α_X) can be calculated using the following equation (1):

$$\alpha_X = \frac{1}{R}\left(\frac{1 - R}{AK_2} + \frac{1 + \alpha_L LK_1}{1 + LK_1}\right) \tag{36}$$

In this equation, R denotes the shift in the arecoline/[^{3}H]NMS competition curve caused by gallamine, and α_X denotes the cooperativity between the binding of gallamine and [^{3}H]NMS. If the estimates of the affinity constants of [^{3}H]NMS (K_1) and gallamine (K_2) and the cooperativity of their interaction are known (α_L), then these values and that of R can be substituted into (36) to estimate α_X. Using this approach for the data in Fig. 7b, an average value of log α_X of –1.97 was estimated.

3. Future Directions

One disadvantage of the radioligand binding methods of analysis described here is that the allosteric interaction must always be studied in the presence of the radioligand. This creates a problem when the allosteric modulator exhibits high negative cooperativity with the radioligand because one must measure binding at very high concentrations of the radioligand to discriminate between competitive inhibition and high negative cooperativity. It may be unfeasible to do this because of high cost and nonspecific binding of the radioligand.

Investigating how an allosteric modulator affects the kinetics of receptor alkylation by a site-directed electrophile that alkylates the orthosteric site does not suffer from these disadvantages (17, 18). The primary interaction is between the orthosteric alkylating agent and the putative allosteric modulator, both of which can be used over a wide range of concentrations, unlike a radioligand. Since the nature of their interaction is preserved through the covalent bond with the receptor, one has only to remove the interacting ligands and estimate the residual unalkylated receptors using a feasible concentration of radioligand. As we have described previously (17, 18), competitive inhibitors and allosteric modulators differ in their pattern of receptor protection.

The method can be adapted to other receptors for which moderately potent, irreversible, orthosteric ligands are available. It should be possible to introduce a reactive electrophilic moiety into the structure of many orthosteric ligands to enable them to bind covalently to their receptor. In addition, mutagenesis could be used to introduce an accessible nucleophilic residue (e.g., aspartic acid or cysteine) into the binding pocket to enable the electrophilic ligand to bind covalently. These methods should have widespread application in drug discovery.

4. Conclusions

Our model for allosteric interactions enables one to measure potential changes in the observed affinity and intrinsic efficacy of a population of receptors. The effect of the modulator is defined by the functions p (change in observed affinity; (2)) and q (change in observed efficacy, (3)). These represent the factors by which agonist affinity and efficacy changes. As the concentration of the modulator increases from zero to receptor saturating, p and q change from their initial value of one to their limiting values of α and β, respectively.

In the analysis of receptor function at a level downstream from receptor activation, it is always possible to obtain accurate estimates of the affinity constant of the modulator (K_2) and product of the modulation in affinity and efficacy (γ), but not necessarily each component (i.e., α and β). In fact, for the examples described in this chapter, it was only possible to make this discrimination for the negative modulator of the mGluR1a receptor. One could estimate α in a radioligand binding assay and then divide the functional estimate of γ by α to obtain β.

With regard to the effect of FTIDC, it caused a complete inhibition of the ability of L-glutamate to activate the mGluR1a receptor, without affecting the affinity of L-glutamate. Suzuki et al. (13) have shown that FTIDC has no effect on the binding of [^{3}H]L-quisqualate to the mGluR1a receptor. These results are consistent with the postulate that FTIDC inhibits agonist efficacy without influencing affinity. The effects of site-directed mutagenesis on the ability of FTIDC to interfere with mGluR1a mediated Ca^{2+} mobilization suggest that FTIDC interacts with the TM domains 6 and 7 near extracellular loop 3. Perhaps it interferes with the coupling of the receptor to G_q without affecting the binding of agonists to the glutamate binding pocket in the large amino terminal domain. If so, the action of FTIDC may not be allosteric in the sense that allosterism is defined as the induction or selection of a conformational change in the receptor that affects the binding of the agonist or the equilibrium between ground and active state of the receptor. Nonetheless, our method for analyzing allosteric interactions is useful for measuring this type of noncompetitive inhibition (5).

We have described a few specific protocols for the analysis of allosterism in functional assays. In cases where the modulator causes a change in the E_{max} or Hill slope of the concentration-response curve, it is best to fit (6) to all of the curves simultaneously using global nonlinear regression analysis. In this way, the model is fitted directly to all of the data. In principle, it is always possible to fit (6) to data obeying the model, but many nonlinear regression applications are not designed to converge on a solution set when the latter is infinite. Consequently, it may be necessary to constrain the value of log α to a reasonable constant and search parameter space to identify the domain of log α where the residual sum of squared deviations is a minimum. This can be done easily as described above.

In situations where the modulator only causes a parallel shift in the concentration-response curve without a concentration-dependent change in the Hill slope, there is no advantage to fitting the data to (6). The reciprocal shift in the concentration-response curve caused by the modulator is equivalent to pq, which is also known as the relative activity of the agonist (RA). In this situation, a modified logistic equation can be used to analyze the data. The modification simply involves introducing the reciprocal of function defined in (22) (RA) as a factor for the EC_{50} value:

$$\text{Response} = \frac{X^n E_{max}}{X^n + \left(\dfrac{1 + AK_2}{1 + A\gamma K_2}\right) EC_{50}{}^n} \qquad (37)$$

The equation can be fitted to all of the concentration-response curves simultaneously by global nonlinear regression analysis to estimate K_2 and γ as well as E_{max}, EC_{50} and the Hill slope (n).

We have always had interest in developing simple methods of analysis that reveal the underlying stimulus and its allosteric modulation in a more intuitive way even though our method may involve fitting a regression equation to calculated parameters instead of the primary data itself. For example, in situations where the Hill slope of the concentration-response curve is equivalent to one, the product of the allosteric change in the affinity and efficacy of agonist (RA) can be easily calculated from the EC_{50} and E_{max} values using (14). A plot of log RA against log A can be analyzed by regression analysis to yield the affinity constant of the modulator and the maximal product of the change in affinity and efficacy (γ). When the transducer function is cooperative ($m > 1$ in the operational modal) an allosteric modulation in the efficacy of the agonist-receptor complex can alter the Hill slope of the concentration-response curve, and the simple calculation of RA has error. The amount of error is usually not substantial as shown by the examples in Figs. 2 and 5. It is possible to obtain an accurate estimate of RA in this case using a nonlinear regression method, and these values can then be analyzed by (22) to obtain estimates of K_2 and γ.

While it may seem that the inability to resolve γ into its α and β components is a limitation in the analysis of receptor function, the parameter γ has unique properties that are more fundamental than either α or β. It represents the ratio of microscopic affinity constants of the allosteric modulator for the active (K_f) and inactive (K_e) states of the receptor ($\gamma = K_f/K_e$) (1). Thus, γ is a pure receptor-dependent property, whereas α and β depend on the receptor as well as the G protein and other factors. If the allosteric modulator selects for a unique active conformation of the receptor that directs signaling through recruitment of arrestin rather than G_i, for example, then estimation of γ is a direct measure of the allosteric modulation in pathway selectivity.

One final issue is that the relationship between an agonist concentration-response curve in the absence and presence of an allosteric modulator is analogous to that between the concentration-response curves of two different agonists. In both cases, the differences between the two concentration-response curves can be attributed to changes in agonist affinity (α) and efficacy (β) or, simply put, to a change in γ. Thus, the methods for the analysis of allosterism in functional assays can also be applied to the analysis of the concentration-response curves of different agonists. In the latter case, the estimate of the relative difference in the product of affinity and efficacy of one agonist expressed relative to that of a standard agonist is known as intrinsic relative activity (RA_i) (19). Like the estimate of γ in allosterism, RA_i also has unique properties at the level of receptor states. It is a relative measure of the microscopic affinity constant of the agonist for the active (K_b) state of the receptor. Thus, it is a pure receptor property and is useful for measuring ligand-directed signaling.

References

1. Ehlert FJ, Griffin MT (2008) Two-state models and the analysis of the allosteric effect of gallamine at the M2 muscarinic receptor. J Pharmacol Exp Ther **325**:1039–1060.
2. Stephenson RP (1956) A modification of receptor theory. Br J Pharmacol **11**:379–393.
3. Furchgott RF (1966) The use of β-haloalkylamines in the differentiation of receptors and in the determination of dissociation constants of receptor-agonist complexes. Adv Drug Res **3**:21–55.
4. Ehlert FJ (2005) Analysis of allosterism in functional assays. J Pharmacol Exp Ther **315**:740–754.
5. Ehlert FJ (1988) Estimation of the affinities of allosteric ligands using radioligand binding and pharmacological null methods. Mol Pharmacol **33**:187–194.
6. Christopoulos A (2000) Quantification of allosteric interactions at G protein-coupled receptors using radioligand binding assays. In: Enna SJ (ed) Current Protocols in Pharmacology. Wiley, New York.
7. Lazareno S, Birdsall NJ (1995) Detection, quantitation, and verification of allosteric interactions of agents with labeled and unlabeled ligands at G protein-coupled receptors: interactions of strychnine and acetylcholine at muscarinic receptors. Mol Pharmacol **48**:362–378.
8. Christopoulos A (2002) Allosteric binding sites on cell-surface receptors: novel targets for drug discovery. Nat Rev Drug Discov **1**:198–210.
9. Christopoulos A, Kenakin T (2002) G protein-coupled receptor allosterism and complexing. Pharmacol Rev **54**:323–374.
10. Stockton JM, Birdsall NJ, Burgen AS et al (1983) Modification of the binding properties of muscarinic receptors by gallamine. Mol Pharmacol **23**:551–557.
11. Black JW, Leff P (1983) Operational models of pharmacological agonism. Proc Roy Soc Lond B **220**:141–162.

12. Marino MJ, Williams DL, O'Brien JA et al (2003) Allosteric modulation of group III metabotropic glutamate receptor 4: a potential approach to Parkinson's disease treatment. Proc Natl Acad Sci USA **100**:13668–13673.
13. Suzuki G, Kimura T, Satow A et al (2007) Pharmacological characterization of a new, orally active and potent allosteric metabotropic glutamate receptor 1 antagonist, 4-[1-(2-fluoropyridin-3-yl)-5-methyl-1H-1,2,3-triazol-4-yl]-N-isopropyl-N- methyl-3,6-dihydropyridine-1(2H)-carboxamide (FTIDC). J Pharmacol Exp Ther **321**:1144–1153.
14. Ehlert FJ (1988) Gallamine allosterically antagonizes muscarinic receptor-mediated inhibition of adenylate cyclase activity in the rat myocardium. J Pharmacol Exp Ther **247**:596–602.
15. Candell LM, Yun SH, Tran LL et al (1990) Differential coupling of subtypes of the muscarinic receptor to adenylate cyclase and phosphoinositide hydrolysis in the longitudinal muscle of the rat ileum. Mol Pharmacol **38**: 689–697.
16. Clark AL, Mitchelson F (1976) The inhibitory effect of gallamine on muscarinic receptors. Br J Pharmacol **58**:323–331.
17. Suga H, Ehlert FJ (2010) Investigating the interaction of McN-A-343 with the M2 muscarinic receptor using its nitrogen mustard derivative. Biochem Pharmacol **79**: 1025–1035.
18. Suga H, Figueroa KW, Ehlert FJ (2008) Use of acetylcholine mustard to study allosteric interactions at the M(2) muscarinic receptor. J Pharmacol Exp Ther **327**:518–528.
19. Tran JA, Chang A, Matsui M et al (2009) Estimation of relative microscopic affinity constants of agonists for the active state of the receptor in functional studies on M2 and M3 muscarinic receptors. Mol Pharmacol **75**: 381–396.

Chapter 15

Experimental Designs for the Study of Receptor–Receptor Interactions

Dennis Paul

Abstract

With the rise of interest in receptor–receptor interaction research, optimization of experimental designs to study the nature of interactions has become increasingly important. In this chapter, traditional experimental designs and their associated statistics are reviewed, and their theoretical and practical limitations are described. The logic and theory behind isobolographic analysis is explained and their practical use is also described. As more and more receptor–receptor interactions are recognized with the use of two or more GPCR drugs, advances in the analysis of the type presented here will become more significant and important for clinical pharmacotherapy.

Key words: Experimental design, Isobolographic analysis, Synergism, Additivity, Interference

1. Introduction

Over the past 3 decades, pharmacologists, cell biologists, and neuroscientists have been very good at delineating the mechanism of signal transduction. Indeed, several of the chapters in this book are devoted to the cell signaling of GPCRs. However, this has always been considered as a first step to the understanding of how cells integrate multiple signals from other cells. With recent interest in dimerization of receptors, lipid raft associations, and other mechanisms of receptor–receptor association, advanced experimental designs that can reveal the nature of functional interactions have become more important. In this chapter, we will describe experimental designs and methods that have been used

Craig W. Stevens (ed.), *Methods for the Discovery and Characterization of G Protein-Coupled Receptors*, Neuromethods, vol. 60, DOI 10.1007/978-1-61779-179-6_15,

in the delineation of mechanisms of treatment interactions, and the advantages and disadvantages of each. These designs have been most commonly used with the application of drugs, but are appropriate for any treatment that produces a variable "dose-response" curve. I have concentrated on the conceptual framework of the designs and avoided the mathematics of the statistical methods. The formulas can be found in the referenced articles and a recent overview (1).

The nature of the interaction between two treatments can reveal a lot about how information is processed. Experiments combining treatments can result in several types of interactions. To avoid confusion, common terms that are used (and often misused) to describe these interactions are first defined.

Additivity: The combined effect of Treatment A and Treatment B is approximately as predicted by the addition of the effects of each treatment alone ($2+2=4$).

Potentiation: The combined effect of Treatment A, which has efficacy, and Treatment B, which has no efficacy, is significantly greater than the effect of Treatment A alone ($2+0>2$).

Antagonism: The combined effect of Treatment A, which has efficacy, and Treatment B, which has no efficacy, is less than the effect of Treatment A ($2+0<2$).

Synergy: The combined effect of Treatment A and Treatment B, both of which have efficacy, is significantly greater than expected by an additive model ($2+2>4$); also known as "super-additivity" or "synergism."

Interference: The combined effect of Treatment A and Treatment B, both of which have efficacy, is less than would be expected by an additive model ($2+2<4$); also known as "sub-additivity."

Each of these possible interactions requires proper experimental designs to determine the nature of the interaction. However, it is rare that the optimal design is used, particularly for interactions other than antagonism. Accordingly, in this chapter we will review standard experimental designs from single-dose studies to shifts in dose-response curves. We will then discuss the use of isobolographic analysis in the determination of the nature of interactions. Although these designs were originated for the study of drug–drug interactions, they are applicable to any stimuli that can be varied to produce a "dose-response" curve. Thus, the term "treatment" instead of "drug" or "concentration" is used, but the term "dose" is still employed to describe the degree of treatment.

2. Methods for the Analysis of Receptor–Receptor Interaction

Tables 1–3 summarize the steps taken for the following analyses which are described further below.

2.1. Antagonism and Potentiation

2.1.1. Single-Dose Designs

Of the possible interactions, antagonism and potentiation are the most straightforward to determine experimentally. In the simplest design to show antagonism, a single dose or concentration of antagonist is combined with a single dose or concentration of an agonist, and degree of blockade, competition, or reversal is reported (Fig. 1a). Similarly, to demonstrate potentiation,

Table 1
Protocol for simple isobologram, relative potency of treatments unknown

Generate a dose-response curves for treatments A and B individually. Determine the ED_{50} ± SEM or CI95
Compute the theoretical additive ED_{50} ± SEM or CI95
Prepare a drug combination at the constant ratio of $ED_{50}A:ED_{50}B$. Prepare serial dilutions of this combination
Generate a dose-response curve for the combination and determine the true ED_{50} ± SEM or CI95

Table 2
Protocol for simple isobologram relative potency known

Estimate the potency ratio of treatments A and B from previous experience with two treatments
Simultaneously generate dose-response curves for the two individual treatments and the isobole combination. Determine ED_{50}s ± SEM or CI95

Table 3
Protocol for 3D interactions

For treatments A, B, and C generate the three possible 2D isobolograms (A:B, A:C, and B:C) as described in Protocol 1 or 2
Construct a 3D isobologram with these three 2D isobolograms
Calculate the *F* values for the surface of additivity and the 2D interaction surface
Generate the dose-response curves for 3+ combinations ratios of the three drugs
Plot the ED_{50}s of the combinations and determine the *F* value for the surface described by these ED_{50}s
Compare the three *F* values

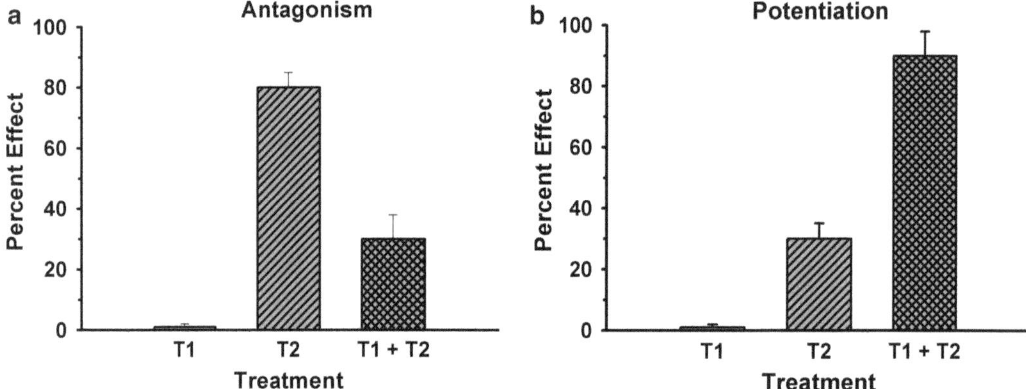

Fig. 1. Single-dose designs for determining antagonism or potentiation. (**a**) A single dose of treatment T1 has no efficacy. A single dose of treatment T2 has 80% efficacy. When added together, treatment T1 reduces the efficacy of treatment T2, and is thus an antagonist. (**b**) Treatment T1 has no efficacy. Treatment T2 has a little efficacy. When added together, treatment T1 increases the efficacy of Treatment T2, and is thus a potentiating agent. Degree of antagonism or potentiation cannot be determined with this design.

single-dose designs are often used. A single dose or concentration of the potentiating agent is added to a single dose or concentration of the agonist and the degree to which the dependent measure increases over agonist treatment alone is reported on a bar chart (Fig. 1b).

Statistical Comparison. These single-dose experiments are typically depicted as a bar graph (Fig. 1a) and compared using a *t*-test, or 95% confidence intervals (CI95) can be calculated, and the overlap of error bars determined.

Limitations. Like a thermometer, every experimental model is only useful within a range of sensitivity, with a lower and upper limit. When approaching these limits, effects are less predictable. It is obvious that you would not try to demonstrate antagonism of a drug that has only 10% efficacy, or show potentiation of a drug that already has 90% efficacy. Thus, to show an antagonist effect, it is best to use a dose of the agonist that produces 60–80% of maximum effect. If a dose that produces 100% is used, then it may not be clear if the agonist dose is at the lowest dose that produces 100% or a much higher dose. In the latter case, the antagonist dose may not be sufficient to produce sufficient blockade of effect to bring the agonist effect below 100%, and thus, the antagonist effect may be missed. If the agonist dose is below 60%, it may be difficult to show a statistically significant antagonism (Fig. 2).

The degree that this affects the results is dependent on the difference between A and A′ (A→A′). For example, in studies of pain perception, for ethical reasons, it is typical to limit the time that a rat is exposed to a painful stimulus. Therefore, A→A′ may be large. In many functional assays, the number of spare receptors may determine A→A′.

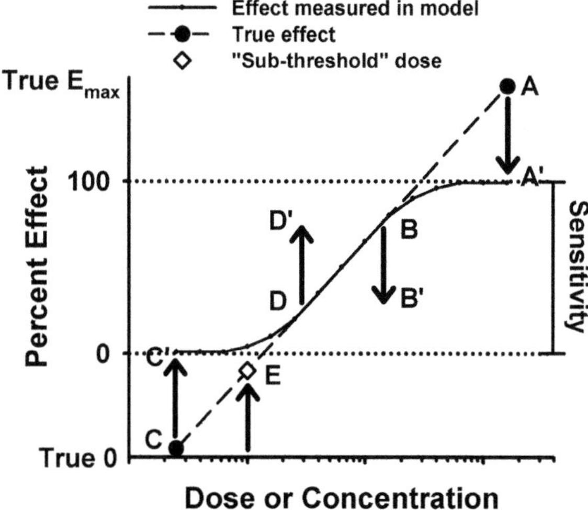

Fig. 2. Sensitivity range of an assay vs. the true dose-response effect. A→A′: Activity of an antagonist above the assay's range of sensitivity. B→B′: Activity of an antagonist within the assay's range of sensitivity. C→C′: Activity of a potentiating agent below the assay's range of sensitivity. D→D′: Activity of a potentiating agent within the assay's range of sensitivity. E: Activity of a "sub-threshold" dose of an agonist below the assay's range of sensitivity.

At the lower end of the sensitivity of the model, there is the issue of whether "0" is truly "0." In the thermometer analogy, one of the thermometers that we use to measure the temperature in our laboratory is effective from 0 to 100°C. Thus, its sensitivity begins at 273°C above absolute 0. Likewise, "0" in a binding or functional assay may not represent all that is happening biochemically or physiologically. In pharmacological studies, we define a threshold dose, below which we observe no functional effect. However, the threshold dose can be dependent upon the assay parameters and background activity. Moreover, we cannot be certain that the "sub-threshold" dose is doing nothing. It may be just doing less than what is detectable in our model (Fig. 2e). As a consequence, with single-dose experiments, we can only say that the treatment is an antagonist. As for potentiation, we cannot definitively determine whether a treatment is either a potentiating agent or an agonist that is additive or synergistic.

2.1.2. Dose-Response Curve Shift

A more effective design to determine antagonism or potentiation is to compare dose-response curves with and without the addition of the antagonist or potentiating agent. Antagonism is classically represented by a rightward shift in a dose-response curve, using doses spaced on a log scale, with the addition of a treatment with no efficacy (Fig. 3). Similarly, potentiation is depicted as a leftward shift in a dose-response curve with the addition of a treatment with no efficacy. For example, we have demonstrated that increasing the intracellular concentration of divalent cations by stimulating 5HT-3 receptors (a nonselective ligand-gated cation channel) will shift the potency of the opioid agonist DPDPE to

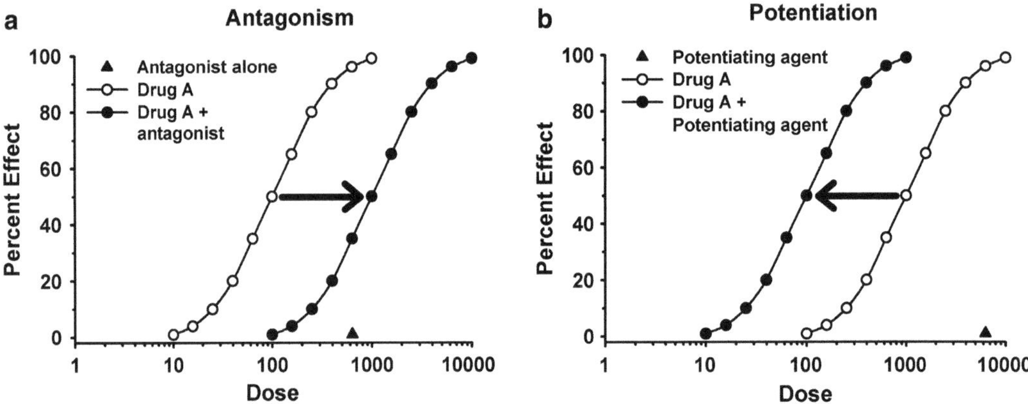

Fig. 3. Shifts in dose-response curves. (**a**) Classic rightward shift produced by a competitive antagonist in a dose-response curve for Treatment A. (**b**) A classic leftward shift of the dose-response curve for Treatment A by a potentiating agent.

inhibit forskolin-stimulated cAMP production tenfold to the left (unpublished).

Statistical Comparison. Analysis of Variance (ANOVA) is often used to compare the two dose-response curves. However, because ANOVA is a "vertical" comparison of doses, this is only statistically appropriate when the exact same doses of the agonist are used for both curves. In this case, power is lost in the extremes, as the two curves merge. Moreover, if the antagonist or potentiating agent shifts the dose-response curve tenfold, typically only half of the doses would overlap. Statistical comparison of dose-response curves is more appropriately done by comparing the ED_{50}s (or EC_{50}s or IC_{50}s) of the two curves. In principle, this is a *t*-test between ED_{50}s. Graphically, it is a horizontal comparison, rather than a vertical comparison.

Limitations. Dose-shift designs are most informative when working with two drugs working through the same receptor. If the drugs are competitive, the shift is parallel. However, when comparing the interaction of two drugs acting through different receptors, the interaction is less predictable. It may produce a shift, a change in slope or both. In other words, models of allosterism are more appropriate (see Chap. 14). In addition, with dose-response curve shift designs, if we want to know the true nature of the interaction, we must be certain that an antagonist is not an inverse agonist (the effects may appear to be an antagonist, when in fact they are additive), or that a potentiating agent has no effect on its own (e.g., sub-threshold agonist). Inverse agonists are rare; however, sub-threshold agonists are frequently used. This type of experiment may be deceptive when using a sub-threshold dose of a treatment that normally has efficacy at higher doses. As with the single-dose designs, it is the sensitivity range of the assay that is at issue. For example, suppose we inject a sub-analgesic threshold dose of morphine combined with a full dose-response curve of morphine. Would it be correct to say that the sub-threshold dose of morphine potentiates the effect of the supra-threshold dose? (This question was initially posed to me by Byron Yoburn, Ph.D., at St. John's University). Clearly, this would not be the case. In this example, the sensitivity of the assay at the lower range limits the ability to detect efficacy.

2.2. Interaction of Two Agonists: Additivity, Synergy, and Interference

With pairings of different receptor types (e.g., heterodimers, two distinct receptors on the same cell, or receptors on different cells) examining the interaction of two endogenous neurotransmitters may lead to an understanding of how their signals are integrated to produce an observed response. As with antagonism and potentiation, simple designs are often used, but these tell us little about the nature of the interaction.

2.2.1. Single-Dose Interaction Experiments

Frequently, drug–drug interactions are tested with single doses of each treatment and represented as a bar graph. The results of theoretical experiments are depicted in Fig. 4. In the first experiment, Treatment T1 and T2 each produce a significant effect. When combined, they produce less than would be expected in an additive model. There is no doubt that this is interference. The only question is whether the degree of interference is properly represented for reasons discussed earlier. In the second experiment (Fig. 4b), Treatment T1 produces a 50% effect, as does Treatment T2. When the same doses of these two treatments are administered simultaneously, the combination produces a 90% effect. Because 50 + 50 = 100, there are valid reasons that the data may represent additivity, synergy, or interference, but the experimental design is insufficient to distinguish among them. Figure 4c illustrates a theoretical dose-response curve for the combination of Treatments T1 and T2. As the effect approaches the maximum

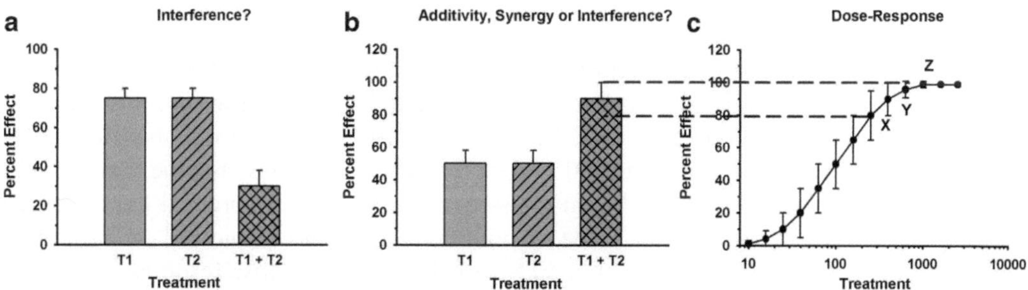

Fig. 4. Interaction of two active treatments using single-dose designs. (**a**) A single dose of either Treatment T1 or T2 has efficacy, but when administered together, efficacy is decreased. (**b**) The selected doses of Treatments T1 and T2 alone each produce moderate efficacy. When simultaneously administered, produces a near asymptotic effect. (**c**) A theoretical dose-response curve for the combination of T1 + T2 from (**b**).

sensitivity of the assay and the dose-effect curve asymptotes, it is apparent that the effect represented by bar "T1 + T2" in Fig. 4b may represent a result anywhere from 80 to 100%. Therefore, the true result may, in fact, be less than additive (*X*; interference), additive (*Υ*), but with a lesser value because of the asymptotic nature of the dose-response curve, or synergistic disguised by the asymptotic nature of the curve (*Z*).

Thus, a single combination of two treatments is insufficient to characterize the nature of an agonist–agonist interaction. Even when extremely low doses of each treatment are used, the nature of the interaction cannot be definitively determined. If a dose of each treatment produces, for example, 20%, and the combination produces 90%, then it is likely that the treatments are synergistic. However, the sensitivity at the lower range of the assay and the nature of "zero" are again at issue, as is the asymptotic nature of the lower end of the dose-response curve. Using the Celsius thermometer analogy again, if we could start at absolute 0, then add enough calories to raise the temperature of the thermometer to the threshold of sensitivity (273 cal/mL H_2O), it would appear that we have done nothing. If we added an additional 273 cal/mL, raising the temperature to 273°C, it would appear that our two manipulations are synergistic, rather than additive.

2.2.2. Single Dose of Treatment A, Dose-Response Curve of Treatment B

Others have investigated the interaction of two treatments by using a single dose of Treatment A and a full dose-response curve of Treatment B. Thus, the results are equivalent to an experiment demonstrating potentiation, except that both treatments have efficacy. Typically, if the effect of Treatment A alone plus the effect of Treatment B at multiple doses is equal to the dose-response curve for Treatment A + Treatment B, then the interaction is said to be additive. If the curve is greater than expected, then it is said to be synergistic, and if it is less than expected, then it is said to be interference.

Statistical Comparison. The dose-response curve for the combination should be compared to a theoretically additive dose-response curve, as well as the control curve. The theoretical curve is generated by simple addition of the effect of the single dose of Treatment A to the Treatment B curve. However, this design is limited by the efficacy of the dose of Treatment A. For doses of A that produce 50% effect or greater, one must assume that the shift is parallel because less than half of the theoretical curve can be generated. Dose-response curves would be plotted on a standard semi-log graph. ED_{50}s of the predicted dose-response curve and the combination dose-response curve would be compared using a *t*-test or CI95. The dose-response curve for the combination should be compared to a theoretically additive dose-response curve, rather than the control curve. The theoretical curve is generated by simple addition of the effect of the single dose of

Treatment A to the Treatment B curve. However, this design is limited by the efficacy of the dose of Treatment A. For doses of A that produce 50% effect or greater, one must assume that the shift is parallel because less than half of the theoretical curve can be generated.

Limitations. This design gives us the nature of the interaction of only one dose of Treatment A with Treatment B. Moreover, in this experimental design, because the dose of Treatment A remains constant as the dose of Treatment B changes, the dose ratio is different for each point on the combination dose-response curve. This can produce nonparallel shifts in the dose-response curves.

2.3. Isobolographic Analysis

With the limitations of these other designs in mind, the method of isobolographic analysis was developed (2). This method uses a dose-response curve of a constant ratio combination of two treatments to determine the nature of the interaction. Initially, dose-response curves for Treatment A and Treatment B are generated and ED_{50} determined. Then one or more dose-response curves for a combination of the two treatments are generated, each at a constant dose ratio. When a single ratio is used, it is often the ratio of the ED_{50} of Treatment A to the ED_{50} of Treatment B (2–4). Typically, this is accomplished by diluting the two treatments, each to the concentration that would give us the ED_{50}, then doing serial dilutions of this combination in 1/2–1/4 log unit steps to complete the dose-response curve. The concentration of each treatment in the ED_{50} of this dose-response curve (Treatment A vs. Treatment B) would then be plotted on linear scale Cartesian coordinates with concentration of Treatment A on the *X*-axis and concentration of Treatment B on the *Y*-axis (Fig. 5). A diagonal line from the ED_{50} of Treatment A to the ED_{50} of Treatment B is drawn and labeled as the *line of additivity*. This graph is called an

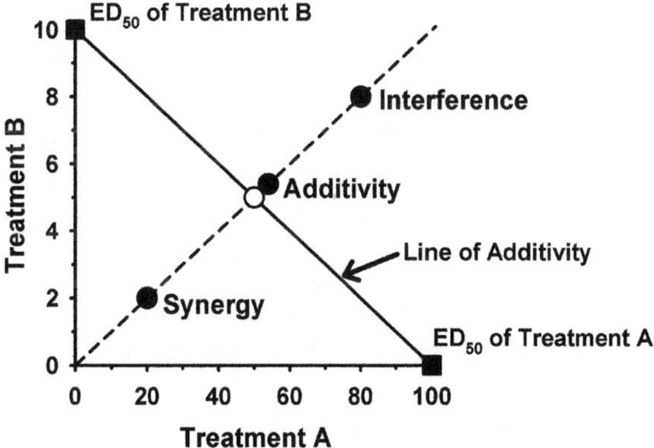

Fig. 5. A theoretical isobologram. *Open circle*: Predicted (theoretical) isobole for the combination of Treatments A and B. *Filled circle*: Three possible outcomes for experimental isoboles.

isobologram, and the experimental ED_{50} is called an *isobole*. Because the dose ratio is constant, this isobole can be plotted using either of the treatments as if it were alone, and then multiplying the ED_{50} by the ratio to get the content of the other treatment in the isobole. If a drug–drug interaction is additive, the isobole of any ratio of the combination of these two treatments should fall on the line of additivity (Fig. 5). The predicted (or *theoretical*) isobole for the tested ratio can be plotted on the line of additivity. Alternatively, the SEM or CI95 for each single-treatment ED_{50} can be plotted on the predicted combination isobole (3, 5, 6). In addition to plotting a predicted isobole, the individual single treatment ED_{50}s ± SEM or CI95 can be plotted along the axes as the *X* and *Y* intercepts for the line of additivity (7).

Then, the combination isobole determined experimentally is plotted on the radial line that extends from the origin through the predicted isobole. The SEM for the experimental isobole is calculated in the usual manner, and can be plotted on the radial line (45° dashed line in Fig. 5), or both vertically and horizontally from the interaction ED_{50}. As the name implies, if the experimental isobole falls on or near the line of additivity, it is said to be additive. If the interaction is synergistic, the isobole should fall closer to the origin than the line of additivity. Conversely, if the nature of the combination is interference, the isobole would be plotted further from the origin than the line of additivity.

Statistical Comparison. In Loewe's original analysis, significance was determined graphically. The 95% confidence intervals (computed as the geometric mean of the CIs of the single treatment ED_{50}s for the predicted isobole) were plotted along a radial axis from the origin through the isobole, and if the error bars of the predicted and experimental ED_{50}s did not overlap, the synergistic or interfering interaction was said to be significant (2). Through the efforts of several investigators, most notably Tallarida (1, 3–6), comparison of isoboles has advanced with a stronger grounding in statistical theory. The simplest of these uses the formulas of the student's *t*-test to compare the predicted and experimental ED_{50}s. This is appropriate when only one ratio of the two agonists is tested. With multiple combination dose ratios, evidence for synergy, additivity, or interference is determined by ANOVA. For testing variable dose ratios, surface analysis can be used (8).

Limitations. By working within the linear part of the single treatment and combination dose-response curves and maintaining a constant ratio between Treatment A and Treatment B, isobolographic analysis avoids the limitations of classic experimental designs. In its simplest form, isobolographic analysis assumes that the dose-response curves for treatments A and B are parallel, and that the two treatments have similar efficacy (the same maximum effect). Derivative methods have been developed to account for violations of these assumptions (9).

2.4. Comparing Isobole Shifts After Experimental Manipulation

To this point, we have listed experimental designs that are descriptive only. That is, we have provided designs that determine whether an interaction is additive, sub-additive (interference), or supra-additive (synergistic). However, just as with the classic designs, isobolographic designs can be used for hypothesis testing. The interaction of two receptors may be compared, manipulating a factor that is thought to mediate the interaction. For example, Overland et al. (10) have examined the role of protein kinase C in the synergistic interaction of δ opioid receptors and α-2 adrenoceptors. Pharmacological blockade or depletion of the enzyme changed analgesic synergy of stimulation of these two receptors to an additive interaction.

For hypothesis testing, parallel experiments are done with treatment vs. no treatment, antisense or siRNA knock-down vs. missense or scrambled RNA control, or in transgenic vs. wild-type mice. Time-course data can also be analyzed by comparing multiple isoboles.

Statistical Comparison. Two isobolograms can be compared in one of two ways. In a method suggested by Ossipov et al. (11), the doses of Treatment A and Treatment B can be added to give the dose-response curves for the two interactions. The ED_{50} of each interaction can be determined and compared as in dose-shift experiments. This method has the advantage of being able to visually determine if the interaction is equal throughout the dose range (it produces a parallel shift).

Alternatively, the two isobolograms can be compared using their *interaction indexes.* Tallarida (12) proposed that an interaction index (γ) can be calculated by triangulating the location of the experimental and theoretical interaction isoboles. Conceptually, this index is the ratio of the distance between the origin and the experimentally determined interaction isobole, divided by the distance between the origin and the theoretical additive isobole. Brodkin and Shannon (13) had previously proposed that inverse of this index be used, so that synergy would be represented by a number greater than 1. However, Tallarida's index has been more frequently used.

An index for synergistic treatments would be less than 1, whereas an index for interfering treatments would be greater than 1. This ratio can be calculated by dividing the total dose for treatments A and B for the experimentally determined combination isobole by the total dose for treatments A and B for the theoretical additive isobole. The SEM or CI95 can be determined for control and experimental indices, and the two compared using the *t*-statistic or comparison of the CI95s (see Fig. 6b).

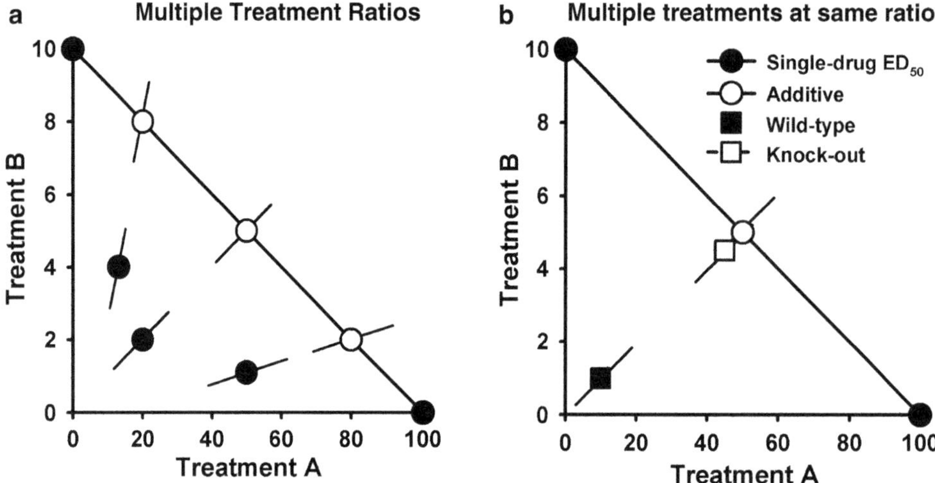

Fig. 6. Multiple ratios and multiple treatments. (**a**) Multiple treatment ratios can be used to verify that a combination has the same interaction. (**b**) Isoboles from experimental manipulations that do not change the individual treatment ED_{50}s can be plotted on the same isobologram.

If the experimental manipulation does not affect the single treatment dose-response curves, then the significance of the manipulation can be determined by a *t*-test or CI95s directly between the two isoboles. However, if the manipulation does affect the ED_{50}s of either or both of the single treatment dose-response curves, then two separate isobolograms should be generated, and the significance of the manipulation should be tested using the interaction indexes. If more than two isobolograms are generated, then an ANOVA can be used for the comparison. This may be useful for comparison or multiple doses of a potentiating agent, antagonist, or time-course data.

2.5. Three-Treatment Interactions

Integration of cell signaling often involves more than two receptors or factors. We have proposed a method for evaluating the nature of interaction of three different treatments. In an application of ANOVA to the method proposed by Gershwin and Smith (14) in which three isobolograms are combined to form a three-dimensional (3D) Cartesian coordinate graph, we reasoned that the surface defined by the ED_{50}s for three or more dose-ratio combinations could be compared to the surface defined (predicted) by the surface defined (predicted) by the ED_{50}s of the single treatment dose-response curves, or to ED_{50}s of the two-dimensional (2D) interactions using the *F* statistics of each surface in a process analogous to a two-way ANOVA. We have used multiple dose-ratio combinations to define a 3D interaction surface (Fig. 7). For example, we have used the following ratios: (1) ED_{50}:ED_{50}:ED_{50}; (2) ED_{60}:ED_{20}:ED_{20}; (3) ED_{20}:ED_{60}:ED_{20}; and (4) ED_{20}:ED_{20}:ED_{60}. A second surface (triangle) is defined by the lines of additivity of three 2D isoboles, and a third surface by

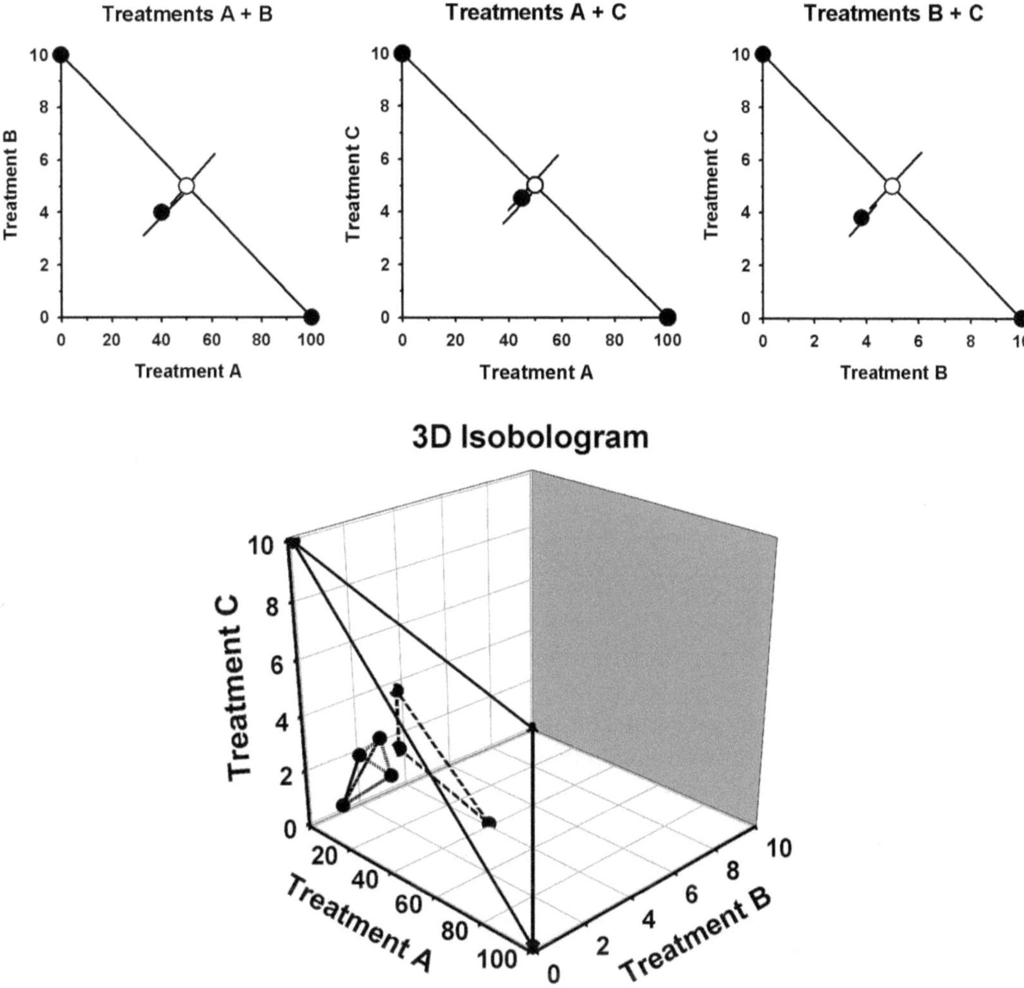

Fig. 7. A three-dimensional (3D) isobologram constructed from three two-dimensional (2D) isobolograms. The 3D isobologram depicts the surface of additivity with *solid lines* connecting the three single-treatment ED_{50}s, the 2D interaction surface with *dashed lines* connecting the three 2-treatment isoboles (embedded in the *walls* of the graph), and four 3-treatment isoboles with *dotted lines* (closer to the origin than the 2D interaction surface). In this example, although the 2D interaction surface is not significant than the surface of additivity, the 3-treatment interaction is super-additive (greater than would be expected by the 2-treatment interactions).

the three ED_{50}s of the treatments. The *F* value for each of these surfaces would be determined, weighted by the inverse of the variance ($1/SD^2$). By weighting in this manner, the *F* value is adjusted for the variability of each individual dose-response curve. Likewise, the ED_{50}s for the three 2D isoboles, weighted for the inverse of the variance, determined the 3D interaction surface predicted by the 2D interactions, and the ED_{50}s for the three single treatment dose-response curves, weighted for the inverse of the variance, determined the experimental 3D interaction surface. The *F* values can then be used for determination of statistically significant differences by ANOVA.

3. Future Directions

It seems that the more we understand about the integration of cell signaling, the more complex we see that it is. Therefore, we may have to use progressively more complex experimental designs. The interaction index can be extended to 3D isoboles. 2D or 3D interactions over time could be examined using surface analysis. 3D Isoboles with a fourth manipulation, such as knockdown or knockout of a protein, are conceivable. However, it must be noted that as each of these designs takes on another dimension, we lose power. Therefore, the number of interactions that we measure in a single design is limited by the extent of potentiation, synergy, or interference.

4. Conclusions

The nature of functional interactions between two or more treatments can be an important signpost to the underlying mechanisms connecting receptors. In this chapter, we have tried to show how an ineffective experimental design can lead us to false conclusions about such interactions, and with the proper designs we can be more definitive about our interpretations. Although the focus of this chapter has been the use of these designs and associated statistics in the study of receptor–receptor interactions, each is applicable to any biological (or physical) interactions.

References

1. Tallarida RJ (2006) An overview of drug combination analysis with isobolograms. J Pharmacol Exp Ther **319**:1–7
2. Loewe S (1953) The problem of synergism and antagonism of combined drugs. Arzneim-Forsch **3**:385–390
3. Tallarida RJ, Porreca F, Cowan A (1986) Statistical analysis of drug-drug and site-site interactions with isobolograms. Life Sci. **45**:947–961
4. Tallarida RJ (1992) Statistical analysis of drug combinations for synergism. Pain **49**:93–97
5. Tallarida RJ, Murray RB (1987) Manual of Pharmacologic Calculation with Computer Programs, 2nd ed, Springer, New York
6. Tallarida RJ, Stone DJ, Raffa RB (1997) Efficient designs for studying synergistic drug combinations. Life Sci **61**:PL417–425
7. Stone LS, MacMillan LB, Kitto KF et al (1997) The α2a adrenergic receptor subtype mediates spinal analgesia evoked by α2 agonists and is necessary for spinal adrenergic-opioid synergy. J Neurosci **17**:7157–7165
8. Tallarida RJ, Stone DJ Jr, McCary JD et al (1999) Response surface analysis of synergism between morphine and clonidine. J Pharmacol Exp Ther **289**:8–13
9. Grabovsky, Y, Tallarida, RJ (2004) Isobolographic analysis for combinations of a full and partial agonist: Curved isoboles. J Pharmacol Exp Ther **310**:981–986
10. Overland AC, Kitto KF, Chabot-Doré AJ et al (2009) Protein kinase C mediates the synergistic interaction between agonists acting at α2-adrenergic and *delta* opioid receptors in spinal cord. J Neurosci **29**:13264–13273

11. Ossipov MH, Lopez Y, Bian D et al (1997) Synergistic antinociceptive interactions between *alpha* 2-adrenergic receptor antagonists. Anesthesiology **86**:1–9
12. Tallarida RJ (2002) The interaction index: a measure of drug synergism. Pain **98**: 163–168.
13. Brodkin J, Shannon HE (2000) Communicating synergism in drug combination studies. Arzneimittelforschung. 2000 **50**:765–767
14. Gershwin ME, Smith NT (1973) Interaction between drugs using three-dimensional isobolographic interpretation. Arch Int Pharmacodyn Ther. **201**:154–161

Part IV

The Regulation of G Protein-Coupled Receptors

Chapter 16

Elucidating Agonist-Selective Mechanisms of G Protein-Coupled Receptor Desensitization

Chris P. Bailey and Eamonn Kelly

Abstract

In pharmacology, a central tenet of receptor theory has been that different agonists acting at a particular G protein-coupled receptor subtype produce the same profile of cellular responses. In recent years, advances in molecular pharmacology and the availability of diverse cell signaling assays have indicated that this idea is not sufficient to explain all the data obtained, and that agonists can produce different response profiles following binding to a receptor subtype in a cell. This has been termed biased agonism or functional selectivity, and is thought to be due to the ability of agonists to stabilize different active conformations of the receptor. Logically, there is no reason why this idea cannot also be extended to receptor regulatory mechanisms, since different receptor conformations could exhibit differential affinities for regulatory elements such as the kinases involved in receptor phosphorylation and desensitization. Nevertheless, great care must be taken when analyzing agonist response and regulatory pathways, since other factors such as differences in agonist efficacy need to be considered as contributing factors to agonist-dependent regulation. In the case of the μ-opioid receptor (MOPr), we have shown that two agonists, morphine and the peptide agonist DAMGO, can induce MOPr desensitization by different mechanisms involving largely protein kinase C (PKC) and G protein-coupled receptor kinase/arrestin respectively. This could explain why opioid agonists have variable clinical profiles and liabilities to induce tolerance and dependence. Here we describe the experimental approaches that can be used to investigate mechanisms of MOPr desensitization with a particular focus on endogenous MOPr in neurons. In addition, we discuss the role that agonist efficacy might play in desensitization and describe methods to estimate agonist efficacy for responses downstream of receptor activation, including arrestin recruitment which can be regarded as both a regulatory and a signaling mechanism.

Key words: Desensitization, Tolerance, Functional selectivity, Biased agonists, G protein-coupled receptors, Opioid receptors, MOPr, Internalization

1. Introduction

1.1. GPCR Desensitization

Most, if not all, G protein-coupled receptors (GPCRs) can undergo agonist-induced desensitization. That is, once a receptor has been activated by an agonist, various intracellular mechanisms

Craig W. Stevens (ed.), *Methods for the Discovery and Characterization of G Protein-Coupled Receptors*, Neuromethods, vol. 60, DOI 10.1007/978-1-61779-179-6_16,

take place to affect that receptor and render it nonfunctional (1). The process of GPCR desensitization therefore reduces the efficacy of GPCR agonists over time, and so is a major contributor to drug tolerance caused by repeated administration of GPCR agonists.

The classical model of GPCR desensitization consists of the activated receptor becoming a substrate for a specific class of kinases, G protein-coupled receptor kinases (GRKs). Although phosphorylation of the receptor by GRK itself leads to minimal loss of receptor function, critically it allows arrestin to bind to the GRK-phosphorylated, agonist-bound receptor with high affinity, rendering it desensitized (2, 3). Although clearly important, the GRK/arrestin-dependent process is by no means the only mechanism by which agonist-induced GPCR desensitization can occur. In many systems, second messenger-dependent kinases such as protein kinase A (PKA) and protein kinase C (PKC) are critically involved in agonist-induced GPCR desensitization, and can either occur in the absence of, or synergistically with, GRK/arrestin-dependent mechanisms (4–6).

The fact that agonist-induced GPCR desensitization can take place by distinct intracellular mechanisms raises some interesting possibilities for the heterogeneity of GPCR desensitization. First, cell-specific desensitization could occur, since if different cells have a different intracellular milieu (e.g., different amounts or subtypes of GRKs, PKAs, or PKCs) then desensitization of a specific GPCR will depend on the cell in which it is located (6, 7). Second, it allows for the phenomenon of agonist-specific GPCR desensitization.

1.2. Functional Selectivity

One of the key advances in GPCR pharmacology in recent years is the concept of functional selectivity. Classical receptor theory postulates that when an agonist has bound to the receptor, it stabilizes an active conformational state of the receptor able to modulate G protein-dependent signaling. As different ligands have different efficacies, this "linear" model of receptor activation can account for a range of ligand–receptor interactions from antagonist, through partial agonist, to full agonist (a range of ligands that stabilize the active conformational state of the receptor with increasing efficacies), as well as inverse agonists (ligands that stabilize the *in*active conformational state of the receptor).

However, this "linear" model of receptor activation cannot account for several experimental findings, whereby different agonists, at a given GPCR, in a given system, can cause different functional responses. This is often termed functional selectivity (also sometimes called "biased agonism" or "collateral efficacy"). Although this was first described over 20 years ago, many recent studies have refocussed on this aspect of GPCR signaling (8, 9). In terms of fundamental receptor theory, in order to account for functional selectivity, more than one active conformational state

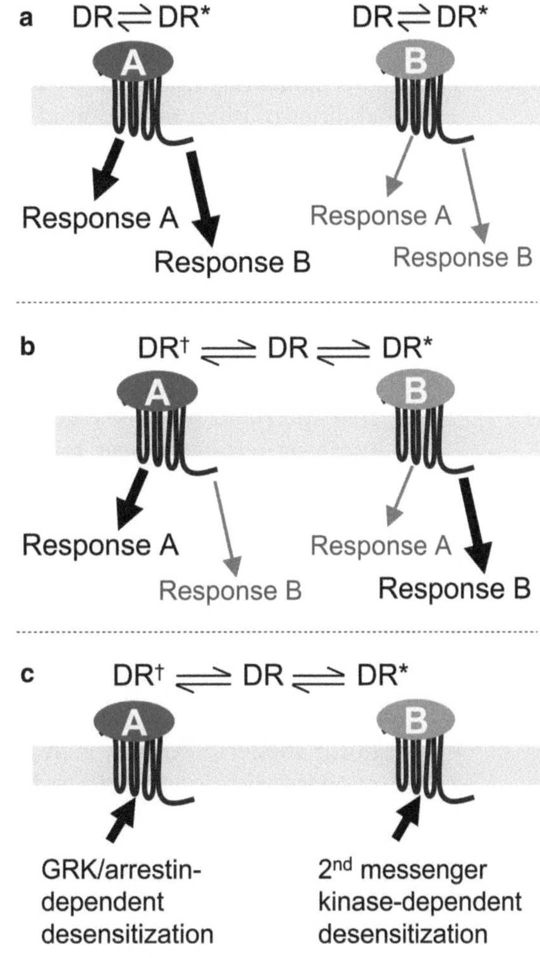

Fig. 1. Functional selectivity of G protein-coupled receptor (GPCR) signaling and desensitization. (**a**) Linear model of GPCR activation. Once an agonist (D) has bound to a receptor (R), it activates the receptor (DR*) with a given efficacy. Higher efficacy drugs (agonist A) produce greater functional responses than lower efficacy drugs (agonist B). (**b**) Functional selectivity of GPCR activation. Different agonists can stabilize different conformational states of the receptor (DR* and DR†). Agonist A stabilizes DR†, which favors signaling response A. Agonist B stabilizes DR*, which favors signaling response B. (**c**) Functional selectivity of GPCR desensitization. Different agonists stabilize different conformational states of the receptor (DR* and DR†). DR† then allows G protein-coupled receptor kinase (GRK)/arrestin-dependent desensitization, whereas DR* allows second messenger kinase-dependent desensitization. NB, functional selectivity at the level of desensitization can occur even if the signaling caused by DR* and DR† is the same.

of the receptor must be incorporated into the model (Fig. 1). Although functional selectivity is most often studied in terms of cellular signaling, some groups have also uncovered functional selectivity in terms of GPCR desensitization.

1.3. Functional Selectivity and GPCR Desensitization

As outlined above, studies have shown that agonist-dependent desensitization can take place via GRK/arrestin-dependent or second messenger kinase-dependent mechanisms. One of the first studies to show that different agonists, acting at the same receptor, might cause desensitization by different intracellular mechanisms, was shown by Keith et al. (10), who examined the abilities of different μ-opioid receptor (MOPr) agonists to cause arrestin-dependent internalization of MOPr. Generally speaking, if a receptor is desensitized by the GRK/arrestin pathway, it is rapidly internalized and then either rapidly recycled or down-regulated. The study found that etorphine and DAMGO, but not morphine, could cause arrestin-dependent internalization of MOPr. Although internalization and desensitization are not necessarily linked (in fact, it is important to note that as MOPrs rapidly recycle following internalization, there is good evidence that internalization is a key contributor to REsensitization of receptor responsiveness rather than DEsensitization) (11–14), this work was followed up by a number of papers, using a variety of different cell types (e.g., HEK cells, AtT20 cells, and rat locus coeruleus neurons) showing that indeed morphine, unlike other MOPr agonists such as DAMGO, etorphine, fentanyl, and methadone, causes minimal GRK/arrestin-dependent MOPr desensitization (11, 15–22).

As the GRK/arrestin-dependent mechanism of GPCR desensitization requires agonist-induced activation of the receptor, the linear model of GPCR activation would lead us to expect that agonists with lower efficacy have less capacity to induce GRK/arrestin-dependent desensitization (23). Indeed, as morphine has lower intrinsic efficacy at MOPr than agonists such as DAMGO, etorphine, fentanyl, and methadone, it is to be expected that, out of these agonists, morphine elicits the least GRK/arrestin-dependent MOPr desensitization. However, more recently we and others (24–32) have shown that different agonists can cause MOPr desensitization by different mechanisms. For example, DAMGO-induced MOPr desensitization is largely GRK/arrestin mediated, and morphine-induced MOPr desensitization is largely PKC-dependent. This suggests that functional selectivity can take place at the level of GPCR desensitization. Again, to fit this experimental finding to receptor theory, the most plausible explanation is that there are more than one active conformational states of the receptor, which can in turn be desensitized by different mechanisms (Fig. 1).

Although agonist-selective desensitization has also been demonstrated for other GPCRs such as the D1 dopamine and neurokinin 1 receptors (33, 34), the bulk of research into this phenomenon has been conducted on the MOPr. Partly this is in an effort from a number of laboratories to decipher the cellular mechanisms underlying opioid tolerance, a clinical problem that complicates the use of opioids as analgesics, but partly it is because such experiments using MOPrs as the GPCR are facilitated by the extensive range of available ligands and assays that can be used.

The rest of this chapter discusses experimental approaches to examine agonist-selective MOPr desensitization. It is not possible to cover, in detail, all of the possible assays that could be used in such studies. Rather, we will focus on the techniques that a number of laboratories (including ours) have used, discussing what we feel are the specific advantages of those approaches, along with possible pitfalls we have encountered that may apply to any study of GPCR desensitization.

2. Functional Assays to Elucidate Agonist-Selective Mechanisms of GPCR Desensitization

As GPCR desensitization is so ubiquitous, virtually any assay where GPCR activation can be measured could be used to study agonist-selective GPCR desensitization. This includes (^{35}S)GTPγS assays (mainly for $G_{i/o}$-coupled receptors, but can also be used for G_s- or G_q-coupled receptors), cAMP ($G_{i/o}$ or G_s) or IP_3 (G_q) assays, and calcium mobilization assays (particularly for G_q-coupled receptors, but also for $G_{i/o}$ or G_s-coupled receptors using mutant G proteins). However, generally speaking, these assays need to be performed either in isolated membrane preparations or in cultured cells. A priority in our laboratory, partly determined by our focus to translate findings on a cellular level with tolerance in the whole animal, has been to study native receptors in mammalian neurons whenever possible. To this end, we have made extensive use of electrophysiological recording techniques in brain slices.

2.1. MOPr Desensitization in Locus Coeruleus Neurons

One of the cellular consequences of activating $G_{i/o}$-coupled GPCRs, such as MOPr, is the opening of a certain class of K^+-channel, the G protein-coupled inwardly rectifying potassium channel (GIRK or Kir3). This type of receptor–effector coupling is particularly noticeable in the soma and dendrites of neurons, and can be used to study neuronal MOPr function. The details of cutting and maintaining brain slices for electrophysiology, as well as the techniques of intracellular and whole-cell patch-clamp recordings are well covered elsewhere (e.g., (35)). In terms of using electrophysiology to study MOPrs, by far the most widely used brain region is the locus coeruleus (LC). The reasons for this are, first, out of the three subtypes of opioid receptor (μ, δ, κ), only MOPr are present in LC neurons (both in the rat and the mouse). Although there are some relatively selective MOPr agonists, such as DAMGO, many agonists, particularly many clinically useful agonists such as morphine, do have agonist activity at other opioid receptor types. So, to have a preparation where there are no δ or κ-opioid receptors is a specific advantage when only MOPrs are to be studied. Second, the population of LC neurons is remarkably homogeneous; unusual for mammalian brain areas, each LC neuron is virtually identical (in terms of size, electrical properties, and receptor expression), and no inhibitory interneurons are present within the LC. A further advantage is that the amplitude of the MOPr-activated GIRK response in LC neurons is quite large, particularly in rats, which improves the signal–noise ratio of the recording and greatly facilitates experimentation.

Desensitization of MOPrs in LC neurons has been studied for many years by electrophysiological methods. Activation of MOPrs results in the opening of GIRKs that causes hyperpolarization of the membrane potential. Recordings can be performed either by intracellular recording techniques (whereby changes in membrane potential are directly measured) or by whole-cell patch-clamp recordings (whereby the membrane potential is "clamped" near its resting membrane potential, and the current flow through the GIRK channel is measured). Either way, the recording of membrane potential or GIRK current gives a real-time measure of MOPr activation. This is another key advantage that this technique has over many other cell-based assays, as it is straightforward to look at the time-course of MOPr desensitization. Using this technique, various groups have shown that agonists such as Met-Enkephalin, DAMGO, fentanyl, and etorphine cause far greater, and more rapid, MOPr desensitization than agonists such as morphine or oxycodone ((15, 19–21, 26, 29–31, 36); Fig. 2).

2.2. MOPr Desensitization in Locus Coeruleus Neurons: Procedural Concerns

The decision of whether to perform intracellular or whole-cell patch-clamp recordings is largely a matter of personal choice, but each has some advantages and disadvantages. The main disadvantage for whole-cell patch-clamp recordings is that key components of the intracellular fluid can dialyze out of the cell into the pipette. Potentially, this can have serious ramifications. For example, if a particular signaling protein that is critical for MOPr desensitization is removed from the intracellular milieu by dialysis, its effects would never be seen. In our hands, this has not been a problem, and can be confirmed by comparing data using whole-cell patch-clamp with perforated-patch experiments (28), indeed, it could be argued that the perforated patch should replace whole-cell patch-clamp for these types of experiments. The key advantage in using intracellular recordings over whole-cell patch-clamp recordings is that dialysis of cell contents will never be a problem. A potential disadvantage in using intracellular recordings is that very large MOPr-induced changes in membrane potential could be underestimated, as when the membrane potential approaches the potassium equilibrium potential, any further membrane potential changes would appear truncated. However, again this does not appear to have been a problem in practice (19).

The fact that in whole-cell patch-clamp recordings, the pipette solution dialyses with the cell contents can also be used as an advantage. First, it permits the use of non-membrane-permeable inhibitors,

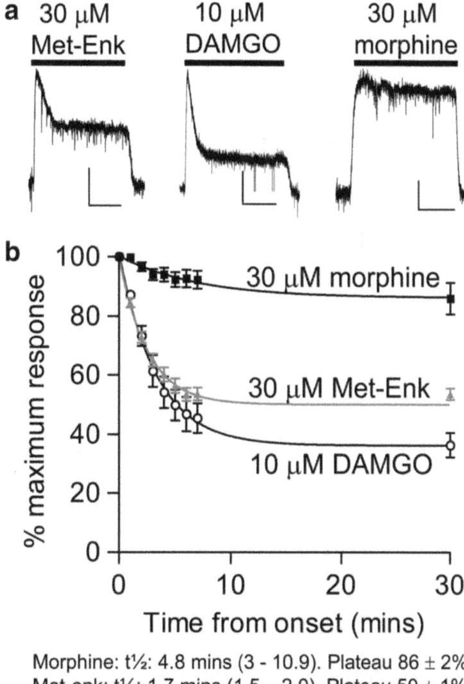

Fig. 2. Rapid agonist-induced μ-opioid receptor (MOPr) desensitization in rat locus coeruleus neurons. MOPr-evoked potassium currents recorded from rat locus coeruleus neurons in acute brain slices using whole-cell patch-clamp recordings. DAMGO and Met-Enkephalin produced far greater rapid desensitization than morphine. All agonists were applied at approximately receptor-saturating concentrations. (**a**) Recordings from three individual neurons, scale bars: 50 pA, 10 min. (**b**) Pooled data of percentage maximum response plotted against time after maximum response was attained. Mean data were fitted to a single exponential; half-life shown as mean values with 95% confidence intervals in *parentheses*, plateau shown as mean ± S.E.M. Figure adapted from (21).

while secondly by a subtle change in the recording protocol, even very small MOPr-induced GIRK-currents can be recorded, and the signal–noise ratio increased. As the K^+-channel is inwardly rectifying, K^+ ions more efficiently pass through the channel from outside the cell to inside, yet, under physiological conditions, K^+ ion flow is outward. However, K^+ ion flow can be made inward by altering the conditions so that the K^+ reversal potential is less negative than the holding membrane potential, either by increasing the concentration of extracellular K^+ ions, or by making the holding potential more negative, or both ((14, 28, 29, 37); Fig. 3).

Another advantage in recording GIRK currents in native neurons, when examining GPCR desensitization, is that it is relatively straightforward to prove that any desensitization that is seen is homologous. As with any assay where the readout is downstream of G protein activation (i.e., all assays except (^{35}S)GTPγS), it is always possible that any reduction in response is only apparent receptor desensitization, caused by run-down of the effector

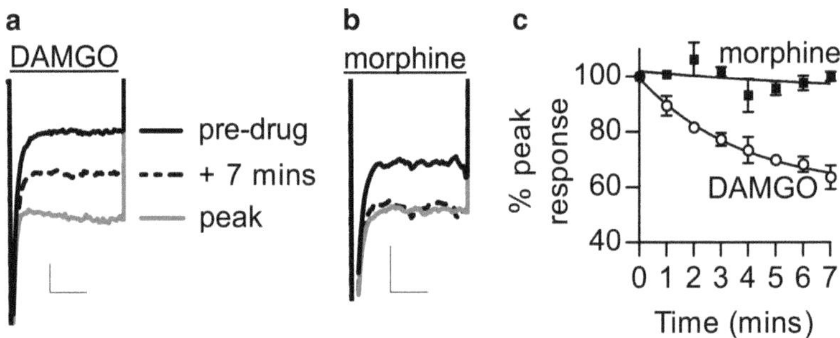

Fig. 3. Rapid agonist-induced MOPr desensitization in mouse locus coeruleus neurons. MOPr-evoked potassium currents are generally smaller in mouse LC neurons compared with rat LC neurons. Using whole-cell voltage-clamp recordings with standard K^+ concentrations, Vh = –60 mV, the maximum morphine-evoked current can often be as little as 20 pA. In order to facilitate recordings, slices can be bathed in high extracellular K^+ (in this case, 10 mM) and the holding potential is hyperpolarized (in this case, Vh was stepped from –60 to –130 mV for 60 ms). The currents during the final 10 ms of the voltage step were analyzed. (**a**) Rapid desensitization caused by 10 μM DAMGO. In this example, the peak current was approximately 600 pA which had declined to approximately 300 pA after 7 min. (**b**) 30 μM Morphine elicited a peak current of approximately 200 pA which did not decline over 7 min. (**c**) Pooled data of recordings similar to those in (**a**) and (**b**). *Scale bars*: 20 ms, 200 pA. *Traces* show data obtained from PKCα knockout mice, *graph* shows data obtained from wildtype mice. Figure adapted from (29).

(a form of heterologous desensitization), rather than "true" desensitization of the receptor itself. This can easily be resolved by examining the current elicited by activation of a different $G_{i/o}$-coupled receptor, in the same cell as it appears to be a general phenomenon that if more than one $G_{i/o}$-coupled receptor is located in the same cell body, then they are coupled to the same population of GIRK channels (38, 39). In the case of studying MOPr desensitization in LC neurons, homologous desensitization can be confirmed by examining currents elicited by activating either the α_2 adrenoceptor, somatostatin SST_{2A} receptor or the ORL1 receptor (15, 40). An example of how interpreting experimental data might be confounded by run-down of the channel, rather than true MOPr desensitization, is that methadone, at high concentrations, causes direct GIRK blockade (20, 41). This makes methadone-induced MOPr desensitization difficult to study using this technique, as an apparent time-dependent decrease in methadone response consists of both MOPr desensitization, as well as of GIRK blockade.

The most widely used approach for studying agonist-selective MOPr desensitization, in recent years, has been by measuring K^+-currents in LC neurons, but, many other cell types could be used, for example, dorsal root ganglion and trigeminal ganglion neurons, and neurons from the periaqueductal gray or raphe nuclei (42–44). Generally speaking, these experiments are more challenging than in the LC, as the neuronal populations in these regions are usually more heterogeneous, they often express other opioid receptor types in addition to MOPrs, and currents are often smaller. As well as measuring MOPr activation by the opening of GIRK channels, MOPrs close certain types of calcium channels which offer an alternative, but technically more challenging, method of assessing MOPr activation (e.g., (11)). Furthermore, although researchers are often driven to examine receptor desensitization in native neurons wherever possible, the experimental approach outlined above can equally be used in heterologous expression systems. For example, HEK293 cells (transfected to express both MOPr and GIRK channels), AtT20 cells (transfected to express MOPr but recording endogenous Ca^{2+}- or K^{+}-currents), and xenopus oocytes (transfected to express MOPr, GIRK, GRK, and arrestin) have all been used to study rapid MOPr desensitization (11, 17, 28).

2.3. Agonist-Selective MOPr Desensitization

Using studies outlined above, numerous groups have shown that lower efficacy MOPr agonists such as morphine can cause less desensitization compared with higher efficacy agonists such as DAMGO (15–22). As discussed, if agonist-induced MOPr desensitization can only occur by a GRK/arrestin-dependent mechanism, then this finding is not surprising, as GRK and arrestin desensitize the activated receptor, and higher efficacy agonists cause greater receptor activation. However, more recently, it has been shown that the relative inability of some agonists to trigger GRK/arrestin-dependent regulation allows those agonists to cause MOPr desensitization by a PKC-dependent mechanism. Moreover, these two mechanisms of MOPr desensitization are entirely separate.

In rat LC neurons, although morphine causes minimal MOPr desensitization, this is greatly enhanced if PKC is activated, either directly using the phorbol ester, 12-myristate 13-acetate (PMA) or indirectly by activating G_q-coupled receptors such as the M3 muscarinic receptor. There is an array of cell-permeable, small-molecule PKC inhibitors available that can be used to inhibit PKC and negate this effect, such as GF109203X, Go6976, chelerythrine, as well as cell-permeable PKC-isoform-specific inhibitors known as RACK (Receptors for Activated C-Kinase) inhibitors (45), and non-cell permeable inhibitors such as the PKC pseudosubstrate PKC(19-31) that can be effective by loading into the recording pipette during whole-cell patch-clamp recordings (28, 29). Although basal PKC levels in rat LC neurons, and presumably in other neurons where minimal rapid morphine-induced MOPr desensitization has been observed, are too low to demonstrate robust rapid morphine-induced MOPr desensitization, if LC neurons are incubated with morphine for a prolonged length of time (>4 h), PKC does not need to be exogenously activated to see profound morphine-induced MOPr desensitization. On the other hand, in both HEK293 cells and AtT20 cells, rapid morphine-induced MOPr desensitization can be readily observed (11, 28),

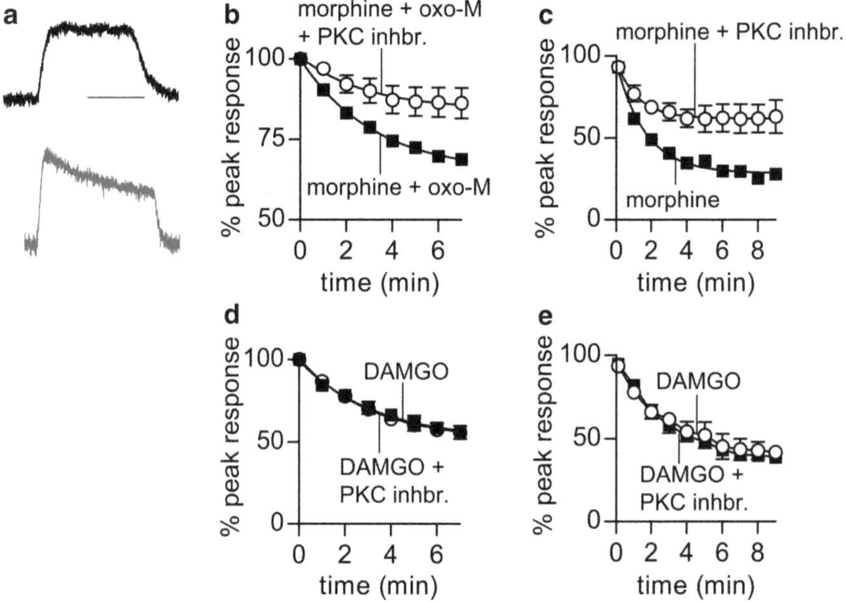

Fig. 4. Agonist-selective mechanisms of MOPr desensitization in rat LC neurons and HEK293 cells. (**a**) *Upper trace* showing MOPr evoked K^+-current elicited by 30 μM morphine in a rat LC neuron; minimal desensitization is observed. In the *lower trace*, morphine elicits much greater rapid MOPr desensitization when protein kinase C (PKC) is activated (by activation of M3 muscarinic receptors with 10 μM oxotremorine-M (oxo-M)). *Scale bar* 5 min. (**b**) In rat LC neurons, in the presence of oxo-M, morphine-induced MOPr desensitization is blocked by inhibition of PKC (using 3 μM chelerythrine). (**c**) In HEK293 cells, morphine produces robust MOPr desensitization, even without exogenous activation of PKC, which is reduced when PKC is inhibited (by 1 μM PKC(19-31) in the recording pipette). DAMGO-induced MOPr desensitization is not affected by PKC inhibition, either in rat LC neurons ((**d**) using 1 μM Go6976) or in HEK293 cells ((**e**) using 1 μM PKC(19-31) in the recording pipette). DAMGO applied at 10 μM. Figure adapted from (26, 28, 29).

without exogenous PKC activation, which can be attenuated by PKC inhibition (Fig. 4; (28); Chen, McPherson, and Henderson, 2011, personal communication). Whether or not this is due to higher PKC activity in these cells remains to be determined.

Inhibiting GRKs is generally more challenging than inhibiting most other kinases, as there are no truly effective small-molecule inhibitors. Of the GRK isoforms, GRK2 and 3 have been shown to mediate MOPr desensitization induced by certain agonists. A non-membrane-permeable GRK2 inhibitor peptide (W643-S670) is available and can be used when applied in the recording pipette during whole cell recordings (31, 44, 46), although this may be ineffective because of slow diffusion of large molecules from pipettes into large cells ((47); Bailey and Henderson, unpublished data). Similarly, there is a small-molecule GRK inhibitor (methyl-5-(2-(5-nitro-2-furyl)vinyl)-2-furoate, also known as β-ARK1 inhibitor) that can be effective when applied through the patch pipette (Fig. 5; (48)). In heterologous expression systems, it is relatively straightforward to inhibit specific GRK isoforms by expressing kinase-deficient dominant negative mutant (DNM) variants of the full-length proteins, for

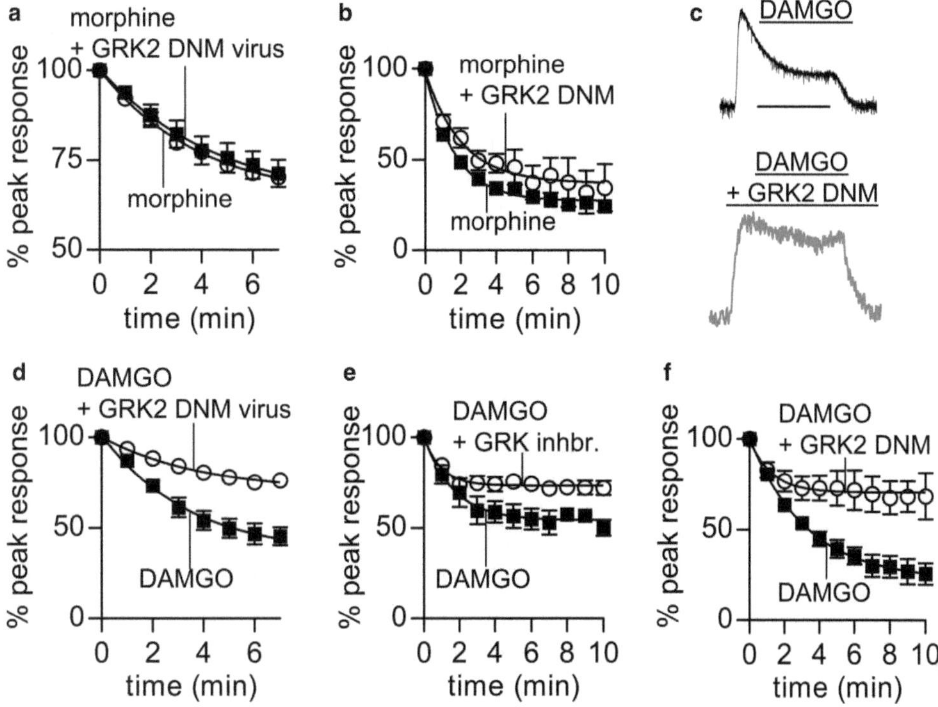

Fig. 5. Inhibition of GRK2 reduced DAMGO- but not morphine-induced MOPr desensitization. Expression of the kinase-deficient dominant negative mutant of GRK2 (GRK2 DNM) has no effect on morphine-induced MOPr desensitization, either in rat LC neurons ((**a**) morphine-induced desensitization shown in the presence of enhanced PKC activity using 1 μM phorbol ester, 12-myristate 13-acetate (PMA); GRK2 DNM expressed by viral-mediated gene transfer) or in HEK293 cells (**b**). By contrast, DAMGO-induced MOPr desensitization is significantly reduced by GRK inhibition. In rat LC neurons, GRK2 DNM was expressed using viral-mediated gene transfer ((**c**) representative traces; scale bar; 5 min, (**d**) or inhibited using β-ARK1 inhibitor (100–300 μM in the recording pipette (**e**)). GRK2 DNM was expressed in HEK293 cells (**f**)). Morphine was applied at 30 μM, DAMGO was applied at 10 μM. Figure adapted from (28, 29, 48).

example GRK2-K220R (Fig. 5; (28, 49)). Using this approach in native neurons is more difficult, but can be achieved using viral-mediated gene transfer, wherein viruses such as the adenovirus can be transformed so as to express the protein of interest (e.g., GRK2-K220R), bicistronically with a marker protein (e.g., Green Fluorescent Protein (GFP)) so as to identify infected cells. Viruses are injected directly into the brain region of interest and, using fluorescence microscopy, neurons expressing GFP (and therefore also GRK2-K220R) can be identified and recorded from. This is an approach we have taken recently to show that GRK2, but not GRK6, mediates DAMGO-induced MOPr desensitization in rat LC neurons (Fig. 5; (29)). One complicating factor in the application of this technique is the choice of promoter used, for example, the widely used CMV promoter is not effective in driving protein expression in monoaminergic neurons such as LC neurons, although this problem could be turned into an advantage by the use of cell-type specific promoters such as PRSx8 (for monoaminergic neurons; (50)). A final approach is to use knockout mice. Although the GRK2 knockout mouse is embryonically lethal (51), the GRK3 knockout mouse has been used to show that fentanyl- but not morphine-induced MOPr desensitization is mediated by GRKs (22). There is also the arrestin-3 knockout mouse, although there is some discrepancy between groups as to whether MOPr desensitization is affected in this mouse, perhaps because of redundancy issues (31, 52–54).

Overall, these studies suggest that agonist-induced MOPr desensitization can occur by at least two separate mechanisms: one is mediated by PKC and the other is mediated by GRK/arrestin. It is interesting to note that in native neurons, unlike in heterologous cell systems *in vitro*, PKC generally needs to be exogenously activated in order to see the PKC-mediated component, although *in vivo* sufficient ongoing PKC activation is present in order to see PKC-mediated tolerance (30, 48). Although our work has focused on the different roles of PKC and GRK/arrestin, there is also evidence to link other intracellular signaling mechanisms, such as ERK (Extracellular signal-Regulated Kinase) and CaMKII (Calcium/calmodulin-dependent protein Kinase II), to agonist-selective MOPr desensitization (31, 55).

2.4. MOPr Desensitization: Selecting the Probe and Quantifying Desensitization

One of the key advantages in using electrophysiological techniques to study GPCR desensitization is the ability to examine receptor desensitization in real-time (e.g., in Fig. 2). However, of the three traces shown in Fig. 2a, each of them shows MOPr desensitization *elicited* by three different agonists (Met-Enkephalin, DAMGO, and morphine), but also "probed" using each of those three different agonists. By taking this approach, it is difficult to directly compare the *amount* of receptor desensitization that each agonist produces. Often this is not a problem, if the experiment is only concerned with examining differences in the mechanisms of desensitization between two agonists (e.g., in Fig. 4), however, this approach is not appropriate when trying to quantify the amount of desensitization that each agonist has produced, and could lead to erroneous conclusions.

In pharmacological terms, GPCR desensitization is assumed to be similar to application of an irreversible antagonist, in that, the process of desensitization decreases the number of available receptors that an agonist can activate, without changing the intrinsic properties (affinity and intrinsic efficacy) of that agonist. Figure 6a shows modeled data of a DAMGO concentration–response curve in rat LC neurons under basal conditions, and following removal of 25, 50, 75, and 90% of available receptors. As DAMGO is a full agonist, with a large receptor reserve, there is little change in the concentration–response curve until in excess of 50% of the available receptors have been removed. So, if 10 μM DAMGO was applied to these cells for, say, 30 min, and caused 50% desensitization (defined here as a 50% reduction in total receptor function) in that time, the response to 10 μM DAMGO

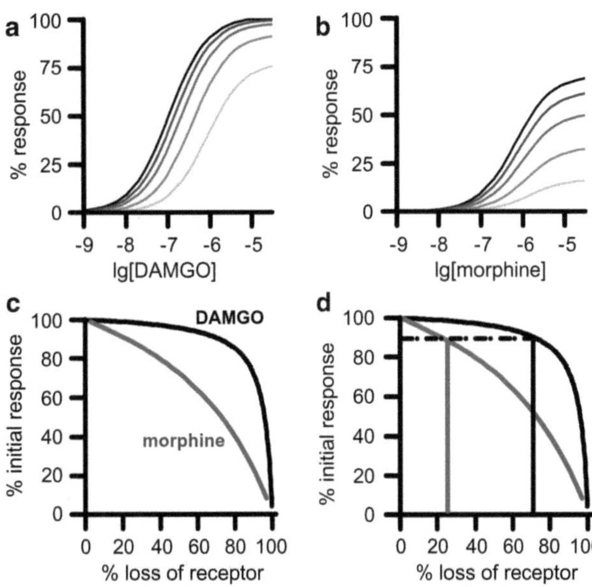

Fig. 6. Quantification of desensitization-induced loss of receptor function. Simulated concentration response curves to (**a**) DAMGO and (**b**) morphine under control conditions (*black lines*) and following 25, 50, 75, and 90% reduction in number of functional receptors (successive *gray lines* to the *right*). (**c**, **d**) Percentage decrease in response to 30 μM morphine and 10 μM DAMGO, as a function of percentage loss of receptor function. Simulations based on experimental data obtained from rat locus coeruleus neurons (30), and using the operational model of agonism (4, 57, 58). (**d**) Also shows that in order to observe a 10% loss of initial response, a far greater loss of receptor function is required for DAMGO than morphine.

would not have decreased appreciably over those 30 min; in effect, it would look like no desensitization had taken place, clearly a "false-negative" conclusion.

This is potentially a problem for any experiment designed to study agonist-selective GPCR desensitization, and has been extremely well described previously (56), but, in practice this can be overcome in a number of ways. First, a submaximal concentration of a "probe" agonist can be applied at the end of, or during, application of the "desensitizing" agonist. For a full agonist a submaximal concentration of the agonist will be more sensitive to a reduction in functional receptor number than a saturating concentration (Fig. 6a). Many studies in locus coeruleus brain slices have used Met-Enkephalin as this ligand (14, 15, 19), as it has the advantage of washing out of the slice very rapidly compared to most other agonists. Or, a partial agonist could be applied after the desensitizing agonist. As a partial agonist has no receptor reserve, any decrease in the number of functional receptors will cause a decrease in response to that partial agonist, whatever concentration is used (Fig. 4b). For example, in locus coeruleus brain slice experiments, morphine can be used (30).

For similar reasons, when directly comparing the ability of agonists to induce GPCR desensitization, it is necessary to apply concentrations of the desensitizing agonist at equivalent receptor occupancy levels. By far the easiest way to achieve this is to apply the agonists at approximately receptor-saturating concentrations (e.g., a concentration of 100× the K_D value results in approximately 99% receptor occupancy).

Figure 6a, b was derived from original experimental data transformed using the operational model of agonist action ((57, 58); and see Sect. 3.1). This model can in turn be used to quantify the loss of receptor function following agonist-induced GPCR desensitization. The operational model is explained in more detail in Sect. 3.1, but, briefly, concentration response curves can be fitted to this equation, and a value of τ can be derived. This value is dependent on both the intrinsic efficacy of the agonist, and the number of functional receptors. So, if we assume that receptor desensitization only causes a decrease in the number of functional receptors, one can derive a τ value under baseline conditions (τ_1) and again following desensitization (τ_2), and the percentage loss of receptor function (f) would be $f = 100 \times (1 - (\tau_2 / \tau_1))$ (4, 30, 59). Figure 6c, d shows modeled data using this approach, derived using original parameters taken from rat LC neuron experiments. Any decrease in the actual, experimental response to the "probe" agonist (in this case 30 μM morphine and 10 μM DAMGO) can be converted into a % loss of receptor function. What is evident here is that the lower the efficacy of the "probe" agonist used, the more linear is the relationship between reduction in response and reduction in receptor function. For example, if DAMGO is used as the probe, in excess of 72% of loss of receptor function must have taken place even to see just a 10% decrease in actual response (whereas for morphine only 24% loss of receptor function would be required; Fig. 6d). Again this states the case for the use of a lower efficacy agonist as the "probe" agonist when wanting to directly compare the amount of desensitization caused by different agonists.

3. Is Agonist-Selective MOPr Desensitization Based on Agonist Efficacy?

From these studies outlined above, we were led to the conclusion that there are at least two distinct mechanisms of agonist-induced MOPr desensitization. In particular, agonists such as DAMGO cause MOPr desensitization almost exclusively by a GRK/arrestin-dependent mechanism, whereas agonists such as morphine cause MOPr desensitization largely by a PKC-dependent mechanism. As stated earlier, using a linear model of GPCR activation, one might expect DAMGO (a higher efficacy agonist than morphine) to be more efficient than morphine at causing GRK/arrestin-dependent desensitization, but this cannot account for morphine's ability to cause MOPr desensitization by a *different* mechanism. To our mind, this can only be explained as a form of functional selectivity where morphine preferentially stabilizes a different active conformational state of the receptor to DAMGO.

If the different mechanisms of desensitization of MOPr by morphine and DAMGO are due to different agonist efficacy, then it is clear that irrespective of the assay used to measure MOPr function, reliable estimates of agonist efficacy need to be obtained.

3.1. Measuring Agonist Efficacy

MOPr agonist efficacy has in the past been estimated by comparing agonist responses following addition of a saturating concentration of agonist or by comparing maximum responses from concentration–effect curves (e.g., (60)). However this is actually a measure of intrinsic activity (maximum response of the agonist compared to maximum response of a full agonist) rather than an absolute measure of efficacy. Intrinsic activity varies depending upon the level of receptor expression and, in addition, intrinsic activity is the same (i.e., 1) for all agonists that are full agonists in a particular system. In fact, the term efficacy includes both drug and tissue-dependent components, and so it is better to use the term intrinsic efficacy (61, 62), which refers specifically to the drug-dependent component of efficacy. In addition relative intrinsic efficacy can be described as the relative abilities of agonists to produce a response resulting from the occupation of the same number of receptors in a tissue. There have been attempts to measure the relative intrinsic efficacy of opioid agonists at MOPr; in a study by Selley et al. (63) agonist efficacy was estimated by the method of Ehlert (64), which is based in part upon the difference between the K_D and EC_{50} values of the agonist. Other studies to quantify relative intrinsic efficacy (11) have involved the receptor inactivation method pioneered by Furchgott and Bursztyn (65), while we (30) and others (59) have used operational analysis (57, 58) to determine values of relative intrinsic efficacy, and also used this approach to assess functional desensitization of MOPr responses (see Sect. 2.4). For this analysis, untransformed concentration–effect curves are fitted to the following equation:

$$E = EmT^n[A]^n/(K_D + [A])^n + T^n[A]^n,$$

where E is response, T is τ which is the operational efficacy, (A) is the concentration of agonist, K_D is the equilibrium dissociation constant of the agonist, Em is the theoretical maximum response of the system, and n represents a slope factor (57, 58). Such analysis can be conveniently performed using, for example,

GraphPad Prism. For operational analysis, it is also necessary to determine K_D values for the agonists concerned. This can be achieved by radioligand binding assay, preferably using membranes prepared from the same cells that are used for the functional assay, or by the receptor inactivation method using the MOPr antagonist β-funaltrexamine (30).

In theory at least, any assay of MOPr activity can be used to determine agonist efficacy as well as used to study MOPr desensitization. Apart from the electrophysiological approach described above, we and others have used a range of assays to study MOPr coupling and desensitization including (^{35}S)GTPγS assays in membrane preparations (28), or cAMP assays in intact cells or membrane preparations (66). In addition ERK activation is also used as a marker for MOPr function (67). We have used the (^{35}S)GTPγS assay in broken cell membranes to determine agonist intrinsic efficacy. An advantage of this method (68) is that the response is closely linked to receptor occupancy, and is generally more straightforward than cAMP assays since agonist activation of MOPr leads to increased (^{35}S)GTPγS binding whereas in the cAMP assay the level of cAMP must be raised in order for MOPr-mediated inhibition of cAMP to be observed. A consideration with the (^{35}S)GTPγS assay is that cytosolic regulatory proteins such as arrestins will be largely lost during membrane preparation, and it is not clear how this might influence measurements of agonist efficacy. In these assays, (^{35}S)GTPγS associated with membranes is usually separated from free radionucleotide by filtration, but we have used a Scintillation Proximity Assay (SPA) to assay bound (^{35}S)GTPγS, which involves the centrifugation of multiwell plates and is suited to a high-throughput approach rather than filtration of individual samples and the addition of scintillation fluid. Example concentration–effect curves for DAMGO and morphine-stimulated (^{35}S) GTPγS binding to membranes of HEK293 cells stably expressing the rat MOPr are shown in Fig. 7a.

3.2. Measuring Agonist-Induced Arrestin Translocation

Given that nonvisual arrestins appear to play an important part in the regulation of MOPr by some agonists but not others, it seems important that the recruitment of arrestins by the MOPr can be quantified. In addition, since arrestins can form signaling scaffolds, arrestin recruitment potentially forms an important further measure of MOPr signaling. Receptor internalization has been used as an indirect measure of arrestin recruitment; however, there are arrestin-independent pathways for GPCR trafficking (69) and so this is not a reliable means to quantify arrestin interaction. The translocation of nonvisual arrestins, usually tagged with GFP, from cytosol to plasma membrane has been widely used as a marker for GPCR activation as well as regulation by an arrestin-dependent process (70). While this can produce convincing live cell images of arrestin translocation (71), a drawback of this approach has been

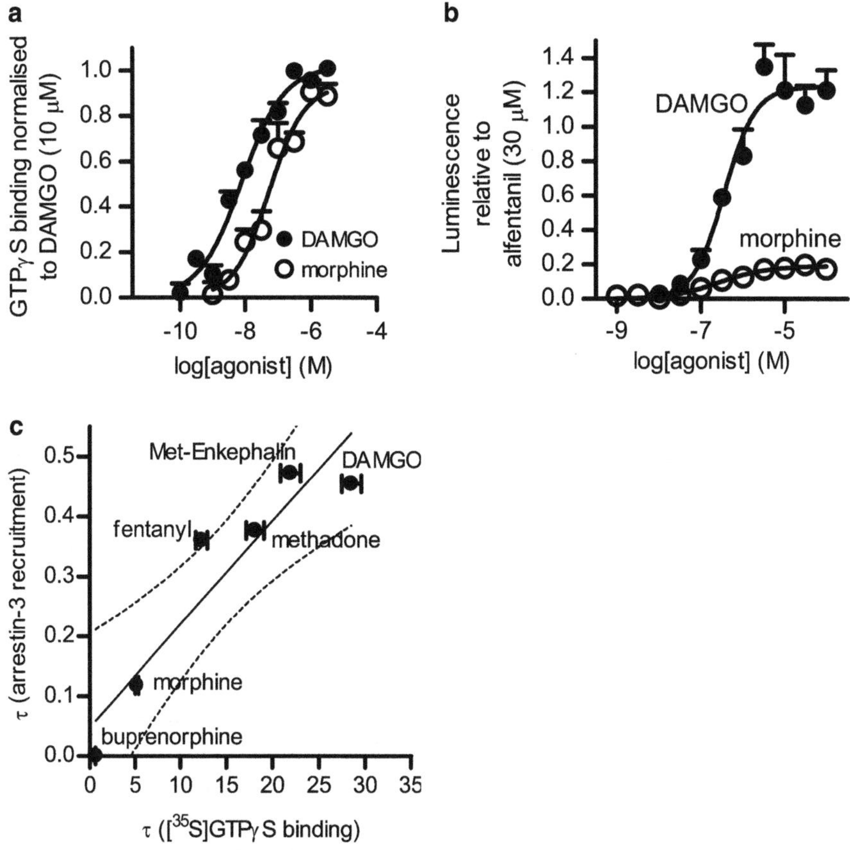

Fig. 7. Quantification of the operational efficacy of MOPr agonists for G protein activation and arrestin recruitment. (**a**) DAMGO and morphine-induced (^{35}S)GTPγS binding to membranes of HEK293 cells stably expressing MOPr. (**b**) DAMGO and morphine-induced arrestin-3 recruitment quantified by an enzyme complementation assay. (**c**) Correlation of operational efficacy (τ) values for G protein activation and arrestin-3 recruitment for a series of MOPr agonists. Unpublished data provided by the authors and colleagues.

the difficulty in quantifying the agonist-induced responses recorded in such images, particularly with regard to establishing agonist concentration–response relationships. The coimmunoprecipitation of arrestins with receptor has also been used to quantify receptor–arrestin interactions (72); however, this involves gel electrophoresis followed by Western blotting and is labor-intensive. Fluorescence Resonance Energy Transfer (FRET) has also been used to study real time changes in GPCR-arrestin interactions (73); in this approach the receptor can be tagged at the C-terminus with Cyan Fluorescent Protein (CFP) and the arrestin with Yellow Fluorescent Protein (YFP), with FRET occurring only when the CFP and YFP are in very close proximity, i.e., when the arrestin is recruited to the GPCR. Although this has the advantage of recording real time interactions in an intact cell, this approach does require advanced imaging equipment and the FRET signals obtained can be quite small.

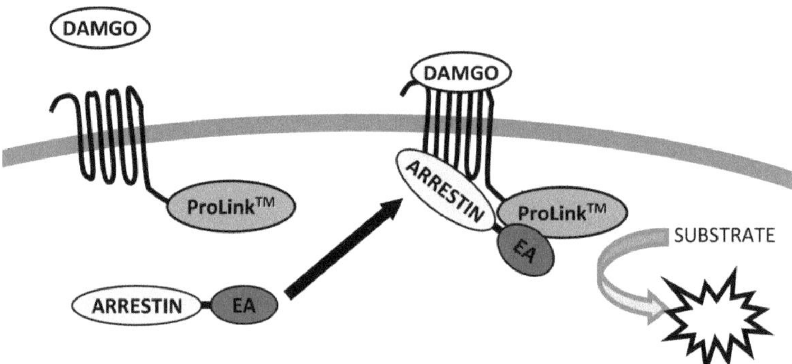

Fig. 8. Principle of the enzyme complementation assay for quantifying arrestin recruitment to a GPCR. The receptor is tagged at the C-terminus with a ProLink™ moiety and the arrestin is tagged with an Enzyme Acceptor (EA) moiety. Both are expressed in a suitable cell system such as HEK293 cells. When agonist binds to the receptor, arrestin is recruited to the receptor and the ProLink™ and EA moieties come into close contact and reconstitute functional β-galactosidase activity, which is quantified by adding a reagent which forms a chemiluminescent product.

In an attempt to provide a more reliable assay for arrestin recruitment, enzyme complementation assays have been developed (74) that are amenable to high-throughput technology. In one such approach (http://www.discoverx.com/gpcrs/arrestin.php), cells are stably transfected with the GPCR of interest coupled at the COOH-terminus to a ProLink™ moiety, and also with nonvisual arrestin coupled to an N-terminal deletion mutant of β-galactosidase (the Enzyme Acceptor, or EA), such that if the arrestin is recruited by the GPCR in response to agonist, then the EA and the ProLink™ come together to form functional β-galactosidase activity, which can be readily quantified by adding a proprietary reagent and assessing chemiluminescence in a standard plate reader (Fig. 8). The advantage of this method is that it can be used to rapidly generate full concentration–effect curves for agonists (Fig. 7b), which previously had not been a realistic option for most laboratories. There are issues still to consider with this methodology, such as the length of time required to obtain measurable agonist responses (many minutes to hours), and in addition it is not clear how reversible the receptor–arrestin interaction is under these conditions, and also whether or not the receptor–arrestin complex trafficks normally when tagged with ProLink™ and the EA moieties, and how this might affect the signal obtained. Nevertheless, this approach provides for the first time a means to rapidly obtain quantitative data for arrestin recruitment to a GPCR. Figure 7b shows concentration–effect curves for DAMGO- and morphine-stimulated arrestin-3-EA recruitment to MOPr-ProLink™ in HEK293 cells.

Once agonist concentration–effect curves have been constructed, then they can be fitted to the operational model and values of operational efficacy (τ) obtained. If this type of analysis is also performed on another MOPr response, and for a sufficient number of agonists, then a correlation between agonist operational efficacies for the two responses can be made. We have done this with values of operational efficacy for (^{35}S)GTPγS and arrestin-3 recruitment measured in HEK293 cells, and the results are shown in Fig. 7c. This type of analysis therefore should be able to detect agonists which display functional selectivity with regard to G protein coupling and arrestin recruitment. There is no reason why other MOPr responses (e.g., internalization, phosphorylation, ERK activation) cannot be analyzed in this way and further correlations carried out.

4. Future Directions

There remains much controversy over the mechanisms of MOPr desensitization, with evidence from various studies to indicate that GRKs and arrestins, PKC, CaMKII, and ERKs can all contribute to MOPr desensitization under various conditions. However, the precise involvement of these kinases in MOPr desensitization in different neuronal types, and in response to different agonists, has yet to be explored in depth. The increasing availability of more selective kinase inhibitors as well as the means to introduce mutant GRKs and arrestins as well as siRNA constructs into neurons (29) should accelerate progress in this field. Whether or not a unifying picture of MOPr desensitization in neurons will emerge is an interesting question, particularly in light of the fact that the morphine-occupied MOPr recruits arrestins in some neurons but not others (7).

The now compelling evidence for the role of PKC in morphine-induced MOPr desensitization and morphine tolerance is particularly intriguing as there is little evidence to indicate that agonist activation of the $G_{i/o}$-coupled MOPr leads to PKC activation. However, a recent study reports that PKCε is selectively translocated to the MOPr "signaling complex" following morphine addition (32), while we have provided evidence for a role of PKCα in morphine-induced MOPr desensitization in neurons (29). However, it is still unknown how PKC isoforms are recruited to the morphine-occupied MOPr but not, for example, the DAMGO-occupied MOPr. Convenient assays for the activation/recruitment of individual PKC isoforms would be very useful in exploring such mechanisms in greater depth.

A key question for the future is the identity of the amino acid residues in MOPr that are phosphorylated in response to agonist (and potentially under basal conditions also). Many previous studies have tried to address this issue using point mutants or deletion mutants of the receptor, but these have led to variable results and conclusions (75). A better approach would be to identify phosphoacceptor residues by mass spectrometry of receptor peptide sequences derived from receptors expressed in, and isolated from, agonist-treated cells. This has been achieved for some GPCRs already (76), but the success of the approach depends upon whether or not sufficient receptor protein can be isolated to take forward for enzymatic digestion and mass spectrometry. It will be extremely interesting to know whether the phosphorylation of MOPr in response to various agonists differs in quantity and/or quality, the latter providing good evidence for agonist-dependent desensitization mechanisms. This type of approach would also facilitate the production of antiphosphoreceptor antibodies (as with the Ser^{375} antiphosphoreceptor antibody; (12)), and also the development of knock-in mice with specific mutations that could test the relationship between MOPr phosphorylation, desensitization, and *in vivo* tolerance to opioid agonists.

With regard to different active conformations of MOPr, studies with other GPCRs suggest ways ahead for the MOPr. These include the use of fluorescent probes to monitor ligand-induced changes in receptor protein conformation (77), and the use of intra-receptor FRET to detect distinct GPCR conformations when the receptor is occupied by different agonist ligands (78). Such approaches have yet to be seriously applied to the MOPr, but would answer the important question of whether morphine stabilizes a truly independent active conformation of MOPr, or whether morphine, due to its low efficacy, instead stabilizes a conformation of the MOPr that is an intermediate and transitory conformation of the receptor activated by high-efficacy agonists such as DAMGO.

5. Conclusions

It is now clear that opioid agonists can induce MOPr desensitization by different molecular mechanisms irrespective of whether the receptors in question are heterologously expressed in a cell line or are endogenously expressed in neurons. Whether this is due to multiple active conformations of MOPr, or is also related to agonist efficacy, remains to be fully resolved, as does the precise role of MOPr desensitization in the development of opioid tolerance *in vivo*. However, recent developments in the area of molecular pharmacology suggest that it will not be too long before answers to these important questions are found.

6. Note

Since this Chapter was written, we have published a study where we have used some of the assays described here to correlate G protein activation with arrestin recruitment for a large number of MOPr agonists (79).

References

1. Hausdorff WP, Caron MG, Lefkowitz RJ (1990) Turning off the signal: desensitization of beta-adrenergic receptor function. FASEB J **4**:2881–2889
2. Benovic JL, Strasser RH, Caron MG et al (1986) *Beta*-adrenergic receptor kinase:identification of a novel protein kinase that phosphorylates the agonist-occupied form of the receptor. Proc Natl Acad Sci USA **83**:2797–2801
3. Lohse MJ, Benovic JL, Codina J et al (1990) *Beta*-Arrestin:a protein that regulates *beta*-adrenergic receptor function. Science **248**:1547–1550
4. Lohse MJ, Benovic JL, Caron MG et al (1990) Multiple pathways of rapid *beta* 2-adrenergic receptor desensitization: Delineation with specific inhibitors. J Biol Chem **265**:3202–3211
5. Pitcher J, Lohse MJ, Codina J et al (1992) Desensitization of the isolated *beta* 2-adrenergic receptor by *beta*-adrenergic receptor kinase, cAMP-dependent protein kinase, and protein kinase C occurs via distinct molecular mechanisms. Biochemistry **31**:3193–3197
6. Kelly E, Bailey CP, Henderson G (2008) Agonist-selective mechanisms of GPCR desensitization. Br J Pharmacol **153 Suppl 1**: S379–388
7. Haberstock-Debic H, Kim KA, Yu YJ et al (2005) Morphine promotes rapid, *arrestin*-dependent endocytosis of *mu*-opioid receptors in striatal neurons. J Neurosci **25**:7847–7857
8. Raymond JR (1995) Multiple mechanisms of receptor-G protein signaling specificity. Am J Physiol **269**:F141–158
9. Urban JD, Clarke WP, von Zastrow M et al (2007) Functional selectivity and classical concepts of quantitative pharmacology. J Pharmacol Exp Ther **320**:1–13
10. Keith DE, Murray SR, Zaki PA et al (1996) Morphine activates opioid receptors without causing their rapid internalization. J Biol Chem **271**:19021–19024
11. Borgland SL, Connor M, Osborne PB et al (2003) Opioid agonists have a different efficacy profiles for G-protein activation, rapid desensitization, and endocytosis of *mu*-opioid receptors. J Biol Chem **278**:18776–18784
12. Schulz S, Mayer D, Pfeiffer M et al (2004) Morphine induces terminal *mu*-opioid receptor desensitization by sustained phosphorylation of serine-375. EMBO J **23**:3282–3289
13. Koch T, Widera A, Bartzsch K et al (2005) Receptor endocytosis counteracts the development of opioid tolerance. Mol Pharmacol **67**:280–287
14. Arttamangkul S, Torrecilla M, Kobayashi K et al (2006) Separation of *mu*-opioid receptor desensitization and internalization: endogenous receptors in primary neuronal cultures. J Neurosci **26**:4118–4125
15. Harris GC, Williams JT (1991) Transient homologous *mu*-opioid receptor desensitization in rat locus coeruleus neurons. J Neurosci **11**:2574–2581
16. Yu Y, Zhang L, Yin X et al (1997) *Mu* opioid receptor phosphorylation, desensitization, and ligand efficacy. J Biol Chem **272**: 28869–28874
17. Kovoor A, Celver JP, Wu A et al (1998) Agonist induced homologous desensitization of *mu*-opioid receptors mediated by G protein-coupled receptor kinases is dependent on agonist efficacy. Mol Pharmacol **54**:704–711
18. Whistler JL, von Zastrow M (1998) Morphine-activated opioid receptors elude desensitization by *beta*-arrestin. Proc Natl Acad Sci USA **95**:9914–9919
19. Alvarez VA, Arttamangkul S, Dang V et al (2002) *Mu*-opioid receptors: Ligand-dependent activation of potassium conductance, desensitization, and internalization. J Neurosci **22**:5769–5776
20. Blanchet C, Sollini M, Lüscher C (2003) Two distinct forms of desensitization of G-protein coupled inwardly rectifying potassium currents evoked by alkaloid and peptide *mu*-opioid receptor agonists. Mol Cell Neurosci **24**: 517–523
21. Bailey CP, Couch D, Johnson E et al (2003) *Mu*-opioid receptor desensitization in mature

rat neurons: lack of interaction between DAMGO and morphine. J Neurosci **23**: 10515–10520

22. Terman GW, Jin W, Cheong YP et al (2004) G-protein receptor kinase 3 (GRK3) influences opioid analgesic tolerance but not opioid withdrawal. Br J Pharmacol **141**:55–64
23. Clark RB, Knoll BJ, Barber R (1999) Partial agonists and G protein-coupled receptor desensitization. Trends Pharmacol Sci **20**: 279–286
24. Mandyam CD, Thakker DR, Christensen JL et al (2002) Orphanin FQ/nociceptin-mediated desensitization of opioid receptor-like 1 receptor and *mu* opioid receptors involves protein kinase C:a molecular mechanism for heterologous cross-talk. J Pharmacol Exp Ther **302**:502–9
25. Mandyam CD, Thakker DR, Standifer KM (2003) *Mu*-opioid-induced desensitization of opioid receptor-like 1 and *mu*-opioid receptors: differential intracellular signaling determines receptor sensitivity. J Pharmacol Exp Ther **306**:965–972
26. Bailey CP, Kelly E, Henderson G (2004) Protein kinase C activation enhances morphine-induced rapid desensitization of *mu*-opioid receptors in mature rat locus ceruleus neurons. Mol Pharmacol **66**:1592–1598
27. Bailey CP, Smith FL, Kelly E et al (2006) How important is protein kinase C in *mu*-opioid receptor desensitization and morphine tolerance? Trends Pharmacol Sci **27**:558–565
28. Johnson EA, Oldfield S, Braksator E et al (2006) Agonist-selective mechanisms of *mu*-opioid receptor desensitization in human embryonic kidney 293 cells. Mol Pharmacol **70**:676–685
29. Bailey CP, Oldfield S, Llorente J et al (2009) Involvement of PKC*alpha* and G-protein-coupled receptor kinase 2 in agonist-selective desensitization of *mu*-opioid receptors in mature brain neurons. Br J Pharmacol **158**:157–164
30. Bailey CP, Llorente J, Gabra BH et al (2009) Role of protein kinase C and *mu*-opioid receptor (MOPr) desensitization in tolerance to morphine in rat locus coeruleus neurons. Eur J Neurosci **29**:307–318
31. Dang VC, Napier IA, Christie MJ (2009) Two distinct mechanisms mediate acute *mu*-opioid receptor desensitization in native neurons. J Neurosci **29**:3322–3327
32. Chu J, Zheng H, Zhang Y et al (2010) Agonist-dependent *mu*-opioid receptor signaling can lead to heterologous desensitization. Cell Signal **22:**684–696
33. Lewis MM, Watts VJ, Lawler CP et al (1998) Homologous desensitization of the D1A dopamine receptor: efficacy in causing desensitization dissociates from both receptor occupancy and functional potency. J Pharmacol Exp Ther **286**:345–353
34. Simmons MA (2006) Functional selectivity of NK1 receptor signaling:peptide agonists can preferentially produce receptor activation or desensitization. J Pharmacol Exp Ther **319**:907–913
35. Walz W (ed) (2007) Neuromethods, Vol. 38: Patch-clamp analysis: Advanced techniques, Second Edition. Humana, Totowa
36. Dang VC, Williams JT (2005) Morphine-induced *mu*-opioid receptor desensitization. Mol Pharmacol **68**:1127–1132
37. Ingram S, Wilding TJ, McCleskey EW et al (1997) Efficacy and kinetics of opioid action on acutely dissociated neurons. Mol Pharmacol **52**:136–143
38. North RA, Williams JT (1985) On the potassium conductance increased by opioids in rat locus coeruleus neurones. J Physiol **364**: 265–280
39. Connor M, Vaughan CW, Chieng B et al (1996) Nociceptin receptor coupling to a potassium conductance in rat locus coeruleus neurones *in vitro.* Br J Pharmacol **119**:1614–1618
40. Christie MJ, Williams JT, North RA (1987) Cellular mechanisms of opioid tolerance:studies in single brain neurons. Mol Pharmacol **32**:633–638
41. Rodriguez-Martin I, Braksator E, Bailey CP et al (2008) Methadone: does it really have low efficacy at *mu*-opioid receptors? Neuroreport **19**:589–593
42. Bagley EE, Chieng BC, Christie MJ et al (2005) Opioid tolerance in periaqueductal gray neurons isolated from mice chronically treated with morphine. Br J Pharmacol **146**:68–76
43. Johnson EE, Chieng B, Napier I et al (2006) Decreased *mu*-opioid receptor signalling and a reduction in calcium current density in sensory neurons from chronically morphine-treated mice. Br J Pharmacol **148**:947–955
44. Li AH, Wang HL (2001) G protein-coupled receptor kinase 2 mediates *mu*-opioid receptor desensitization in GABAergic neurons of the nucleus raphe magnus. J Neurochem **77**: 435–444
45. Schechtman D, Mochly-Rosen D (2002) Isozyme-specific inhibitors and activators of protein kinase C. Methods Enzymol **345**: 470–489
46. Koch WJ, Hawes BE, Inglese J et al (1994) Cellular expression of the carboxyl terminus of a G protein-coupled receptor kinase attenuates G *beta gamma*-mediated signaling. J Biol Chem **269**:6193–6197

47. Pusch, M. and Neher, E (1988) Rates of diffusional exchange between small cells and a measuring patch pipette. Pflugers Arch. **411**, pp. 204–211
48. Hull LC, Llorente J, Gabra BH et al (2010) The effect of PKC and GRK inhibition on tolerance induced by *mu*-opioid agonists of different efficacy. J Pharmacol Exp Ther **332**:1127–1135
49. Kong G, Penn R, Benovic JL (1994) A *beta*-adrenergic receptor kinase dominant negative mutant attenuates desensitization of the *beta* 2-adrenergic receptor. J Biol Chem **269**: 13084–13087
50. Hwang DY, Carlezon WA, Isacson O et al (2001). A high-efficiency synthetic promoter that drives transgene expression selectively in noradrenergic neurons. Hum Gene Ther **12**:1731–1740
51. Jaber M, Koch WJ, Rockman H et al (1996) Essential role of *beta*-adrenergic receptor kinase 1 in cardiac development and function. Proc Natl Acad Sci USA **93**:12974–12979
52. Bohn LM, Gainetdinov RR, Lin FT et al (2000) *Mu*-opioid receptor desensitization by *beta*-arrestin-2 determines morphine tolerance but not dependence. Nature **408**:720–723
53. Dang VC, Christie MJ (2006) *Beta*-arrestin-2-independent regulation of *mu* opioid receptor. Soc Neurosci Abstr **32**:426.11
54. Walwyn W, Evans CJ, Hales TG (2007) *Beta*-arrestin2 and c-Src regulate the constitutive activity and recycling of *mu* opioid receptors in dorsal root ganglion neurons. J Neurosci **27**:5092–5104
55. Koch T, Kroslak T, Mayer P et al (1997) Site mutation in the rat *mu*-opioid receptor demonstrates the involvement of calcium/calmodulin-dependent protein kinase II in agonist-mediated desensitisation. J Neurochem **69**:1767–1770
56. Connor M, Osborne OB, Christie MJ (2004) *Mu*-opioid receptor desensitization:is morphine different? Br J Pharmacol **143**:685-696
57. Black JW, Leff P (1983) Operational models of pharmacological agonism. Proc R Soc Lond B Biol Sci **220**:141–162.
58. Black JW, Leff P, Shankley NP et al (1985) An operational model of pharmacological agonism: the effect of E/(A) curve shape on agonist dissociation constant estimation. Br J Pharmacol **84**:561–571
59. Osborne PB, Williams JT (1995) Characterization of acute homologous desensitization of *mu*-opioid receptor-induced currents in locus coeruleus neurones. Br J Pharmacol **115**:925–932
60. Clark MJ, Furman CA, Gilson TD et al (2006) Comparison of the relative efficacy and potency of *mu*-opioid agonists to activate $Galpha_{i/o}$ proteins containing a pertussis toxin insensitive mutation. J Pharmacol Exp Ther **317**: 858–864
61. Clarke WP, Bond RA (1998) The elusive nature of intrinsic efficacy. Trends Pharmacol Sci **19**:270–276
62. Christopoulos A, El-Fakahany EE (1999) Qualitative and quantitative assessment of relative agonist efficacy. Biochem Pharmacol **58**:735–748
63. Selley DE, Liu Q, Childers SR (1998) Signal transduction correlates of *mu*-opioid agonist intrinsic efficacy: Receptor-stimulated (^{35}S) GTP*gamma*S binding in mMOR-CHO cells and rat thalamus. J Pharmacol Exp Ther **285**:496–505
64. Ehlert FJ (1985) The relationship between muscarinic receptor occupancy and adenylate cyclase inhibition in the rabbit myocardium. Mol Pharmacol **28**:410–421
65. Furchgott RF, and Bursztyn P (1967) Comparison of dissociation constants and of relative efficacies of selected agonists acting on parasympathomimetic receptors. Ann NY Acad Sci **144**:882–893
66. Law P-Y, Erickson LJ, El-Kouhen R et al (2000) Receptor density and recycling affect the rate of agonist-induced desensitization of *mu*-opioid receptor. Mol Pharmacol **58**: 388–398
67. Zheng H, Loh HH, Law PY (2008) *Beta*-arrestin-dependent *mu*-opioid receptor-activated extracellular signal-regulated kinases (ERKs) translocate to nucleus in contrast to G protein-dependent ERK activation. Mol Pharmacol **73**:178–190
68. Harrison C, Traynor JR (2003) The (^{35}S) GTP*gamma*S binding assay: approaches and applications in pharmacology. Life Sci **12**: 489–508
69. Van Koppen CJ, Jakobs KH (2004) Arrestin-independent internalization of G protein-coupled receptors. Mol Pharmacol **66**:365–367
70. Kallal L, Benovic JL (2000) Using green fluorescent proteins to study G-protein-coupled receptor localization and trafficking. Trends Pharmacol Sci **21**:175–180
71. Mundell SJ, Matharu AL, Pula G et al (2001) Agonist-induced internalization of the metabotropic glutamate receptor 1a is arrestin- and dynamin-dependent. J Neurochem **78**: 546–551
72. Mundell SJ, Pula G, McIlhinney RA et al (2004) Desensitization and internalization of metabotropic glutamate receptor 1a following activation of heterologous Gq/11-coupled receptors. Biochemistry **43**:7541–7551

73. Krasel C, Bunemann M, Lorenz K et al (2005) *Beta*-arrestin binding to the *beta*2-adrenergic receptor requires both receptor phosphorylation and receptor activation. J Biol Chem **280**:9528–9535

74. van Der Lee MM, Bras M, van Koppen CJ et al (2008) *Beta*-arrestin recruitment assay for the identification of agonists of the sphingosine 1-phosphate receptor EDG1. J Biomol Screen **13**:986–998

75. Johnson EE, Christie MJ, Connor M (2005) The role of opioid receptor phosphorylation and trafficking in adaptations to persistent opioid treatment. Neurosignals **14**:290–302

76. Busillo JM, Armando S, Sengupta R et al (2010) Site-specific phosphorylation of CXCR4 is dynamically regulated by multiple kinases and results in differential modulation of CXCR4 signaling. J Biol Chem **285**: 7805–7817

77. Kobilka BK, Gether U (2002) Use of fluorescence spectroscopy to study conformational changes in the *beta* 2-adrenoceptor. Methods Enzymol **343**:170–182

78. Zurn A, Zabel U, Vilardaga JP et al (2009) Fluorescence resonance energy transfer analysis of *alpha* 2a-adrenergic receptor activation reveals distinct agonist-specific conformational changes. Mol Pharmacol **75**:534–541

79. McPherson J, Rivero G, Baptist M et al (2010) *mu*-Opioid receptor internalization: correlation of agonist operational efficacy for G protein activation with ability to activate processes leading to internalization. Mol Pharmacol 78: 756–766

Chapter 17

Detecting the Role of Arrestins in G Protein-Coupled Receptor Regulation

Laura M. Bohn and Patricia H. McDonald

Abstract

G protein-coupled receptors (GPCRs) are the major sites of actions for the body's endogenous hormones and neurotransmitters which make them ideal targets for pharmaceutical development with the goal of either mimicking the normal transmitter response or tempering it. In recent years, targeting GPCRs has become more complicated as we realize that drug action at receptors is "context dependent" such that activation and inhibition is limited to the response evaluated and agonist and antagonist become terms that reflect a particular condition of the experimental or physiological output. Therefore, the composition of the receptor's immediate environment may determine activation profiles as posttranslational modifications of the receptor or of the binding partners can ultimately lead to regulation of the responsiveness of the receptor.

Key words: Arrestins, Receptor regulation, Phosphorylation, Translocation assay

1. Introduction

One means of receptor regulation involves the phosphorylation by GPCR kinases (GRKs) and subsequent interactions with βarrestins (1, 2). GRKs and βarrestins are ubiquitously expressed when examined on the organ level with some tissues expressing up to five types of nonvisual GRKs. The two isoforms of βarrestins, βarrestin1 and 2, are considered for the most part to be ubiquitously expressed. However, in brain, it appears that βarrestin1 expression is greater than βarrestin2 expression and relative levels of each βarrestin form appear to vary within distinct brain regions (3–5). In cellular models, modifying the expression levels of particular GRKs and βarrestins can shift potency profiles for G protein-coupling in a manner demonstrating a role of these regulatory proteins in desensitizing this particular signaling

Craig W. Stevens (ed.), *Methods for the Discovery and Characterization of G Protein-Coupled Receptors*, Neuromethods, vol. 60, DOI 10.1007/978-1-61779-179-6_17, © Springer Science+Business Media, LLC 2011

pathway. Moreover, genetic deletion or RNA interference of GRK and βarrestin expression enhances G protein-coupling pathways downstream of a number of GPCRs, including opioid receptors, serotonin receptors, adrenergic receptors, and chemokine receptors (5–8).

In addition to affecting receptor coupling to G proteins, the GRKs and βarrestins also greatly impact receptor endocytosis via clathrin-coated pits, such that overexpression of these components in cellular cultures can enhance receptor internalization; and genetic ablation of the βarrestins can prevent trafficking of certain receptors (9–12). The important lesson to gain from these observations is that the presence, and perhaps even the levels, of the regulatory units can ultimately affect the function of the receptor in response to a drug. Therefore, the responsiveness of a receptor to a particular drug can be defined by its association with certain regulatory proteins.

β-Arrestin represent a fulcrum for G protein-coupled receptor (GPCR) signaling. As a scaffolding protein, it can determine what other proteins can interact, such as JNK and Src (13, 14) respectively, and what proteins cannot (such as Gα proteins) (15, 16). The recruitment of arrestins to a receptor can be induced by agonist activation of the receptor. The downstream effect, whether initiating a signaling pathway or preventing pathway engagement determines whether βarrestins function as GPCR signaling molecules or as GPCR dampeners. Therefore, GPCR signaling is defined by the context of the cellular function, and signaling is relative to function observed, such that a βarrestin can be seen as a desensitizing agent when it disrupts further G protein-coupling and as a signaling molecule when it facilitates the activation of downstream signaling cascades.

Regardless of the cellular outcome, whether stimulation or inhibition of a particular cascade of signaling events, the recruitment of βarrestin molecules occurs in response to ligand binding. Agonist and antagonist are relative terms referring to activation or termination of subsequent signaling cascades and with respect to this particular GPCR signaling event, agonists induce βarrestin recruitment and antagonists prevent and even displace agonist-induced βarrestin recruitment. In this regard, the interaction between a GPCR and a βarrestin can reveal functional information regarding the ability of a ligand to engage a particular receptor.

2. Methods to Elucidate Arrestin Interactions with GPCRs

Much of what we know about how βarrestins interact with GPCRs comes from cell-based functional assays. Typically, a receptor is expressed at high concentrations in a particular cell line (historically, HEK-293 cells) along with a βarrestin protein. Generally, at least one of the proteins is tagged with a reporter or a fluorescent tag. An early assay developed to examine βarrestin–GPCR interactions, was the βarrestin-green fluorescent protein (GFP) translocation assay. This was first described by Lawrence Barak with Marc Caron and Robert J. Lefkowitz at Duke University. The assay entailed tagging βarrestin with a GFP at the N-terminus (early studies showed that either end of the βarrestin could suffice for detection, but the N-terminus proved to produce more robust signals) and visualizing transfected cells using confocal fluorescent microscopy (17, 18).

In this assay, either of the two ubiquitously expressed βarrestin molecules, βarrestin1 or βarrestin2, can be tagged. Expression of βarrestin2-GFP in HEK-293 cells can be seen in the cytosol and is excluded from the nucleus; βarrestin1-GFP can be seen in both the cytosol and the nucleus. Cotransfection of the βarrestin-GFP with any GPCR of interest allows the user to visualize translocation of the βarrestin-GFP from the cytosol to the plasma membrane upon addition of an agonist. The translocation event can range in intensity from the development of puncta on the surface, to intense green spots outlining the membrane, to a continuous green ring around the cell, and, in some cases, the co-internalization of the receptor and the βarrestin-GFP into intracellular domains (19). Antagonist screens can also be developed by assessing whether the test ligand can prevent the appearance of the membrane punctae when treated with a known agonist.

The most beneficial aspect of this GPCR activation assay is that it is independent of G protein-directed second messenger cascades. Classically, to assess agonist activation of a GPCR, one needs first to know to what Gα protein the receptor couples most preferentially. For example, if the receptor couples to Gαs, then one would measure cAMP accumulation as an indication of adenylyl cyclase activation. For Gαi/o coupling one would typically assess inhibition of forskolin-induced adenylyl cyclase activity or alternatively, adenylyl cyclase that is activated by an endogenously expressed Gαs coupled receptor. Gαq coupling entails measuring phospholipase C activity (PLC) by measuring agonist-stimulated phosphoinositide turnover or changes in intracellular calcium levels. Coupling to G proteins can be directly assessed using radioisotope labeled GTP, however, this approach is most robust for the Gαi/o coupled receptors. Downstream kinases can also be assessed but these endpoints are more distal to the receptor activation event and thereby allow for potential of compound interference with signaling events downstream of the receptor. The βarrestin interaction with a GPCR occurs upon agonist binding regardless of the class of G protein-coupling to the receptor. In a sense, the βarrestin interaction may be even more proximal to receptor activation than the G protein-coupling event.

This proximity and universality of the βarrestin translocation provides an important benefit when trying to identify ligands for orphan receptors where typically the G protein partner is unknown, as one can utilize a single experimental G protein-independent readout to determine activity profiles. Collaboration with Drs. Paul Taghert and Erik Johnson, led to the use of this approach to successfully deorphanize several GPCRs. Starting with 15 orphan Drosophila GPCR cDNA constructs and ten Drosophila neuropeptides, the βarrestin2-GFP translocation assay was used to pair six GPCRs with their cognate neurotransmitters (20, 21).

After extensively using the βarrestin translocation assay to understand GPCR trafficking and regulation (17, 18, 22–24), Barak and his team developed the βarrestin translocation assay into a high content screening platform that has been used to screen small molecule libraries for novel ligands (25–28). This technology is currently supplied by Molecular Devices (Sunnyvale, CA). Since these early studies at the turn of the century (early 2000), the βarrestin2- translocation assay has evolved into several powerful high throughput screening (HTS) platforms, each of which is based on measuring the effect of ligand binding to GPCR to affect βarrestin interactions.

The bioluminescent resonance energy transfer (BRET) technique has become a very popular method for studying protein–protein interactions (29, 30) and this approach has been developed into a high throughput compatible assay to monitor interactions between GPCRs and βarrestin (31, 32). This methodology involves heterologous co-expression of genetically fused proteins that link the GPCR of interest to an acceptor fluorophore (such as GFP) and the βarrestin to bioluminescent donor enzyme (such as *Renilla* luciferase-RLuc). The tags can also be interchanged, such that the fluorescent protein resides on the βarrestin and the RLuc tag resides on the receptor. If the fusion proteins are in close proximity, resonance energy will be transferred from the donor to the acceptor molecule and subsequent fluorescence from the acceptor can be detected at a characteristic wavelength. Co-expression of the GPCR and βarrestin fusion constructs in live cells enables their interaction to be monitored in real time. Following agonist stimulation, cells are treated with a luciferase substrate such as coelantrazine, which emits light at ~400 nm upon *Renilla* luciferase catalyzed oxidation. The emission spectra of the substrate overlap with the excitation spectra of the acceptor molecule. The assay relies on a multimodal plate reader capable of dual-filter luminometry to measure the light emitted in both spectra as the RLuc emission serves as an internal control and the GFP emission determines the excitation. The BRET signal is determined by the energy emitted by the acceptor relative to that emitted by the donor and is usually reported as a ratio of excitation of the two emissions (GFP/RLuc). There are currently three generations of BRET ($BRET^1$, $BRET^2$ and eBRET) available, the three derivations are technically similar but differ according to their spectral properties and luciferase substrate. Other modifications to the GPCR:βarrestin BRET assay have been described that improve the scalability of the assay (33).

Two commercial HTS assays are available through DiscoveRx (PathHunter™) (34–36) and Invitrogen (Tango™) (37, 38). These assays measure ligand-activated GPCR-βarrestin2 recruitment using either enzyme fragment complementation or transcriptional reporter gene activity, respectively. The DiscoveRx PathHunter™ platform utilizes the non-covalent interaction of two polypeptide enzyme fragments of β-galactosidase, one fused to the receptor and the other fused to βarrestin. The use of EFC to monitor GPCR–βarrestin interactions was first described by Yan et al. (39). The agonist mediated receptor–βarrestin interaction leads to measurable reconstituted enzyme activity which can be quantified on addition of a chemiluminescence substrate. The two enzyme fragments have low affinity and this principle discourages basal complementation in the absence of βarrestin and receptor collisions.

Invitrogen offers the Tango™ assay, this involves tagging the receptor with a transcription factor tethered by a linker containing a protease cleavage site. Upon GPCR activation, the βarrestin translocation presents the protease to the receptor where the transcription factor is cleaved from its C terminus. The transcription factor then translocates to the nucleus where it binds to a GAL4 promoter driving expression of the reporter gene β-lactamase. Enzyme activity is assessed by detection of product which involves a termination of a fluorescence resonance energy transfer (FRET) to reveal emission in another spectrum (37, 38, 40, 41).

Other approaches have also been taken to assess βarrestin-receptor interactions that are not yet commercially available but may be amenable to HTS. Fluorescent protein complementation involves tagging the C-terminus of the GPCR with half of the fluorescent protein molecule (such as YFP) and the βarrestin with the other half. Upon activation the two molecules should come together to produce a functional fluorescent protein which would be detectable via a fluorescence imaging plate reader. In this same vein, FRET can also be used, wherein each molecule, the βarrestin and the receptor, are tagged with fluorescent proteins. The excitation of the one protein emits in a spectrum that excites the second protein. Like BRET, FRET also measures the emission of the first and second protein and a ratio is calculated.

The duration of time to measure the βarrestin translocation event at the receptor is very different between these large-scale methodologies. BRET, FRET, Fluorescent protein complementation, and the Transfluor assay, all measure the interaction that occurs immediately following addition of agonist and do not necessarily rely on downstream secondary cellular events to produce a signal. As such, these assays can be used to measure the early dynamic events involved in GPCR–βarrestin interactions as a function of time. The PathHunter™ system as well as the Tango™ assay are both read hours following stimulation and therefore give less information regarding the temporal dynamics of the interaction, although DiscoveRx recommends testing earlier timepoints for optimization of the assay. Further, reporter assays that measure single receptor activation events (such as the one time cleavage of the transcription factor from the receptor) do not give information as to the transient nature of the interaction which may prove to be fundamentally important in determining the role that the βarrestin may play in the interaction. Finally, it should be noted that most of the HTS platforms that have been developed have focused on the βarrestin2 interaction. For the most part, tagged βarrestin2 seems to have higher affinity than tagged βarrestin1 for many GPCRs in these label-based assays; however, the ability to recruit a particular βarrestin may be as pharmacologically relevant as whether the receptor recruits a βarrestin at all (19, 42).

3. Future Directions

The high throughput βarrestin platforms make it relatively easy to assess a veritable multitude of compounds in a relatively short time frame in order to identify putative novel ligands for GPCRs, independent of the second messenger cascade. An important question to consider in such an approach is how applicable will such a cell-based screening system be towards elucidating the pharmacological and physiological relevancy of the GPCR–arrestin interactions. Further, when designing a cell-based assay, particularly for seeking ligands to neurotransmitter receptors the question arises whether the cells that are amenable to HTS will truly recapitulate an environment that would enable the receptor signaling such that relevant βarrestin coupling would be revealed. In most cell-based drug discovery efforts, neuron specific scaffold proteins are not present in the cell type of choice and therefore assessment of receptor function may not reflect the desired endogenous state. Further, without the relevant scaffolding proteins, signaling events optimized in cell culture may fail to truly mimic the endogenous state. For example, will a ligand promote a βarrestin coupling event in cells that would never occur in a neuron? Interactions with scaffolding proteins such as PSD-95 and spinophilin, expressed in neurons but not in most immortalized lines, may determine the permissivity of such interactions in neurons. Receptor–protein interactions ultimately dictate receptor conformations and, by facilitating or preventing binding to other scaffolding proteins, ultimately dictate receptor functional signaling. These interactions may also determine the effect a particular ligand has on a receptor and may serve as a point to bifurcate ligand-directed signaling events (43, 44).

While βarrestins and GRKs are somewhat ubiquitously expressed, other regulatory and scaffolding components are only expressed in certain cell types. The best example of this are proteins that make up the postsynaptic densities expressed by neuronal cells, including members of the membranes associated guanylate kinases (MAGUK) family such as PSD-95; the Shank/proSAP family, Stargazin, and Homer (for review see (45)). PSD-95 is a large PDZ domain containing scaffolding protein that regulates AMPA receptors and may represent a functional link between GPCRs and ligand gated ion channels such as AMPA and NMDA. It also directly interacts with GPCRs to regulate their function. PSD-95 modulates D1 dopamine receptor function by decreasing surface expression and dampening agonist-induced cAMP signaling (46). Serotonin 2A and 2C receptor expression is decreased in the absence of PSD-95; further, downstream signaling is also disrupted which correlates with a decrease in receptor-mediated behavioral responses in PSD-95 knockout mice (47, 48). Conversely, PSD-95 is important for internalizing the serotonin receptors. Interestingly, βarrestin2 also plays an important role in internalizing the serotonin 2A receptor and the events that occur due to or subsequent to internalization may be critical in determining the normal function of this receptor *in vivo* (43, 49).

The concept that receptor signaling is contextual becomes very important when considering how to assess arrestin regulation of GPCRs in cellular models. For example, if the receptor is in complex with a protein expressed only in neurons, and that interaction facilitates the agonist-induced βarrestin recruitment, then assessing βarrestin-translocation to the receptor in a non-neuronal cell type may not represent normal function of the agonist at the receptor. While it is not practical to do HTS for ligands in primary neuronal cultures, it is essential to keep in mind that the ligands identified using recombinant systems may not fully recapitulate the functional selectivity of the GPCR activation event *in vivo*. Ultimately, the function of a ligand at a receptor is determined by the receptor environment and therefore, it remains of great importance to determine the signaling cascades that underlie normal biological function.

Ligand-directed signaling (or "biased agonism") has been a topic of considerable interest among pharmacologists in the last several years (49–51). The concept is based on the idea that the chemical characteristics of the ligand may alter the conformation of the receptor such that it will interact preferentially with certain cellular proteins to mediate distinct biological responses. Therefore, the nature of the receptor–protein interactions will dictate the signal transduction pathway based on the properties of the ligand. The βarrestins may be key in determining receptor responsiveness to ligand in a particular system as they have multifaceted roles. These regulatory proteins have been shown to mediate receptor desensitization and internalization extensively in cell culture model systems and *in vivo* (52–58). βArrestins can also mediate GPCR signaling and this has been demonstrated in cell culture models for vasopressin, angiotensin, and β-adrenergic receptors; this list continues to grow (59–64). βArrestins can mediate signaling *in vivo* as well, and this has been demonstrated for D2 dopamine and serotonin 2A receptors in βarr2-KO mice (43, 65). βArrestin-mediated GPCR signaling may ultimately prove to be as essential as (or even more important than) G protein-mediated signal transduction for some receptors. This consideration becomes imperative as the current advances in molecular pharmacology reveal that diverse agonists can direct specific receptor signaling complexes (49, 66–68).

Such ligand-directed signaling may determine drug responsiveness and may underlie diverse physiological responses generated by different agonists at a particular receptor. For example, the *mu* opioid receptor (MOR) binds to morphine, methadone, and fentanyl and each agonist can, at different doses, produce equiefficacious G protein-coupling. Further, each agonist, at different doses, can stimulate a similar extent of analgesia. However, the three ligands differ in their propensity to regulate the receptor. While methadone and fentanyl robustly promote receptor phosphorylation, βarrestin recruitment, and receptor internalization, morphine weakly promotes these events. We have previously shown that morphine induces βarrestin-MOR interactions and that this interaction is preferential for βarr2 (69). Accordingly, βarr2-KO mice show greatly altered responses to morphine wherein morphine induces enhanced analgesia (54–56). Methadone and fentanyl, which robustly recruit both βarr1 and βarr2, produce the same analgesic response in both genotypes (69). Interestingly, while morphine tolerance is not observed in βarr2-KO mice in the hot plate analgesia test, tolerance to methadone and fentanyl is not affected by deletion of this regulatory protein (unpublished observations). These studies suggest that the distinct agonists at the MOR differentially depend upon βarr2 to mediate the same physiological response.

Using the same mouse model, the impact that the cellular environment makes on receptor responsiveness can also be demonstrated. βArrestin2 acts as a negative regulator of MOR signaling in pain pathways as discussed earlier; however, morphine-induced constipation and respiratory suppression are attenuated in this mouse model suggesting that βarrestin2 may be functioning as a facilitator of MOR signaling in centers that control these physiological functions. Other physiological symptoms induced by morphine are differentially affected by the deletion of βarrestin2 in response to morphine: hypothermia is enhanced; locomotor hyperactivity is suppressed; dopamine release and reinforcement are enhanced; and drug reinforcement is also enhanced. While it is tempting to assign roles of βarrestin2 as a signaling or dampening regulator at MOR for each of these physiological outputs in response to morphine, we must consider that *in vivo* pharmacology represents the concerted effects of a single drug treatment which may impact on a complex interplay of neurotransmitter responsiveness. In this way, while it may appear that βarrestin2 functions as a facilitator of MOR signaling underlying locomotor hyperactivity, it must also be considered that MOR activation induces the release of dopamine which in turns acts at dopamine receptors which are also regulated by βarrestin2. In fact, postsynaptic D2 dopamine receptors have been shown to signal via βarrestin2 in striatal neurons (65), which would imply that while we see enhanced striatal dopamine release following morphine challenge in the βarr2-KO mice, hyperactivity is actually attenuated.

4. Conclusions

The use of genetically modified mouse models, while very useful, must be approached with caution and careful experimental design. It should also be done with the full assumption that by knocking out a particular protein, disruption of any particular system is more likely than conservation of biological processes. While the use of conditional knockouts may help to overcome some developmental adaptations, the fact that multiple systems are at play when a drug is administered is not overcome. Likewise, while targeted deletions may help to rule out interplay between organ systems or brain regions, the contribution of other neurotransmitter systems is not eliminated.

The ultimate determination of whether a particular receptor system has been impacted upon by the deletion of the regulatory element must be shown biochemically. Biochemical assessments of GPCR activation and scaffolding are difficult as stimulation windows are generally less than twofold and agonist specificity can make it difficult to accurate test a single receptor population. To assess βarrestin–GPCR interactions *in vivo* is very difficult primarily because the levels of the signaling partners are very low in the cell (as most of our technology has been optimized for cell-based recombinant systems). Further, to demonstrate the direct interaction between endogenous proteins, the current technology relies on antibody recognition of the proteins of interest. For GPCRs, the antibodies should recognize regions of the protein that are not involved in signaling complexes (whether intracellular scaffolding proteins such as βarrestins or regions associated with other receptors forming dimers). For βarrestins, the antibodies must recognize the species of interest, which for mouse models, is not commercially available.

Despite the challenges, the most meaningful information that can be gained regarding GPCR–arrestin interactions should be garnered from the endogenous setting. As it becomes increasingly clear that the cellular environment can impact on the ligand properties and ligand properties can direct engagement with the cellular environment, the need to determine drug actions in target tissues becomes a top priority.

References

1. Lefkowitz RJ and Whalen EJ (2004) *Beta*-arrestins: traffic cops of cell signaling. Curr Opin Cell Biol **16**:162–168.
2. Lefkowitz RJ (2004) Historical review: a brief history and personal retrospective of seven-transmembrane receptors. Trends Pharmacol Sci **25**:413–422.
3. Gurevich EV, Benovic JL and Gurevich VV (2004) Arrestin2 expression selectively increases during neural differentiation. J Neurochem **91**:1404–1416.
4. Ahmed MR, Gurevich VV, Dalby KN et al (2008) Haloperidol and clozapine differentially affect the expression of arrestins, receptor kinases, and extracellular signal-regulated kinase activation. J Pharmacol Exp Ther **325**:276–283.
5. Violin JD, Ren XR and Lefkowitz RJ (2006) G-protein-coupled receptor kinase specificity for *beta*-arrestin recruitment to the *beta*2-adrenergic receptor revealed by fluorescence resonance energy transfer. J Biol Chem **281**:20577–20588.
6. Drake MT, Violin JD, Whalen EJ et al (2008) *Beta*-arrestin-biased agonism at the *beta*2-adrenergic receptor. J Biol Chem **283**: 5669–5676.
7. Galandrin S and Bouvier M (2006) Distinct signaling profiles of *beta*1 and *beta*2 adrenergic receptor ligands toward adenylyl cyclase and mitogen-activated protein kinase reveals the pluridimensionality of efficacy. Mol Pharmacol **70**:1575–1584.
8. Zidar DA, Violin JD, Whalen EJ et al (2009) Selective engagement of G protein coupled receptor kinases (GRKs) encodes distinct functions of biased ligands. Proc Natl Acad Sci USA **106**:9649–9654.
9. Kohout TA, Lin, FS, Perry SJ et al (2001) *Beta*-arrestin 1 and 2 differentially regulate heptahelical receptor signaling and trafficking. Proc Natl Acad Sci USA **98**:1601–1606.
10. Kohout TA and Lefkowitz RJ (2003) Regulation of G protein-coupled receptor kinases and arrestins during receptor desensitization. Mol Pharmacol **63**:9–18.
11. Vines CM, Revankar CM, Maestas, DC et al (2003) N-formyl peptide receptors internalize but do not recycle in the absence of arrestins. J Biol Chem **278**:41581–41584.
12. DeFea KA, Zalevsky J, Thoma, MS et al (2000) *Beta*-arrestin-dependent endocytosis of proteinase-activated receptor 2 is required for intracellular targeting of activated ERK1/2. J Cell Biol **148**:1267–1281.
13. McDonald PH, Chow, CW, Miller WE et al (2000) *Beta*-arrestin 2: a receptor-regulated MAPK scaffold for the activation of JNK3. Science **290**:1574–1577.
14. Luttrell LM, Ferguson SS, Daaka Y et al (1999) *Beta*-arrestin-dependent formation of *beta*2 adrenergic receptor-Src protein kinase complexes. Science **283**:655–661.
15. Freedman NJ and Lefkowitz RJ (1996) Desensitization of G protein-coupled receptors. Recent Prog Horm Res **51**:319–351.
16. Tohgo A, Choy EW, Getsy-Palmer D et al (2003) The stability of the G protein-coupled receptor-*beta*-arrestin interaction determines the mechanism and functional consequence of ERK activation. J Biol Chem **278**:6258–6267.
17. Barak LS, Warabi K, Feng X et al (1999) Real-time visualization of the cellular redistribution of G protein-coupled receptor kinase 2 and *beta*-arrestin 2 during homologous desensitization of the substance P receptor. J Biol Chem **274**:7565–7569.
18. Barak LS, Zhang J, Ferguson SS et al (1999) Signaling, desensitization, and trafficking of G protein-coupled receptors revealed by green fluorescent protein conjugates. Methods Enzymol **302**:153–171.
19. Oakley RH, Laporte SA, Holt JA et al (2000) Differential affinities of visual arrestin, *beta* arrestin1, and *beta* arrestin2 for G protein-coupled receptors delineate two major classes of receptors. J Biol Chem **275**:17201–17210.
20. Johnson EC, Bohn LM, Barak LS et al (2003) Identification of Drosophila neuropeptide receptors by G protein-coupled receptors-*beta*-arrestin2 interactions. J Biol Chem **278**(52): 52172–52178.

21. Johnson EC (2003) Identification and characterization of a G protein-coupled receptor for the neuropeptide proctolin in *Drosophila melanogaster*. Proc Natl Acad Sci USA **100**: 6198–6203.
22. Oakley RH, Laporte SA, Holt JA et al (1999) Association of *beta*-arrestin with G protein-coupled receptors during clathrin-mediated endocytosis dictates the profile of receptor resensitization. J Biol Chem **274**:32248–32257.
23. Wilbanks AM, Laporte SA, Bohn LM et al (2002) Apparent loss-of-function mutant GPCRs revealed as constitutively desensitized receptors. Biochemistry **41**:11981–11989.
24. Claing A, Laporte SA, Caron MG et al (2002) Endocytosis of G protein-coupled receptors: roles of G protein-coupled receptor kinases and *beta*-arrestin proteins. Prog Neurobiol **66**:61–79.
25. Oakley RH, Hudson CC, Cruickshank RD et al (2002) The cellular distribution of fluorescently labeled arrestins provides a robust, sensitive, and universal assay for screening G protein-coupled receptors. Assay Drug Dev Technol **1**: 21–30.
26. Hudson CC, Oakley RH, Sjaastad MD et al (2006) High-content screening of known G protein-coupled receptors by arrestin translocation. Methods Enzymol **414**:63–78.
27. Ghosh RN, DeBiasio R, Hudson CC et al (2005) Quantitative cell-based high-content screening for vasopressin receptor agonists using transfluor technology. J Biomol Screen **10**:476–484.
28. Oakley RH, Hudson CC, Sjaastad MD et al (2006) The ligand-independent translocation assay: an enabling technology for screening orphan G protein-coupled receptors by arrestin recruitment. Methods Enzymol **414**:50–63.
29. Bertrand L, Parent S, Caron MG et al (2002) The BRET2/arrestin assay in stable recombinant cells: a platform to screen for compounds that interact with G protein-coupled receptors (GPCRS). J Recept Signal Transduct Res **22**:533–541.
30. Hamdan FF, Percherancier Y, Breton B et al (2006) Monitoring protein-protein interactions in living cells by bioluminescence resonance energy transfer (BRET). Curr Protoc Neurosci **5**:5–23.
31. Vrecl M, Jorgensen R, Pogacnik A et al (2004) Development of a BRET2 screening assay using *beta*-arrestin 2 mutants. J Biomol Screen **9**: 322–333.
32. Hamdan FF, Audet M, Garneau P et al (2005) High-throughput screening of G protein-coupled receptor antagonists using a bioluminescence resonance energy transfer 1-based *beta*-arrestin2 recruitment assay. J Biomol Screen **10**:463–475.
33. Heding A (2004) Use of the BRET 7TM receptor/*beta*-arrestin assay in drug discovery and screening. Expert Rev Mol Diagn **4**:403–411.
34. van Der Lee MM, Bras M, van Koppen CJ et al (2008) *Beta*-Arrestin recruitment assay for the identification of agonists of the sphingosine 1-phosphate receptor EDG1. J Biomol Screen **13**:986–998.
35. Zhao X, Jones A, Olson KR et al (2008) A homogeneous enzyme fragment complementation-based *beta*-arrestin translocation assay for high-throughput screening of G-protein-coupled receptors. J Biomol Screen**13**:737–747.
36. McGuinness D, Maliksay A, Visconti R et al (2009) Characterizing cannabinoid CB2 receptor ligands using DiscoveRx PathHunter *beta*-arrestin assay. J Biomol Screen **14**:49–58.
37. Wetter JA, Revankar C, Hanson BJ (2009) Utilization of the Tango *beta*-arrestin recruitment technology for cell-based EDG receptor assay development and interrogation. J Biomol Screen **14**:1134–1141.
38. Doucette C, Vedik K, Koepnick E et al (2009) *Kappa* opioid receptor screen with the Tango *beta*-arrestin recruitment technology and characterization of hits with second-messenger assays. J Biomol Screen **14**:381–394.
39. Yan YX, Boldt-Houle DM, Tillotson BP et al (2002) Cell-based high-throughput screening assay system for monitoring G protein-coupled receptor activation using *beta*-galactosidase enzyme complementation technology. J Biomol Screen **7**:451–459.
40. van der Lee MM, Blomenrohr M, van der Doelen AA et al (2009) Pharmacological characterization of receptor redistribution and *beta*-arrestin recruitment assays for the cannabinoid receptor 1. J Biomol Screen **14**:811–823.
41. Hanson BJ, Wetter J, Bercher MR et al (2009) A homogeneous fluorescent live-cell assay for measuring 7-transmembrane receptor activity and agonist functional selectivity through *beta*-arrestin recruitment. J Biomol Screen **14**:798–810.
42. Laporte SA, Oakley RH, Holt JA et al (2000) The interaction of *beta*-arrestin with the AP-2 adaptor is required for the clustering of *beta* 2-adrenergic receptor into clathrin-coated pits. J Biol Chem **275**:23120–23126.
43. Schmid CL, Raehal KM, Bohn LM (2008) Agonist-directed signaling of the serotonin 2A receptor depends on *beta*-arrestin-2 interactions *in vivo*. Proc Natl Acad Sci USA **105**: 1079–1084.
44. Kinzer-Ursem TL, Linderman JJ (2007) Both ligand- and cell-specific parameters control

ligand agonism in a kinetic model of G protein-coupled receptor signaling. PLoS Comput Biol **3**:e6.

45. Kennedy MJ, Ehlers MD (2006) Organelles and trafficking machinery for postsynaptic plasticity. Ann Rev Neurosci **29**:325–362.
46. Zhang J, Vinuela A, Neely MH et al (2007) Inhibition of the dopamine D1 receptor signaling by PSD-95. J Biol Chem **282**: 15778–15789.
47. Xia Z, Gray JA, Compton-Toth BA et al (2003) A direct interaction of PSD-95 with 5-HT2A serotonin receptors regulates receptor trafficking and signal transduction. J Biol Chem **278**:21901–21908.
48. Abbas AI, Yadav PN, Yao WD et al (2009) PSD-95 is essential for hallucinogen and atypical antipsychotic drug actions at serotonin receptors. J Neurosci **29**:7124–7136.
49. Urban JD, Clarke WP, von Zastrow M et al (2007) Functional selectivity and classical concepts of quantitative pharmacology. J Pharmacol Exp Ther **320**:1–13.
50. KenakinT (2007) Functional selectivity through protean and biased agonism: who steers the ship? Mol Pharmacol **72**:1393–1401.
51. Rajagopal S K, Rajagopal K, Lefkowitz RJ (2010) Teaching old receptors new tricks: biasing seven-transmembrane receptors. Nat Rev Drug Discov **9**:373–386.
52. Pierce KL, Luttrell LM, Lefkowitz RJ (2001) New mechanisms in heptahelical receptor signaling to mitogen activated protein kinase cascades. Oncogene **20**:1532–1539.
53. Luttrell LM, Lefkowitz RJ (2002) The role of *beta*-arrestins in the termination and transduction of G-protein-coupled receptor signals. J Cell Sci **115**:455–465.
54. Bohn LM, Lefkowitz RJ, Gainetdinov RR et al (1999) Enhanced morphine analgesia in mice lacking *beta*-arrestin 2. Science **286**: 2495–2498.
55. Bohn LM, Gainetdinov RR, Lin FT et al (2000) *Mu*-opioid receptor desensitization by *beta*-arrestin-2 determines morphine tolerance but not dependence. Nature **408**:720–723.
56. Bohn LM, Lefkowitz RJ, Caron MG (2002) Differential mechanisms of morphine antinociceptive tolerance revealed in *beta*-arrestin-2 knock-out mice. J Neurosci **22**:10494–10500.
57. Gainetdinov RR, Premont RT, Bohn LM et al (2004) Desensitization of G protein-coupled receptors and neuronal functions. Ann Rev Neurosci **27**:107–144.
58. Raehal KM, Walker JK, Bohn LM (2005) Morphine side effects in *beta*-arrestin 2 knockout mice. J Pharmacol Exp Ther **314**: 1195–1201.
59. Ren XR, Reiter E, Ahn S et al (2005) Different G protein-coupled receptor kinases govern G protein and *beta*-arrestin-mediated signaling of V2 vasopressin receptor. Proc Natl Acad Sci USA **102**:1448–1453.
60. Charest PG, Oligny-Longpre G, Bonin H et al (2007) The V2 vasopressin receptor stimulates ERK1/2 activity independently of heterotrimeric G protein signalling. Cell Signal **19**:32–41.
61. Ahn S, Nelson CD, Garrison TR et al (2003) Desensitization, internalization, and signaling functions of *beta*-arrestins demonstrated by RNA interference. Proc Natl Acad Sci USA **100**:1740–1744.
62. Tohgo A, Pierce KL, Choy EW et al (2002) *Beta*-arrestin scaffolding of the ERK cascade enhances cytosolic ERK activity but inhibits ERK-mediated transcription following angiotensin AT1a receptor stimulation. J Biol Chem **277**:9429–9436.
63. Shenoy SK, Drake MT, Nelson CD et al (2006) *Beta*-arrestin-dependent, G protein-independent ERK1/2 activation by the *beta*2 adrenergic receptor. J Biol Chem **281**:1261–1273.
64. Luttrell LM, Roudabush FL, Choy EW et al (2001) Activation and targeting of extracellular signal-regulated kinases by *beta*-arrestin scaffolds. Proc Natl Acad Sci USA **98**:2449–2454.
65. Beaulieu JM, Sotnikova TD, Marion S et al (2005) An Akt/*beta*-arrestin 2/PP2A signaling complex mediates dopaminergic neurotransmission and behavior. Cell **122**:261–273.
66. Wei H, Ahn S, Shenoy SK et al (2003) Independent *beta*-arrestin 2 and G protein-mediated pathways for angiotensin II activation of extracellular signal-regulated kinases 1 and 2. Proc Natl Acad Sci USA **100**:10782–10787.
67. Kohout TA, Nicholas SL, Perry SJ et al (2004) Differential desensitization, receptor phosphorylation, *beta*-arrestin recruitment, and ERK1/2 activation by the two endogenous ligands for the CC chemokine receptor 7. J Biol Chem **279**:23214–23222.
68. Abbas A and Roth BL (2008) Arresting serotonin. Proc Natl Acad Sci USA **105**:831–832.
69. Bohn LM, Dykstra LA, Lefkowitz RJ et al (2004) Relative opioid efficacy is determined by the complements of the G protein-coupled receptor desensitization machinery. Mol Pharmacol **66**:106–112.

Chapter 18

Characterizing Molecular Mobility and Membrane Interactions of G Protein-Coupled Receptors

Vladana Vukojević, Yu Ming, and Lars Terenius

Abstract

Drugs targeting the opioid neurotransmission system have been used for centuries recreationally and for medical purposes. In spite of this vast experience and competence in opioid pharmacotherapy, fine details about the cellular and molecular mechanisms underlying opioid receptor physiology remain unknown. We present here two methods with single-molecule sensitivity, confocal laser scanning microscopy with avalanche photodiode (APD) detectors (APD imaging) and fluorescence correlation spectroscopy (FCS) suitable for nondestructive study of molecular interactions and intracellular transporting processes in living cells. These high-resolution methods provide functional readouts, giving measures of concentration, mobility and affinity, for the investigated molecules and enable us to monitor changes in these properties in living cells in real time. We have used these methods to study early events in opioid receptor activation with specific and nonspecific ligands, and discuss the new insights obtained by these approaches.

Key words: Opioid receptors, Opiates, Alcohol, Protein–lipid interactions, APD imaging, Fluorescence correlation spectroscopy (FCS)

1. Introduction

The plasma membrane is the natural environment where cell-surface receptors reside and perform their biological function. Throughout years, our view on the structural organization of the plasma membrane and its impact on cell-surface receptor signaling has changed, gradually evolving from the classic Singer–Nicolson model describing the plasma membrane as a two-dimensional lipid fluid in which proteins diffuse freely (1) to the current view of the plasma membrane as a complex and dynamic mosaic of lipids, proteins, and carbohydrates, where surface receptors are self-organized into protein- and lipid-enriched structural and

Craig W. Stevens (ed.), *Methods for the Discovery and Characterization of G Protein-Coupled Receptors*, Neuromethods, vol. 60, DOI 10.1007/978-1-61779-179-6_18, © Springer Science+Business Media, LLC 2011

functional units (2–6). The notion that protein–protein and protein–lipid interactions in the plasma membrane have intricate direct and indirect effects on surface receptor reactivity and, hence on their biological function prompts the application and further development of sophisticated biophysical methods to study these interactions nondestructively in the living cell. The great complexity of the problem – low number of molecules of interest in comparison to the surrounding matrix molecules (concentrations of the order 0.1–100 nM), fast molecular interactions (characterized by on-off rates of the order 10^5–10^8 $M^{-1}s^{-1}$), short life time of transient molecular complexes (10^{-6}–10^{-3} s), and rapid molecular movement (diffusion coefficients of the order 10^{-13}–10^{-10} m^2s^{-1}) – requires ultrafast methods with high spatial (sub-micrometer) and temporal (sub-millisecond) resolution and single-molecule sensitivity.

The aim of this chapter is to introduce two such methods: fluorescence imaging with avalanche photodiode detectors (APD imaging) (7) and fluorescence correlation spectroscopy (FCS) (8–11). These complementary methods enable us to visualize the sparse receptor molecules in the plasma membrane, measure their local concentrations, quantitatively characterize their mobility, and identify the underlying modes of motion, i.e. to distinguish free Brownian diffusion from directed movement or motion impeded by molecular interactions. Local surface density and lateral mobility are, in addition to chemical affinities, critical determinants of surface receptor reactivity. According to collision theory, the rate of chemical reactions is directly proportional to the number of collisions between the reacting molecules. Thus, slower moving or confined molecules interact less efficiently with one another and their corresponding ligands than the fast and freely moving ones. Since receptor reactivity is inseparable from its biological function, APD imaging and FCS provide the functional read out at the level of individual surface receptor molecules in real time in the living cell. We therefore refer to the combined APD imaging/FCS approach as to functional fluorescence microscopy imaging (fFMI). This approach contrasts with other approaches, such as fluorescence recovery after photobleaching (FRAP), that provide macroscopic, temporally or ensemble averaged values. FCS is also more versatile than Förster resonance energy transfer (FRET) for the study of molecular interactions because it is less distance dependent. FRET is limited to distances between chromophores of several nanometers whereas such requirement does not exist for FCS where molecular interactions can be verified through co-motion of interacting molecules by multicolor cross-correlation measurements, i.e., fluorescence cross-correlation spectroscopy (FCCS).

For many years, our group is involved in investigating the function of the opioid neurotransmission systems at different levels of organization of an organism; cellular, tissue, organ, and the whole organism (12–21). The four opioid receptors; *mu* (MOP), *delta* (DOP), *kappa* (KOP), and nociceptin (NOP), are G protein-coupled receptors (GPCR) (22–24). Together with their corresponding endogenous peptide ligands they control the nociceptive pathways, modulate affective behavior, regulate neuroendocrine physiology and control respiration, thermoregulation, blood pressure, and gastrointestinal motility (25). Opioid receptors are also a very important pharmacological target. Natural alkaloid opioids like morphine, or synthetic opioids like methadone are extremely potent pain-killers, but also highly addictive drugs. Opioid antagonists such as naloxone and naltrexone are shown to reduce alcohol craving, and used in the clinics for treatment of relapse to alcoholism (26). In spite of the great relevance of the opioid systems and their long-time use for the treatment of severe pain and heroin addiction, the mechanisms underlying opioid receptor function are not fully understood at the cellular and molecular level.

2. Background on Functional Fluorescence Microscopy Imaging

2.1. Fluorescence Correlation Spectroscopy

Principles of FCS were formulated about 40 years ago (27–34) in the framework of a physical method called fluctuation correlation analysis. Pioneers in the FCS field developed a theoretical basis and built state-of-the-art equipment to perform fluorescence fluctuation measurements, setting cornerstones not only for FCS but also for fluorescence photobleaching recovery (35–40), multiphoton FCS (41, 42), and multiphoton microscopy (43–45). The early use of FCS after its introduction in the 1970 involved large excitation volumes and long correlation times, causing photo-destruction. Only after introducing technological improvements (31, 33, 34) did FCS acquired its present characteristics of single-molecule detection sensitivity and short measurement times. This allowed for studies of fast dynamic processes in living cells even if reporter molecules are present at low levels. Recent advances and development of new techniques that enable protein labeling within the cells by genetic fusion with fluorescent proteins or by small organic fluorophores (46–49) further speeded up the rapid expansion of FCS applications in live cell studies.

In FCS, the time course and the amplitudes of fluorescence intensity fluctuations occurring in a very small, diffraction limited volume element are analyzed using statistical methods in order to derive molecular numbers, conventional diffusion transport coefficients, and chemical rate constants. For more comprehensive reviews on FCS, see for example (50–52). We describe here briefly the most important elements of an FCS instrument and the most important steps in FCS analysis.

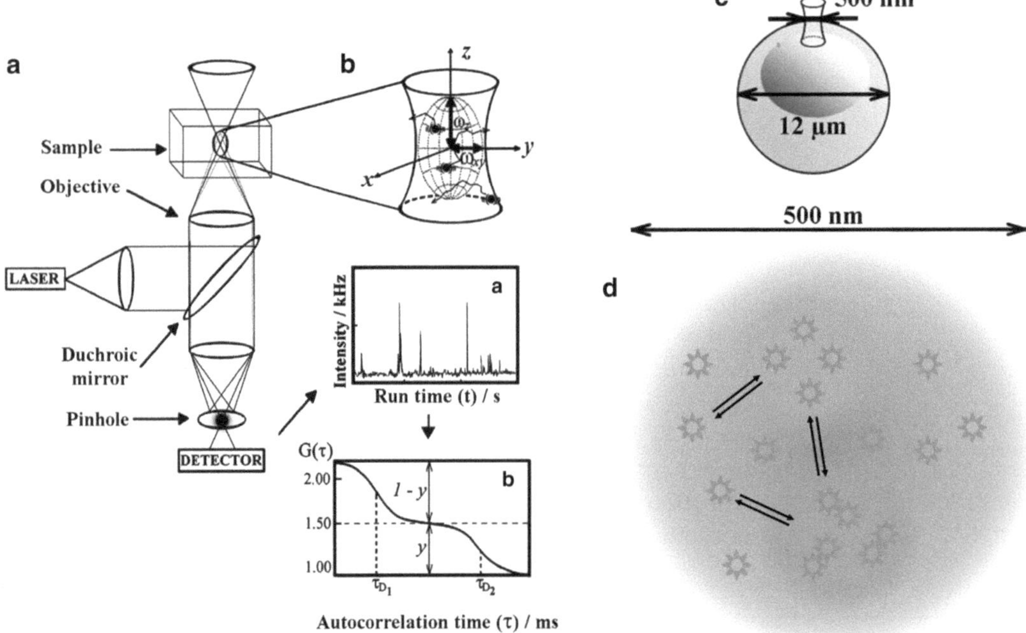

Fig. 1. Schematic presentation of the instrumentation for CLSM/FCS. (**a**) To induce fluorescence, the sample is illuminated by incident laser light. The irradiating laser beam is reflected by a dichroic mirror and is sharply focused by the objective to form a diffraction limited volume element. (**b**) A confocal aperture, set in the image plane to reject the out-of-focus light, further reduces the volume from which fluorescence is detected. This is crucial for providing an ellipsoidal observation volume element and enabling sub-micrometer resolution and quantitative analysis. After the absorption of energy, fluorescent molecules lose energy through photon emission. Light emitted by fluorescing molecules passing through the confocal volume element (magnified in the inset) is separated from the exciting radiation and the scattered light by a dichroic mirror and barrier filter and transmitted to the detector. The number of pulses originating from the detected photons, recorded during a specific time interval, corresponds to the measured light intensity. *Inset* (a) Fluorescence intensity fluctuations recorded in real time. *Inset* (b) normalized autocorrelation curve $G(\tau)$ showing that two components with diffusion times τ_{D1} and τ_{D2} are being detected. The experimental autocorrelation curve is fitted using the autocorrelation function derived for an underlying model. In the example given, a model describing free three-dimensional diffusion of two components was used. (**c**) Schematic presentation of a PC12 cell with the observation volume element focused at the apical plasma membrane. (**d**) Schematic presentation of possible GPCR organization in the plasma membrane.

To induce fluorescence, the sample is illuminated by incident light of a certain wavelength delivered by a laser (Fig. 1a). The laser beam is reflected by a dichroic mirror and sharply focused by the objective in the sample. Following the absorption of energy, fluorescent molecules lose energy through photon emission. The emitted light is separated from the exciting radiation and the scattered light by a dichroic mirror and barrier filter, and transmitted to the detector, which responds with an electrical pulse to each detected photon. The volume from which fluorescence is detected is further reduced by a pinhole (confocal aperture) in the image plane, to reject out-of-focus light. This is crucial to achieve a small and very well-defined observation volume element (typically about 2×10^{-16} L – roughly the size of a bacterium, Fig. 1b), as well as quantitative and low background analysis. The number of pulses originating from the detected photons, recorded during a specific time interval, corresponds to the measured light intensity. Thus, in one experimental run changes in fluorescence intensity in time are registered (Fig. 1a, inset a).

Statistical methods of data analysis are applied to detect non-randomness in the data. Typically, this is done by temporal autocorrelation analysis, but other ways of data analysis such as higher order autocorrelation functions (53, 54), fluorescence intensity distribution analysis (FIDA) (55, 56), photon-counting histograms (PCH) (57, 58), and recently developed fluorescence cumulant analysis (FCA) (59) can also be applied. In temporal autocorrelation analysis we first derive the intensity autocorrelation function $C(t)$. $C(t)$ gives the correlation between the fluorescence intensity, $I(t)$, measured at a certain time, t, and its intensity measured at a later time $t+\tau$, $I(t+\tau)$. The intensity autocorrelation function may be defined as an ensemble average:

$$C(\tau) = \langle I(t)I(t+\tau)\rangle \tag{1a}$$

or, alternatively, as a time average of the product $I(t)I(t+\tau)$ measured over a certain accumulation time T:

$$C(\tau) = \lim_{T\to\infty}\frac{1}{T}\int_0^T I(t)I(t+\tau)\mathrm{d}t. \tag{1b}$$

Since the unprocessed data in FCS are essentially fluorescence fluctuations over the mean fluorescence intensity $<I>$, it is also possible to express the autocorrelation function through fluctuations of light intensities $\delta I(t) = I(t) - <I>$ and $\delta I(t+\tau) = I(t+\tau) - <I>$, at times t and $t+\tau$, respectively. In this way, the intensity autocorrelation function is defined as:

$$C(\tau) = \langle I\rangle^2 + \langle \delta I(t)\delta I(t+\tau)\rangle. \tag{1c}$$

Regardless of the form of expression (1a)–(1c), as they are all equivalent, it is not convenient to use the intensity correlation function in practice because its value depends on properties of the applied experimental setup (9). Therefore, instead of using the intensity autocorrelation function, it is more convenient to use the so-called normalized autocorrelation function, $G(\tau)$, defined as:

$$G(\tau) = 1 + \frac{\langle \delta I(t)\delta I\ (t+\tau)\rangle}{\langle I\rangle^2}, \tag{2a}$$

which is independent of the properties of the experimental setup, such as laser intensity, detection efficiency of the instrumentation, and fluorescence quantum yield of the fluorophore.

For further analysis, the normalized autocorrelation function $G(\tau)$ has to be plotted for different time lags, i.e. for different autocorrelation times τ. In molecular systems undergoing only stochastic fluctuations, we observe random variations of $G(\tau)$ around the value $G(\tau)=1$. For processes that are not random, an autocorrelation curve is determined (Fig. 1, inset b). Typically, one observes a maximal limiting value of $G(\tau)$ as $\tau \to 0$, decreasing to the value of $G(\tau)=1$ at long times, indicating that correlation between the initial and the current value has been lost. Very often only the so-called nonuniform part of the normalized autocorrelation function is analyzed. In this case, one observes a maximal limiting value of $G(\tau)$ as $\tau \to 0$ that decreases to the value of $G(\tau)=0$ at long times. The limiting value of $G(\tau)$, as $\tau \to 0$, is then inversely proportional to the absolute concentration of the fluorescing molecules, as we shall show later.

Thereafter, the experimentally obtained autocorrelation curve has to be compared to an autocorrelation function that is derived for a corresponding model system. For example, if we are studying binding of a small, fluorescently labeled molecule to a cell-surface receptor the autocorrelation curve (Fig. 1, inset b) is compared to the autocorrelation function that describes free diffusion of the unbound ligand and the diffusion of the ligand bound to a surface (33):

$$G(\tau) = 1 + \frac{1}{N}\left(\frac{1-y}{\left(1+\frac{\tau}{\tau_{D1}}\right)\sqrt{1+\frac{w_{xy}^2}{w_z^2}\frac{\tau}{\tau_{D1}}}} + \frac{y}{\left(1+\frac{\tau}{\tau_{D2}}\right)} \right) \tag{3a}$$

In (3a), N is the average number of ligand molecules in the observation volume; y is the fraction of bound ligand molecules; w_{xy} and w_z are radial and axial parameters, respectively, related to spatial properties of the detection volume element (Fig. 1b), τ is the autocorrelation; τ_{D1} and τ_{D2} are characteristic decay times of the autocorrelation curve, in this case called the diffusion times because they present the average lateral transition time of the unbound/bound ligand molecule through the volume element. Spatial properties of the detection volume, represented in (3a) by the square of the ratio of the radial and axial parameters ($(w_{xy}/w_z)^2$), are determined experimentally in calibration measurements (60). The calibration is performed *in vitro*, by using a solution of a dye for which the diffusion coefficient is known. The autocorrelation curve derived in the calibration experiments is fitted by the autocorrelation function describing free diffusion of the referent dye molecules:

$$G(\tau) = 1 + \frac{1}{N} \cdot \frac{1}{\left(1 + \frac{\tau}{\tau_D}\right)\sqrt{1 + \frac{w_{xy}^2}{w_z^2}\frac{\tau}{\tau_D}}}, \tag{3b}$$

The diffusion time, τ_D, of the investigated component is determined from the autocorrelation function (3a) that matches best the actual, experimentally determined, autocorrelation curve using the value of $(w_{xy}/w_z)^2$ determined from the calibration experiments. This parameter, derived directly from FCS measurements, is related to the translation diffusion coefficient D:

$$\tau_D = \frac{w_{xy}^2}{4D}, \tag{4}$$

which is related to the size of the fluorescent molecules *via* the Stokes–Einstein equation:

$$D = \frac{kT}{6\pi\eta R}. \tag{5}$$

In (5) k is the Boltzmann constant, T is absolute temperature, η is the solvent viscosity, and R is the hydrodynamic radius of a hypothetical compact sphere in a viscous medium. If the investigated molecule is of globular shape, its molecular mass (M) can be estimated from FCS measurements:

$$\frac{1}{D} \propto \frac{6\pi\eta}{kT} \cdot \sqrt[3]{M}. \tag{5a}$$

As can be seen from (3a) and (3b), the limiting value of $G(\tau)$ as $\tau \rightarrow 0$ is related to the average number of molecules in the observation volume (N), i.e. it can be used to determine the absolute concentration (c) of the fluorescing molecules. For example, if $G(\tau) = 2.25$ at $\tau = 0$ (Fig. 1a, inset b) and the observation volume is $V = 2.0 \times 10^{-19}$ m^3, the concentration of the fluorescent molecules in the sample is $c = 6.6 \times 10^{-9}$ mol dm^{-3}.

Although the measured fluctuations are utterly stochastic by themselves, their average rate of relaxation to the equilibrium value is not stochastic. Rather, it is constrained by macroscopic properties of the sample. And it is exactly this interrelation that makes it possible to apply fluctuation analysis to obtain information about mobility coefficients, local concentration, apparent hydrodynamic radius, chemical reaction constants and rates, association, and dissociation and equilibrium binding constants. In cases other than the analyzed example of free diffusion of a single fluorescent species, the autocorrelation function attains forms different from the one given in (3a) and (3b) because all processes leading to statistical fluctuations in the fluorescence signal will generate a characteristic decay time in the autocorrelation curve (8).

Autocorrelation analysis, when applied to systems with more than one component, has a practical detection limit that depends critically on the quantum yields, concentrations, and the differences in diffusion times of the investigated molecules (61). For example, to distinguish two molecules by diffusion rate using the previously described autocorrelation analysis, their diffusion times have to differ by a factor of 1.6 for bright fluorophores, or even more if they are less bright, meaning that their masses should differ about 5–8 times. By the previously mentioned data analysis techniques such as FIDA and PCH, distinction of fluorescent species by differences in their molecular brightness can be achieved (62).

2.2. APD Imaging

FCS instrumentation uses the optical layout of a confocal microscope to generate a diffraction limited illumination volume element in the sample and a pinhole in front of the detector to reduce the field of detection, thereby reaching single-molecule sensitivity under optimum conditions. Therefore, it is only natural to integrate FCS with confocal laser scanning microscopy (CLSM), in one instrumental setup. However, in commercial FCS/CLSM instrumentation the optical pathways for CLSM imaging and FCS are typically not fully integrated and the instrumentation is used as two separate methods in one system. In the imaging mode, the photomultiplier tubes (PMTs) are used as detectors, whereas the avalanche photodiodes (APDs) that are characterized by superior quantum yield and collection efficiency (about 70% in APDs as compared to 15–25% in PMTs), higher gain, faster response time, negligible dark current, and better efficiency in the red part of the spectrum, are used in the FCS mode. This dichotomy in instrumental design creates double negative effects; in order to visualize the molecules of interest by CLSM imaging one needs to overexpress them, thereby introducing biological artifacts and "killing" the FCS analysis. As explained earlier and shown in (3a) and (3b), the higher the concentration, i.e., the number of molecules in the observation volume element (N), the smaller are the fluorescence intensity fluctuations and the lower is the sensitivity of FCS. Therefore, we took the initiative to integrate fully the optical pathways and improve the detection efficiency of CLSM imaging by using APD detectors. Thus, the only difference between FCS and APD imaging is at the level of the readout. In classical FCS, the readout is in the form of fluorescence intensity fluctuations recorded in one stationary spot, giving changes in fluorescence intensity over time (Fig. 2, upper time trace), whereas the APD imaging readout is in the form of a digital image, showing spatial and temporal fluorescence intensity fluctuations (Fig. 3). Such instruments are now commercially available.

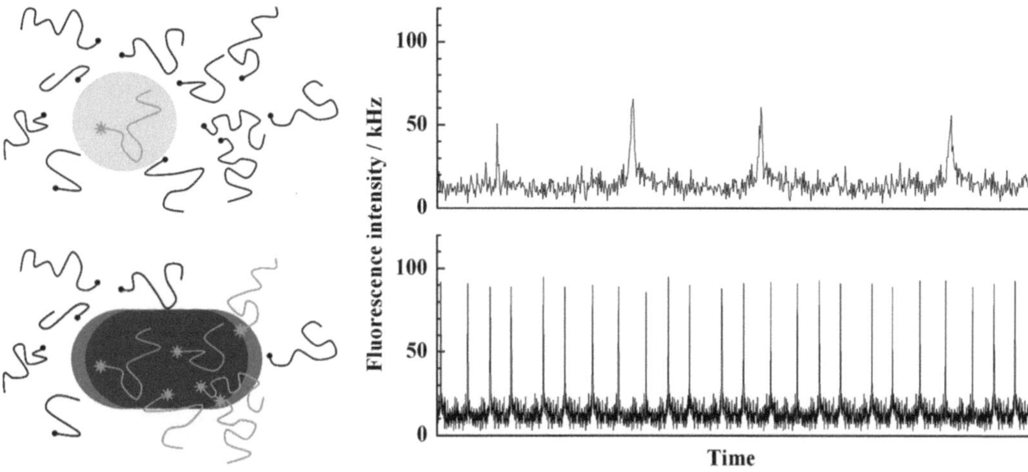

Fig. 2. Scanning FCS. In classic FCS, the observation volume element is stationary. Fluorescence bursts from molecules of interest are recorded only for a short period of time, during their passage across the observation volume element. In between fluorophore encountering events, background signal is collected over extended time (*upper row*). In the scanning mode, the encountering frequency of fluorescent molecules is higher because of the translation of the observation volume element. In this way, a larger number of molecules can be inspected for the same measurement time. Accordingly, the contribution of the background is reduced (*lower row*). The signal intensity is further enhanced by avoiding the transition of the fluorophore to the triplet state (see Fig. 2 in (7)).

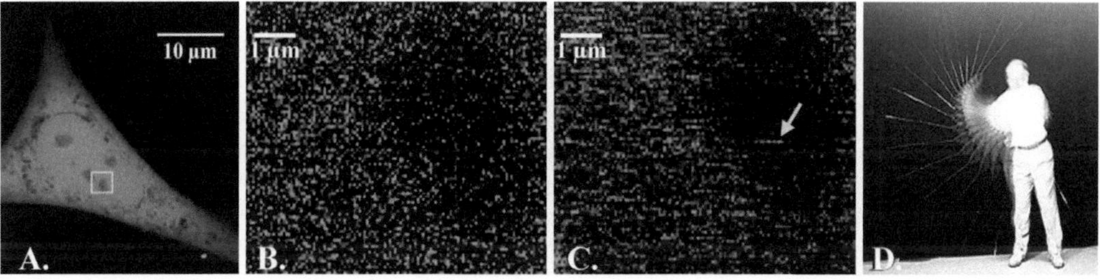

Fig. 3. Dissecting molecular movement by CLSM imaging. (**a**) Image of a live neuroblastoma cell expressing the enhanced green fluorescent protein (eGFP). (**b**) Detail of the image shown in (**a**) acquired by fast scanning; the movement of the scanning head $\tau_{Scan}=3.5$ μs/pixel is faster than eGFP movement by diffusion, $\tau_D=35$ μs/pixel. (**c**) The same detail acquired by slow scanning, $\tau_{Scan}=52$ μs/pixel $>\tau_D$. In scanning, imaging of mobile molecules is coupled with the linear movement of the scanning head. Therefore, molecular movement is observed as long stretches of illuminated pixels (*arrow*). eGFP movement cannot be observed in the fast acquired image (**b**), where EGFP molecules appear to be stationary. (**d**) Movement captured in a series of superimposed stroboscopic images is shown for comparison. Objects moving at different speeds have different "readouts" – the slowly moving body is confined to the same place, the arm movement gives a smeared continuous trace, and the fast moving golf club is captioned in many different positions. (Image supplied and used with permission from Andrew Davidhazy, Rochester Institute of Technology, Rochester, NY, USA).

The considerably improved detection efficiency achieved by a simple replacement of the PMTs by APDs enabled us now to reach out and extend the FCS principles to imaging (7). Classic, stationary FCS probes molecular interactions in a small region of the cell at a time (Fig. 1c). This enables us to study molecular mechanisms underlying complex cellular processes in great detail in a very well defined location (Fig. 1d). However, this approach is not best suited for quantitative study of processes that are occurring in an area that is larger than the FCS observation volume element. Stationary FCS also has limitations when it comes to studying molecular assemblies that are characterized by low mobility. Immobile molecules do not produce fluorescence intensity fluctuations. They are rather contributing to the background, thereby reducing the signal-to-noise ratio (SNR) and obstructing FCS measurements. In addition, molecular assemblies characterized by very low mobility are preferentially subjected to photo-destruction, i.e., irreversible photobleaching of the fluorophore. These disadvantages of classic FCS can be efficiently resolved by scanning the observation volume element over the sample (63) or by introducing multiple observation sites as in multifocal FCS (64).

Implementation of the APD detectors for imaging also enables fast scanning in the imaging mode (7). Fast scanning has been suggested as a possible way to increase signal intensity in CLSM (65–68), but has not been systematically pursued. A contributing factor is that with increasing scanning speed the number of detected photons becomes lower. With low photon counts, detector properties become increasingly relevant and the internal noise of the detector considerably limits the quality of the image. However, fast scanning is also advantageous. It increases the fluorescence yield by abolishing intersystem crossing (7), and enables inspection of a larger number of molecules for the same measurement time, thereby increasing the SNR (Fig. 2). The significantly improved detection efficiency by APD detection now enables imaging with single-molecule sensitivity, and even visual dissection of differences in molecular movement (Fig. 3) (7).

3. Molecular Mechanisms Underlying Opioid Receptor Function

3.1. Opioid Receptor Activation by Specific Ligands

3.1.1. Single-Color FCS Study

Using APD imaging, we could now visualize live PC12 cells stably transformed to express MOP receptor molecules genetically fused with the enhanced green fluorescent protein (eGFP) to yield a fluorescent construct MOP-eGFP (Fig. 4a). We found that MOP expression levels differ strongly between cells, with local receptor densities ranging from 1 to 1,000 molecules in the observation volume element. (Assuming homogenous MOP distribution, the average number of receptors per cell was estimated to vary between 5×10^2 and 5×10^5) (20).

Two fractions of MOP were identified at the plasma membrane that could be distinguished by differences in lateral mobility. The majority of MOP, 80–90%, showed fast lateral mobility, $\tau_{MOP,1} = (250 \pm 150)\,\mu s$. The slowly moving fraction, $\tau_{MOP,2} = (2.5 \pm 1.5)\,ms$, dissipated upon cholesterol depletion and was enriched in detergent insoluble cellular extracts, suggesting

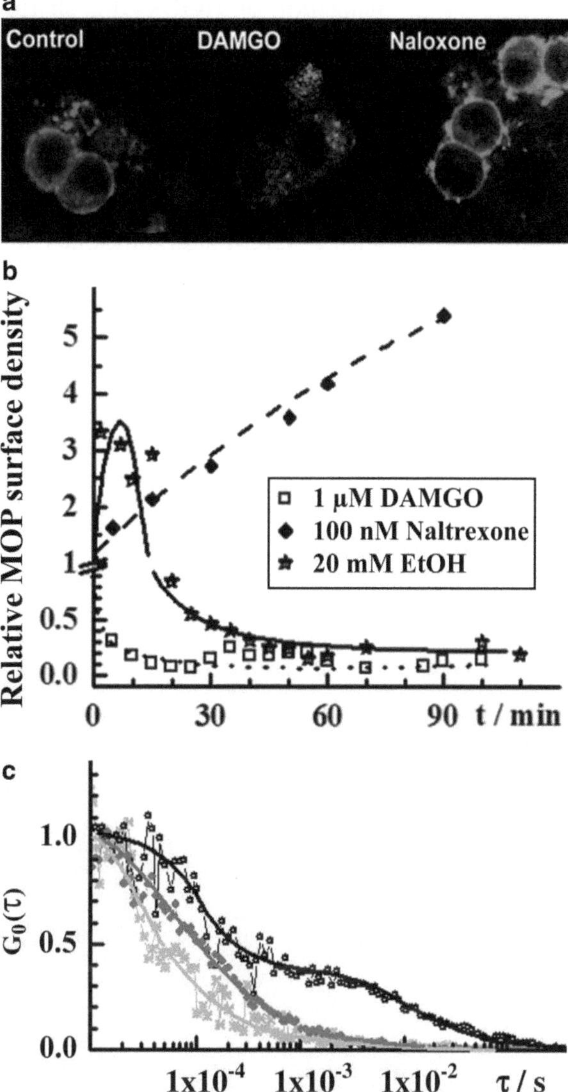

Fig. 4. Effect of ethanol and opioid receptor agonists and antagonists on MOP surface density and dynamics in PC12 cells. (**a**) Confocal fluorescence images showing subcellular localization of MOP-eGFP in PC12 cells under control conditions (*left*), and upon 3 h treatment with DAMGO (1.0 μM; *middle*) or naloxone (200 nM; *right*). (**b**) Autocorrelation curves obtained by FCS analysis showing increased lateral mobility of MOP upon stimulation with ethanol and the opposite effect caused by naltrexone. (**c**) Relative changes in MOP-eGFP surface density under stimulation with ethanol (*light gray*) and naltrexone (*dark gray*), as compared to MOP dynamics in control cells (*black*).

that this fraction may correspond to MOP associated with lipids and other components in detergent insoluble protein/lipid-rich micro-domains (20).

Selected ligands at MOP receptors elicit specific responses (Fig. 4). The *mu*-selective agonist DAMGO caused rapid internalization of MOP-eGFP, sharply reducing its surface density. In contrast, the opioid antagonists naloxone and naltrexone caused a monotonous increase in MOP-eGFP surface density and stalled its mobility in the plasma membrane, as evident from the appearance of a second component in the autocorrelation curve, characterized by longer residence times (Fig. 4c).

Similarly, two fractions of the *kappa* opioid receptor (KOP) with different lateral mobility were observed in PC12 cells stably transformed to express KOP-eGFP (Fig. 5). As compared to MOP-eGFP, KOP-eGFP shows slower lateral mobility and higher degree of association in larger molecular assemblies in PC12 cells (Table 1). KOP lateral mobility in NGF treated PC12 cells showing a neuronal-like cell soma (Fig. 5, red line) was even slower than in the control, untreated cells (Fig. 5, black line), showing both a more populated slowly moving fraction and longer residence times (Table 1).

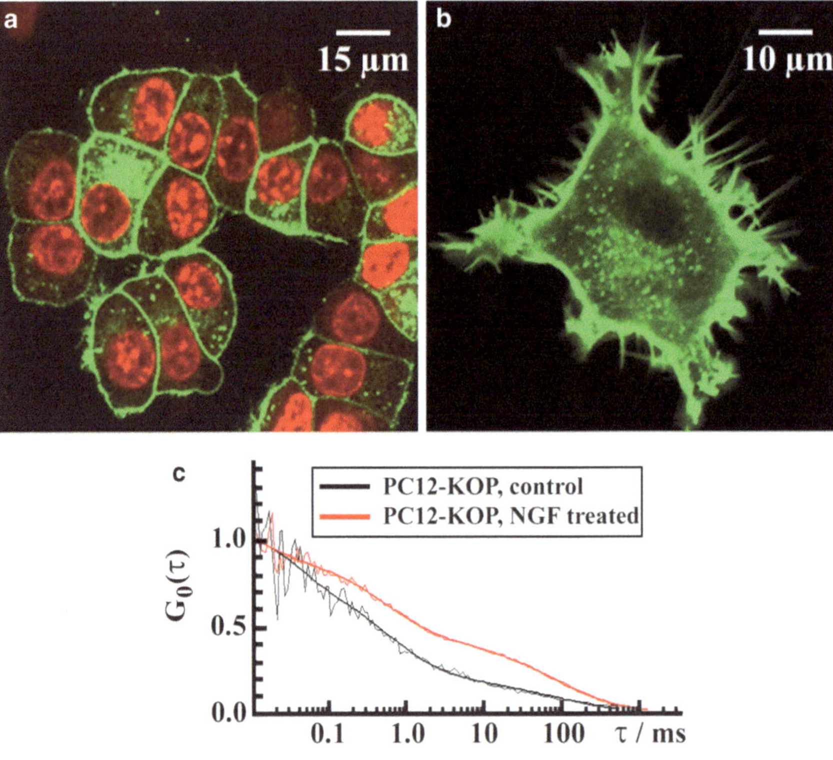

Fig. 5. Confocal images showing KOP-eGFP distribution in PC12 cells. (**a**) Undifferentiated PC12 cells expressing KOP-eGFP. Cell nuclei were visualized using a vital nuclear Vybrant® DyeCycle™ Ruby stain with near-infrared emission. (**b**) Soma growth and morphological changes are observed in PC12 cells stably expressing KOP-eGFP in response to nerve growth factor (NGF) treatment. (**c**) KOP lateral mobility in NGF treated PC12 cells is slower (*top trace*) than in the control, not treated cells (*bottom trace*). The slowly moving fraction is both more populated, $x_{2,\mathrm{NGF}} = (0.4 \pm 0.2)$ compared to $x_{2,\mathrm{control}} = (0.25 \pm 0.15)$ and characterized by longer residence times, $\tau_{2,\mathrm{NGF}} = (0.10 \pm 0.02)$ s as compared to $\tau_{2,\mathrm{control}} = (0.05 \pm 0.02)$ s.

Table 1
Opioid receptor lateral mobility in PC12 cells

	MOP	KOP	KOP_{NGF}
x_1	0.90 ± 0.05	0.80 ± 0.05	0.5 ± 0.1
$\tau_{D1}/\mu s$	250 ± 150	500 ± 200	500 ± 200
x_2	0.10 ± 0.05	0.20 ± 0.05	0.5 ± 0.1
τ_{D2}/ms	2.50 ± 1.5	5 ± 2	10 ± 5

3.1.2. Dual-Color FCCS study

Simultaneous labeling of the ligand and receptor enables quantitative studies of ligand-receptor interactions by FCCS. We used this approach to monitor interactions of an endogenous enkephalin derivative, a hepta-enkephalin peptide (hEnk; Tyr-Gly-Gly-Phe-Met-Arg-Phe) fluorescently labeled with tetramethylrhodamine isothiocyanate (TRITC) (20). Interactions between TRITC-hEnk peptide and the MOP-eGFP could be studied using multicolor cross-correlation measurements, enabling the observation of ligand-receptor complexes at the plasma membrane of a living cell.

Dual-color FCCS was performed by exciting both fluorophores simultaneously or using sequential FCCS (Fig. 6). These experiments also suggested that two pools of ligand-receptor complexes exist that can be distinguished by their average residency times (Fig. 6e). Because of a small cross-talk between detectors, visible as the positive correlation in the triplet range (Fig. 6e), sequential FCCS was applied to confirm the formation of ligand-receptor complexes. In the sequential mode the fluorophores were excited alternatingly, as described in the Experimental section, minimizing the cross-talk between the detectors. Sequential FCCS confirmed that after 60 min incubation with 50 nM TRITC-labeled hEnk, ligand-receptor complexes characterized with correlation $\tau_{D2} = (4.0 \pm 2.0)$ ms had formed (Fig. 6e, inset), showing that the observed cross-correlation is not due to the cross-talk between channels. From the FCCS measurements, the equilibrium binding constant for hEnk-MOP was estimated to be $K_b = (5.0 \pm 1.0) \times 10^{-9}$ M^{-1}. This finding is in good agreement with MOP receptor binding constants for several other enkephalin derivatives reported in the literature.

3.2. Opioid Receptor Activation by Nonspecific Ligands

Ethanol is a nonspecific substance known to modulate the opioid system activity. Interactions between alcohol and the CNS opioid signaling system are well established and documented in basic research and clinical practice (69, 70). However, the mechanisms that mediate alcohol-opioid interactions are not yet fully resolved.

We are using fluorescence imaging and FCS to investigate cellular and molecular mechanisms of ethanol action at the level

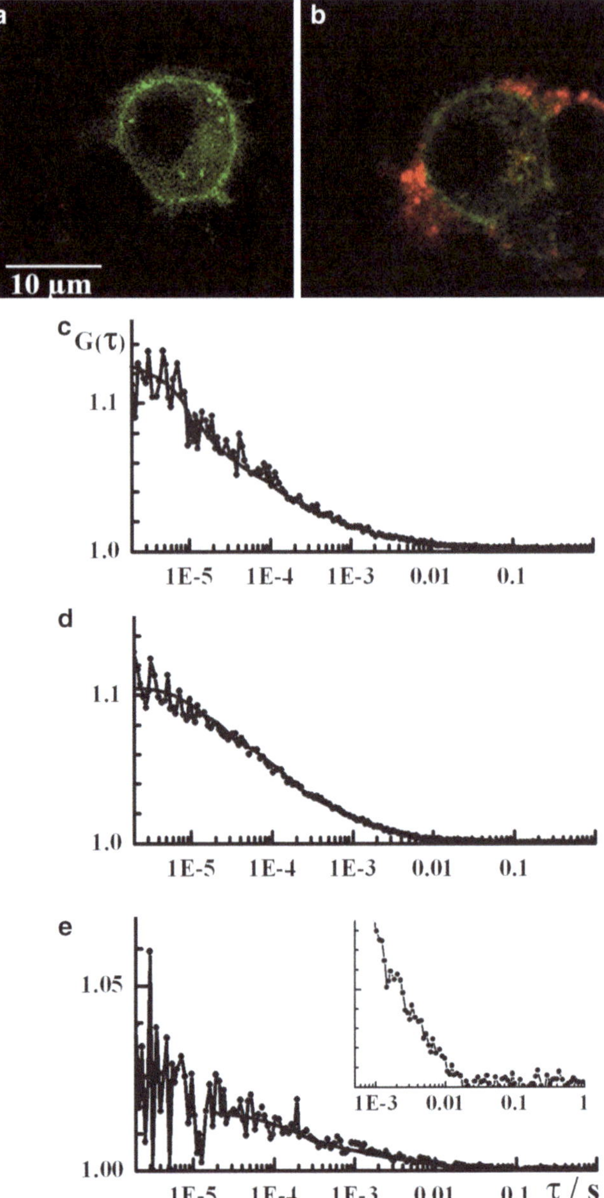

Fig. 6. Confocal images showing the expression of MOP-eGFP in undifferentiated PC12 cells before (**a**) and after 110 min incubation with 50 nM TRITC-labeled hEnk (**b**). TRITC and eGFP fluorescence were excited using the 488 nm line of the Ar laser. (**c**–**e**) Detection of ligand-receptor complexes in living cells using temporal dual-color cross-correlation analysis. Single-color autocorrelation curves for MOP-eGFP (**c**) and TRITC-labeled hEnk (**d**), and dual-color cross-correlation curves obtained in the continuous (**e**) and the switching mode (magnified in the insert). FCCS measurements were performed and analyzed as described in the caption to Fig. 2. The two-component model for free two-dimensional-diffusion ((2a), without the triplet term) was applied to evaluate the cross-correlation curve, confirming existence of ligand-receptor complexes with lateral mobility $\tau_{D1} = (250 \pm 150)\,\mu\text{s}$ and $\tau_{D2} = (4.0 \pm 2.0)$ ms, respectively. From the FCCS measurements, the hEnk binding constant at MOP was estimated to be $(5.0 \pm 1.0) \times 10^9\ \text{M}^{-1}$. Image adapted from Vukojević et al. (20).

of the opioid receptor and the plasma membrane lipids. In the live cell studies conducted in (21), we observed a complex, biphasic action of physiologically relevant concentrations of ethanol (10–40 mM) on MOP surface density and mobility (Figs. 4 and 7). As seen in Fig. 4b, ethanol induces an abrupt transient increase in MOP-eGFP surface density, followed by a state that is characterized by reduced MOP surface density. The time course of ethanol-induced changes in MOP surface density is also markedly different from that induced by the specific ligands; ethanol-induced internalization half-time, $t_{1/2,\text{ethanol}} = 25$ min, is markedly different from the agonist-induced internalization half-time, $t_{1/2,\text{agonists}} = 2.5$ min. In addition, the lateral mobility of MOP also changed with the slowly moving fraction of MOP diminished upon exposing the PC12 cells to ethanol (Fig. 7a).

Based on these findings, we have proposed an alternative hypothesis on ethanol-induced euphoria, shifting the focus from its effect on endogenous opioid ligands to its effect on opioid receptors (21). The previous hypothesis asserted that ethanol increases the activity of the endogenous opioid system through the release of opioid peptides and that anticraving/antihedonic

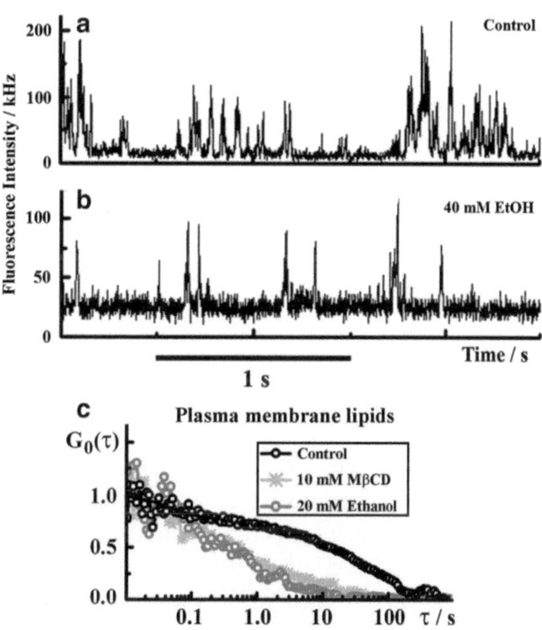

Fig. 7. Dynamics of plasma membrane lipids under MOP stimulation with naltrexone and ethanol. (**a**) Fluorescence intensity fluctuations recorded in control cells, labeled with DiIC_{18} (5). The corresponding temporal autocorrelation curve is shown in Fig. 5a (*black curve*). (**b**) Fluorescence intensity fluctuations showing the effect of 40 mM ethanol on the dynamics of plasma membrane lipids recorded after 15 min exposure to ethanol. (**c**) Normalized autocorrelation curves showing changes in the dynamics of plasma membrane lipids, caused by 20 mM ethanol (*dark gray*) or 10 mM MßCD (*light gray*) as compared to the dynamics in control cells (*black curve*).

effects of naltrexone (Revia®), used clinically to prevent relapse in alcoholism are achieved through antagonizing the effect of opioid peptides released by ethanol that are acting at the MOP receptor. However, attempts to measure ethanol-induced increase in endogenous opioids levels have failed so far. In addition, animal model studies showed that neither null-mutation of preproenkephalin, nor homozygous knockout of proopiomelanocortin (the precursor of β-endorphin) affects the voluntary intake of ethanol in mice (71–73), whereas MOP knockout mice do not self-administer alcohol (74).

3.3. Opioid Receptor-Lipid Interactions

We are also able to study in live cells the dynamics and structural organization of plasma membrane lipids. Using a fluorescent lipid marker $DiIC_{18}$ (5) (1,1′-dioctadecyl-3,3,3′,3′-tetramethylindocarbocyanine perchlorate) as an indicator, we observed a complex temporal dynamics of the plasma membrane lipids in PC12 cells (Fig. 7). This was reflected by compound fluorescence intensity bursts, consisting of alternating high–low-intensity spurts separated by prolonged intervals (about 0.7 s) of low-intensity fluorescence fluctuations. Marked changes in the lipid dynamics were observed under treatment with 10 mM methyl-β-cyclodextrin (MßCD), a compound known to extract cholesterol from the plasma membrane. The fluorescence intensity fluctuation pattern changed its complex form to regular fluorescence intensity fluctuations and the characteristic decay time shifted towards shorter times (Fig. 7).

4. Future Directions and Conclusion

We have introduced two methods with single-molecule sensitivity, fluorescence imaging with avalanche photodiode detectors (APD imaging) and FCS that enable us to characterize quantitatively the receptor dynamics at the plasma membrane of living cells. Receptor status, determined by receptor level at the membrane surface and receptor associations in functional complexes, plays a central role in determining receptor availability, function, and the downstream consequences of receptor activation. Therefore, this methodology may become a method of choice in future studies of GPCR activation in live cells and opens a way to study important aspects of the role of lipids in cell-surface receptor activation. The prospect to monitor in real time and with single-molecule sensitivity changes in the surface density and mobility of the activated receptor, and observe concomitantly changes in the structural organization and dynamics of lipids in the plasma membrane is a challenging task not yet easily achieved with other methods.

Using the opioid receptors as an example, we demonstrated how these methodologies can be used to study molecular mechanisms of surface receptor action nondestructively, in live cells. Working out detailed molecular mechanisms is invaluable for our understanding of the etiology of complex psychiatric diseases and may help designing new treatment strategies.

Acknowledgements

The authors acknowledge support from The Knut and Alice Wallenberg Foundation, The Swedish Research Council and The Swedish Brain Foundation.

References

1. Singer SJ, Nicolson GL (1972) The fluid mosaic model of the structure of cell membranes. Science **175:**720–731.
2. Allen JA, Halverson-Tamboli RA et al (2007) Lipid raft microdomains and neurotransmitter signalling. Nat Rev Neurosci **8:**128–140.
3. Pike LJ (2003) Lipid rafts: bringing order to chaos. J Lipid Res **44:**655–667.
4. Simons K, Toomre D (2000) Lipid rafts and signal transduction. Nat Rev Mol Cell Biol **1:**31–39.
5. Hanyaloglu AC, von Zastrow M (2008) Regulation of GPCRs by endocytic membrane trafficking and its potential implications. Ann Rev Pharmacol Toxicol **48:**537–568.
6. Hanson MA, Cherezov V, Griffith MT et al (2008) A specific cholesterol binding site is established by the 2.8 A structure of the human *beta*2-adrenergic receptor. Structure **16:**897–905.
7. Vukojević V, Heidkamp M, Ming Y et al (2008) Quantitative single-molecule imaging by confocal laser scanning microscopy. Proc Natl Acad Sci USA **105:**18176–18181.
8. Elson EL (2004) Quick tour of fluorescence correlation spectroscopy from its inception. J Biomed Optics **9:**857–864.
9. Vukojević V, Pramanik A, Yakovleva T et al (2005) Study of molecular events in cells by fluorescence correlation spectroscopy. Cell Mol Life Sci **62:**535–550.
10. García-Sáez AJ, Schwille P (2007) Single molecule techniques for the study of membrane proteins. Appl Microbiol Biotechnol **76:**257–266.
11. Vukojević V, Papadopoulos DK, Terenius L et al (2010) Quantitative study of synthetic Hox transcription factor–DNA interactions in live cells. Proc Natl Acad Sci USA **107:** 4087–4092.
12. Terenius L (1973) Stereospecific interaction between narcotic analgesics and a synaptic plasma-membrane fraction of rat cerebral-cortex. Acta Pharm Toxicol **32:**317–320.
13. Terenius L (1973) Characteristics of the "receptor" for narcotic analgesics in synaptic plasma membrane fraction from rat brain. Acta Pharmacol Toxicol (Copenh) **33:** 377–384.
14. Terenius L. (1977) Opioid peptides and opiates differ in receptor selectivity. Psychoneuroendocrinology **2:**53–58.
15. Gunne LM, Lindström L, Terenius L (1977) Naloxone-induced reversal of schizophrenic hallucinations. J Neural Transm **40:**13–19.
16. Terenius L. (1978) Endogenous peptides and analgesia. Ann Rev Pharmacol Toxicol **8:** 189–204.
17. Vincent SR, Dalsgaard CJ, Schultzberg M et al (1984) Dynorphin-immunoreactive neurons in the autonomic nervous-system. Neurosci **11:**973–987.
18. Nylander I, Hyytia P, Forsander O et al (1994) Differences between alcohol-preferring (AA) and alcohol-avoiding (ANA) rats in the prodynorphin and proenkephalin systems. Alcohol Clin Exp Res **18:**1272–1279.
19. Kuzmin A, Sandin J, Terenius L et al (2003) Acquisition, expression, and reinstatement of ethanol-induced conditioned place preference in mice: Effects of opioid receptor-like 1 receptor agonists and naloxone. J Pharmacol Exp Ther **304:**310–318.

20. Vukojević V, Ming Y, D'Addario C et al (2008) *Mu*-opioid receptor activation in live cells. FASEB J **22:**3537–3548.
21. Vukojević V, Ming Y, D'Addario C et al (2008) Ethanol/naltrexone interactions at the *mu*-opioid receptor. CLSM/FCS study in live cells. PLoS ONE **3:**e4008.
22. Pierce KL, Premont RT, Lefkowitz RJ (2002) Seven-transmembrane receptors. Nat Rev Mol Cell Biol **3:**639–650.
23. Wolfe BL, Trejo J (2007) Clathrin-dependent mechanisms of G protein-coupled receptor endocytosis. Traffic **8:**462–470.
24. Jalink K, Moolenaar WH (2010) G protein-coupled receptors: the inside story. BioEssays **32**:13–16.
25. Molina PE (2006) Opioids and opiates: analgesia with cardiovascular, haemodynamic and immune implications in critical illness. J Intern Med **259**:138–154.
26. Tambour S, Quertemont E. (2007) Preclinical and clinical pharmacology of alcohol dependence. Fund Clin Pharmacol **21**:9–28.
27. Magde D, Webb WW, Elson E (1972) Thermodynamic fluctuations in a reacting system - measurement by fluorescence correlation spectroscopy. Phys Rev Lett **29:**705–708.
28. Elson EL, Magde D (1974) Fluorescence correlation spectroscopy. Conceptual basis and theory. Biopolymers **13:**1–27.
29. Ehrenberg M, Rigler R (1974) Rotational Brownian motion and fluorescence intensity fluctuations. Chem Phys **4:**390–401.
30. Koppel DE (1974) Statistical accuracy in fluorescence correlation spectroscopy. Phys Rev A **10:**1938–1945.
31. Koppel DE, Axelrod D, Schlessinger J et al (1976) Dynamics of fluorescence marker concentration as a probe of mobility. Biophys J **16:**1315–1329.
32. Ehrenberg M, Rigler R (1972) Polarized fluorescence and rotational Brownian motion. Chem Phys Lett **14:**539–544.
33. Rigler R, Mets Ü, Widengren J et al (1993) Fluorescence correlation spectroscopy with high count rate and low background. Eur Biophys J **22:**169–175.
34. Eigen M, Rigler R (1994) Sorting single molecules: application to diagnostics and evolutionary biotechnology. Proc Natl Acad Sci USA **91:**5740–5747.
35. Axelrod D, Koppel DE, Schlessinger J et al (1976) Mobility measurement by analysis of fluorescence photobleaching recovery kinetics. Biophys J **16:**1055–1069.
36. Elson EL, Schlessinger J, Koppel DE et al (1976) Measurement of lateral transport on cell surfaces. Prog Clin Biol Res **9:**137–147.
37. Schlessinger J, Koppel DE, Axelrod D et al (1976) Lateral transport on cell membranes: Mobility of concanavalin A receptors on myoblasts. Proc Natl Acad Sci USA **73:** 2409–2413.
38. Schlessinger J, Axelrod D, Koppel DE et al (1977) Lateral transport of a lipid probe and labeled proteins on a cell membrane. Science **195:**307–309.
39. Jacobson K, Elson E, Koppel D et al (1982) Fluorescence photobleaching in cell biology. Nature **295:**283–284.
40. Yechiel E, Edidin M (1987) Micrometer-scale domains in fibroblast plasma membranes. J Cell Biol **105:**755–760.
41. Berland KM, So PTC, Gratton E (1995) Two-photon fluorescence correlation spectroscopy: Method and application to the intracellular environment. Biophys J **68:**694–701.
42. Schwille P, Haupts U, Maiti S et al (1999) Molecular dynamics in living cells observed by fluorescence correlation spectroscopy with one- and two-photon excitation. Biophy. J 77:2251–2265.
43. Denk W, Strickler JH, Webb WW (1990) Two-photon laser scanning fluorescence microscopy. *Science* **248:** 73–76
44. Zipfel WR, Williams RM, Christie A et al (2003) Live tissue intrinsic emission microscopy using multiphoton-excited native fluorescence and second harmonic generation. Proc Natl Acad Sci USA **100:**7075–7080.
45. Zipfel WR, Williams RM, Webb WW (2003) Nonlinear Magic: Multiphoton microscopy in the biosciences. Nature Biotech **21:** 1369–1377.
46. Miyawaki A, Sawano A, Kogure T (2003) Lighting up cells: labelling proteins with fluorophores. Nat. Cell Biol. Suppl: S1–S7.
47. Giepmans BN, Adams SR, Ellisman MH et al (2006) The fluorescent toolbox for assessing protein location and function. Science **312:**217–224.
48. Marks KM, Nolan GP (2006) Chemical labeling strategies for cell biology. Nat Methods **3:**591–596.
49. Terasaki M, Jaffe LA (2004) Labeling of cell membranes and compartments for live cell fluorescence microscopy. Methods Cell Biol **74:**469–489.
50. Liu P, Ahmed S, Wohland T (2008) The F-techniques: advances in receptor protein studies. Trends Endocrinol Metab **19**: 181–190.
51. Haustein E, Schwille P (2007) Fluorescence correlation spectroscopy: novel variations of an established technique. Annu Rev Biophys Biomol Struct **36**:151–169.

52. Elson EL. (2001) Fluorescence correlation spectroscopy measures molecular transport in cells. Traffic. **2**:789–796.
53. Thompson, N L (1991) Fluorescence correlation spectroscopy. In: Lakowicz JR (ed) Topics in Fluorescence Spectroscopy, Volume 1: Techniques. Plenum Press, New York.
54. Qian H, Elson EL (1990) On the analysis of high order moments of fluorescence fluctuations. Biophys J **57**:375–380.
55. Kask P, Palo K, Ullmann D et al (1999) Fluorescence-intensity distribution analysis and its application in biomolecular detection technology. Proc Natl Acad Sci USA **96:** 13756–13761.
56. Kask P, Palo K, Fay N et al (2000) Two-dimensional fluorescence intensity distribution analysis: theory and applications. Biophys J **78:**1703–1713.
57. Chen Y, Müller JD, So PTC et al (1999) The photon counting histogram in fluorescence fluctuation spectroscopy. Biophys J **77:** 553–567.
58. Hillesheim LN, Müller JD (2003) The photon counting histogram in fluorescence fluctuation spectroscopy with non-ideal photodetectors. Biophys J **85:**1948–1958.
59. Müller JD (2004) Cumulant analysis in fluorescence fluctuation spectroscopy. *Biophys J* **86:**3981–3992.
60. Weisshart K, Jüngel V, Briddon SJ (2004) The LSM 510 META-ConfoCor 2 system: an integrated imaging and spectroscopic platform for single-molecule detection. Curr Pharm Biotechnol **5:**135–154.
61. Meseth U, Wohland T, Rigler R et al (1999) Resolution of fluorescence correlation measurements. Biophys J **76:**1619–1631.
62. Chen Y, Wei LN, Müller JD (2003) Probing protein oligomerization in living cells with fluorescence fluctuation spectroscopy. Proc Natl Acad Sci USA **100:**15492–15497.
63. Petrásek Z, Hoege C, Mashaghi A et al (2008) Characterization of protein dynamics in asymmetric cell division by scanning fluorescence correlation spectroscopy. Biophys J **95:** 5476–5486.
64. Blancquaert Y, Gao J, Derouard J et al (2008) Spatial fluorescence cross-correlation spectroscopy by means of a spatial light modulator. J Biophotonics **1**:408–418.
65. Eigen M, Rigler R (1994) Sorting single molecules: application to diagnostics and evolutionary biotechnology. Proc Natl Acad Sci USA **91:**5740–5747.
66. Rigler R, Mets Ü, Widengren J et al (1993) Fluorescence correlation spectroscopy with high count rate and low background: analysis of translational diffusion. Eur Biophys *J* **22:**169–175.
67. Koppel DE, Morgan F, Cowan AE et al (1994) Scanning concentration correlation spectroscopy using the confocal laser microscope. Biophys J **66:**502–507.
68. Donnert G, Eggeling C, Hell SW (2007) Major signal increase in fluorescence microscopy through dark-state relaxation. Nat Methods **4:**81–86.
69. Herz A (1997) Endogenous opioid systems and alcohol addiction. Psychopharmacology (Berl) **129:**99–111.
70. Herz A (1998) Opioid reward mechanisms: a key role in drug abuse? Can J Physiol Pharmacol **76**:252–258.
71. Koenig HN, Olive M (2002) Ethanol consumption patterns and conditioned place preference in mice lacking preproenkephalin. Neurosci Lett **325**:75–78.
72. Griesel JE, Mogil JS, Grahame NJ et al (1999) Ethanol oral self-administration is increased in mutant mice with decreased *beta* -endorphin expression. Brain Res **835**:62–67.
73. Grahame NJ, Low MJ, Cunningham CL (1998) Intravenous self-administration of ethanol in *beta*-endorphin-deficient mice. Alcohol Clin Exp Res **22**:1093–1098.
74. Roberts AJ, McDonald JS, Heyser CJ et al (2000) *Mu*-opioid receptor knockout mice do not self-administer alcohol. J Pharmacol Exp Ther **293**:1002–1008.

Chapter 19

Using RNA Interference to Downregulate G Protein-Coupled Receptors

Philippe Sarret, Louis Doré-Savard, Pascal Tétreault, Valérie Bégin-Lavallée, Marc-André Dansereau, and Nicolas Beaudet

Abstract

Technologies developed to interfere with gene transcripts were developed back in the 1980. However, it was not before the last decade that light was shed on the underlying mechanisms of what is now known as RNA interference. From then, RNAi was propelled to the forefront as a revolutionizing approach for basic research and clinical therapy. In the present chapter, we will present an overview of RNAi and its mechanisms with a focus on GPCRs applied to neuroscience. We will briefly detail the steps to move from the *in vitro* assessment to the *in vivo* proof-of-principle to further ensure appropriate clinical transfer. Examples of successful experiments will be given in each section. Advances, drawbacks, and future directions will be discussed with RNAi as an exciting new technology that can be used to treat GPCR-related neuropathologies intractable with commonly-used pharmacological agents.

Key words: siRNA, shRNA, DsiRNA, ODN, Clinical trials, Silencing

1. Introduction

1.1. History of RNAi

Pharmacological targeting of GPCR was a key concept that drove industrial drug development into the modern era (1). Despite the fact that the role of these invaluable targets is still investigated nowadays in various pathologies, it has become increasingly more difficult to discover, develop, and launch new individual molecules in recent years. Long-term declines in the productivity of pharmaceutical industry is supported by the drastic reduction in the number of new drug products approved by the FDA, going from 53 in 1996 to only 18 in 2006 (2). Yet, classical pharmacology faces a recent slowing down that has driven the development of

Craig W. Stevens (ed.), *Methods for the Discovery and Characterization of G Protein-Coupled Receptors*, Neuromethods, vol. 60, DOI 10.1007/978-1-61779-179-6_19,

alternative techniques to study, validate, and target GPCRs in the hope to design new blockbusters. The first approach that comes in mind is genetic engineering through the use of transgenic mice to knockout GPCRs expression. Although really insightful, this technique remains criticized for its main use limited to rodent and for reported developmental lethality or compensation mechanisms occurring early in the developmental stages (3). However, recent discoveries in the field of protein expression lead to a brand new approach circumventing those drawbacks: RNA interference (RNAi). Indeed, exogenous RNA could transiently prevent GPCRs expression without intervening directly in early gene expression steps.

The first endogenous RNAi-like processes were discovered in 1984 in *Escherichia coli* (4). It was demonstrated through genetic engineering that small RNA could bind to and inhibit the translation of a complementary RNA sequences. Reduction in mRNA expression caused by another mRNA was also reported in plants in the early 1990 (5). The mechanisms underlying these phenomena were unknown (not to say mystical) at the time, and it was not before 1997 that the idea of a well-defined RNAi process emerged (6). One year later, Fire and Mellow confirmed their hypothesis and laid the foundations for a mechanistic explanation of RNAi through an elegant work on *Caenorhabditis elegans* twitching behaviors and double-stranded RNA (dsRNA) (7). This mechanism presented enormous potential for selective repression of protein expression both *in vitro* and *in vivo*; a potential that was acknowledged in 2006, when Fire and Mellow earned the Nobel Prize in Physiology and Medicine for that discovery. In the early 2000, this animated field took another giant step forward when different groups unveiled the existence of microRNAs (miRNA): endogenous untranslated short dsRNAs that regulate protein synthesis (8). It was these miRNA molecules that served as template to develop the exogenous RNAi technologies as we know and use them today.

1.2. Cell Biology of RNAi Pathway

RNAi belongs, with antisense oligonucleotides (ODNs) and rybozymes, to the ODN-based technologies (9). ODNs are short, single-stranded products that affect gene expression upon hybridization to complementary sequences by multiple mechanisms, including translational inhibition by physical blockage and mRNA cleavage by recruitment of RNase H. On the other hand, ribozymes are catalytic RNAs that cleave target mRNAs by sequence complementation. Although antisense oligodeoxynucleotides were frequently used in the 1990 as an RNA silencing research tool, the past decade has seen an explosive development in the application of RNAi to modulate gene expression using exogenous dsRNAs. In the following section, we present a brief overview of their mechanism of action.

1.2.1. Endogenous Pathway

Micro RNAs are a class of posttranscriptional regulators used by the cell to downregulate gene expression. The disruption of this endogenous pathway can initiate pathologies (10). The synthesis of miRNA originates in the nucleus, where the enzyme polymerase III transcribes the miRNA encoding parts of the genome in precursors called pri-miRNAs (Fig. 1, step 1). These primary transcripts consist of a ~80-nucleotides stem-loop structure. The first step of miRNA biogenesis is the nuclear excision of the upper part of this hairpin structure by Drosha, a class II RNase-III family enzyme. The product of this processing yields pre-miRNAs with 2-nt 3′ overhangs. Afterwards, the karyopherin Exportin-5 (Exp-5) mediates their nuclear export to the cytoplasm. At this point, pre-miRNAs are recruited and processed by a member of the class III-RNAse-III family called Dicer into miRNA duplex of ~21 nucleotides of length. Following cleavage by Dicer, miRNA molecules are then incorporated into a multi-protein RNA-induced silencing complex (RISC) (Fig. 1).

1.2.2. Exogenous Pathway

By contrast, exogenous gene silencing strategies using small interfering RNA (siRNA) and dicer-substrate small interfering RNAs (DsiRNA) take place directly in the cytoplasm after their cellular uptake following *in vitro* incubation or *in vivo* administration. siRNAs are synthetic 21 nucleotide-long duplexes that mimic the natural product of Dicer enzyme processing. Those compounds enter into the cellular interference pathway directly via the RISC complex (Fig. 1, step 2). In comparison to siRNA, several groups demonstrated that slightly longer synthetic RNAs, namely DsiRNAs, were taken in charge by Dicer upon cellular entry thus yielding a greater efficiency *in vitro* and *in vivo* (11, 12) (Fig. 1, step 3). Indeed, the presence of longer strands not exceeding 30 bp met greater silencing efficiency at lower concentration than did siRNAs (13). Those findings lead to the development of Dicer-substrate siRNAs consisting of 27 nucleotide-long double-strands, modified in such a way to direct the Dicer cleavage for a specific 21 nt end-product (14). The pre-association of DsiRNA with Dicer facilitates the loading of DsiRNA products into the RISC complex (12). Another strategy for RNAi that is well characterized is the use of a plasmid or vector that expresses short hairpin RNA (shRNA). Inside the nucleus, polymerase III transcribes the information encoded in the expression system in a structure homologous to pre-miRNA. As in the miRNA pathway, Drosha and Dicer proceed to the biogenesis and maturation of shRNA into siRNA and the target degradation is ensured by RISC complex.

All the previous dsRNA therapeutics described earlier enter RISC. Many conserved proteins have been identified as essential across species. In human cells, there are three major players; Dicer, Argonaute (AGO2), and dsRNA-binding protein TRBP (15). The transfer of miRNA into RISC is facilitated by the binding of

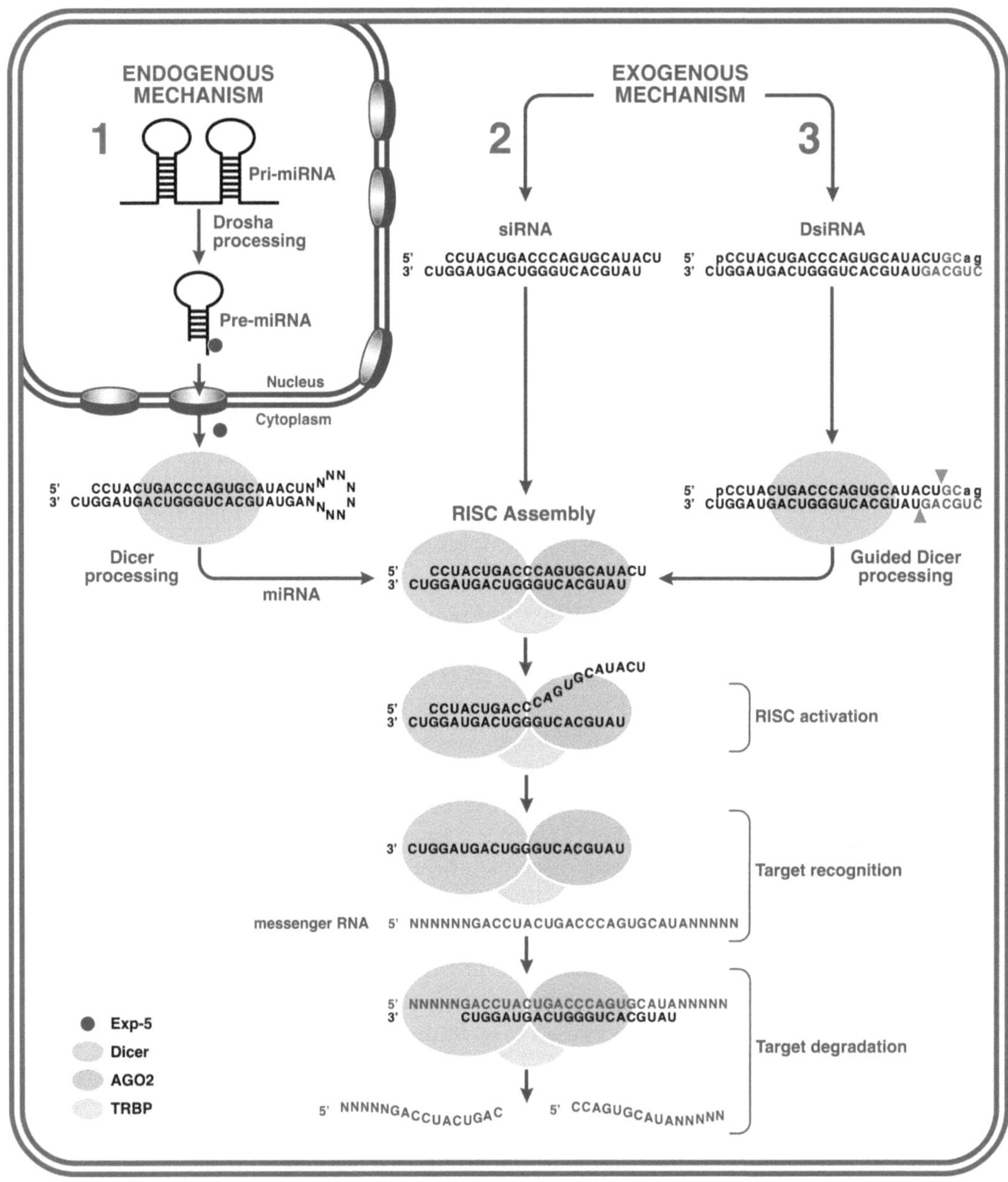

Fig. 1. Endogenous and exogenous RNA interference mechanism. (1) Biogenesis of microRNA (miRNA). Inside the nucleus, pri-miRNAs are first cleaved by Drosha to yield in pre-miRNAs stem-loop structure. Exportin-5 (Exp5) transports pre-miRNAs into the cytoplasm. Those transcripts are then processed in miRNA duplexes by Dicer. Finally, miRNAs are loaded into RISC. (2) Small interfering RNA molecules (siRNA) are synthetic molecules that mimic Dicer products and upon cellular entry are directly loaded into RISC. (3) By contrast, the more recent generation of siRNA consisting of 27 nucleotides entitled Dicer-substrate siRNA (DsiRNA) benefit from pre-association with the enzyme Dicer to facilitate loading into RISC. The convergent point of the three pathways is the loading of Dicer products into RISC. During RISC assembly, one of the two siRNA strands, referred to as the passenger strand, is cleaved and released, whereas the other strand (guide strand) is incorporated to the argonaute protein AGO2 component of RISC (RISC activation). The remaining single-stranded siRNA then guides the RISC complex to recognize its target mRNA by perfect base-pairing (target recognition). Finally, the target is degraded by the endonucleolytical activity located in the AGO2 protein (target degradation).

Dicer to both AGO2 and TRBP. During RISC assembly, one of the two dsRNAs strands referred to as the guide strand is loaded onto AGO2 at the core of the RISC complex. The other strand called the passenger strand will be cleaved and trigger RISC activation. The guide strand thus directs the sequence specificity of RISC to RNAi silencing. The incorporated strand that serves as the guide strand is generally the one whose 5′ terminus is at the thermodynamically less stable end of the duplex (16). AGO2 uses the guide strand to recognize its target by hybridizing perfectly the complementary mRNA sequence and then catalyzes the splicing. Finally, the cleavage of mRNA leads to its degradation by cellular exonucleases.

Both siRNA and shRNA are designed to mimic similar intermediates of miRNA pathway and require common cellular mechanisms for gene silencing. However, the availability of proteins involved in the endogenous pathway complex remains essential for appropriate function of miRNA. Indeed, high doses of sh/siRNA can interfere by overloading the different RISC complexes and thus produce toxicity (17, 18). This competition with endogenous RNAi machinery can cause important side effects and lead to lethality (19). Thus, RNAi therapeutic transfer to human requires close surveillance since despite giving good results, these mechanisms present drawbacks.

1.3. Comparison of RNA Interference with Conventional Antisense Oligonucleotides

ODNs represent the most mature therapeutic entities with more than 39 trials currently in development (20). However, RNAi has been reported to be a highly efficient process, exceeding the potency of conventional antisense approaches (21, 22). Indeed, it has been estimated that required concentration of siRNA to attain half-maximal inhibition levels are 100- to 1,000-fold lower than for ODNs directed against the same target, both *in vitro* and *in vivo* (23, 24). Since much higher doses of ODNs are required to achieve significant knockdown of the gene of interest, this may also explain the lower level of *in vivo* toxicity observed with siRNA. Another advantage that RNAi has over ODN technology is that it appears to be more selective to the target gene, thus avoiding potential side effects related to nonspecific interference (25). This last feature may also allow the use of multiple RNAi targets at the same time in a single treatment (26). In addition, siRNAs were demonstrated to be more resistant to nuclease degradation as compared with ODNs and, therefore have longer therapeutic effects than antisense therapy (27). Furthermore, as opposed to the ODN approach, shRNAs and miRNA may be integrated in vector systems for nonviral and viral delivery (28, 29). Thus, a single administration can trigger long-term expression of the therapeutic RNAi and consequently, the use of RNAi technology can reduce the high dosing schedule frequency required with ODN therapy. The potential toxicities that have been observed following persistent expression of shRNA may perhaps be overcome by using inducible vector systems.

2. Methods to Knockdown the Expression of GPCRs

In the following sections, we will review the recent methods described in the literature to knockdown GPCRs *in vitro* and *in vivo*, thus displaying their potential in treating diverse health issues.

2.1. Silencing GPCRs In Vitro

As cited previously, recent RNAi technologies can be divided in three main categories: siRNA (for short interfering RNA), shRNA (for short hairpin RNA), and miRNA (for micro RNA). Taking the latest advances in this research field as a starting point, this section will briefly describe the elaboration of an efficient experiment using RNAi *in vitro* and expose examples for GPCR transcripts interference in the central nervous system (CNS).

2.1.1. An Adequate Preparation is Key to RNAi Success

A meticulous design, rigorous controls, and the choice of a representative cell assay are the basis to optimize the RNAi validation towards a further use in *in vivo* research. Various design methods exist, but in all cases, there are basic rules to follow. Even if those rules are observed, the potency of a specific siRNA still remains unpredictable. Thus, it is highly recommended to design multiple siRNA candidates for a same transcript of interest and to double-check high score candidates in more than one of the available algorithms in order to choose the sequences with the best potential (30). In the last decade, plenty of design algorithms have emerged in the literature and are now freely accessible on the web while proposing embedded tools to facilitate the design (see Table 1).

Few strategies involving miRNA are commercially available at this point. Some companies have developed artificial miRNAs mimicking endogenous mature miRNAs. However, tools mainly exist to: (1) evaluate the activity of a specific miRNA (via a reporter assay) or (2) modulate the activity of miRNA (with a selective inhibitor or enhancer). Indeed, as we reported in a previous section, exogenous siRNA can interfere with the miRNA functions. In this regard, many databases exist to identify endogenous miRNA sequences in the goal of avoiding possible overlaps. Furthermore, some companies have developed a vast variety of products and resources for miRNA assessment and modulation (Table 1).

2.1.2. Strategic Engineering to Penetrate Endogenous Defenses

RNAi molecules have encountered a rapid evolution since their discovery. The main reason ensues to the importance of ameliorating RNA molecules to help them evade from nuclease

Table 1
Web-based tools for the design of RNAi studies

siRNA design tools list	http://itb.biologie.hu-berlin.de/~nebulus/sirna/siDesign.htm
DsiRNA design tool	http://www.idtdna.com/Scitools/Applications/RNAi/RNAi.aspx
shRNA guidelines	www.genelink.com/sirna/shRNAi.asp
miRNA resources	www.mirbase.org/, http://escience.invitrogen.com/ncode
miRNA products	www.ambion.com/techlib/resources/miRNA/index.html

degradation while decreasing their capacity to induce off-target effects. In this regard, many strategies to avoid nucleases involve inserting chemical substitutions (such as fluoride or methyl groups) in the antisense strand (31). Keeping the length of siRNA under 30 bp also eludes the off-target effects and activation of innate immune response through toll-like receptors (32). Also, avoiding a particular seed region at the 5′ end of the antisense strand decreases risks of interaction with the miRNA pathway (33). These are just a few examples of structural modifications that have been proposed to increase the half-life of RNAi molecules or decrease the concentrations required to produce a strong knockdown, thus attenuating risks of off-target effects. Many reviews in the literature are dedicated to this topic (31).

2.1.3. Selecting the Appropriate Cellular Model

Once the Dsi/si/shRNA candidates are designed and manufactured, validation will need to be performed on a cell line representative of the hypothesis scope and the tissue aimed in future *in vivo* studies or clinical therapy. In the case of neuropathologies involving GPCRs, neuronal or glial cell lines or primary cultures are a priority. However, if the selected cell type has a low expression level of the gene of interest which is hardly inducible or detectable with classical biochemical techniques, it will be necessary to use a transgenic expression cell line. One has to keep in mind that artificial overexpression of a gene in transfected cell lines can exaggerate the real potential of a candidate siRNA. This bias has to be taken into consideration when transferring to ex vivo or *in vivo* validation. Consequently, it is essential to identify at least three potent *in vitro* candidates in case of weak *in vivo* hit.

2.1.4. Getting the Experimental Design Ready for In Vitro Testing

The validation process of candidate RNAi technologies involves rigorous controls in order to attribute the consequence of silencing to the target mRNA and not to an off-target effect. A minimum of four controls are required for this purpose, even if they are not used in all studies: (1) cells without treatment to evaluate basal gene expression level; (2) cells and transfection reagent (or empty viral vector for shRNA) per se to assess possible toxicity and/or undesired effects of the administration route on gene expression; (3) cells, transfection agent and negative control siRNA (i.e. scrambled siRNA, mismatch, etc.) or viral vector with negative control shRNA to assess off-target effects; (4) cells, transfection agent and positive control siRNA or viral vector with positive control shRNA (e.g. HPRT housekeeping gene) to ensure the efficiency of the silencing and the assay performance.

Molecular validation (western blot and/or qPCR) represents a direct measure of silencing potency and needs to be evaluated. Note that the choice of transfection agent is as crucial *in vitro* as it is *in vivo* to ensure an appropriate silencing. Cationic lipids, polymers, viral vectors, and many others are being tested to improve the delivery of RNAi molecules into nervous cells and tissues. From our observations (unpublished data), an optimal transfection agent *in vitro* can yield poor delivery efficiency and/or raise adverse effects *in vivo*. Thus, penetration efficiency must be verified at every step. A cellular uptake control is accurate in RNAi cases of hard-to-transfect cells such as neurons. Fluorophores tagged to an RNAi molecule allow the experimenter to visualize the penetrating efficiency of RNA formulations in cells. This step can be observed on fixed cells or in live acquisition with a confocal spinning disc microscope. However, in all cases, performing this set of controls is a prerequisite for validating functional RNAi technique and increasing the chances of success for further *in vivo* applications.

2.1.5. Example of Effective GPCRs Knockdown In Vitro

Screening of the actual literature yields approximately 20 *in vitro* studies on RNAi technologies focused on GPCR silencing in the neuroscience field. We have listed the different GPCRs according to their respective class of ligands (see Table 2) and will review the methods for our selection of studies.

GPCRs Class A: Peptide Family

Among the peptide class A, chemokine receptors are implicated in a number of neurological disorders and are a growing interest in the GPCR research community. Since chemokines are known to be selective for more than one receptor (34), RNAi appears the most specific approach to study each chemokine receptor function. In a recent paper from Agrawal and colleagues (35), the authors looked at the role of CCR3 and CCR5 in HIV-1 infection of brain microglia. They used several cultures of primary human brain microglia and transfected them with Lipofectamine 2000 (Invitrogen, Carlsbad, CA, USA) complexed with either 21-nt siRNA with 2-nt deoxythymidine 3′-overhangs (0–10 nM) or 63-nt plasmid cloned siRNA (5 μg). They found that CCR3, as it was previously reported for CCR5, is a potential co-receptor for some subtypes of HIV-1. Indeed, by specifically silencing CCR3 or CCR5, a decrease in HIV-1 replication occurred.

Table 2
G protein-coupled receptors targeted in in vitro neuroscience research

GPCR class	Receptor type	Associated disease	Cell line	RNAi technology	Transfection agent	Reference
Class A peptidic	Formylpeptide (FPR)	Glioma	U87 (Glioblastoma)	siRNA (n/a)	n/a	(37)
	Formylpeptide (FPR)	Glioma	U87 (Glioblastoma)	siRNA (21 nt)	n/a	(38)
	Gastrin-releasing peptide (GRPR)	Neuroblastoma (NB)	SK-N-SH, BE(2)-C and LAN-1, (NB)	siRNA (n/a)	Electroporation	(68)
	Chemokines (CCR3 and CCR5)	HIV-1	U87-CD4-CCR3 and HOS- CD4-CCR3 (brain microglia) and MDM (monocyte-derived macrophages)	siRNA (modified 21 or 63 nt in plasmid)	Lipofectamine 2000	(35)
	Chemokine (CCR5)	Neuronal cells differentiation	Murine cortical neuronal cell	siRNA (n/a)	n/a	(69)
	Chemokine (CXCR2)	Alzheimer's disease (AD)	HEKsw (kidney)	siRNA (n/a)	Lipofectamine 2000	(36)
	Chemokine (CXCR4)	Uveal melanoma	Human uveal melanoma cells	siRNA (n/a)	n/a	(70)
	Chemokine (CXCR4)	Focal cebrebral ischemia	Rat neural progenitor cells (subven-tricular zone)	siRNA (n/a)	Retroviral vector	(71)
	Chemokine (CXCR4)	Diabetic retinopathy	HRMEC (retinal microvascular endothelial cells)	siRNA (modified 21 nt)	Targefect	(72)
	Neurokinin (NK-1)	Metastatic neuro-blastoma (NB)	CHP212 (NB cell lines) and SY5Y (NB from the bone marrow)	siRNA (in plasmid)	Superfect	(73)
Class A amine	Adrenoceptor (β_1-AR)	Memory modulatory processes	HEK293 (kidney)	siRNA (modified 21 nt)	Lipofectamine 2000	(74)
	Dopamine (DRD1)	Nicotine dependence	HEK293 (kidney)	miRNA (n/a)	Lipofectamine 2000	(75)
	Serotonin (5-HTR$_{1B}$)	Complex behavioral disorders	HeLa (cervical cancer)	miRNA (n/a)	Lipofectamine 2000	(40)
	Serotonin (5-HTR)	Mood disorders (humans) synaptic plasticity (Aplysia)	Aplysia cultured neurons or ganglia	dsRNA (n/a)	Microinjection	(39)

(continued)

Table 2 (continued)

GPCR class	Receptor type	Associated disease	Cell line	RNAi technology	Transfection agent	Reference
Class B hormone	Corticotropin-releasing factor receptor type 1 and 2 (CRF-R1/2)	Pituitary function (secretion of gonadotropin)	Rat anterior pituitary cells	siRNA (n/a)	Adenovirus	(41)
Class C hormone	Metabotropic glutamate (mGlu1)	Neurotoxicity	Rat cerebellar neurons	siRNA (21 nt)	Transit-TKO	(76)
	Metabotropic glutamate (mGlu5)	Neurotocixity	BV2 (microglia)	siRNA (n/a)	Lipofectamine 2000	(42)
	Gamma-aminobutyric acid ($GABA_{B1}$)	Ethanol effect	Rat primary cortical neuron	siRNA (n/a)	Lipofectamine 2000	(77)
	Gamma-aminobutyric acid ($GABA_{B1}$)	Neurodegenerative disease	Rat primary cortical neuron	siRNA (21 nt)	Lipofectamine 2000	(78)

Furthermore, CXCR2 was studied for its potential role in Alzheimer's disease (AD) (36). In this study, their model was based on HEK_{SW} cells stably expressing the amyloid precursor protein (APP) comprising the Swedish mutation. Ambion's proprietary 21-mer siRNAs (120 pmol) were transfected with Lipofectamine 2000 into the cells, and β-amyloid protein Aβ40 and Aβ42 concentrations were analyzed at 64, 72, 96, and 120 h. They showed that downregulation of CXCR2 receptor reduced Aβ levels. Given the fact that CXCR2 is upregulated in AD brains, the authors propose its potential as a promising therapeutic target.

In two recent studies, Bian and colleagues focused on the class A GPCR formylpeptide receptor (FPR) (37, 38). In these studies, they transfected 20 μM of FPR siRNA (21 nt) with Oligofectamine (Invitrogen). They managed to partially abolish FPR expression in the human glioblastoma cell line U87, as they qualitatively showed by reverse-transcriptase PCR (RT-PCR) (37). More precisely, they demonstrated the capacity of FPR to: (1) promote glioblastoma cell proliferation and (2) stimulate tumors production of two important angiogenic factors (IL-8 and VEGF). They further conclude that the knockdown of this receptor significantly suppresses the aggressive phenotype of FPR-KO glioma cells injected in nude mice.

GPCRs Class A: Amine Family

In regard to the class A GPCR amine family, we will mention two studies on the serotonin receptor (5-HTR). In a recent paper from the group of Kaang (39), they targeted 5-HTR_{apAC1} with dsRNA (of≈200 bp) manufactured with the MEGAscript RNAi kit (Ambion, Austin, TX, USA). The microinjection of 500 ng/μL of 5-HTR dsRNA resulted in the blockade of cAMP production, decrease in the serotonin-induced elevation in membrane exitability and in the spike duration in primary *Aplysia* sensory neurons. Therefore, synaptic facilitation decreased in nondepressed or partially depressed sensory-to-motor neuron synapses.

Interestingly, Jensen and colleagues reported that the regulation of the serotonin receptor 1B (5-HTR_{1B}) by an innate miRNA (miR-96) was associated with aggressive human behaviors (40). Following this observation, they transfected the commercially available miR-96 (Ambion) with Lipofectamine 2000 in the human cervical cancer extruded cell line HeLa. They monitored an increase in the suppressive effect of the ancestral A-element, which is a specific polymorphism of the 5-HTR_{1B}, known to be implicated in the regulation directed by endogenous miR-96 and consequently modulating human behaviors.

GPCRs Class B: Hormone Family

Only one paper studied the GPCR class B (secretin-like family) in a neuroscience RNAi paradigm. Nemoto et al. (41) worked on type 1 and 2 corticotropin-releasing factor receptors (CRF-R1/2) for the purpose of elucidating the secretion mechanisms of gonadotropin that regulate pituitary function. To achieve their goal, they used freshly prepared rat anterior pituitary cells and transfected them with an adenovirus containing either siRNA for CRF-R1 or CRF-R2. Their results showed that inhibiting CRF-R2 increased the secretion and mRNA expression of gonadotropins, concluding that paracrine activation of CRF-R2 on gonadotrophs tonically inhibits the expression and secretion of gonadotropins.

GPCRs Class C: Hormone Family

The GPCR class C is more represented than class B with regards to RNAi studies targeting neurological premises, with three studies. In one of them, Faden's group (42) assessed neurotoxicity in the murine microglial cell line BV2 by targeting the metabotropic glutamate receptor 5 (mGluR5). Indeed, using Lipofectamine 2000 as the transfection reagent, the mouse mGluR5 siRNA (100 nM; Santa Cruz, CA, USA) silencing effect was assessed through the levels of TNF-α release following a bacterial endotoxin LPS stimulation. They conclude that mGluR5 negatively regulates the release of microglial-associated inflammatory mediators in response to LPS and thus prevents microglial-mediated neuronal cell death.

2.2. Silencing GPCRs In Vivo

Once the initial knockdown validations are conclusive *in vitro* and the best RNAi candidates retained, experiments can be designed for *in vivo* evaluation. We have found 24 *in vivo* studies in the last 10 years in which GPCRs were silenced by a dsRNA approach in a neuroscience paradigm (see Table 3). We will review, in the following sections, those RNAi studies conducted in insects as well as in rodent spinal and supraspinal structures.

2.2.1. Neuroscience RNAi Lessons Taught by Insects

Neuroscience research is not limited to higher mammals like rodents and nonhuman primates. Insects are often used for their rapid generation time, easy genetic modifications through conventional techniques and fast development to adult stage, as it is the case for flies or crickets (43). *Drosophila melanogaster* is an insect frequently used in research for studying genetics and most useful to elucidate GPCR functions at larval or mature stages.

Indeed, RNAi has thus been first used in the fruit fly embryo to target neuronal receptors. More recently, microinjection of RNAi constructs in the adult fruit fly allowing a specific silencing in the nervous system came from the group of Manev and colleagues (44), who knocked-down γ-aminobutyric acid B receptor 1 ($GABA_BR1$). Briefly, a regular unmodified siRNA (21–22 nt) was injected directly into the brain using concentrations of 10 and 100 ng/μL (0.2 μL per fly; 10% of the body fluids). Long strands of dsRNA (819 bp) gave similar results in this model. The observed knockdown was first validated by RT-PCR at 24, 48, and 72 h following the injection. Despite the fact that no quantification of the mRNA decrease was conducted, the qualitative fading of the bands

Table 3
G protein-coupled receptors used as *in vivo* proof-of-principle

GPCR class	Receptor type	Associated disease	Species	RNAi technology	Reference
Class A	Angiotensin (AT1A)	Receptor functions	Mice	shRNA	(79)
Class A	Angiotensin (AT1B)	Brainstem function	Mice	shRNA	(80)
Class A	Melanocortin (MC_4R)	Neg results	Rats	siRNA	(49)
Class A	Dopamine-like (AmDOP2)	Locomotion	Honey bees	siRNA	(47)
Class A	Dopamine-like (DD2R)	Locomotor function	Drosophila	pUAS-ds-DD2R	(45)
Class A	Dopamine (DopR)	Human neuronal pathogenic mechanisms	Cricket, *Gryllus bimaculatus*	siRNA	(46)
Class A	Dopamine (DRD1)	Neg results	Rat	siRNA	(48)
Class A	Dopamine (DRD3)	Psychiatric disorders and drug dependence	Rat	shRNA - lentivirus	(81)
Class A	Tachykinins	Olfactory behaviors	Drosophila	RNAi	(82)
Class A	Opioid (MOR)	Ethanol consumption	Mice	shRNA	(51)
Class A	Muscarinic acetylcholine (CHRM2,3,4)	Nociception	Rat	siRNA	(54)
Class A	Neurotensin (NTS2)	Nociception	Rat	DsiRNA	(11)
Class A	Adrenoreceptor (ADRA1b)	Memory modulatory processes	Rat	siRNA	(74)
Class A	Adrenoreceptor (ADRA2a)	Anxiety-related behavior	Rat	siRNA	(50)
Class A	Octopamine	Memory processes	Honey bees	dsRNA	(83)
Class A	Serotonin	Depression, anorexia, and autism	*C. elegans*	dsRNA	(84)
Class A	Chemokine (CCR5)	Neuronal cells differentiation	Rat	siRNA	(69)
Class A	Tachykinin (NK3R)	Peripheral osmotic challenge	Rat	siRNA	(85)
Class A	Opioid (DOR)	Nociception	Rat	siRNA	(53)
Class C	Metabotropic (GABA-B)	Locomotor activity	Drosophila	DsRNA	(86)

(continued)

Table 3 (continued)

GPCR class	Receptor type	Associated disease	Species	RNAi technology	Reference
Class C	Metabotropic (GABA-B)	Development	Drosophila	RNAi	(87)
Class C	Metabotropic (GABA-B)	Behavior response to alcohol	Drosophila	RNAi	(44)
Class C	Metabotropic (GABBR1)	Neurodegenerative disease	Rat	RNAi	(78)
Class C	Metabotropic ($mGluR_7$)	Anxiety	Mice	siRNA	(88)
Putative	Gustatory (GR68a)	Olfactive and behavior	Drosophila	siRNA	(89)

was observed. Accordingly, $GABA_B$ agonist-induced inhibition of locomotor activity was significantly reduced as well as the number of catatonia-like episodes in RNAi treated flies. Sequence-independent off-target effects and efficient penetration of the construct in the brain were assessed by a GFP-coupled 750 bp sequence. Several studies used similar models to evaluate the role of dopamine-like receptors in insects (e.g. flies and honeybees) with efficiencies ranging from 35 to 60% inhibition (45–47).

As an example, the study of Hamada et al. depicts the effects of an injection of a long interfering dsRNA targeting a homologous dopamine receptor (DopR) in the cricket, *Gryllus bimaculatus* (46). In this study, 0.7 μL of a PCR-generated (372–660 bp) dsRNA (20 μM) was microinjected in the nymph, while the effects visualized in the adult cricket upon development. Inhibition of the targeted mRNA was robustly verified using qPCR. Two different RNAi templates induced 60% inhibition in the insects and modified their call song patterns. A *DsRed*-dsRNA was used as a control to eliminate the possibility of off-target effects. Taken together, these results obtained in different "lower organisms" give interesting options for CNS-associated GPCR research. Indeed, *Drosophila* and other insects are relatively cost-efficient in comparison to rodents and even to cell culture, while allowing behavioral observations to extrapolate knockdown effects to other "higher organisms."

2.2.2. Rodents Putting Their Brain to RNAi Improvement

Nonetheless, rodents remain the common gold standard in preclinical research, especially when it comes to neurological disease models and behavioral observations. Consequently, rats and mice are increasingly used to validate RNAi strategies targeting GPCRs in the CNS. At a supraspinal level, one typically has to microinject the dsRNAs directly in the structure where the receptor is expressed. In fact, intracerebroventricular administration of naked siRNA yielded poor efficacy in several studies (48, 49). In contrast, siRNAs injected directly in brain specific regions have met with greater success. Dygalo et al. injected 120 ng/5 μL of unmodified siRNA targeting the *alpha*2A-adrenoreceptor directly in the brain (1 mm caudal to lambda and 5.5 mm deep) for three consecutive neonatal days (50). They observed a successful knockdown in the cortex and brainstem on the day following the treatment as evaluated by qualitative RT-PCR (63 and 47% respectively). Consequently, the binding properties assessed by ligand radiolabeling significantly decreased. Following *alpha*2A receptor downregulation, the animals increased their activity on the elevated-plus maze test, generally used for anxiety assessment. Scrambled-nucleotides sequences were used to assess off-target effects. It is worth mentioning that siRNA performance was superior to the tested ODNs in this context.

As mentioned earlier in this chapter, the use of transfection agents to formulate dsRNA and facilitate their penetration in tissues and cells while protecting them from degradation is a preferred approach in comparison to naked dsRNA. Accordingly, numerous research groups use viral vectors to target the CNS in rodents. Lasek and colleagues downregulated *mu* opioid receptor (MOR) with a lentivirus loaded with a shRNA vector thus inducing high efficiency silencing in the ventral tegmental area (VTA) of mice (51). In brief, lentivirus expressing MOR or scrambled shRNA were produced in 293FT cells following expression and efficiency validation of the construct *in vitro*. The virus was then microinjected in the VTA at a concentration of 3×10^7 pg/mL (1 μL). The virus was tagged with GFP to allow visualization of the effective infection on dissected brain sections. Lentiviral delivery partially abolished the binding capacity of radiolabeled agonist to MOR for a period lasting for up to 6 weeks in the VTA. The knockdown resulted in decreased ethanol consumption by as much as 25% in the treated animals over 1 month.

2.2.3. Rats Showing the Way for Intrathecal RNAi Analgesic Therapy

The spinal cord is the first neural center for the processing and modulation of nociceptive information. It is increasingly studied and targeted in RNAi research elucidating GPCRs roles in nociception and analgesia (11, 52–54). Since siRNA delivered via intrathecal administration ends up in the cerebrospinal fluid, doses required for efficient knockdown in spinal and ganglia neurons with naked siRNA are very high (52, 55). However, cationic lipids and chitosan nanoparticles were shown to significantly reduce the effective siRNA concentrations (53, 56). Our recent study combined the use of cationic lipids to an advanced generation of siRNAs to knockdown the neurotensin receptor 2 (NTS2) in the context of pain research (11). The formulation of highly efficient DsiRNA to the cationic lipid i-Fect enabled the use of very low intrathecal doses (1 μg) to induce appropriate silencing. A 10 μL volume of the complex (i-Fect 5:1 saline, mixed 5 min before administration) was injected at the lumbar level for 2 consecutive days resulting in an efficient knockdown of the NTS2 receptor for 3 days following the treatment. The penetration of the complex in dorsal root ganglia (DRG) and spinal cord (SC) tissues was validated with scrambled complex tagged with Texas-Red fluorophores. A significant decrease of the targeted mRNA was observed 24 h after the treatment by quantitative real-time PCR (qPCR) in DRG and SC (up to 85 and 70%, respectively). In addition, two different siRNAs directed against distinct sequences targeting the receptor transcript were used to validate the potency and rule out potential off-target effects. The efficient knockdown of NTS2 resulted in the complete abolition of the antinociceptive effect of the NTS2-selective agonist JMV-431 at the tail-flick test. This proof of concept demonstrates the role of the NTS2 GPCR in pain modulation at the spinal level.

Chitosan nanoparticles have been evaluated for many years as potential carriers and were recently applied to RNAi intrathecal delivery (54, 57). A recent study discriminated between different subtypes of muscarinic acetylcholine receptors (M_2–M_4) in the mechanisms of thermal nociception (54). This elegant study demonstrates that M_4 subtype is the most important for the modulation of nociception. Chitosan was dissolved in acetate buffer and adjusted to pH 4.6 before muscarinic receptor (MR) subtypes siRNAs were added and stirred until final concentration was reached. An intrathecal administration of a low dose of 5 μg/5 μL of a chitosan-MR siRNA was then performed for three consecutive days in adult rats. The effective uptake of siRNA by neuronal cells was verified by using Alexa Fluor 488-coupled chitosan-siRNA nanoparticles. Visualization of DRG and SC sections showed chitosan/siRNA-488 uptake while naked siRNA-488 sections uptake was absent. The efficiency of the knockdown was verified at the mRNA level by qPCR (M_2: 60%, M_3: 35%, M_4: 50%). Moreover, the inhibition of the expression of M_2 and M_4, but not M_3, receptors significantly reduced the analgesic effect induced by the administration of a nonselective agonist of the muscarinic receptors 3 days after the siRNA treatment.

Taken together, the scientific approach and the numerous controls used in the last two described studies represent state-of-the-art RNAi approaches targeting CNS GPCRs *in vivo*. With similar appropriate controls and an additional immune response array, dsRNA therapeutics targeting pro-nociceptive targets represent a high potential in treating chronic pain in the clinic. Intrathecal injections, although an invasive technique, thus represent a realistic route of administration for patients coping with intractable pain. In summary, the rapid expansion of *in vivo* RNAi technologies for CNS applications in the last 3 years is encouraging for upcoming clinical translation. We will discuss those future directions in the next section.

3. Translation from In Vitro to In Vivo: A Challenge

3.1. RNAi Status for Treating Clinical Neuropathologies Involving GPCRs

Despite the major advances of *in vitro* and *in vivo* RNAi applications in the last decade, only nine clinical trials involving dsRNA are presently under way (20) (Table 4). There are numerous limitations to explain the lack of translation to human applications when comparing to its broad use in molecular biology. First and foremost, safety issues and efficient delivery to target tissues are limits to genetic approaches in the clinic. However, the RNAi packaging, as detailed in the previous section, has made rapid advancements thus decreasing concerns and accelerating clinical transfer as observed in the last 3 years.

RNAi clinical trials have been initiated through instillation or various topical administrations. Among those clinical trials, only two are neuroscience-focused. Quark Pharmaceuticals undertook phase I trials to ultimately manage optic nerve atrophy or ischemic optic neuropathy via the intravitreal application of a 19 nucleotides long siRNA inhibiting the RTP801L gene mRNA (Table 4). Degeneration of axons in the optic nerve head can be caused in part by mechanisms leading to elevated intraocular pressure or by insufficient blood supply. These symptoms are among the world leading causes of blindness (Patent WO/2007/141796). This first RNAi neuroscience clinical trial is feasible due to the ease of accessibility to the retinal ganglion neurons with intravitreal administration. The neuroprotective treatment was shown, in pre-clinical studies, to prevent the progressive loss of retinal ganglion cells from the start of the experiment as well as two weeks after the degeneration was initiated (58).

It is noteworthy that despite more than 15 RNAi clinical studies performed involving dsRNA, none targeted GPCRs. This fact is surprising, considering that GPCRs account for more than 50% of all marketed prescription drugs, thus pinpointing their crucial role in pathology management (59). Overall, everything remains to be done in regards to RNAi therapies targeting neuropathologies involving GPCR dysregulation. Pharmaceutical companies specialized in clinical RNAi therapeutical solutions present pipelines dedicated to CNS pathologies such as acute and chronic neurodegenerative diseases, spinal cord injury, amyotrophic lateral sclerosis, and Alzheimer's disease among others. However, none of them clearly identify GPCRs as potential targets.

Table 4
Clinical trials involving RNAi against GPCR targets in 2009–2010

Investigator	Therapy	Pathology	Target	Site of administration	Phase	Status
Allergan and Sirna Therapeutics	AGN211745	Age-related macular degeneration	n/d	Intravitreal	II	Failed
Calando Pharmaceuticals	CALAA-01	Solid tumor cancers	M2 ribonucleotide reductase subunit	Intravenous	I	Active
Duke University Medical Center	n/d	Melanoma	Immunoproteasomes LMP2, LMP7, MECL1	Intradermal	I	Recruiting
Opko Health	Bevasiranib	Diabetic macular edema	VEGF	Intravitreal	II	Completed
Opko Health	Bevasiranib	Age-related macular degeneration	VEGF	Intravitreal	III	Interrupted
PC Project and Transderm	TD101	Pachyonychia congenita	Keratin K6a (N171K mutation)	Intraplantar (callus)	Ib	Completed
Quark Pharmaceuticals	QPI-1007	Chronic optic nerve atrophy	Retinal ganglion cells	Intravitreal	I	Recruiting
Quark Pharmaceuticals	QPI-1007	Acute non-arteritic anterior ischemic optic neuropathy	Retinal ganglion cells	Intravitreal	I	Recruiting
Quark Pharmaceuticals	QPI-1002	Acute kidney injury	p53	Intravenous	I, IIa	Active
Quark Pharmaceuticals	QPI-1002	Delayed graft function	p53	Intravenous	I, II	Active
Silence Therapeutics AG	Atu027	Advanced solid cancer	n/d	Intravenous	I	Recruiting
Tekmira Pharmaceuticals Corporation	PRO-040201	High cholesterol	n/d	Intravenous	I	Terminated
Quark Pharmaceuticals and Pfizer	PF-04523655	Diabetic macular edema	RTP801	Intravitreal	II	Recruiting
Quark Pharmaceuticals and Pfizer	PF-04523655	Age-related macular degeneration	RTP801	Intravitreal	II	Recruiting
Alnylam Pharmaceuticals	ALN-RSV01	Respiratory syncytial virus	Nucleocapsid N	Inhalation	IIb	Recruiting
Gradalis	FANG	Solid tumors	Furin	Intradermic	I	Recruiting

3.2. Advantages and Disadvantages of RNAi over Classical Pharmacology

In the last decade, microarray and proteomic approaches have led to the discovery of hundreds of potential molecular targets for a wide range of diseases or pathological disorders. *In vivo* validation of these novel targets via classical pharmacology is particularly limited by the lengthy process of synthesizing agonists or antagonists selective for the protein product of the target gene. Furthermore, the use of different predicting methods reveals that most of the disease-associated targets that have been identified are not "druggable" with conventional therapeutics, such as small molecules, proteins, or monoclonal antibodies (60). Proteins with conformations not amenable to small molecule binding or the identification of compounds selective for one target over closely related family members represent a major obstacle to the development of classical pharmacological tools. In this context, RNAi constitutes a powerful alternative approach offering the major advantages that; (1) sequences are rapidly designed for highly specific inhibition of the target of interest and, (2) targets can comprise any molecular class, including non-druggable proteins. An interesting aspect of RNAi is also its potentiality to silence allele-specific genes (27, 61). Many familial neurodegenerative diseases, such as dystonia, Huntington's disease, congenital myasthenic syndrome, amyotrophic lateral sclerosis, frontotemporal dementia, or spinocerebellar ataxia are inherited disorders caused by a dominant negative action of the mutated protein. Therefore, targeting mutated alleles specifically by RNAi, without affecting the expression of the wild-type protein carrying normal biological function holds promises for developing fine-tuned therapy for many neurological disorders currently untreatable with traditional biotherapeutic drugs.

In the field of GPCR research, receptor cell membrane expression, desensitization, internalization, and downregulation represent different steps that can limit the effectiveness of drug therapy. Indeed, agonists and antagonists, when administered acutely or chronically, can induce changes in receptor number and/or sensitivity (c.f. Chap. 20). For instance, drugs of abuse affecting receptor activity are believed to be responsible for some of the biochemical and cellular processes related to tolerance and dependence. In that context, success in designing therapeutic compounds therefore requires a good understanding of the underlying mechanisms that govern GPCR trafficking and function. All of this detailed information on the life cycle of GPCR is not necessarily important when using RNAi technology. Furthermore, in term of efficacy, it is generally believed that over 60% of receptor occupancy is needed for an antagonist to produce significant therapeutic effects. For example, the classical antipsychotics like haloperidol are clinically active only at 80% receptor occupancy. Therapeutic use of RNAi has revealed that some siRNAs inducing limited knockdown of GPCR mRNA (less than 30%) were nevertheless able to produce significant behavioral effects (62). These examples provide evidence of the impressive efficacy of siRNA in inhibiting gene function. However, this statement has to be verified; what may work for some target genes may not be applicable for others.

Even if the exploitation of gene silencing via RNAi has revolutionized the way in which gene function is now elucidated, it is important to bear in mind that one of the current challenges of using RNAi as a therapeutic approach is to deliver and release sufficient siRNAs to the intracellular compartment of the target cells (shRNAs and primary miRNAs facing the additional challenge that they need to access the nucleus). Viral and nonviral delivery systems have been developed to improve their capacity to reach the desired target cells in a highly specific manner and after the administration of the lowest possible doses (9). Unfortunately, the downside of using delivery systems is the potential toxicity of the delivery agent itself and its capacity to secure a prolonged release (63). Thus, since GPCRs represent a class of proteins located on the cell surface, RNAi will probably not replace the more classical pharmacopeia (i.e. small molecules or antibodies).

4. Future Directions

In the field of neuroscience, the development of effective treatment for CNS diseases remains a great challenge in clinical medicine. With the exciting data emerging from the first clinical trials, it is now more likely that RNAi will provide another potent class of therapeutic agents. However, the main challenges for the translation of RNAi-based therapy from cell culture to *in vivo* therapeutics in animals and humans are the delivery and the proper release of dsRNA to the target tissue and cells (64). This step is even more crucial in neuroscience applications since the *in vivo* administration of RNAi technology is limited by the incapacity of RNA molecules to cross the blood–brain barrier (BBB). Currently, invasive techniques are thus required to locally administer RNAi treatment in the brain or the spinal cord, making this approach less attractive for treating humans. Thus, considerable effort is being made to develop appropriate vehicle to optimize the delivery and stability for siRNA (63). To increase *in vivo* biodistribution, several nonviral based RNAi delivery methods have been used, including liposome and lipids, cationic polymers, cholesterol conjugation, RNA aptamers, and peptide transduction domains (65). In this regard, recent advances in the Trojan Horse Liposome technology applied to the transvascular nonviral gene therapy of brain disorders present a promising solution to the shRNA delivery across the BBB (66). Another aspect that will help in establishing future RNAi-based clinical trials is going to be our capacity to investigate the pharmacodynamic and pharmacokinetic properties of siRNA. Visualization and monitoring of siRNA by different *in vivo* noninvasive imaging techniques, including optical imaging, radionuclide-based imaging (SPECT and PET), and MRI should indeed facilitate the transformation of RNAi into a powerful therapeutic modality in the clinic (67).

5. Conclusion

The discovery of RNAi was very timely for the drug discovery and development field since there has been a sharp decline in research productivity in the pharmaceutical industry over the last decade. RNAi-based drugs are already in clinical trials and it is hopeful that siRNA will receive FDA approval in the future for brain disorders and GPCR targets. However, carefully planned preclinical analyses of therapeutic RNAi are required to prevent premature failures in human trials that could delay or dampen maturation of the field.

References

1. Davey J (2004) G-protein-coupled receptors: new approaches to maximise the impact of GPCRS in drug discovery. Expert Opin Ther Targets **8**:165–170.
2. Owens J (2007) 2006 drug approvals: finding the niche. Nat Rev Drug Discov **6**:99–101.
3. Cazzin C, Ring C J (2009) Recent advances in the manipulation of murine gene expression and its utility for the study of human neurological disease. Biochim Biophys Acta **1802**:796–807.
4. Mizuno T, Chou MY, Inouye M (1984) A unique mechanism regulating gene expression: translational inhibition by a complementary RNA transcript (micRNA). Proc Natl Acad Sci USA **81**:1966–1970.
5. Napoli C, Lemieux C, Jorgensen R (1990) Introduction of a chimeric chalcone synthase gene into petunia results in reversible co-suppression of homologous genes in trans. Plant Cell **2**:279–289.
6. Rocheleau CE, Downs WD, Lin R et al (1997) Wnt signaling and an APC-related gene specify endoderm in early C. elegans embryos. Cell **90**:707–716.
7. Fire A, Xu S, Montgomery MK et al (1998) Potent and specific genetic interference by double-stranded RNA in *Caenorhabditis elegans.* Nature **391**:806–811.
8. Pasquinelli AE (2002) MicroRNAs: deviants no longer. Trends Genet **18**:171–173.
9. Sepp-Lorenzino L, Ruddy M (2008) Challenges and opportunities for local and systemic delivery of siRNA and antisense oligonucleotides. Clin Pharmacol Ther **84**: 628–632.
10. Martino S, di Girolamo I, Orlacchio A et al (2009) MicroRNA implications across neurodevelopment and neuropathology. J Biomed Biotechnol **2009**:654346.
11. Dore-Savard L, Roussy G, Dansereau MA et al (2008) Central delivery of Dicer-substrate siRNA: a direct application for pain research. Mol Ther **16**:1331–1339.
12. Kim D-H, Behlke MA, Rose SD et al (2005) Synthetic dsRNA Dicer substrates enhance RNAi potency and efficacy. Nat Biotechnol **23**:222–226.
13. Kubo T, Zhelev Z, Bakalova R et al (2007) Enhancement of gene silencing potency and nuclease stability by chemically modified duplex RNA. Nucleic Acids Symp Ser (Oxf), 407–408.
14. Rose SD, Kim D-H, Amarzguioui M et al (2005) Functional polarity is introduced by Dicer processing of short substrate RNAs. Nucleic acids research **33**:4140–4156.
15. Wang HW, Noland C, Siridechadilok B et al (2009) Structural insights into RNA processing by the human RISC-loading complex. Nat Struct Mol Biol **16**:1148–1153.

16. Schwarz D, Hutvágner G, Du T et al (2003) Asymmetry in the assembly of the RNAi enzyme complex. Cell **115**:199–208.
17. Snove O, Jr., Rossi J J (2006) Toxicity in mice expressing short hairpin RNAs gives new insight into RNAi. Genome Biol 7, 231.
18. Hutvagner G, Simard MJ, Mello CC et al (2004) Sequence-specific inhibition of small RNA function. PLoS Biol **2**:E98.
19. Grimm D, Streetz KL, Jopling CL et al (2006) Fatality in mice due to oversaturation of cellular microRNA/short hairpin RNA pathways. Nature **441**:537–541.
20. ClinicalTrials.gov. (2010) Information on Clinical Trials and Human Research Studies.
21. Grunweller A, Wyszko E, Bieber B et al (2003) Comparison of different antisense strategies in mammalian cells using locked nucleic acids, 2′-O-methyl RNA, phosphorothioates and small interfering RNA. Nucleic Acids Res **31**:3185–3193.
22. Kretschmer-Kazemi Far R, Sczakiel G (2003) The activity of siRNA in mammalian cells is related to structural target accessibility: a comparison with antisense oligonucleotides. Nucleic Acids Res **31**:4417–4424.
23. Bertrand JR, Pottier M, Vekris A et al (2002) Comparison of antisense oligonucleotides and siRNAs in cell culture and *in vivo*. Biochem Biophys Res Commun **296**:1000–1004.
24. Miyagishi M, Hayashi M, Taira K (2003) Comparison of the suppressive effects of antisense oligonucleotides and siRNAs directed against the same targets in mammalian cells. Antisense Nucleic Acid Drug Dev **13**:1–7.
25. Tan P H, Yang L C, Ji R R (2009) Therapeutic potential of RNA interference in pain medicine. Open Pain J **2**:57–63.
26. Grimm D, Kay MA (2007) Combinatorial RNAi: a winning strategy for the race against evolving targets? Mol Ther **15**:878–888.
27. Genc S, Koroglu TF, Genc K (2004) RNA interference in neuroscience. Brain Res Mol Brain Res **132**:260–270.
28. Castanotto D, Rossi JJ (2009) The promises and pitfalls of RNA-interference-based therapeutics. Nature **457**:426–433.
29. Pardridge WM (2007) shRNA and siRNA delivery to the brain. Adv Drug Deliv Rev **59**:141–152.
30. Amarzguioui M, Lundberg P, Cantin E et al (2006) Rational design and *in vitro* and *in vivo* delivery of Dicer substrate siRNA. Nat Protoc **1**:508–517.
31. Behlke MA (2008) Chemical modification of siRNAs for *in vivo* use. Oligonucleotides **18**:305–319.
32. Kurreck J (2009) RNA interference: from basic research to therapeutic applications. Angew Chem Int Ed Engl **48**:1378–1398.
33. John M, Constien R, Akinc A et al (2007) Effective RNAi-mediated gene silencing without interruption of the endogenous microRNA pathway. Nature **449**:745–747.
34. Murphy PM, Baggiolini M, Charo IF et al (2000) International union of pharmacology. XXII. Nomenclature for chemokine receptors. Pharmacol Rev **52**:145–176.
35. Agrawal L, Maxwell CR, Peters PJ et al (2009) Complexity in human immunodeficiency virus type 1 (HIV-1) co-receptor usage: roles of CCR3 and CCR5 in HIV-1 infection of monocyte-derived macrophages and brain microglia. J Gen Virol **90**:710–722.
36. Bakshi P, Margenthaler E, Laporte V et al (2008) Novel role of CXCR2 in regulation of gamma-secretase activity. ACS Chem Biol **3**:777–789.
37. Chen DL, Ping YF, Yu SC et al (2009) Downregulating FPR restrains xenograft tumors by impairing the angiogenic potential and invasive capability of malignant glioma cells. Biochem Biophys Res Commun **381**:448–452.
38. Yao Y, Wang C, Varshney RR et al (2009) Antisense makes sense in engineered regenerative medicine. Pharm Res **26**:263–275.
39. Lee YS, Choi SL, Lee SH et al (2009) Identification of a serotonin receptor coupled to adenylyl cyclase involved in learning-related heterosynaptic facilitation in Aplysia. Proc Natl Acad Sci USA **106**:14634–14639.
40. Jensen KP, Covault J, Conner TS et al (2009) A common polymorphism in serotonin receptor 1B mRNA moderates regulation by miR-96 and associates with aggressive human behaviors. Mol Psychiatry **14**:381–389.
41. Nemoto T, Yamauchi N, and Shibasaki T (2009) Novel action of pituitary urocortin 2 in the regulation of expression and secretion of gonadotropins. J Endocrinol **201**:105–114.
42. Loane DJ, Stoica BA, Pajoohesh-Ganji A et al (2009) Activation of metabotropic glutamate receptor 5 modulates microglial reactivity and neurotoxicity by inhibiting NADPH oxidase. J Biol Chem **284**:15629–15639.
43. Belles X (2010) Beyond *Drosophila*: RNAi *in vivo* and functional genomics in insects. Annu Rev Entomol **55**:111–128.
44. Dzitoyeva S, Dimitrijevic N, and Manev H (2003) *Gamma*-aminobutyric acid B receptor 1 mediates behavior-impairing actions of alcohol in *Drosophila*: adult RNA interference and pharmacological evidence. Proc Natl Acad Sci USA **100**:5485–5490.

45. Draper I, Kurshan P T, McBride E et al (2007) Locomotor activity is regulated by D2-like receptors in *Drosophila*: an anatomic and functional analysis. Dev Neurobiol **67**:378–393.
46. Hamada A, Miyawaki K, Honda-sumi E et al (2009) Loss-of-function analyses of the fragile X-related and dopamine receptor genes by RNA interference in the cricket *Gryllus bimaculatus*. Dev Dyn **238**:2025–2033.
47. Mustard JA, Pham PM, Smith BH (2009) Modulation of motor behavior by dopamine and the D1-like dopamine receptor AmDOP2 in the honey bee. J Insect Physiol **56**:422–430.
48. Isacson R, Kull B, Salmi P et al (2003) Lack of efficacy of 'naked' small interfering RNA applied directly to rat brain. Acta Physiologica Scandinavica **179**:173–177.
49. Senn C, Hangartner C, Moes S et al (2005) Central administration of small interfering RNAs in rats: a comparison with antisense oligonucleotides. Eur J Pharmacol **522**:30–37.
50. Dygalo NN, Kalinina TS, Shishkina GT (2008) Neonatal programming of rat behavior by downregulation of *alpha*2A-adrenoreceptor gene expression in the brain. Ann N Y Acad Sci **1148**:409–414.
51. Lasek AW, Janak PH, He L et al (2007) Downregulation of *mu* opioid receptor by RNA interference in the ventral tegmental area reduces ethanol consumption in mice. Genes Brain Behav **6**:728–735.
52. Dorn G, Patel S, Wotherspoon G et al (2004) siRNA relieves chronic neuropathic pain. Nucleic Acids Res **32**:e49.
53. Luo MC, Zhang DQ, Ma SW et al (2005) An efficient intrathecal delivery of small interfering RNA to the spinal cord and peripheral neurons. Mol Pain **1**:29.
54. Cai YQ, Chen SR, Han HD et al (2009) Role of M2, M3, and M4 muscarinic receptor subtypes in the spinal cholinergic control of nociception revealed using siRNA in rats. J Neurochem **111**:1000–1010.
55. Altier C, Dale CS, Kisilevsky AE et al (2007) Differential role of N-type calcium channel splice isoforms in pain. J Neurosci **27**:6363–6373.
56. Howard KA (2009) Delivery of RNA interference therapeutics using polycation-based nanoparticles. Adv Drug Deliv Rev **61**:710–720.
57. Zhang HM, Chen SR, Cai YQ et al (2009) Signaling mechanisms mediating muscarinic enhancement of GABAergic synaptic transmission in the spinal cord. Neuroscience **158**: 1577–1588.
58. Today MN (2009) Quark Pharmaceuticals Announces Data Indicating Potential Utility of QPI-1007 for Treatment of Glaucoma. MediLexicon International Ltd.
59. Heilker R, Wolff M, Tautermann CS et al (2009) G-protein-coupled receptor-focused drug discovery using a target class platform approach. Drug Discov Today **14**:231–240.
60. Lopez-Fraga M, Martinez T, Jimenez A (2009) RNA interference technologies and therapeutics: from basic research to products. BioDrugs **23**:305–332.
61. Miller GG, Voronina TA (2005) (Perspective technologies for drug design). Antibiot Khimioter **50**:52–63.
62. Hoyer D, Thakker DR, Natt F et al (2006) Global down-regulation of gene expression in the brain using RNA interference, with emphasis on monoamine transporters and GPCRs: implications for target characterization in psychiatric and neurological disorders. J Recept Signal Transduct Res **26**:527–547.
63. Whitehead KA, Langer R, Anderson DG (2009) Knocking down barriers: advances in siRNA delivery. Nat Rev Drug Discov **8**: 129–138.
64. Jones D (2009) Teaming up to tackle RNAi delivery challenge. Nat Rev Drug Discov **8**: 525–526.
65. Eguchi A, Dowdy SF (2009) siRNA delivery using peptide transduction domains. Trends Pharmacol Sci **30**:341–345.
66. Boado RJ, Pardridge WM (2009) Comparison of blood-brain barrier transport of glial-derived neurotrophic factor (GDNF) and an IgG-GDNF fusion protein in the rhesus monkey. Drug Metab Dispos **37**:2299–2304.
67. Hong H, Zhang Y, Cai W (2010) *In vivo* imaging of RNA interference. J Nucl Med **51**: 169–172.
68. Ishola TA, Kang J, Qiao J et al (2007) Phosphatidylinositol 3-kinase regulation of gastrin-releasing peptide-induced cell cycle progression in neuroblastoma cells. Biochim Biophys Acta **1770**:927–932.
69. Park MH, Lee YK, Lee YH et al (2009) Chemokines released from astrocytes promote chemokine receptor 5-mediated neuronal cell differentiation. Exp Cell Res **315**:2715–2726.
70. Li H, Yang W, Chen PW et al (2009) Inhibition of chemokine receptor expression on uveal melanomas by CXCR4 siRNA and its effect on uveal melanoma liver metastases. Invest Ophthalmol Vis Sci **50**:5522–5528.
71. Liu XS, Chopp M, Santra M et al (2008) Functional response to SDF1 *alpha* through over-expression of CXCR4 on adult subventricular zone progenitor cells. Brain Res **1226**: 18–26.
72. Yu K, Zhuang J, Kaminski JM et al (2007) CXCR4 down-regulation by small interfering RNA inhibits invasion and tubule formation of

human retinal microvascular endothelial cells. Biochem Biophys Res Comm **358**:990–996.

73. Mukerji I, Ramkissoon SH, Reddy KK et al (2005) Autocrine proliferation of neuroblastoma cells is partly mediated through neurokinin receptors: relevance to bone marrow metastasis. J Neurooncol **71**:91–98.
74. Fu AL, Yan XB, Sui L (2007) Down-regulation of *beta*1-adrenoceptors gene expression by short interfering RNA impairs the memory retrieval in the basolateral amygdala of rats. Neurosci Lett **428**:77–81.
75. Huang W, Li MD (2009) Differential allelic expression of dopamine D1 receptor gene (DRD1) is modulated by microRNA miR-504. Biol Psychiatry **65**:702–705.
76. Pshenichkin S, Dolinska M, Klauzinska M et al (2008) Dual neurotoxic and neuroprotective role of metabotropic glutamate receptor 1 in conditions of trophic deprivation - possible role as a dependence receptor. Neuropharmacology **55**:500–508.
77. Lee HY, Li SP, Park MS et al (2007) Ethanol's effect on intracellular signal pathways in prenatal rat cortical neurons is GABA-B1 dependent. Synapse **61**:622–628.
78. Naseer MI, Lee HY, Ullah N et al (2010) Ethanol and PTZ effects on siRNA-mediated GABA-B1 receptor: down regulation of intracellular signaling pathway in prenatal rat cortical and hippocampal neurons. Synapse **64**:181–190.
79. Chen Y, Chen H, Hoffmann A et al (2006) Adenovirus-mediated small-interference RNA for *in vivo* silencing of angiotensin AT1a receptors in mouse brain. Hypertension **47**:230–237.
80. Lin Z, Chen Y, Zhang W et al (2008) RNA interference shows interactions between mouse brainstem angiotensin AT1 receptors and angiotensin-converting enzyme 2. Exp Physiol **93**:676–684.
81. Bahi A, Boyer F, Bussard G et al (2005) Silencing dopamine D3-receptors in the nucleus accumbens shell *in vivo* induces changes in cocaine-induced hyperlocomotion. Eur J Neurosci **21**:3415–3426.
82. Ignell R, Root CM, Birse RT et al (2009) Presynaptic peptidergic modulation of olfactory receptor neurons in Drosophila. Proc Natl Acad Sci USA **106**:13070–13075.
83. Farooqui T, Vaessin H, Smith BH (2004) Octopamine receptors in the honeybee (*Apis mellifera*) brain and their disruption by RNA-mediated interference. J Insect Physiol **50**: 701–713.
84. Carre-Pierrat M, Baillie D, Johnsen R et al (2006) Characterization of the *Caenorhabditis elegans* G protein-coupled serotonin receptors. Invert Neurosci **6**:189–205.
85. Jensen D, Zhang Z, Flynn FW (2008) Trafficking of tachykinin neurokinin 3 receptor to nuclei of neurons in the paraventricular nucleus of the hypothalamus following osmotic challenge. Neuroscience **155**:308–316.
86. Dimitrijevic N, Dzitoyeva S, Satta R et al (2005) *Drosophila* GABA(B) receptors are involved in behavioral effects of gamma-hydroxybutyric acid (GHB). Eur J Pharmacol **519**:246–252.
87. Dzitoyeva S, Gutnov A, Imbesi M et al (2005) Developmental role of GABA-B(1) receptors in *Drosophila*. Brain Res Dev Brain Res **158**:111–114.
88. Fendt M, Schmid S, Thakker DR et al (2008) mGluR7 facilitates extinction of aversive memories and controls amygdala plasticity. Mol Psychiatry **13**:970–979.
89. Thorne N, Bray S, Amrein H (2005) Function and expression of the *Drosophila gr* genes in the perception of sweet, bitter and pheromone compounds. Chem Senses **30 Suppl 1**:i270–272.

Chapter 20

Upregulating G Protein-Coupled Receptors with Receptor Antagonists

Ellen M. Unterwald

Abstract

The phenomenon of antagonist-induced receptor upregulation is common to many G protein-coupled receptors (GPCRs) such as adrenergic, muscarinic, opioid, cannabinoid, histamine, GABA(B), serotonin, and dopamine receptors. This chapter reviews data that support antagonist-induced upregulation specifically of opioid receptors but many of the principles apply to other GPCRs as well. It is well documented that chronic exposure to opioid receptor antagonists reliably produces increases in binding to opioid receptors when the antagonists are administered *in vivo* or applied *in vitro* to cell culture systems. Antagonist exposure increases receptor number and is associated with functional supersensitivity to subsequent agonist administration. For example, the analgesic potency of morphine is increased following prior administration of opioid receptor antagonists. Likewise, coupling of opioid receptors to G proteins is increased following antagonist exposure, as is the ability of opioid agonists to regulate adenylyl cyclase activity. The most common approach used to measure receptor upregulation is radioligand receptor binding. This chapter includes methods to measure receptor number by radioligand binding and by immunohistochemical approaches. Also included are methods to assess alterations in receptor function following antagonist exposure. The methods can be applied to tissue or cell homogenates or to *in situ* preparations in order to increase the anatomical specificity of the resulting data.

Key words: Receptor upregulation, Receptor antagonists, Receptor binding, Quantitative receptor autoradiography, Opioid receptor regulation

1. Introduction

A characteristic common to many G protein-coupled receptors (GPCRs) is their ability to upregulate in response to antagonist exposure. Antagonist-induced upregulation has been reported for numerous GPCRs including, but not limited to, dopamine receptors (1–3), muscarinic receptors (4), adrenergic receptors (5), serotonin receptors (6), histamine receptors (7), GABA(B)

Craig W. Stevens (ed.), *Methods for the Discovery and Characterization of G Protein-Coupled Receptors*, Neuromethods, vol. 60, DOI 10.1007/978-1-61779-179-6_20, © Springer Science+Business Media, LLC 2011

receptors (8), and opioid receptors (9). Antagonist-induced upregulation is robust and reproducible. This chapter will focus on opioid receptors to highlight the response of GPCRs to antagonist exposure. Methods that have been used to measure receptor upregulation are provided at the end of the chapter.

Opioid receptor upregulation following exposure to opioid receptor antagonists is a well-documented phenomenon. It was first characterized over 3 decades ago in nervous tissue of rodents. One of the earliest reports of opioid receptor upregulation following opioid receptor antagonist administration came from Loh and colleagues (10) as part of a larger study on the regulation of opioid receptor binding by morphine. Binding to opioid receptors in mouse brain was significantly increased 3 days after implantation of pellets containing the opioid receptor antagonist naloxone. Chronic administration of opioid receptor antagonists also produces functional opioid receptor supersensitivity, and this was first reported by Tang and Collins (11) who demonstrated that chronic administration of naloxone results in enhanced morphine-induced analgesia, which is accompanied by an increase in the number of opioid binding sites (12). Shortly thereafter, Herz and colleagues (13) found that chronic exposure of guinea pigs to naloxone for 1–2 weeks caused an increase in the sensitivity to opioids in the electrically stimulated longitudinal muscle-myenteric plexus ileum preparation. Once again, the enhanced inhibitory properties of opioid agonists were associated with elevations in the number of opioid receptors as measured by radioligand binding in both the guinea pig ileum and brainstem.

These early studies documenting the increase in opioid receptor number and function following opioid receptor antagonist exposure have been since extended to describe fully the pharmacological characteristics of this phenomenon, and potential molecular mechanisms have been suggested. In the next section, the characteristics of antagonist-induced opioid receptor upregulation are reviewed. Methods to assess receptor upregulation are presented in the last section.

2. Opioid Receptor Upregulation Following Opioid Receptor Antagonist Exposure

2.1. In Vivo Receptor Regulation Studies

The initial reports of antagonist-induced opioid receptor supersensitivity and upregulation were followed by more detailed characterization of this phenomenon. Antagonist-induced opioid receptor upregulation occurs following chronic administration of the nonselective opioid receptor antagonists naloxone or naltrexone, and the antagonists are most often administered to rodents by implanting drug-containing pellets or minipumps into the subcutaneous (sc) space. It is interesting to note that antagonist-induced upregulation of opioid receptor binding sites is observed consistently and robustly following continuous sc infusion of antagonist, but less consistently after intermittent sc injections (14).

Radioligand binding is the most commonly used technique to measure opioid receptor upregulation and is performed on tissue homogenates or on tissue sections using quantitative receptor autoradiography. In all cases, increases in receptor number (B_{max}) rather than increases in receptor affinity (K_d) are apparent (15–20). The three types of opioid receptors, *mu*, *delta*, and *kappa*, are regulated to different extents in brain following chronic administration of naloxone or naltrexone. Receptor binding studies using selective radioligands indicate that *mu* opioid receptors are most affected by these antagonists, followed by *delta* opioid receptors (22–26). Naltrexone administration increases the density of *mu* opioid receptors by 80–100% and *delta* receptors by about 30% in preparations of whole brain minus cerebellum (21, 22). *Kappa* opioid receptors are more resistant to regulation by naloxone or naltrexone (19, 24, 25, 60). Thus, the degree of upregulation to any given dose of antagonist is greatest for *mu*, followed by *delta*, and lowest for *kappa* opioid receptors (17, 19, 23–26). This may be due to differences in the molecular mechanisms involved in the regulation of the three types of opioid receptors or, alternatively, due to the relative affinity of naloxone and naltrexone for the three opioid receptors. Naloxone and naltrexone have higher affinity for *mu* opioid receptors than the other two opioid receptors, although their affinity for *kappa* receptors is generally reported to be greater than for *delta* receptors (27).

Antagonist-induced *mu* receptor upregulation has also been measured using immunohistochemistry (28). Adjacent brain sections from rats exposed continuously for 7 days to naltrexone were processed for measurement of *mu* opioid receptors by immunohistochemistry and by receptor autoradiography. In agreement with other receptor autoradiography studies (23, 25, 29), increased binding to *mu* opioid receptors was widespread and occurred in many brain regions including the central gray, hypothalamus, interpeduncular nucleus, ventral tegmental area, amygdala, thalamus, hippocampus, and globus pallidus. By contrast, significant increases in *mu* receptor immunoreactivity were limited to the interpeduncular nucleus, amygdala, hippocampus, and thalamus. The results from this study indicate that chronic naltrexone exposure produces robust changes in *mu* receptor binding in many brain areas, but relatively smaller increases in *mu* opioid receptors as measured by immunoreactivity. Increases in immunoreactivity were more modest in magnitude than the increases in receptor binding, suggesting that chronic naltrexone increases the fraction of *mu* opioid receptors capable of binding ligand, without a large change in the total levels of receptor protein (28).

2.2. In Vitro Receptor Regulation Studies

Exposure of cells expressing opioid receptors to opioid receptor antagonists in culture produces receptor upregulation, similar to what occurs following *in vivo* antagonist exposure. For example, human embryonic kidney 293 (HEK293) cells stably expressing the murine *mu* opioid receptor and exposed to naloxone for 18 h show significant increases in surface *mu* receptors using flow cytometric analysis (30). Likewise, naloxone increases both *mu* and *delta* opioid receptor densities in a dose-dependent manner in SH-SY5Y cells (31). Similar results have been reported in cell lines expressing D2 receptors (2, 3) and cultured neuronal cells expressing muscarinic (4) and 5HT-1A receptors (6). These studies demonstrate that GPCRs expressed by cells in culture undergo upregulation in response to application of receptor antagonists *in vitro*, with similar pharmacological properties to that seen in the whole animal.

2.3. Functional Receptor Supersensitivity

Chronic exposure to opioid receptor antagonists not only produces an increase in the number of opioid receptor binding sites but also increases the subsequent response to opioid receptor agonists. For example, long-term exposure to naloxone or naltrexone results in supersensitivity to morphine analgesia as demonstrated by a leftward shift in the morphine analgesic dose–response curve (17, 20, 26, 32). In agreement with the receptor binding data, supersensitivity to morphine on analgesic tests returns to baseline levels 6 days after cessation of naltrexone administration (17, 22). Other physiological responses to opioid agonists are exaggerated following antagonist administration. Six injections of naloxone over 3 days are sufficient to produce an increase in locomotor response to subsequent morphine administration (33). Likewise, the hyperthermic response to acute morphine is enhanced following chronic naltrexone (20). Neurons in the locus coeruleus of chronic naltrexone-treated rats exhibit enhanced inhibitory responses to morphine (34). Augmented morphine withdrawal signs are seen when the morphine treatment is preceded by chronic naloxone (35). The lethality of morphine is increased 2.5-fold following chronic naltrexone treatment in the mouse (36).

As described above, opioid receptor antagonists can upregulate receptor number and can increase subsequent pharmacological responses to agonists. Receptor upregulation has been shown to be accompanied by increases in the ability of opioid receptors to activate G proteins and subsequently inhibit cAMP production. Thus, activation of G proteins by *mu* opioid receptor agonists as measured by [^{35}S]GTPγS binding is augmented in mouse spinal cord following 7 days of naloxone injections (37), indicating enhanced receptor G protein coupling. Likewise, chronic naltrexone has been shown to augment the efficacy of opioid receptor agonists to inhibit adenylyl cyclase activity (38). Taken together, these results indicate that chronic exposure to opioid receptor antagonists leads to increases in opioid receptor number, enhanced coupling to intracellular effector systems, and augmentation in functional responses.

2.4. Antagonist-Induced Opioid Receptor mRNA Regulation

As described above, chronic exposure to opioid receptor antagonists leads to increases in opioid receptor number as measured by radioligand binding and immunoreactivity. Studies have been performed to determine if the increases in opioid binding sites are associated with increases in the steady-state levels of opioid receptor mRNA. Many studies have demonstrated that opioid receptor upregulation occurs in the absence of changes in the levels of opioid receptor mRNA ((39–42), but see (43)). For example, we have found that 7-day infusion of naltrexone significantly upregulated *mu* opioid receptor binding in rat brain; however, *mu* opioid receptor mRNA levels were not significantly altered in any brain region investigated (39). Therefore, it appears that changes in the steady-state level of opioid receptor mRNAs do not occur in response to chronic antagonist exposure, as a result of either increased transcription or decreased degradation of the receptor transcripts. Similar findings have been reported for D2 dopamine (2) and histamine H2 (7) receptors, in that antagonist treatment leads to receptor upregulation in the absence of changes in receptor mRNA levels. By contrast, changes in protein and mRNA levels have been reported for the m2- and m3-muscarinic acetylcholine receptors (4).

Potential mechanisms underlying antagonist-induced opioid receptor upregulation have been reviewed recently (9). Support for the idea that some antagonists can act as pharmacological chaperones has grown. Evidence from several GPCRs suggests that some antagonists can act intracellularly by binding to and stabilizing newly synthesized receptors, thereby promoting proper protein folding, maturation, exit from the endoplasmic reticulum, and trafficking to the cell surface (42, 44–47). Other studies suggest that antagonists can block constitutive receptor downregulation and reduce the rate of degradation of receptor binding sites (7, 42, 48, 49), thus increasing functional receptors.

3. Methods for Measuring GPCR Upregulation Following Receptor Antagonist Exposure

Several methods have been used to measure the increase in binding sites that occurs with chronic exposure to receptor antagonists. The most common of these is radioligand binding to whole cells or homogenates prepared from brain tissues or cell membranes. Both receptor number and affinity can be determined.

3.1. Radioligand Binding

Radioligand binding has been used to document antagonist-induced receptor upregulation of opioid receptors (10, 12, 17, 39, 42), *beta* 2-adrenoceptors (5), D2 dopamine receptors (2), and

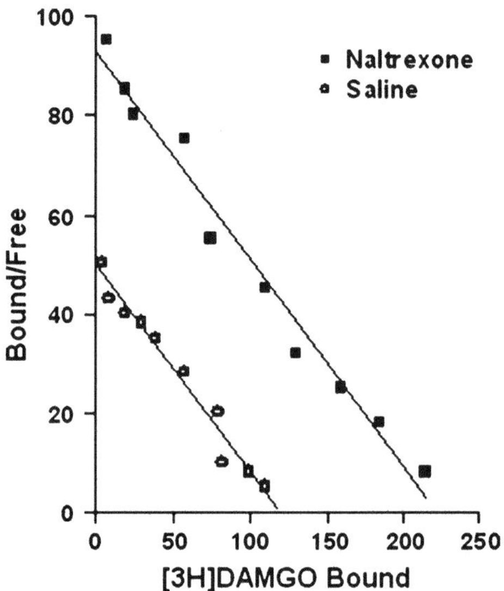

Fig. 1. Scatchard analysis of the binding of [^{3}H]-DAMGO to *mu* opioid receptors in whole brain (minus cerebellum) of rats exposed to saline (*open circles*) or naltrexone (8 mg/kg/day; *closed squares*) by osmotic minipumps for 7 days. Binding to *mu* receptors was 85% higher in brains from naltrexone-treated as compared to saline-treated rats (B_{max} = 120 vs. 223 fmol/mg protein), without a difference in receptor affinity (K_d).

histamine H2 receptors (7). The methods described below for opioid receptor binding are based on our published work (39) and were used to generate the data in Fig. 1 which shows the binding of [^{3}H]-DAMGO to *mu* opioid receptors in whole brain homogenates of rats infused with saline or naltrexone for 7 days. Results demonstrate an 85% increase in B_{max} (120 vs. 223 fmol/mg protein) following naltrexone administration. Composition of the buffers and the incubation time and temperature can vary with measurement of different GPCRs and with different radioligands.

1. Tissues or cells are homogenized in ice-cold buffer (50 mM Tris HCl, pH 7.4 at 4°C) using a tissue/cell disruptor such as a Polytron for 30 s. The homogenates are centrifuged for 15 min at 30,000 × *g*. The resulting supernatants are discarded and the pellets are resuspended in the same buffer. A second centrifugation is performed, again 15 min at 30,000 × *g*, and the pellets are resuspended in the same buffer.
2. Incubate the homogenate at 37°C in an oscillating water bath for 30 min. This step promotes the dissociation of endogenous ligand or exogenously administered drug at the binding site. It is critical to remove all of the applied or administered antagonist prior to radioligand binding to accurately measure binding density. If antagonist remains in the preparation, it

will displace the radioligand and result in an underestimate of true receptor number.

3. A third centrifugation is performed, 15 min at 30,000 × *g*. The final pellet is resuspended in the assay buffer.
4. Radioligand binding is usually performed in triplicate. A series of test tubes (12 × 75 mm) are prepared containing increasing amounts of radioligand (six tubes of each). About 10 radioligand concentrations are sufficient to allow accurate determination of binding constants. Radioligand concentrations should extend from approximately tenfold below to tenfold above the expected dissociation constant (K_d) at the receptor of interest.
5. A 1,000-fold excess of unlabeled ("cold") ligand is added to half of the tubes at each radioligand concentration (i.e., three tubes). The binding in these tubes is used to define "background" or "nonspecific" binding, while the binding in the other tubes is used to define "total" radioligand binding. Therefore, an average radioligand binding assay would consist of 60 tubes for each tissue or cell homogenate: (3 total + 3 nonspecific binding) × 10 radioligand concentrations.
6. Equal amounts of tissue/cell homogenate are added to each tube. Total sample volumes (usually 1 ml) are kept the same by altering the volume of buffer.
7. Samples are vortexed and incubated at the temperature and time appropriate for each radioligand and GPCR in order for the reaction to reach equilibrium binding. Temperature and time are inversely related in this regard; time to reach equilibrium binding is longer at lower temperatures. Standard incubation conditions for opioid receptor binding range from 45 min at 22°C to 60 min at 4°C, although these can vary.
8. Following incubation, samples are filtered onto Whatman GF/B filters under vacuum, and the tubes and filters rapidly washed 3 times with ice cold buffer. The Brandel or other brands of cell harvesters work well for this purpose. Filtration allows the separation of radioligand bound to receptors from nonbound (or "free") radioligand.
9. Filters are dried and transferred to scintillation vials. The vials are filled with scintillation fluid, vortexed, and subjected to liquid scintillation counting for measurement of radioactivity.
10. Protein concentrations of the homogenates are determined by standard techniques (i.e., Lowry, Bradford, or other protein assays).
11. Data can be analyzed by two methods to determine binding constants: saturation analysis and Scatchard plots. Detailed description of these methods can be found in (50).

12. For saturation analysis, radioligand concentrations are plotted on the abscissa and specific radioligand bound (fmol/mg protein) on the ordinate. Curve fitting statistical programs (e.g., GraphPad Prism, Sigma Plot, LIGAND, BioSoft, etc.) can be used to calculate receptor densities and affinities. The amount of specific binding when the isotherm reaches the plateau approximates the B_{max}. An estimate of the K_d can be derived from the concentration of radioligand that labels half of the receptors (i.e., 50% of B_{max}).
13. Binding data can be linearized using a Scatchard plot. For Scatchard analysis, amount of radioligand bound (fmol/mg protein) is plotted on the abscissa and radioligand bound/free plotted on the ordinate. Linear regression analysis is used to determine the K_d (the negative reciprocal of the slope) and the B_{max} (the intercept of the line with the abscissa). Figure 1 shows a Scatchard plot of radioligand binding to *mu* opioid receptors in rat brain homogenates following chronic exposure to saline or naltrexone.

3.2. Quantitative Receptor Autoradiography

Radioligand receptor binding can also be performed in situ on tissue sections. Although receptor autoradiography is generally more time consuming and labor intensive than homogenate binding methods, receptor autoradiography offers several important advantages including greater anatomical resolution and greater sensitivity. It is also possible to measure several receptors from adjacent sections from the same tissue. Quantitative receptor autoradiography has been used to demonstrate brain region-specific upregulation of opioid receptors (23, 25, 28, 29) and D1 dopamine receptors (1) following antagonist treatment. Figure 2 shows images from autoradiographic films of sagittal brain sec-

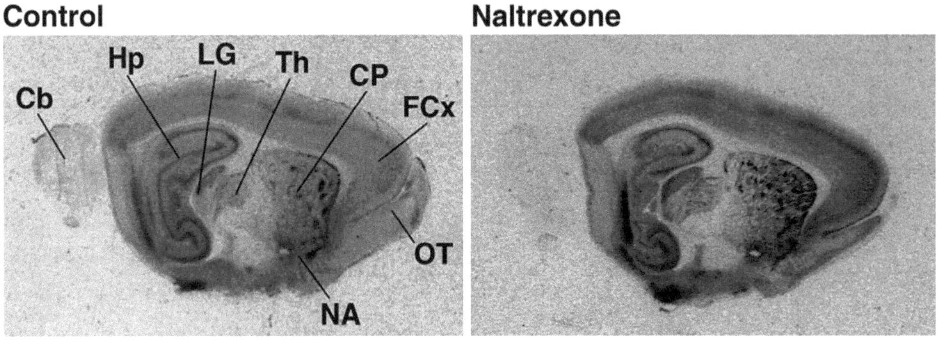

Fig. 2. *Mu* opioid receptor upregulation in rat brain after chronic naltrexone administration. Sagittal brain sections from a saline control (*top*) and a naltrexone-exposed (*bottom*) rat were processed for quantitative *mu* opioid receptor autoradiography by the methods provided. Images were generated by exposure of [^{3}H]DAMGO-labeled brain sections to film and subsequent image analysis. *Mu* opioid receptor binding was increased by 10–200% depending on the brain region in brains exposed chronically to naltrexone (adapted from (28, 51)). *Cb* cerebellum; *Hp* hippocampus; *LG* lateral geniculate; *Th* thalamus; *CP* caudate putamen; *FCx* frontal cortex; *OT* olfactory tubercle; *NA* nucleus accumbens.

tions from rats exposed to saline or naltrexone for 7 days. *Mu* opioid receptors were labeled with [^{3}H]DAMGO and the sections exposed to tritium-sensitive film, as described below. Antagonist administration resulted in an increase in *mu* opioid receptor binding, which ranged from 10 to 200% depending on the brain region (adapted from (28, 51)). These methods are based on our previous publications for measurement of opioid receptors (28, 52).

1. Brains or other tissues are rapidly removed and immediately immersed in isopentane cooled to –30°C on dry ice. After 30 s, brains or tissues are removed and stored at –20°C.
2. Frozen tissues are sectioned on a cryostat at –18°C (10–20 μm thick) and mounted onto treated slides (e.g., gelatin-coated or Superfrost Plus (Fisher) microscope slides). After brief air drying on ice, sections are stored at –20 to 70°C with desiccant.
3. Sections are preincubated in 50 mM Tris HCl buffer, pH 7.4, at 21°C for 30 min. This preincubation facilitates the dissociation of exogenous or endogenous ligands from the receptor sites. Again, this step is crucial to remove the antagonist from the receptor sites and tissue prior to radioligand application.
4. Sections are next incubated in radioligand and 50 mM Tris buffer for 45–60 min at 4°C or room temperature, depending on the radioligand. Sections prepared in this way will be used to measure "total" receptor binding. It is important for quantitative measurement of receptor binding on tissues to use a concentration of radioligand that is at least fivefold higher than the K_d of the specific ligand at the receptor of interest. In this way, binding will approximate B_{max} values and will facilitate the determination of changes in receptor density.
5. Similar sections are incubated in the same solution as above with the addition of 1,000-fold excess of nonlabeled radioligand. Radioligand binding to these sections serves as a measurement of nonspecific binding.
6. All sections (total and nonspecific) are washed in ice-cold buffer to separate bound from free radioligand (6×20 s for opioid ligands). The wash times are optimized empirically for the optimal retention of bound radioligand while removing unbound ligand (i.e., to achieve high specific binding). Sections can be dipped into ice-cold dH_2O after the rinses to remove any salts from the washes. Sections are immediately dried under a stream of cold air.
7. Labeled sections are exposed to film (e.g., Kodak BioMax MR) together with a set of radioactive standards used for

quantification. Films are developed using Kodak D19 developer and fix.

8. Optical densities of the brain regions of interest are determined using a computer-assisted image analysis system. Autoradiograms are illuminated on a light box and the image captured by a videoscanner or camera. The integrated optical densities of the tissue regions of interest are measured. Likewise, the image of the radioactive standards on the film is also analyzed and a standard curve of optical density vs. known amounts of radioactivity is derived from the standards. Optical densities of the regions of interest are converted to fmol/mg tissue by comparison to the standard curve. Nonspecific binding is subtracted from total binding to determine specific binding for each brain region. Specific binding, fmol/mg tissue, can be compared between treatment groups for each brain region of interest.

3.3. Agonist-Stimulated [^{35}S]GTPγS Binding

Agonist-stimulated [^{35}S]GTPγS binding can be used to assess the degree of coupling between specific GPCRs and G proteins. With this method, antagonist-induced upregulation of GPCRs can be evaluated from a functional standpoint. Similar to receptor binding, [^{35}S]GTPγS binding can be performed on tissue or cell homogenates or in situ on fresh frozen tissue sections (see Sect. 3.4). The methods presented here are based on the work of Sim, Selley, and Childers (53), and have been used to demonstrate upregulation of opioid receptor (37, 42) and GABA(B) receptor (8) function following antagonist exposure. It should be noted that this method is best for GPCRs that are coupled to Gi/Go (54, 55).

1. Crude membranes of antagonist-treated and control brain tissues or cells are prepared by homogenizing the tissues or cells in 20 volumes of ice cold 50 mM Tris HCl, 3 mM $MgCl_2$, 1 mM EGTA, pH 7.4. The resulting homogenates are centrifuged at 30,000 × g at 4°C for 15 min. The pellets are resuspended in buffer and centrifuged again at 30,000 × g at 4°C for 15 min. The resulting pellets are resuspended in assay buffer containing 50 mM Tris HCl, 3 mM $MgCl_2$, 0.2 mM EGTA, 100 mM NaCl, pH 7.4, and stored at −80°C until assayed.
2. Receptor agonist-stimulated [^{35}S]GTPγS binding is performed with homogenates in the above assay buffer. Samples of homogenates containing 10 μg of protein are incubated at 30°C for 1 h in assay buffer in the absence or presence of the receptor agonist (0.01–30 μM), GDP (20 μM), and 0.05 nM [^{35}S]GTPγS in 1 ml total volume (in triplicate). Basal binding is assessed in the absence of agonist and presence of GDP, and nonspecific binding assessed in the presence of GDP and 10 μM unlabeled GTPγS.

3. The reaction is terminated by rapid filtration under vacuum through Whatman GF/B glass fiber filters, followed by three washes with 3 ml of ice-cold 50 mM Tris buffer.
4. Bound radioactivity is determined by liquid scintillation spectrophotometry.
5. Nonspecific binding is subtracted from basal and agonist-stimulated binding. Specific agonist-stimulated binding is expressed as a percent of specific basal binding.
6. Agonist-stimulated [^{35}S]GTPγS binding can be expressed as a percent of basal on the ordinate and concentration of agonist on the abscissa. EC_{50} and E_{max} values can be calculated and compared between treatment groups.

3.4. Agonist-Stimulated [^{35}S]GTPγS Binding on Tissue Sections

Similar to quantitative receptor autoradiography, in situ methods can be applied to measurements of agonist-stimulated [^{35}S]GTPγS binding (53, 56, 57). We have used this method to assess the functionality of cocaine-induced *mu* opioid receptor upregulation (58) and the methods below are based on those previously published (53, 58).

1. Brains or other tissues are rapidly removed and immediately immersed in isopentane cooled to −30°C on dry ice. After 30 s, brains are removed and stored at −80°C.
2. Frozen tissues are sectioned on a cryostat at −18°C (10–20 μm thick) and mounted onto treated slides (e.g., gelatin-coated or Superfrost Plus (Fisher Scientific) microscope slides). Slide-mounted sections are briefly air dried on ice and then stored at −80°C with desiccant.
3. Slide-mounted brain sections are incubated in assay buffer composed of 50 mM Tris HCl, 3 mM $MgCl_2$, 0.2 mM EGTA, 100 mM NaCl, pH 7.4, for 10 min at 25°C. Sections are next incubated in the same assay buffer containing 1 mM GDP for 15 min.
4. Sections are then incubated in assay buffer containing [^{35}S]GTPγS (0.04 nM), agonist to the receptor of interest (0–10 μM), and GDP (1 mM) for 2 h at 25°C. Nonspecific binding is assessed by incubating sections in a similar solution of [^{35}S]GTPγS (0.04 nM) and GDP (1 mM) that also contains 10 μM GTPγS.
5. Labeled sections are washed in ice cold 50 mM Tris HCl buffer, dried, and exposed to film together with ^{14}C-standards.
6. [^{35}S]GTPγS binding to brain regions of interest is measured by determining the optical densities on the films as described above for quantitative receptor autoradiography (Sect. 3.2). [^{35}S]GTPγS binding in the absence of agonist is a measure of basal binding. Binding in the presence of agonist is expressed

as a percent of basal binding. Agonist-stimulated [^{35}S]GTPγS binding can be compared between experimental groups exposed to the receptor antagonist vs. controls.

3.5. In Situ Immunolocalization of GPCRs

3.5.1. IHC with Biotinylated Secondary Antibody

Receptor upregulation can also be measured using immunohistochemical (IHC) techniques. IHC has been used to demonstrate antagonist-induced plasticity of 5-HT1A receptors in hippocampal cell culture (6). IHC detects all receptor proteins, whereas receptor binding methods measure only those receptors that are in a confirmation capable of binding the radioligand. Therefore, results of antagonist-induced receptor upregulation obtained by IHC methods might differ somewhat from those methods that rely on receptor binding (28, 42). IHC is generally considered to be a semiquantitative method. IHC procedures vary widely depending on the protein of interest and the antibody used to detect it. Below are methods for the labeling of *mu* opioid receptors in rodent tissues (59).

1. Anesthetized animals are perfused transcardially with 4% paraformaldehyde and 0.2% picric acid in 0.16 M phosphate buffer. The tissues (brain, spinal cord, or others) are removed and postfixed in the same fixative for 90 min at 4°C and then stored in 10% sucrose in 0.01 M phosphate-buffered saline (PBS) containing 0.01% sodium azide and 0.02% bacitracin for 24 h. Tissues sections are cut (14 μm) and mounted onto coated slides.
2. For immunoperoxidase staining, the sections are incubated with an antibody to the *mu* opioid receptor diluted in PBS (pH 7.4) containing 0.3% Triton X-100, sodium azide, and Bacitracin for 48 h at 4°C. After incubation with the primary antibody, the sections are rinsed 2 × 15 min in PBS and incubated with the secondary biotinylated anti-IgG antibody for 4 h at 22°C, followed by avidin–biotin–peroxidase complex for 3 h at 22°C. The sections are incubated for 10 min in a medium containing 60 mg of 3,3′-diaminobenzidine and 330 μl of 30% H_2O_2 in 0.01 M PBS, rinsed with PBS, dried in air, dehydrated in increasing concentrations of ethanol, cleaned in xylene, and mounted with coverslips. Control experiments are conducted by omitting either the primary or the secondary antisera.
3. The sections are examined under a microscope for quantitative analysis of relative levels of receptor-immunoreactivity. The relative gray levels of immunostaining on tissue regions of interest can be determined on digitized images using an image analysis system. Alternatively, the number of cells positive for receptor immunoreactivity can be counted and compared between treatment groups.

3.5.2. IHC with Radiolabeled Secondary Antibody

We have modified standard IHC methods to enhance the quantitative measurement of receptor density by combining IHC with receptor autoradiography techniques (28). Using this method, an antibody selective to the receptor of interest is applied to the tissue section. The primary antibody is detected by a radiolabeled secondary antibody. The labeled sections are exposed to film and the resulting images quantified similar to those generated by radioligand receptor autoradiography described above (Sect. 3.2). This method has been used to examine the regulation of *mu* opioid receptors following opioid receptor antagonist administration and compared with results generated by receptor autoradiography performed on adjacent tissue sections from the same animals (28).

1. Slide-mounted paraformaldehyde-fixed brain sections are washed in PBS overnight, quenched with 3% H_2O_2/methanol, and rinsed with PBS. Sections are incubated with 10% goat serum, 0.3% Tween 20, and 1% bovine serum albumin in PBS for 2 h at 21°C. Primary antibody in 10% goat serum, 0.3% Tween 20, and 1% bovine serum albumin is added to the sections and incubated in a humid chamber for 24 h at 4°C.
2. After washing in PBS/0.3% Tween, sections are incubated in [^{125}I]-secondary antibody in 5% goat serum, 0.3% Tween 20, and 1% bovine serum albumin in PBS for 2 h at 21°C.
3. Sections are then washed in PBS, dried, and exposed to film with a set of [^{125}I]-standards. Films are developed by standard methods. Receptors are quantified by measuring optical densities of the brain regions of interest on the resulting films as described above (Sect. 3.2) for receptor autoradiography.

4. Future Directions

Upregulation of GPCRs by receptor antagonists is well documented. Despite this, the molecular mechanisms underlying antagonist-induced GPCR upregulation have remained elusive and should be an area of future investigation. It may well be that different mechanisms are involved depending on the specific receptor under investigation. Thus, opioid receptor upregulation does not appear to be due to increases in gene expression (39–42), whereas muscarinic acetylcholine receptor upregulation may be (4). Even within the opioid receptor family, different mechanisms maybe involved in receptor upregulation. For example, evidence suggests that opioid receptor antagonists upregulate *mu* opioid receptors by stabilizing the binding site and inhibiting constitutive internalization and downregulation (48). In the case of the *delta* opioid receptor, it appears that opioid receptor antagonists act as pharmacological chaperones in that they enter the cell and facilitate the folding and transport of intracellular receptors to the plasma membrane resulting in increases in receptor binding (9). Data suggests that opioid receptor antagonists upregulate *kappa* opioid receptors by enhancing the rate of receptor maturation through the secretory pathway and protecting the receptor from degradation (42, 48). Future directions should include investigations into the molecular mechanisms underlying antagonist-induced upregulation of other GPCRs. As many widely used pharmacotherapies are GPCR antagonists, elucidation of the effects of these agents on receptor expression and function is important.

5. Conclusions

As discussed in this chapter, exposure to reversible pharmacological antagonists can result in upregulation of specific GPCRs. Receptor upregulation is most commonly quantified using radioligand binding methods, although immunohistochemistry has also been employed. GPCR upregulation often results in increases in responses to agonists at the same receptor resulting in functional supersensitivity. This has been demonstrated by showing that agonist-stimulated GTPγS binding is enhanced following chronic antagonist treatment. Likewise, enhanced physiological responses to agonist administration accompany increases in receptor number following antagonist exposure. Because antagonists at a variety of GPCRs are widely used therapeutics, the potential effects of long-term exposure to these drugs on receptor expression and function should be an important clinical consideration.

References

1. Parashos SA, Barone P, Tucci I et al (1987) Attenuation of D-1 antagonist-induced D-1 receptor upregulation by concomitant D-2 receptor blockade. Life Sci **41**:2279–2284
2. Filtz TM, Guan W, Artymyshyn RP et al (1994) Mechanisms of up-regulation of D2L dopamine receptors by agonist and antagonists in transfected HEK-293 cells. J Pharmacol Exp Ther **271**:1574–1582
3. Boundy VA, Pacheco MA, Guan W et al (1995) Agonists and antagonists differentially regulate the high affinity state of the D2L receptor in human embryonic kidney 293 cells. Mol Pharmacol **48**:956–964
4. Fukamauchi F, Saunders PA, Hough C et al (1993) Agonist-induced down-regulation and antagonist-induced up-regulation of M2- and M3-muscarinic acetylcholine receptor mRNA and protein in cultured cerebellar granule cells. Mol Pharmacol **44**:940–949
5. Brodde OE, Zerkowski HR, Doetsch N et al (1989) Subtype-selective up-regulation of human saphenous vein *beta* 2-adrenoceptors by chronic *beta*-adrenoceptor antagonist treatment. Naunyn Schmiedebergs Arch Pharmacol **339**:479–482
6. Nishi M, Azmitia EC (1999) Agonist- and antagonist-induced plasticity of rat 5-HT1A

receptor in hippocampal cell culture. Synapse **31**:186–195

7. Osawa S, Kajimura M, Yamamoto S et al (2005) Alteration of intracellular histamine H2 receptor cycling precedes antagonist-induced upregulation. Am J Physiol Gastrointest Liver Physiol **289**:G880-889
8. Pibiri F, Carboni G, Carai MA et al (2005) Up-regulation of GABA(B) receptors by chronic administration of the GABA(B) receptor antagonist SCH 50,911. Eur J Pharmacol **515**:94–98
9. Unterwald EM, Howells RD (2008) Upregulation of opioid receptors. In: Dean R, Bilsky EJ, Negus SS (Eds) Opiate receptors and antagonists: From bench to clinic. Humana Press, New York
10. Hitzemann RJ, Hitzemann BA, Loh HH (1974) Binding of ^{3}H-naloxone in the mouse brain: effect of ions and tolerance development. Life Sci **14**:2393–2404
11. Tang AH, Collins RJ (1978) Enhanced analgesic effects of morphine after chronic administration of naloxone in the rat. Eur. J Pharmacol **47**:473–474
12. Lahti RA, Collins RJ (1978) Chronic naloxone results in prolonged increases in opiate binding sites in brain. Eur J Pharmacol **51**:185–186
13. Schulz R, Wuster M, Herz A (1979) Supersensitivity to opioids following the chronic blockade of endorphin action by naloxone. Naunyn Schmiedebergs Arch Pharmacol **306**:93–96
14. Yoburn BC, Purohit V, Patel K et al (2004) Opioid agonist and antagonist treatment differentially regulates immunoreactive *mu*-opioid receptors and dynamin-2 *in vivo*. Eur J Pharmacol **498**: 87–96
15. Tempel A, Zukin RS, Gardner EL (1982) Supersensitivity of brain opiate receptor subtypes after chronic naltrexone treatment. Life Sci **31**:1401–1404
16. Zukin RS, Sugarman JR, Fitz-Syage ML et al (1982) Naltrexone-induced opiate receptor supersensitivity. Brain Res **245**:285–292
17. Yoburn BC, Nunes FA, Adler B et al (1986) Pharmacodynamic supersensitivity and opioid receptor upregulation in the mouse. J Pharmacol Exp Ther **239**:132–135
18. Tempel A, Gardner EL, Zukin RS (1985). Neurochemical and functional correlates of naltrexone-induced opiate receptor up-regulation. J Pharmacol Exp Ther **232**:439–444
19. Attali B, Vogel Z (1990) Characterization of *kappa* opiate receptors in rat spinal cord-dorsal root ganglion co-cultures and their regulation by chronic opiate treatment. Brain Res **517**:182–188
20. Bhargava HN, Matwyshyn GA, Reddy PL et al (1993). Effects of naltrexone on the binding of [^{3}H]D-Ala^2, MePhe^4, Gly-ol^5-enkephalin to brain regions and spinal cord and pharmacological responses to morphine in the rat. Gen Pharmacol **24**:1351–1357
21. Yoburn BC, Shah S, Chan K et al (1995) Supersensitivity to opioid analgesics following chronic opioid antagonist treatment: relationship to receptor selectivity. Pharmacol Biochem Behav **51**:535–539
22. Giordano AL, Nock B, Cicero TJ (1990) Antagonist-induced up-regulation of the putative epsilon opioid receptor in rat brain: comparison with *kappa*, *mu* and *delta* opioid receptors. J Pharmacol Exp Ther **255**:536–540
23. Morris BJ, Millan MJ, Herz A (1988) Antagonist-induced opioid receptor up-regulation. II. Regionally specific modulation of *mu*, *delta* and *kappa* binding sites in rat brain revealed by quantitative autoradiography. J Pharmacol Exp Ther **247**:729–736
24. Yoburn BC, Luke MC, Pasternak GW et al (1988) Upregulation of opioid receptor subtypes correlates with potency changes of morphine and DADLE. Life Sci **43**:1319–1324
25. Lesscher HMB, Bailey A, Burbach JPH et al (2003) Receptor-selective changes in *mu*-, *delta*-, and *kappa*-opioid receptors after chronic naltrexone treatment in mice. Eur J Neurosci **17**:1006–1012
26. Hummel M, Ansonoff MA, Pintar JE et al (2004) Genetic and pharmacological manipulation of *mu* opioid receptors in mice reveals a differential effect on behavioral sensitization to cocaine. Neuroscience **125**:211–220
27. Raynor K, Kong H, Chen Y et al (1994) Pharmacological characterization of the cloned *kappa*-, *delta*-, and *mu*-opioid receptors. Mol Pharmacol **45**:330–334.
28. Unterwald EM, Anton B, To T et al (1998) Quantitative immuno-localization of *mu* opioid receptors: regulation by naltrexone. Neuroscience **85**:897–905
29. Tempel A, Gardner EL, Zukin RS (1984) Visualization of opiate receptor upregulation by light microscopy autoradiography. Proc Natl Acad Sci USA **81**:3893–3897
30. Zaki PA, Keith DE, Brine GA et al (2000) Ligand-induced changes in surface *mu*-opioid receptor number: relationship to G protein Activation. J Pharmacol Exp Ther **292**:1127–1134
31. Zadina JE, Chang SL, Ge LJ et al (1993) *Mu* opiate receptor down-regulation by morphine and up-regulation by naloxone in SHSY5Y human neuroblastoma cells. J Pharmacol Exp Ther **265**:254–262

32. Yoburn BC, Kreuscher SP, Inturrisi CE et al (1989) Opioid receptor upregulation and supersensitivity in mice: effect on morphine sensitivity. Pharmacol Biochem Behav **32**:727–731
33. Volterra BN, DiGiulio AM, Cuomo V et al (1984) Modulation of opioid system in C57 mice after repeated treatment with morphine and naloxone: biochemical and behavioral correlates. Life Sci **34**:1669–1678
34. Bardo MT, Bhatnagar RK, Gebhart GF (1983) Chronic naltrexone increases binding in brain and produces supersensitivity to morphine in the locus coeruleus of the rat. Brain Res **289**:223–234
35. Suzuki T, Fukagawa Y, Misawa M (1990) Enhancement of morphine withdrawal signs in the rat after chronic treatment with naloxone. Eur J Pharmacol **178**:239–242
36. Yoburn BC, Nunes FA, Adler B et al (1986) Pharmacodynamic supersensitivity and opioid receptor upregulation in the mouse. J Pharmacol Exp Ther **239**:132–135
37. Narita M, Mizoguchi H, Nagase H et al (2001) Up-regulation of spinal *mu*-opioid receptor function to activate G-protein by chronic naloxone treatment. Brain Res **913**:170–173
38. Cote TE, Izenwasser S, Weems HB (1993) Naltexone-induced upregulation of *mu* opioid receptors on 7315c cell and brain membranes: enhancement of opioid efficacy in inhibiting adenylyl cyclase. J Pharmacol Exp Ther **267**:238–244
39. Unterwald EM, Rubenfeld JM, Imai Y et al (1995) Chronic opioid antagonist administration upregulates *mu* opioid receptor binding without altering *mu* opioid receptor mRNA levels. Mol Brain Res **33**:351–355
40. Castelli MP, Melis M, Mameli M et al (1997) Chronic morphine and naltrexone fail to modify *mu*-opioid receptor mRNA levels in the rat brain. Mol Brain Res **45**:149–153
41. Jenab S, Kest B, Inturrisi CE (1995) Assessment of *delta* opioid antinociception and receptor mRNA levels in mouse after chronic naltrexone treatment. Brain Res **691**:69–75
42. Wannemacher K, Yadav P, Howells RD (2007) A select set of opioid ligands induce up-regulation by promoting the maturation and stability of the rat *kappa* opioid receptor in human embryonic kidney 293 cells. J Pharmacol Exp Ther **323:**614–625
43. Jenab S, Inturrisi CE (1994) Ethanol and naloxone differentially upregulate *delta* opioid receptor gene expression in neuroblastoma hybrid (NG108-15) cells. Mol Brain Res **27**:95–102
44. Morello J-P, Salahpour A, Laperriere A et al (2000) Pharmacological chaperones rescue cell-surface expression and function of misfolded V2 vasopressin receptor mutants. J Clin Invest **105**:887–895
45. Petaja-Repo UE, Hogue M, Bhalla S et al (2002) Ligands act as pharmacological chaperones and increase the efficiency of *delta* opioid receptor maturation. EMBO J **21**:1628–1637
46. Chaipatikul V, Erickson-Herbrandson LJ, Loh HH et al (2003) Rescuing the traffic-deficient mutants of rat *mu*-opioid receptors with hydrophobic ligands. Mol Pharmacol **64**:32–41
47. Chen Y, Chen C, Wang Y et al (2006) Ligands regulate cell surface level of the human *kappa* opioid receptor (hKOR) by activation-induced down-regulation and pharmacological chaperone-mediated enhancement: differential effects of non-peptide and peptide agonists. J Pharmacol Exp Ther **319**:765–775
48. Li J, Chen C, Huang P et al (2001) Inverse agonist up-regulates the constitutively active D3.49(164)Q mutant of the rat *mu*-opioid receptor by stabilizing the structure and blocking constitutive internalization and down-regulation. Mol Pharmacol **60**:1064–1075
49. Li J, Huang P, Chen C et al (2001) Constitutive activation of the *mu* opioid receptor by mutation of D3.49(164), but not D3.32(147): D3.49(164) is critical for stabilization of the inactive form of the receptor and for its expression. Biochemistry **40**:12039–12050
50. Yamamura HI, Enna SJ, Kuhar MJ (1985) Neurotransmitter receptor binding. Raven Press, New York.
51. Unterwald EM (2008) Naltrexone in the treatment of alcohol dependence. J Addict Med **2**:121–127
52. Unterwald EM, Rubenfeld JM, Kreek MJ (1994) Repeated cocaine administration upregulates *kappa* and *mu*, but not *delta*, opioid receptors. NeuroReport **5**:1613–1616
53. Sim LJ, Selley DE, Childers SR (1995) *In vitro* autoradiography of receptor-activated G proteins in rat brain by agonist-stimulated guanylyl 5'-[*gamma*-[^{35}S]thio]-triphosphate binding. Proc Natl Acad Sci USA **92**:7242–7246
54. Milligan G (2003) Principles: extending the utility of [^{35}S]GTP *gamma*S binding assays. TIPS **24**:87–90
55. Harrison C and Traynor JR (2003) The [^{35}S] GTP *gamma*S binding assay: approaches and applications in pharmacology. Life Sci **74**:489–508
56. Sim-Selley LJ and Childers SR (2002) Neuroanatomical localization of receptor-activated G proteins in brain. Methods Enzymol **344**:42–58

57. Garcia-Jimenez A, Cowbuurn RF, Winblad B et al (1997) Autoradiographic characterization of [[35]35]GTP*gamma*S binding sites in rat brain. Neurochem Res **22**: 1055–1063
58. Schroeder JA, Niculescu M, Unterwald EM (2003) Cocaine alters *mu* but not *delta* or *kappa* opioid receptor-stimulated in situ [^{35}S] GTP*gamma*S binding in rat brain. Synapse **47**:26–32
59. Zhang X, de Araujo LG, Elde R et al (1999) Effect of morphine on cholecystokinin and *mu*-opioid receptor-like immunoreactivities in rat spinal dorsal horn neurons after peripheral axotomy and inflammation. Neuroscience **95**:197–207
60. Millan MJ, Morris BJ, Herz A (1988). Antagonist-induced opioid receptor upregulation. I. Characterization of supersensitivity to selective *mu* and *kappa* agonists. J Pharmacol Exp Ther **247**:721–728

Index

A

B

C

Craig W. Stevens (ed.), *Methods for the Discovery and Characterization of G Protein-Coupled Receptors*, Neuromethods, vol. 60, DOI 10.1007/978-1-61779-179-6, © Springer Science+Business Media, LLC 2011

D

E

F

G

H

R

V

W

X

Y

Z

MIX
Papier aus verantwortungsvollen Quellen
Paper from responsible sources
FSC® C105338

If you have any concerns about our products,
you can contact us on
ProductSafety@springernature.com

In case Publisher is established outside the EU,
the EU authorized representative is:
Springer Nature Customer Service Center GmbH
Europaplatz 3, 69115 Heidelberg, Germany

Printed by Libri Plureos GmbH
in Hamburg, Germany